8th International Conference on 3D Radiation Dosimetry 2014

(IC3DDose)

AA002942

Journal of Physics: Conference Series Volume 573

Ystad, Sweden
4-7 September 2014

ISBN: 978-1-63439-906-7
ISSN: 1742-6588

Printed from e-media with permission by:

Curran Associates, Inc.
57 Morehouse Lane
Red Hook, NY 12571

Some format issues inherent in the e-media version may also appear in this print version.

Printed by Curran Associates, Inc. (2015)

For permission requests, please contact the Institute of Physics
at the address below.

Institute of Physics
Dirac House, Temple Back
Bristol BS1 6BE UK

Phone: 44 1 17 929 7481
Fax: 44 1 17 920 0979

techtracking@iop.org

Additional copies of this publication are available from:

Curran Associates, Inc.
57 Morehouse Lane
Red Hook, NY 12571 USA
Phone: 845-758-0400
Fax: 845-758-2634
Email: curran@proceedings.com
Web: www.proceedings.com

TABLE OF CONTENTS

NOVEL 3D DETECTORS

MRI/OPTICAL DOSIMETRY

DOSIMETRIC CHALLENGES

EVALUATION OF DOSE DISTRIBUTIONS

NOVEL OPTICAL READOUT SYSTEMS

POSTERS

Journal of Physics: Conference Series **573** (2015) 011001 doi:10.1088/1742-6596/573/1/011001

8th International Conference on 3D Radiation Dosimetry (IC3DDose)

Preface

IC3DDose 2014, the 8th International Conference on 3D Radiation Dosimetry was held in Ystad, Sweden, from 4-7 September 2014. This grew out of the DosGel series, which began as DosGel99, the 1st International Workshop on Radiation Therapy Gel Dosimetry in Lexington, Kentucky. Since 1999 subsequent DoSGel conferences were held in Brisbane, Australia (2001), Ghent, Belgium (2004), Sherbrooke, Canada (2006) and Crete, Greece (2008). In 2010 the conference was held on Hilton Head Island, South Carolina and underwent a name-change to IC3DDose. The 7th and last meeting was held in Sydney, Australia from 4–8 November 2012.

It is worth remembering that the conference series started at the very beginning of the intensity modulated radiotherapy era and that the dosimeters being developed then were, to some extent, ahead of the clinical need of radiotherapy. However, since then the technical developments in radiation therapy have been dramatic, with dynamic treatments, including tracking, gating and volumetric modulated arc therapy, widely introduced in the clinic with the need for 3D dosimetry thus endless. This was also reflected by the contributions at the meeting in Ystad.

Accordingly the scope of the meeting has also broadened to IC3DDOSE - *I See Three-Dimensional Dose.* A multitude of dosimetry techniques and radiation detectors are now represented, all with the common denominator: *three-dimensional* or 3D. Additionally, quality assurance (QA) procedures and other aspects of clinical dosimetry are represented.

The implementation of new dosimetric techniques in radiotherapy is a process that needs every kind of caution, carefulness and thorough validation. Therefore, the clinical needs, reformulated as the aims for IC3DDOSE - *I See Three-Dimensional Dose,* are:

- Enhance the quality and accuracy of radiation therapy treatments through improved clinical dosimetry.
- Investigate and understand the dosimetric challenges of modern radiation treatment techniques.
- Provide a forum to discuss the latest research and developments in 3D and advanced radiation dosimetry.
- Energize and diversify dosimetry research and clinical practice by encouraging interaction and synergy between advanced, 3D, and semi-3D dosimetry techniques.

We commend these IC3Dose 2014 conference proceedings to you and strongly believe they include significant contributions to scientific progress in this field.

We would like to express our sincere gratitude to everybody involved in making the conference possible, the Scientific committee for their work on the general planning, paper review and program formulation, the distinguished invited speakers for their contributions and the local organizing committee members for all their hard work on the practical preparation for the meeting.

Journal of Physics: Conference Series **573** (2015) 011001 doi:10.1088/1742-6596/573/1/011001
Lars E. Olsson, Sven Bäck and Sofie Ceberg
Lund University and Skåne University Hospital, Sweden

International Scientific Committee

Sven Bäck, Sweden (chair)
Clive Baldock, Australia
Sam Beddar, USA
Crister Ceberg, Sweden
Yves de Deene, Belgium/Australia
Simon Doran, UK
Geoffrey Ibbott, USA
Andrew Jirasek, Canada
Kevin Jordan, Canada
Martin Lepage, Canada
Daniel Low, USA
Mark Oldham, USA
Tony Popescu, Canada
John Schreiner, Canada
Cheng-Shie Wuu, USA
David Thwaites, Australia

Local Organizing Committee

Sofie Ceberg (chair)
Lars E. Olsson (conference chair)
Fredrik Nordström
Anneli Edvardsson
Anna Karlsson Hauer
Anna Bäck

Journal of Physics: Conference Series **573** (2015) 011002 doi:10.1088/1742-6596/573/1/011002

CONFERENCE PROGRAMME

September 4-7, 2014
Ystad Saltsjöbad, Sweden

Journal of Physics: Conference Series **573** (2015) 011002 doi:10.1088/1742-6596/573/1/011002

IC3DDose
8th International Conference on 3D Radiation Dosimetry

INTRODUCTION AND WELCOME

It is a great pleasure to announce the 8th International conference on three-dimensional dosimetry, IC3DDose'14. Fifteen years ago the first meeting was solely devoted to gel dosimetry (DOSGEL'99). At the time we were two enthusiasts Clive Baldock and Lars E. Olsson (LEO), who eagerly wanted to organize a meeting on this topic, and they liked to locate it in the US in conjunction to an AAPM meeting. However, Clive was working in Australia and LEO was in Sweden, so it was not obvious how such a mission should be accomplished. Luckily, a local organizer was found in Geoffrey Ibbott. At the time he was working in Lexington, KY, which thereby became the site for the first meeting, to be followed by a series of successful conferences around the world. Now, we will host the meeting at Lund University in Sweden.

It is worth to remember that the conference series started in the very beginning of intensity modulated radiotherapy era and that the dosimeters we were developing were to some extent ahead of the clinical need of radiotherapy. However, since then the technical development in radiation therapy has been dramatic with dynamic treatments including eg. tracking, gating and volumetric modulated arc therapy widely introduced in the clinic and the need for 3D dosimetry thus endless.

Accordingly the scope of the meeting has also broadened. A multitude of dosimetry techniques and radiation detectors are now represented, all with the common denominator "three-dimensional", 3D. Additionally, QA procedures and other aspects of clinical dosimetry are represented.

The implementation of new dosimetric techniques in radiotherapy is a process that needs every kind of caution, carefulness and thorough validation. Therefore, the clinical needs reformulated as the aims for IC3DDOSE - I See Three Dimensional Dose - are:

- Enhance the quality and accuracy of radiation therapy treatments through improved clinical dosimetry.
- Investigate and understand the dosimetric challenges of modern radiation treatment techniques.
- Provide a forum to discuss the latest research and developments in 3D and advanced radiation dosimetry.
- Energize and diversify dosimetry research and clinical practice by encouraging interaction and synergy between advanced, 3D, and semi-3D dosimetry techniques

We hope that you will find the program for this meeting attractive and strongly believe that the conference include significant contributions to the scientific progress in this field.

We would like to express our sincere gratitude to everybody involved in making this meeting possible. The Scientific committee for their work on the general planning, abstract review and the program. The distinguished invited speakers for their contributions. The local organizing committee members for all their hard work on the practical preparation for the meeting.

Finally, we want to welcome you to the south of Sweden and to Ystad. We hope that you will enjoy the meeting, the scientific and social interactions and the pleasant settings in this region.

Lars E. Olsson, Sven Bäck and Sofie Ceberg
Lund University and Skåne University Hospital, Sweden

Journal of Physics: Conference Series **573** (2015) 011002 doi:10.1088/1742-6596/573/1/011002

IC3DDose
8th International Conference on 3D Radiation Dosimetry

IC3DDose14 ACKNOWLEDGES

The scientific and financial support from

VETENSKAPSRÅDET THE SWEDISH RESEARCH COUNCIL Cancerfonden MALMÖ CANCER CENTER

the generous support from our sponsors and exhibitors

ScandiDos

SUN NUCLEAR corporation ACCURAY modusQA Accuracy. Confidence.™ MGS

and the endorsement by

ESTRO European Society for RADIOTHERAPY & ONCOLOGY

INTERNATIONAL SCIENTIFIC COMMITTEE

Sven Bäck, Sweden (chair)
Clive Baldock, Australia
Sam Beddar, USA
Crister Ceberg, Sweden
Yves de Deene, Belgium/Australia
Simon Doran, UK
Geoffrey Ibbott, USA
Andrew Jirasek, Canada
Kevin Jordan, Canada
Martin Lepage, Canada
Daniel Low, USA
Mark Oldham, USA
Tony Popescu, Canada
John Schreiner, Canada
Cheng-Shie Wuu, USA
David Thwaites, Australia

LOCAL ORGANIZING COMMITTEE

Sofie Ceberg (chair)
Lars E. Olsson (conference chair)
Fredrik Nordström
Anneli Edvardsson
Anna Karlsson Hauer
Anna Bäck

Journal of Physics: Conference Series **573** (2015) 011002 doi:10.1088/1742-6596/573/1/011002

IC3DDose
8th International Conference on
3D Radiation Dosimetry

PROGRAM

Thursday 4 September 2014

9:00-12:00 **Arrival and registration**

12:00-13:00 **Lunch Break**

13:00-13:30 **Welcome & introduction**: L E Olsson and S Å J Bäck

13:30-15:00 **Session 1: The role for 3D dosimetry in QA**
Chair: L E Olsson, S Å J Bäck

13:30-14:00 Invited Review 1: T Knöös
QA procedures needed for advanced RT techniques and its impact in treatment outcome
14:00-14:15 B Mijnheer, I Olaciregui-Ruiz, R Rozendaal, H Spreeuw, M van Herk, A Mans
Current status of 3D EPID-based in vivo dosimetry in The Netherlands Cancer Institute
14:15-14:30 J Olofsson, M Gustafsson, U Isacsson, M Olevik-Dunder, M Westermark, H Benedek, P Hållström
Strategies for Quality Assurance of Intensity Modulated Radiation Therapy
14:30-14:45 J Götstedt, A Karlsson Hauer, A Bäck
Complexity metric as a complement to measurement based IMRT/VMAT patient-specific QA
14:45-15:00 A Cardin, X Ding, A Kassaee, L Lin, M J Maryanski, S Avery
Validation of high-resolution 3D patient QA for proton PBS and IMPT using laser CT of improved polymer gel dosimeters

15:00-15:30 **Tea/Coffee Break**

15:30-17:00 **Session 2: Clinical dosimetry I**
Chair: C Baldock, J Schreiner

15:30-16:00 Invited Review 2: D Thwaites and G Ibbott
Audits for advanced treatment dosimetry
16:00-16:15 C R Hansen, J R Sykes, J Barber, K West, R Bromley, K Szymura, S Fisher, J Sim, M Bailey, D Chrystal, S Deshpande, I Franji, T B Nielsen, C Brink, D I Thwaites
Multicentre knowledge sharing and planning/dose audit on Flattening Filter Free beams for SBRT lung
16:15-16:30 T Olding, K M Alexander, C Jechel, A T Nasr, C Joshi
Delivery Validation of VMAT Stereotactic Ablative Body Radiotherapy at Commissioning

16:30-17:00 *Poster Express (one min /one slide)*

17:00 - **Exhibition & Poster viewing – Mingle buffet & sparkling**

Journal of Physics: Conference Series **573** (2015) 011002 doi:10.1088/1742-6596/573/1/011002

IC3DDose
8th International Conference on 3D Radiation Dosimetry

PROGRAM

Friday 5 September 2014

8:15-8:55 **Morning lecture I.** Chair: Local committee
J Schreiner: *True 3D chemical dosimetry (gels, plastics): Development and Clinical Role*

9:00-10:30 **Session 3: Clinical dosimetry II**
Chair: T Knöös, Y De Deene

9:00-9:30 Invited Review 3: T Popescu
Patient-specific 4D QC using Monte-Carlo phase-space predictions and EPID dosimetry

9:30-9:45 S Avery, A Cardin, L Lin, M Kirk, A Kassaee, M J Maryanski
3D dose verification with polymer gel detectors of brain-spine match line for proton pencil beam cranio-spinal: A Preliminary Study

9:45-10:00 T Young, A Xing, P Vial, D Thwaites, L Holloway, S Arumugam
Sensitivity of Collapsed Arc QA method for Delivery Errors in Volumetric Modulated Arc Therapy (VMAT)

10:00-10:15 W Ottosson, C F Behrens, C E Andersen
Dose verification of radiotherapy for lung cancer by using plastic scintillator dosimetry and a heterogeneous phantom

10:15-10:30 M Caloz, M Kafrouni, Q Leturgie, S Corde, S Downes, J Lehmann, D Thwaites
Development of phantom and methodology for 3D and 4D dose intercomparisons for advanced lung radiotherapy

10.30-11.00 **Tea/Coffee Break**

11.00-12:30 **Session 4: Novel 3D detectors I**
Chair: D Thwaites, G Ibbott

11:00-11:30 Invited Review 4: S Beddar
Real-time volumetric scintillation dosimetry

11:30-11:45 R D Franich, J R Supple, B Lindsay, U J Yeo, P Lonski, R L Smith, M L Taylor, L Dunn, T Kron
Reproducibility assessment of dynamically deforming DEFGEL in a respiratory motion phantom

11:45-12:00 Y De Deene, R Hill, P S Skyt, J Booth
Flexydos3D: A new deformable anthropomorphic 3D dosimeter readout with optical dual wavelength CT scanning

12:00-12:15 M Martinho Costa, C McErlean, I Rivens, J Adamovics, M O Leach, G ter Haar, S J Doran
PRESAGE® as a new calibration method for high intensity focused ultrasound therapy

12:15-12:30 T Tominaga, M Yoshioka, S Hayashi, S Usui, M Tada
Availability of a container-less polymer gel detector and a gelatin container

12:30-13:30 **Lunch Break**

13:30-15:00 **Session 5: MRI/Optical dosimetry I**
Chair: K Jordan, S Doran

13:30-14:00 Invited Review 5: M Oldham
Radiochromic detectors

Journal of Physics: Conference Series **573** (2015) 011002 doi:10.1088/1742-6596/573/1/011002

IC3DDose
8th International Conference on
3D Radiation Dosimetry

PROGRAM

14:00-14:15	C McErlean, E Bräuer-Krisch, J Adamovics, MO Leach, SJ Doran *High resolution 3D dosimetry for Microbeam Radiation Therapy using Optical CT*
14:15-14:30	K Jordan, K Dekke, J Battista *Stray light reduction in optical computed tomography using a convergent cone-beam source*
14:30-14:45	G Heilemann, D Georg, A Berg *Pushing the boundaries of spatial resolution in dosimetry using polymer gels and radiochromic films*
14:45-15:00	S Usui, MYoshioka, S Hayashi, T Tominaga *Inclusion Type Radiochromic gel dosimeter for three-dimensional dose verification*
15:00-15:30	**Tea/Coffee Break**
15:30-17:00	**Session 6: Dosimetric challenges I** Chair: D Low, A Karlsson Hauer
15:30-16:00	Invited Review 6: A van Esch Dosimetry data analysis in clinical routine: a study case
16:00-16:15	S Arumugam, T Young, A Xing, D Thwaites, L Holloway *Benchmarking the gamma pass score using ArcCHECK for routine dosimetric QA of VMAT plans*
16:15-16:30	B Sun, S Yaddanapudi, S M Goddu, S Mutic *A Self-Sufficient Method for Calibration of Varian Electronic Portal Imaging Device*
16:30-16:45	K M Alexander, C Pinter, J Andrea, G Fichtinger, L J Schreiner *Implementation of an efficient workflow process for gel dosimetry using 3D Slicer*
16:45-17:00	S J Doran, T Gorjiara, J Adamovics, Z Kuncic, A Kacperek, C Baldock *The quenching effect in PRESAGE dosimetry of proton beams: Is an empirical correction feasible?*

IC3DDose
8th International Conference on 3D Radiation Dosimetry

PROGRAM

Saturday 6 September 2014

8:15-8:55	**Morning lecture II.** Chair: Local committee Y De Deene and A Jirasek: *Uncertainty in 3D dosimetry; various detectors and read-out systems*
9:00-10:30	**Session 7: Evaluation of dose distributions** Chair: C Ceberg, F Nordström
9:00-9:30	Invited Review 7: D Low *The importance of 3D dosimetry*
9:30-9:45	S Arumugam, A Xing, T Young,D Thwaites,L Holloway *Validation of 3DVH estimated DVH metrics for Prostate VMAT plans*
9.45-10:00	A Xing, S Arumugam, S Deshpande, A George, P Vial, L Holloway, G Goozee *Evaluation of 3D Gamma index calculation implemented in two commercial dosimetry systems*
10:00-10:15	R Lafratta, G Ibbott, J Adamovics, D Followill *Comparison of 2D and 3D Gamma Calculations for an IMRT QA Phantom*
10:15-10:30	T Hanušová, V Vondráček, K Badraoui-Čuprová, I Horáková and I Koniarová *New method for estimation of fluence complexity in IMRT fields and correlation with gamma analysis*
10.30-11.00	**Tea/Coffee Break**
11.00-12:30	**Session 8: Novel optical readout system** Chair: C Baldock, M Oldham
11:00-11:30	Invited Review 8: E Kristensson *Optical CT 3D imaging of scattering media using structured illumination*
11:30-11:45	D Ramm, T P Rutten *Feasibility of a dual wavelength laser optical CT scanner with in-air gel readout*
11:45-12:00	Y De Deene *Dual wavelength optical CT scanning of anthropomorphic shaped 3D dosimeters*
12:00-12:15	K Jordan *Radiochromic film thickness correction with convergent cone-beam optical CT scanner*
12:15-12:30	Y J Chang *Development of CCD-based optical computed tomography and comparison with single-beam optical CT scanner*
12:30-13:30	**Lunch Break**
13:30-15:00	**Session 9: MRI/Optical dosimetry II** Chair: K Jordan, J Schreiner
13:30-14:00	Invited Review 9: C-S Wuu *Clinical applications of 3-D dosimeters*
14:00-14:15	T Yao, A Gasparini, A G Denkova, J M Warman *Radiation-chemical and optical properties of a radio-fluorogenic gel*

IC3DDose
8th International Conference on
3D Radiation Dosimetry

PROGRAM

14:15-14:30	A T Nasr, K Alexander, L J Schreiner, K B McAuley *Opportunities for improving the performance of LCV micelle gel dosimeters: I. preliminary investigation*
14:30-14:45	A T Nasr, K M Alexander, T Olding, L J Schreiner, K B McAuley *Opportunities for improving the performance of LCV micelle gel dosimeters: II. Recipe Optimization*
14:45-15:00	T Juang, J Adamovics, M Oldham *Characterization of a reusable PRESAGE 3D dosimeter*
15:00-15:30	**Tea/Coffee Break**
15:30-16:45	**Session 10: Dosimetric challenges II** Chair: A Jirasek, A Bäck
15:30-16:00	Invited Review 10: B Raaymakers *Dosimetric commissioning of a hybrid MRI radiotherapy system*
16:00-16:15	M Alqathami, A Blencowe, G Ibbott *An investigation into the potential influence of oxygen on the efficiency of the PRESAGE dosimeter*
16:15-16:30	M Welch, W D Foltz, D A Jaffray *Timing Considerations for MR-Guided Preclinical RT: Effects of Ion Diffusion, SNR and Imaging Times on FXG Gel Calibration*
16:30-16:45	S T Smith, K-S Masters, K Hosokawa, J Blinco, S B Crowe, T Kairn, J V Trapp *A reduction of diffusion in PVA Fricke hydrogels*
18:00-23:30	**Conference dinner at Svaneholm Castle**

8th International Conference on 3D Radiation Dosimetry (IC3DDose) IOP Publishing
Journal of Physics: Conference Series **573** (2015) 011002 doi:10.1088/1742-6596/573/1/011002

IC3DDose
8th International Conference on
3D Radiation Dosimetry

PROGRAM

Sunday 7 September 2014

8:15-8:55	**Morning lecture III.** Chair: Local committee A Karlsson Hauer and A Bäck: *Quasi 3D dosimetry (EPID, conventional 2D/3D detector matrices)*
9:00-10:30	**Session 11: Dosimetric challenges III** Chair: C-S Wuu, T Popescu
9:00-9:15	A Edvardsson, A Ljusberg, C Ceberg, J Medin, L Ambolt, F Nordström, S Ceberg *Verification of motion induced thread effect during tomotherapy using gel dosimetry*
9:15-9:30	Q Li, T Juang, R Beth, S Chang, M Oldham *Verification of micro-beam Irradiation*
9:30-9:45	P Kozubikova, J Solc, J Novotny Jr, K Vavru, J Pipek, J Semnicka *Assessment of Radiochromic Gel Dosimeter Based on Turnbull Blue Dye for Relative Output Factor Measurements of the Leksell Gamma Knife Perfexion™*
9:45:10:00	J F Pavoni, W F P Neves-Junior, M A Silveira, P A M M Ramos, CM K Haddad, O Baffa *Feasibility on Using Composite Gel-Alanine Dosimetry on the validation of a Multiple Brain Metastasis Radiosurgery VMAT Technique*
10:00-10:15	X Qian, J Admovics, C-S Wuu *A Small Animal Image Guided Irradiation System Study Using 3-D Dosimeters*
10:15-10:30	S Bache, J Malcolm, J Adamovics, M Oldham *Investigation of a low-cost optical-CT system with minimal refractive index-matching fluid*
10.30-11.00	**Tea/Coffee Break**
11.00-12:30	**Session 12: Novel 3D detectors II** Chair: S Beddar, S Å J Bäck
11:00-11:30	Invited Review 11: B Pogue *Cerenkov light to measure 3D dose distributions for external radiation therapy*
11:30-11:45	A Hiroki, S Yamashita, M Taguchi *Enhancement in dose sensitivity of polymer gel dosimeters composed of radiation-crosslinked gel matrix and less toxic monomers*
11:45-12:00	B Berndt, P S Skyt, L Holloway, R Hill, A Sankar, Y De Deene *Do saccharide doped PAGAT dosimeters increase accuracy?*
12:00-12:15	K I Penev, K Mequanint *Methyl thymol blue in Fricke gels*
12:15-12:30	P S Skyt, B Berndt, L Holloway, R Hill, Y De Deene *NMR spectroscopy of saccharide-doped PAGAT dosimeters*
12:30-13:00	C Baldock **Summary of conference - Closing remarks & goodbyes**
13:00-14:00	**Lunch Break**

Journal of Physics: Conference Series **573** (2015) 011002 doi:10.1088/1742-6596/573/1/011002

IC3DDose
8th International Conference on
3D Radiation Dosimetry

PROGRAM

Poster presentations

1 A Asena, T Kairn, S B Crowe, S T Smith, J V Trapp
PAGAT gel dosimeters for dose distribution measurements in the vicinity of high-density implants: A preliminary study

2 A Asena, T Kairn, S B Crowe, J V Trapp
Photon beam dose distributions for patients with implanted temporary tissue expanders

3 DU Yi, WANG Xiangang, XIANG Xincheng
Artifact suppression in optical CT for gel dosimeters by iterative reconstruction

4 E J J Samuel, P Sathiyaraj, D Titus, D S Kumar
Particle Size Analysis of PAGAT Gel dosimetry

5 E J J Samuel, P Sathiyaraj, D Titus, D S Kumar
Antioxidant effect of green tea on polymer gel dosimeter

6 S E S Babu, B P Ravindran
Effect of bloom strength on radiochromic gel dosimeters

7 E M Høye, P S Skyt, E S Yates, L P Muren, J B B Petersen, P Balling
A new dosimeter formulation for deformable 3D dose verification

8 H Kawamura, H Sato, T Hamano, M Suda, H Yoshii
Polymer gel dosimetry for neutron beam at the Neutron exposure Accelerator System for Biological Effect Experiments (NASBEE)

9 J Šolc, V Socho, P Kozubíková
Energy dependence of Fricke-xylenol orange gel and gel based on Turnbull blue for low-energy photons

10 M Kozicki, P Maras
The polyGeVero software for fast and easy computation of 3D radiotherapy dosimetry data

11 M Carroll, G Ibbott, J Adamovics
Feasibility of PET/CT 3-D dosimetry for proton-activated PRESAGE dosimeters

12 M Laoues, A S Moussa, R Khelifi
Ocular brachytherapy dosimetry using ophthalmic applicators 90Sr/90Y

13 S T Smith, T Kairn, S B Crowe, A Asena, J V Trapp
Characterisation of the half-field beam penumbra for a variety of blocking set-ups

14 S-i Hayashi, Y Sakurai, R Uchida, M Suzuki, S Usui, T Tominaga
Preliminary study of MAGAT polymer gel dosimetry for Boron-Neutron Capture Therapy

15 T Kairn, M B Kakakhel, H Johnston, A Jirasek, J V Trapp
Assessment of the effects of CT dose in averaged x-ray CT images of a dose-sensitive polymer gel

16 Y De Deene
Computational Simulations of the Influence of Noise in Optical CT reconstruction

17 A Xing, S Deshpande, S Arumugam, A George, L Holloway, P Vial, G Goozee
Is a quasi-3D dosimeter better than a 2D dosimeter for TomoTherapy delivery quality assurance?

18 K I Penev, K Mequanint
Characterization of a commercially-produced chemically stable Fricke gel dosimeter

19 P Mann, M Witte, S Armbruster, A Runz, C Lang, M Breithaupt, M Berge, J Biederer, C P Karger, T Moser
Feasibility of polymer gel dosimetry for validation of gated radiotherapy treatments in a dynamic porcine lung phantom

20 K Jordan
Investigating hydrogel dosimeter decomposition by chemical methods

21 S Arumugam, A Xing, T Young, D Thwaites, L Holloway
Three dimensional dose verification of VMAT plans using the Octavius 4D dosimetric system

Journal of Physics: Conference Series **573** (2015) 011002 doi:10.1088/1742-6596/573/1/011002

View of Ystad Saltsjöbad

IC3DDose
8th International Conference on
3D Radiation Dosimetry
C. CEBERG

Journal of Physics: Conference Series **573** (2015) 011003 doi:10.1088/1742-6596/573/1/011003

Peer review statement

All papers published in this volume of *Journal of Physics: Conference Series* have been peer reviewed through processes administered by the proceedings Editors. Reviews were conducted by expert referees to the professional and scientific standards expected of a proceedings journal published by IOP Publishing.

QA procedures needed for advanced RT techniques and its impact on treatment outcome

T Knöös
Radiation Physics, Skåne University Hospital, Lund, Sweden and Department of Medical Radiation Physics, Clinical Sciences, Lund University, Lund, Sweden.
SE-205 02 Malmö, Sweden

E-mail: tommy.knoos@med.lu.se

Abstract. The radiotherapy process is reviewed briefly and potential risks or pitfalls are identified. The focus is on modern advanced modalities in radiation therapy such as IMRT, VMAT, gating and tracking and also for the unknown to come. Existing methods, or quality controls (QC), or with better word barriers, are introduced at important steps of process with the purpose of prohibiting errors to continue through the process and thus avoiding an unwanted erroneous irradiation of the patient. The soft branch of quality assurance (QA) such as peer-review is also a major component of today's process and its safety. The importance of knowing your QCs is pointed out. The role of dosimetry method i.e. 3D-dosimetry is reviewed. Staff have to be working with awareness and alertness that can reduce most of the risks. Having comprehensive protocols known by all involved together with well-trained staff at the department with dedicated functions and responsibilities will further reduce the risk for unintended irradiations of patient. Having a well-designed QA system with the appropriate barriers have the possibility of producing high quality radiotherapy, which will also result in better outcome for the patients. The international head and neck trial illustrates very well the importance of accurate radiotherapy.

1. Background

Radiotherapy is evolving fast with new technologies and modalities launched frequently by vendors and by research groups. Today we have commercially available dynamic delivery of rotational intensity modulated treatments e.g. RapidArc, VMAT and Tomotherapy. Intensity modulated radiotherapy (IMRT) is today the standard modality at many departments. We also have systems that adjust the beam delivery according to the patients motions i.e. gating systems [1] and soon we will also have commercial system tracking the movements of the tumour [2, 3]. We have also seen the incorporation of computerised tomography during treatment with equipment mounted on the linear accelerator i.e. cone-beam CT (CBCT) which have given us new possibilities of increasing the precision and accuracy of the patients position but also the opportunity to adapt our treatment delivery to the target of the day. Selecting the plan of the day based on imaging prior to irradiation is possible today [4-6]. The latest development is the incorporation of MR-imaging with the treatment units. Research groups in Utrecht, The Netherlands [7], Calgary, Canada [8] and Sydney, Australia [9] are examples of this. A commercial product Viewray® is also available [10] combining three rotating Co-60 sources equipped with MLC with a MRI system for on-line imaging.

The challenge for medical physicists and other professionals within radiation oncology is to cope with this accelerated process of new modalities and technologies. Especially regarding acceptance and commissioning of the new equipment, but also creating new procedures for the daily work to establish a safe environment for these modalities for patients but also for staff. This pushes our quality management systems to its limits where new quality assurance programs and their quality controls have to be set up and organised. How will we assure that these systems will be able to deliver the accurate and precise absorbed dose (dose distribution) as nicely planned and displayed on our treatment planning stations? We do not only have to do it once, we have to do it for all fractions through the treatment course. In short, the right dose at the right place at the right time. Decision-making is also part of this when we start to think about selecting the best plan for the individual patient from our multi-dimensional environment of modalities. Reports regarding improving safety in radiation oncology has been published during the last 10 years [11-13].

2. New approaches for quality assurance

Radiation oncology cannot continue in the same path regarding quality assurance (QA) and especially not how quality controls (QC) are set-up and performed. The all-mighty model has been to check what ever is possible to check e.g. plans, monitor units, chart checks prior, during and after treatment etc… Even patient specific QA measurement programs have been introduced draining the resources of the medical physics department. Have these programs prevented us from accidents in radiotherapy?

Several severe accidents have been reported the last decade or so (even earlier when technology have passed the users) and some examples are given here (details regarding these accidents can also be found at IAEAs Radiation Protection of Patients website [14]):

- UK, 1982-90: incorrect SSD correction (applied it twice due to lack of knowledge and communication) 1045 patients where under-dosed up to 30%, >492 RT failures (Ash and Bates [15]).
- Spain, 1990: Linac 'repair' led to 36MeV e-beam no matter what was programmed. No dosimetry check. 27 patients overdose, 15 deaths [16, 17].
- France, 2004: incorrect MU for dynamic wedge. 23 patients overdosed 20%, 4 deaths [18].
- Glasgow,2006: incorrect calculation of MU. Planner thought TPS calculated MU/Gy and not MU/fraction. 67% overdose results in death [19].
- France, 2006-7: large ion chamber used for SRS. 145 overdoses [20].
- US, NYC, IMRT without MLC control points, overdose 13 Gy/fraction three times, one patient dead [21].

Thus to have a safer radiotherapy environment in the future we have to change our way of thinking and look at what we are doing from a new perspective. We need a more process-oriented approach.

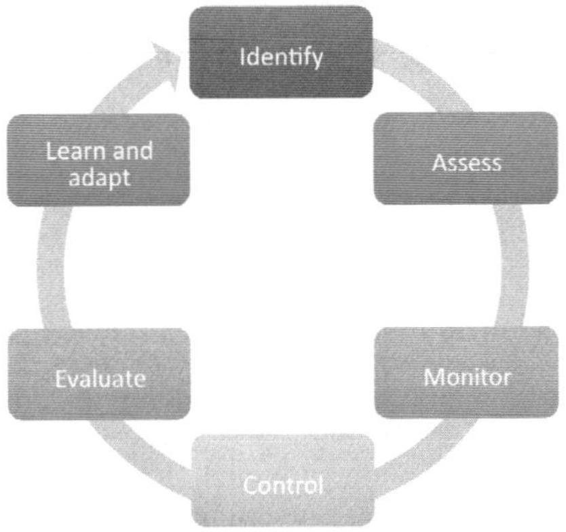

Figure 1. The wheel of improvements for guiding quality management.

Referring to figure 1, we have to look at our radiotherapy process with an approach of identifying potential risks, define and set barriers to assure that potential deviations from the intended treatment plan never reach the patient.

One needs a team made up of representatives of all professionals in radiation oncology who identifies the risks in the radiotherapy process. Identified risks can be assessed and shared within the radiotherapy community. At the identified control points, QC or barriers are set up to prevent errors from disseminate further through down the line. All QCs must be monitored and to have defined tolerances and action levels. Control and evaluation of the effectiveness is mandatory to maintain a potential of improving the radiotherapy process. Finally we have to learn and feedback our experiences etc and fine-tune our QM/QA system.

Effective tools for this do exist in other fields such as aviation, nuclear power, off shore and also from the manufacturing industry. For short we have a) process mapping (identify the main processes and their sub processes), b) identify the processes which are more error prone using established methods as FMEA (Failure Mode and Error Analysis) and RCA (Root Cause Analysis) [22, 23] c) identify measures (QC or barriers) for control [24] and d) using SPC (Statistical Process Control) to determine tolerance and action levels for QC based on measurements [25-27].

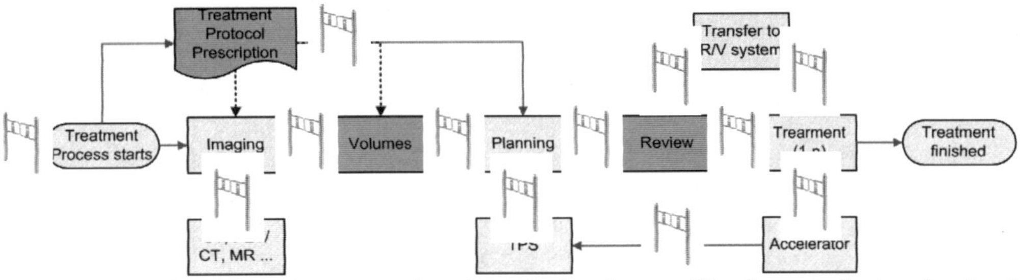

Figure 2. Example of a crude process chart for radiation therapy. The fences indicate safety barriers or QCs between the different sub-processes.

Additionally, one must also have control of the barriers set up in the process and an effective way is to score/report actual and potential deviations thru e.g. an Incident Learning Systems (ILS). Today we have several systems for anonymous reporting thus allows us to get information on risky steps in the process. These are the international system SAFety in Radiation Oncology (SAFRON) [28] and the recently announced US system from ASTRO/AAPM Radiation Oncology Incident Learning System (RO-ILS) [29] and the Canadian National System for Incident Reporting: Radiation Therapy (NSIR-RT) [30]. A lot of national systems exist too e.g. the UK has a very comprehensive reporting system within healthcare including radiation oncology [31] as well as local systems [32, 33]). A predecessor to these were the Radiation Oncology Safety Information System (ROSIS) set up in 2006 within a web environment and which received around 1000 reports in a short time (Cunningham *et al* [34]).

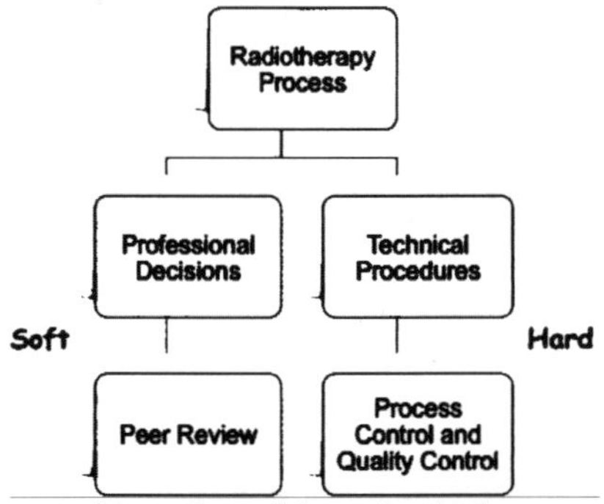

Figure 3. Quality assurance with its two sides - the soft branch which covers all the decisions taken in a patients process and the hard branch with all technical QC as dosimetry checks, plan checks etc.

One must remember that the severe events in radiation oncology that we know so well from literature and from various lectures on safety is only the tip of an iceberg. In daily practice deviations from the intended process occurs frequently, but our QC or barriers catch and corrects them i.e. keeps the process within its limits. This emphasis the use of an ILS for potential advert events too. We have a great opportunity to learn from these.

It is not only the technical part of a QM or QA system that is important, the soft part with an environment that has an open, no-blame, just culture and an organisation that sees all mistakes as an opportunity to learn and improve is even more important.

Peer-review of decisions taken in radiotherapy is another important part of a modern QM system, (figure 3). One can consider this as the soft side of the system compared to all the technical (or hard) QC that exists to assure the dosimetry. This soft part of the system includes e.g. decision to treat (in our case with radiation), prescription, treatment modality, volume(s) delineation, modality, treatment plan etc. All these steps involve a lot of subjective decisions by all professionals especially radiation oncologists. Marks *et al* have discussed this further [35].

Similar to peer-review of the soft branch in radiation oncology it is also of importance for the medical physics to have a system of internal and external audit of the commissioning process. For example, a linear accelerator should never be clinical before an external audit been taken place. For internal audit one should also consider to always having two medical physicists performing acceptance and commissioning [36].

3. The role of 3D dosimetry

Dosimetric QC is one of the major activities the clinical medical physicist is involved in regarding improving the safety within radiation oncology. This can of course be performed in many ways but the topic here is the role of 3D dosimetry. Today we have more or less three methods for 3D dosimetry:

- 3D detectors.
- 2D detectors which are complemented with reconstructions to 3D.
- Calculations based on either pre-planning data or data collected during the treatment or a combination.

The only fully 3D dosimeter in radiotherapy is the radiosensitive gel/plastic system, which are either read-out in a MRI system or in an optical laser scanning CT-like system. These system have displayed application and feasibility in many situations e.g. VMAT, RapidArc, protons, gating, tracking, protons [37-41] However, there is none (to the author's knowledge) reports regarding its role either in commissioning or in standardised and periodical QC measurements.

Regarding 2D systems many efforts have been reported using portal imaging systems integrated with the linear accelerator for dosimetry [42-45]. The concept is to use the transmitted fluence information and after appropriate corrections for different scatter sources and several other methods (e.g. back-projection) calculate the absorbed dose distribution within the patient. One institution that has implemented this fully is the NKI in Amsterdam who today do 3D dosimetry reconstruction of the dose distribution for all patients treated i.e. about 6000 patients annually. This group has also reported what type of deviations etc they have found during clinical practice of 3D-dosimetry (Mans *et al* [46, 47]). The conclusion was that about 4% of the treatments were associated with some kind of deviation from the original plan. Some of the methods also combine this with a 3D cone beam CT during treatment and thus basing the calculations on the information obtained.

The third method for 3D dosimetry is calculations based on the delivered treatment e.g. the 4D Monte Carlo simulations of continuously variable beam configurations by the group from British Columbia [48, 49]. The idea is to use the recorded data of the delivered treatment from the linacs log files. Their system also has the possibility to calculate the planar dose distribution to a detector plane either at the entrance or at the exit side. A system based on these principles may replace today's patient specific measurement based QC as many clinics still are performing (see also figure 4).

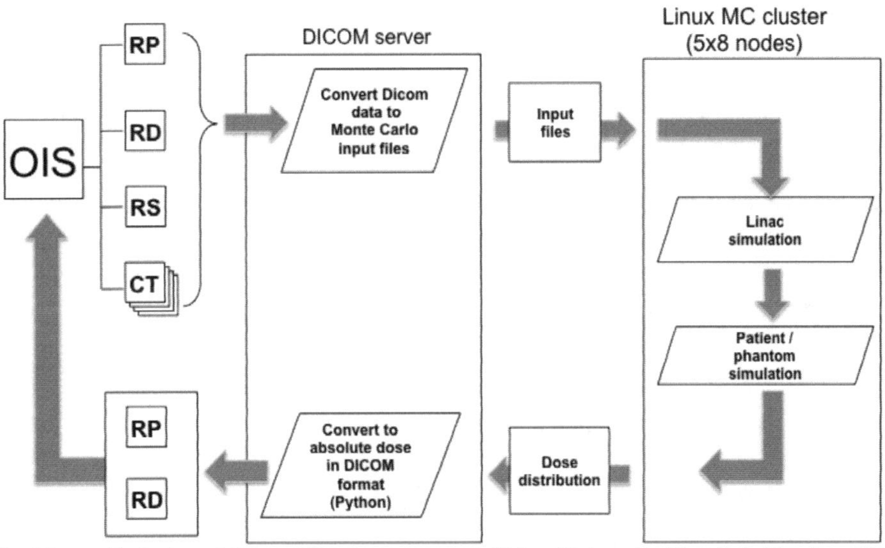

Figure 4. The Monte Carlo based 3D verification system at Skåne University Hospital. The data (DICOM RP – plan, RD - dose, RS - structures and CT slices) from the OIS (Oncological Information System) including all log files from the delivery are exported and converted to input files to the linac simulation package (BEAMnrc) and the resulting phase spaces are communicated to the DOSXYZnrc system to calculate the absorbed dose distribution in the patient. These files are then exported back to the OIS for evaluation (From Cronholm and Nordqvist personnel communication 2014).

4. Quality assurance metrics

The fast adoption of the gamma analysis method for patient specific QC is probably well accepted by the radiation oncology community, however, one must point out that this method is maybe not adequate for all situations it have been applied to. There is a belief that the gamma analysis with its standard criteria of 3 % global dose difference and 3 mm (3%G/3mm) is applicable in many situations. For example using these criteria for commissioning have been nicely summarised to be dis-advantages (Nelms *et al* [50]) Several serious errors may be hidden within this "standard" criterion, which maybe would be obvious if other evaluations methods would have been used. Thus one has to *know and understand the particulars dosimetry system completely, including its limitations, before applying it to a particular validation task* (Schreiner's 1st commandment [51]). A typical example of this are the conclusions from the above-cited paper; "Overreliance on the insensitive metric is counterproductive to quality improvement and can lead to the sense of complacency among the clinical physicists." And "IMRT and VMAT commissioning, along with product validation, would benefit from the retirement of the 3%/3 mm passing rates as a primary metric of performance, and the adoption instead of tighter tolerances, more diligent diagnostics, and more thorough analysis." (Nelms *et al* [39]).

In the future we maybe also see systems that ask for the operators consent before a diagnostic or therapeutic procedure is started. One can just imagine the situation in NYC when the MLC control points were missing if the system would have asked the therapist "Do you want to deliver 13 Gy with these setting on the linac" before the person pressed "BEAM-ON"?

5. Clinical outcome

Several examples show that good radiotherapy matters. For example, the clinical study TROG 0.2.02 HeadStart that compared two different drugs in combination with radiotherapy showed no difference between the two arms neither in overall survival or local control (Rishin *et al* [52]). More than 800 patients were included from 89 institutes in 16 countries. After re-evaluation of each patient's treatment plan regarding the compliance to the treatment protocol one excellently showed that significant difference existed between outcome (overall survival and/or local control) and the compliancy to the treatment-plan protocol (Peters *et al* [53]).

The QA paper of the Swedish study of accelerated radiotherapy for head and neck cancers (ARTSCAN) showed the importance of having consensus meetings with all involved professionals. (Johansson *et al* [54]). This is in agreement with the Schreiner's 2nd commandment - "*Engage in the clinical exchange of ideas and knowledge through publication in scientific journals, and, perhaps more importantly, through regular communication, meetings and workshops with colleagues locally, nationally and internationally*"[40]. It was also concluded that even if the participating department worked at a high technical level deviations were discovered by the QC procedures and directed the treatment into the correct treatment. The QA team's conclusion was "the omission of a comprehensive QA procedure is risky"[43] From the report of the clinical outcome of this study one can read "*In this study every effort was made to assure a similar quality of the radiotherapy throughout the study and at all centres. It is also possible that a stricter QA-process lead to an improved outcome for the conventional treatment* (Rishin et al [41]). *This may, in turn, have lead to less potential for improvement in the experimental arm of the study.*" (Zackrisson *et al* [55])

An example where maybe more extensive risk analysis and full three-dimensional end-2-end dosimetry test may have revealed that the margins used for an image-guided treatment protocol for prostate were too optimistically reduced is the paper by Engels *et al* 2009 [56].

Weber *et al* [57] have published a compilation of results from several clinical studies where they concluded "*These QA data stemming from prospective clinical trials show undisputedly that non adherence to protocol specified RT requirements is associated with reduced survival, local control and potentially increased toxicity.*"

Recently a simulation of various QA levels for an head and neck study was published. This study concluded firstly that increased QA level should lead to better patient outcome. Secondly, from a health care perspective it is also cost effective with a higher level of QA since the less number of patients treated for recurrence [58].

6. Summary

The 3D dosimetry systems available at various clinical practical levels will in the future play a larger role due to its possibility of integrating the dose over the whole studied volume. The gel/plastic dosimeter is best suitable for end-to-end tests, at least as long as the read-out system requires a MRI-system. Probably the future will give us clinical systems based on light-based CT scanning systems. The planar Epid based systems are already today suitable for on-line patient in-vivo dosimetry. An interesting approach is the Monte Carlo simulations based on the recorded log files from the treatment unit. The most extensive model uses MC calculations of the given treatment and predict the dose in the patient CT-set as well as the planar dose distribution to a detector either at the entrance or at the exit plane.

There is undisputable evidence in the literature that a well-designed QA programme with appropriate and effective QC have the potential of resulting in high quality radiotherapy for the patients. An effective system should be designed where the soft QCs e.g. peer-review are included routinely. This gives radiation oncology the possibility to reach the level of clinical outcome that is possible. In other words we are giving the patients the maximum potential to get the most out of the offered treatment. Thus high quality is synonymous with high patient safety.

7. References

1. Korreman S S *et al* 2005 *Radiother. Oncol.* **76** 311-8
2. Poulsen P R *et al* 2012 *Int. J. Radiat. Oncol. Biol. Phys.* **83** e265-71
3. Falk M *et al* 2012 *Med. Phys.* **39** 1588-94
4. Vestergaard A *et al* 2014 *Acta Oncol.* **53** 997-1004
5. Leinders S M *et al* 2013 *Int. J. Radiat. Oncol. Biol. Phys.* **87** 1016-21
6. Heijkoop S T *et al* 2014 *Int. J. Radiat. Oncol. Biol. Phys.* (in press)
7. Lagendijk J J *et al* 2014 *Semin. Radiat. Oncol.* **24** 207-9
8. Fallone B G *et al* 2014 *Semin. Radiat. Oncol.* **24** 200-2
9. Keall P J *et al* 2014 *Semin. Radiat. Oncol.* **24** 203-6
10. Web address as of Sept 20 2014: http://www.viewray.com
11. Dunscombe P 2012 *Front. Oncol.* **28** 129
12. Royal College of Radiologist 2008 https://www.rcr.ac.uk/publications.aspx?PageID=149&PublicationID=281
13. American Society for Radiation Oncology 2012 Safety is no accident: A Framework for Quality Radiation Oncology and Care,
14. Web address of Sept 22 2014: https://rpop.iaea.org/RPOP/RPoP/Content/AdditionalResources/Training/1_TrainingMaterial/AccidentPreventionRadiotherapy.htm
15. Ash D and Bates T 1994 *Clinical Oncology* **6** 214-25
16. Sociedad Española de Física Médica 1991 The Accident of the Linear Accelerator in the "Hospital Clínico de Zaragoza", SEFM, Madrid
17. Medical Management of Radiation Accidents, Second Edition, edited by Igor Gusev, Angelina Guskova, Fred A. Mettle, ISBN 0-8493-7004-3, 2001, CRC Press LLC.
18. Web address as of 12 Oct 2006 http://www.asn.gouv.fr/sections/rubriquesprincipales/actualites/communiques-presse/2006/accidents-radiotherapie
19. Unintended overexposure of patient Lisa Norris during radiotherapy treatment at the Beatson Oncology Centre, Glasgow in January 2006, Report of an investigation by the Inspector appointed by the Scottish Ministers for The Ionising Radiation (Medical Exposures) Regulations 2000.
20. Borius P Y *et al* 2010 *Neurochirurgie* **56** 368-73
21. Web address as of Sept 20 2014 http://www.nytimes.com/2010/01/24/health/24radiation.html?pagewanted=all&_r=0
22. Huq M S *et al* 2008 *Int. J. Radiation Oncology Biol. Phys.* **71** S170-3
23. Ford E C *et al* 2009 *Int. J. Radiation Oncology Biol. Phys.* **74** 852-8
24. Noel C E *et al* 2014 *Med. Phys.* **41** 081717
25. Rath F 2008 *Int. J. Radiation Oncology Biol. Phys.* **71** S187-90
26. Nordström F *et al* 2012 *Radiother. Oncol.* **102** 364-70
27. Pawlicki T *et al* 2008 *Radiother. Oncol.* **89** 330-7

28. Web address as of Sept 20 2014 https://rpop.iaea.org/RPOP/RPoP/Modules/login/safron-register.htm
29. Web address as of Sept 20 2014 https://www.astro.org/Clinical-Practice/Patient-Safety/ROILS/Intro.aspx
30. Web address as of Sept 20 2014 http://www.cpqr.ca/programs/national-incidents-reporting/
31. Web address as of Sept 20 2014 http://www.nrls.npsa.nhs.uk/report-a-patient-safety-incident/about-reporting-patient-safety-incidents/
32. Yeung T K *et al* 2005 *Radiother. Oncol.* **74** 283-91
33. Ford E C *et al* 2012 *Int. J. Radiation Oncology Biol. Phys.* **84** e263-9
34. Cunningham J *et al* 2010 *Radiother. Oncol* **97** 6017
35. Marks L B *et al* 2013 *Practical Radiation Oncology* **3** 149-56
36. Knöös T (in preparation)
37. Ceberg S *et al* 2010 *Phys. Med. Biol.* **55** 4885-98
38. Ceberg S *et al* 2008 *Phys. Med. Biol.* 53 N387-96
39. Gustavsson H *et al* 2004 *Phys Med Biol.* **49** 3847-55
40. Juang T *et al* 2014 *Med. Phys.* **41** 071706
41. Baldock C *et al* 2010 *Phys. Med. Biol.* **55** R1-63
42. van Zijtveld M *et al* 2007 *Radiother. Oncol.* **82** 201-7
43. van Zijtveld M *et al* 2009 *Med. Phys.* **36** 946-52
44. Mijnheer B *et al* 2013 *Med. Phys.* **40** 070903
45. Wendling M *et al* 2009 Med. Phys. **36** 3310-21
46. Mans A *et al* 2010 *Med. Phys.* **37** 2638-44
47. Rozendaal R A *et al* 2014 *Radiother. Oncol.* S0167-8140
48. Asuni G *et al* 2013 *Phys. Med. Biol.* **58** 3535-50
49. Popescu T 2014 International Workshop on Monte Carlo Techniques in Medical Physics. Quebec City, June 2014
50. Nelms B E *et al* 2013 *Med. Phys.* **40** 111722-15
51. Schreiner L J 2011 *Med. Phys.* **36** 189-91
52. Rischin D *et al* 2010 *J. Clin. Oncol.* **28** 2989-95
53. Peters L J *et al* 2010 *J. Clin. Oncol.* **28** 2996-3001
54. Johansson K A *et al* 2008 *Radiother. Oncol.* **87** 290-9
55. Zackrisson B *et al* 2011 *Radiother. Oncol.* **100** 41-8
56. Engels B *et al* 2009 *Int. J. Radiation Oncology Biol. Phys.* **74** 388-91
57. Weber D C *et al* 2012 *Radiother. Oncol.* **105** 4-8
58. Weber D C *et al* 2014 *Radiother. Oncol.* **111** 393-9

Audits for advanced treatment dosimetry

G S Ibbott[1] and D I Thwaites[2]
[1]Department of Radiation Physics, University of Texas MD Anderson Cancer Center, 1515 Holcombe Blvd., Houston, TX 77030, USA

E-mail: gibbott@mdanderson.org

[2]Institute of Medical Physics, School of Physics, University of Sydney, Camperdown, Sydney, NSW 2006, Australia

E-mail: david.thwaites@sydney.edu.au

Abstract. Radiation therapy has advanced rapidly over the last few decades, progressing from 3D conformal treatment to image-guided intensity modulated therapy of several different flavors, both 3D and 4D and to adaptive radiotherapy. The use of intensity modulation has increased the complexity of quality assurance and essentially eliminated the physicist's ability to judge the validity of a treatment plan, even approximately, on the basis of appearance and experience. Instead, complex QA devices and procedures are required at the institutional level. Similarly, the assessment of treatment quality through remote and on-site audits also requires greater sophistication. The introduction of 3D and 4D dosimetry into external audit systems must follow, to enable quality assurance systems to perform meaningful and thorough audits.

1. Background

In recent years the approach to treating cancer patients with radiation therapy has evolved considerably. In the past, it was common for clinicians to base treatment decisions on their individual prior experiences, but now it is more common to employ "evidence-based medicine" and derive prescriptions from data obtained through clinical trials [1]. The highest-level evidence for any type of therapy originates from systematic reviews of randomized clinical trials.

As an example, according to data from the Imaging and Radiation Oncology Core-Houston (IROC-H - formerly the Radiological Physics Center, RPC), nearly 70% of all US radiation therapy centers participate to some degree in cooperative group clinical trials [2]. About 25% of the centers collectively enrol more than 1,000 patients per year on protocols managed by one or more of the clinical trials groups. For a treatment center to join and continue membership in the cooperative groups, the institution's physics staff must perform specific quality assurance (QA) procedures. For protocols that require or permit radiation therapy, there are particular radiation therapy quality assurance and data submission requirements for each patient entered into the trial. Participation in certain advanced technology protocols such as those requiring three-dimensional (3-D) conformal treatment, intensity-modulated radiation therapy (IMRT), volumetric modulated arc therapy (VMAT), stereotactic radiation therapy, or brachytherapy require a substantial physics effort to qualify the institution to enter patients. Further physics effort often is necessary to submit the required data for each patient treated under protocol.

The quality assurance procedures required for participation in clinical trials establish a baseline for institutional QA and as one component they include external audits of dosimetry, planning and delivery. As advanced technologies and treatments are often involved in leading edge clinical trials, significant audit development for such treatments has often been in the context of trial QA. It is critical that these are built on previous 'simpler' levels of audit, as those levels must be fully verified before moving on to more complex levels. However increasingly such advanced treatment audits have moved away from auditing single parameters, or sets of parameters, towards comprehensive 'end-to-end' testing, considering the whole treatment chain as experienced by a patient and using phantoms that are more representative of patients. Clinical trial audits have often also helped to widen adoption of advanced treatments, as this QA has provided confidence in their implementation and delivery. The auditing used in clinical trial QA can augment a hospital's own QA program by bringing and imposing an independent review of the work performed by clinical medical physicists. A recent review of dosimetry for audits and clinical trials can be found in Kron et al [3] and more generally for QA of clinical trials in Ibbott et al [4], some of which has been used as the basis for parts of this summary. It may be noted that dosimetry is taken here in its widest sense to encompass dose measurement, dose performance of the equipment and the whole treatment process, dose determination (eg.treatment planning dosimetry) and dose delivery, i.e. all areas of clinical dosimetry.

More generally, intercomparisons and audits have been developed over at least the last 5 decades at the local, national or international level, not specifically linked to trials, but rather to support the development and improvement of quality and safety in radiotherapy, to avoid errors, to provide confidence in the accuracy of treatment dosimetry generally and to provide a process for centres to ensure new machines and processes can be independently verified before wider clinical use [5]. To varying degrees, such systems have developed through time to audit more complex dosimetry and treatments, with the aim of providing tests and verification that reflect the development of clinical practice, although in general audit sophistication lags behind practice [3, 6-9].

2. Role of audits in radiation therapy clinical trials and link to credentialing for trials

Multi-center clinical trials operated through co-operative clinical trial groups (CCTGs) such as the Radiation Therapy Oncology Group (RTOG, now part of the NRG Oncology Group), European Organization for Research and Treatment of Cancer Radiation Oncology Group (EORTC-ROG), Trans-Tasman Radiation Oncology Group (TROG) etc. generally have the greatest impact on changing clinical practice as these groups set high standards for trial design, QA, and management, resulting in high quality data to support trial results that can be reproduced in the clinic. Although the process of developing a clinical trial may vary among the groups, the same general principles apply which in turn provides an opportunity for clinical trial groups to collaborate to increase patient recruitment [10]. QA programs should be set up to meet the specific needs of the trial and considering the effect of predicted intercenter variability in treatment delivery [11] on the trial analysis. It may be noted that, whilst hospital participation in clinical trials varies, it has been shown that patients treated at hospitals that contribute larger numbers of patients to trials do better than those with low contribution rates [12], implying a wider effect on quality that may well also affect routine clinical practice in those centres. Thus participation in clinical trials and associated QA may affect the way many patients, locally, nationally, and on the international scale, are treated.

A clinical trial may incorporate a credentialing program in an effort to minimize the number of protocol violations and improve the overall quality of the trial [13] and one part of that is satisfactory achievement of the relevant dosimetry audit requirements. Credentialing offers the opportunity to confirm investigators have read and understood the trial protocol and have the necessary resources to treat the patients in all arms of the trial in accordance with the trial's specifications. When trials allow the introduction of a new technique, credentialing is also an opportunity to provide education, peer review and an external formal assessment that the new technique has been appropriately commissioned prior to clinical implementation. For example, the UK trial PARSPORT allowed many centers to implement IMRT for head and neck cancers as a result of participation in the trial [14].

Human error has been identified as one of the major contributing factors in patient treatments and therefore some clinical trial groups will require process maps to be developed and submitted for approval prior to a center opening a clinical trial [15].

Clinical trials that require the use of advanced technologies such as IMRT and prostate brachytherapy are considered sufficiently challenging that institutions are required to demonstrate their ability to use these technologies before being permitted to register patients [16-18]. The IROC-H and other QA offices participate in the credentialing process for clinical trials through several cooperative groups. Credentialing for such clinical trials generally involves an evaluation of most if not all of the following aspects:

- Attestation to use of the particular advanced technology to treat patients previously
- Facility Questionnaire
- Knowledge Assessment Questionnaire
- Benchmark case or phantom
- Patient-specific plan QA review
- Electronic data submission
- QA & dosimetry review
- Clinical review by radiation oncologist
- Reviews of Patient Treatment Records

3. External audits of general dosimetry and QA performance

3.1 Remote audits of treatment machine output

Many national or international organizations conduct regular independent audits of treatment machine output calibration with mailed dosimeters. This is a critical first step for all audits, whether of simpler or advanced treatments. The main systems providing international coverage include:

- The International Atomic Energy Agency (IAEA) based in Vienna, which began its TLD service in 1969, aimed at supporting a wide range of lower income per capita UN member states [19, 20];
- IROC-H based in Houston, Texas, which has the largest program and monitors all of the institutions (>2000 institutions as of early 2014) that participate in NCI sponsored clinical trials, both within the USA and internationally [13, 21]. The IROC-H initiated its TLD program for photon beams in 1977 [22]. In 1982 electron beams were included, and in 2007, measurements of proton beams were initiated. In 2010, the IROC-H adopted the use of optically-stimulated luminescence dosimeters (OSLDs) and largely discontinued the use of TLDs for photons and electrons [23].
- The EQUAL-ESTRO service (European Quality Assurance Laboratory-ESTRO) located in Paris, now a commercial service not formally linked to ESTRO, although growing out of scientific and audit pilot studies aimed at supporting quality in radiotherapy in Europe that began within the ESTRO infrastructure in the early 1990s [24-29].

In addition, there is a wide range of national or regional systems. Many of these were initially supported by IAEA, or linked, developmental projects [7, 9] and using similar TLD methods; others developed independently, eg the Section of Outreach Radiation Oncology and Physics at the National Cancer Center of Japan, in Tokyo, using glass dosimeters [30]. Descriptions of many of these can be found, for example, in proceedings of IAEA dosimetry symposia, available at http://www.iaea.org/Publications/, eg [31].

Most external audit systems for beam output verification are relatively simple, eg the IAEA's simple-assembly TLD stand to be placed in a water phantom [19, 20]. The IROC-H's OSLD system is described here as an example. Each year, institutions receive a package containing a lightweight platform and acrylic mini-phantoms containing several OSLD "nanoDots" (Landauer, Inc., Glenwood, IL) for irradiation with each radiation beam. Instructions are supplied that explain the irradiation procedure and ask the institution to describe their calibration procedure. The blocks and other equipment are returned to the IROC-H where the OSLDs are analyzed. The IROC-H applies corrections for the differences in scatter between the institution's calibration conditions and the OSLD irradiation, and for fading, dose linearity and energy dependence of the OSLD system [23, 32].

The uncertainty of the system for measuring output of accelerators remotely has been evaluated to be 1.5% [32], expressed as the standard deviation of measurements of dose. Consequently, the IROC-H's measurement of an institution's output can be stated at an uncertainty of less than 5% using a 99% confidence interval and this (5%) has been established as the threshold for acceptability. When the OSLD measurement disagrees with an institution's stated dose by more than 5%, the IROC-H initiates further steps to resolve the discrepancy. If the discrepancy cannot be resolved through telephone call discussion and the review of procedures and documentation, an on-site dosimetry visit is scheduled.

Remote audits of machine calibration have been described by many other investigators [3, 7-9,19,20 24-28,30,31,33,34]. For example, the experience at the International Atomic Energy Agency (IAEA) in Vienna is comparable to that of the IROC-H in terms of numbers of separate institutions audited, although the frequency of audits is typically lower [19, 20]. Williams *et al* described the early experience of the Australian (ACDS) dosimetry audits using TLDs and OSLDs [35]. Other systems have used dosimeters made of a thermoluminescent glass rods [18, 34], alanine [36], lithium formate [37], etc. It may be noted that some audit systems have developed methods to use remote auditing for a much wider range of dosimetry parameters than just beam output, eg the IAEA-supported system of linked national audit networks has piloted, and in some cases routinely implemented, remote audits of field size output variation, beam quality, field symmetry, asymmetric outputs, tests of small fields, tests of dose changes under inhomogeneities and other planning system tests etc. [7-9, 19, 20].

3.2 Results of remote audits of machine calibration

As one of the two largest examples, the IROC-H has described results from a large series of annual calibration audits [2, 38]. Consistently each year, 5% to 6% of the US megavoltage beams audited with TLD have fallen outside of the IROC-H's ±5% dose or 5 mm electron depth dose criteria on the first measurement. The analysis indicated that the incorrectly-calibrated beams are distributed among approximately 10% to 20% of the institutions monitored by the IROC-H. This observation has been confirmed by on-site visit measurements using ion chambers; among institutions visited by the IROC-H, approximately 10% to 15% had one or more beams outside of the IROC-H's criteria on an annual basis that required an investigation by the IROC-H [2, 38].

The precedent for performing TLD audits annually was established by the IROC-H early in its development and all current trial data and results are based on having this level of QA. A review of EORTC trial results indicated that decreases in tumour control probability were associated with discrepancies in the beam calibration, as measured by a TLD audit program [11]. At the same time, increases in normal tissue morbidity were associated with discrepant high TLD measurements. The article also indicated that sequential TLD audits improved the uniformity of the clinical outcome and that small deviations in beam output might lead to clinically important variations in outcome. Mailed TLD audits were deemed to be an integral part of quality assurance for trials. Other studies have shown the importance of independent audits to assure clinical trial quality [12, 39].

The IROC-H data also showed that 41% of the institutions monitored by the IROC-H had exactly one discrepancy detected by the TLD program during the 10 years between 1998-2008 [38, 40]. However, thanks at least in part to the audit intervention, a much smaller percentage had two or more discrepancies during this period, a feature of most audit reporting of experience over time and over repeated audits. As indicated above, while calibration discrepancies were detected for 15 - 20% of the major contributors to clinical trials in any one year, these institutions did not consistently have such discrepancies. Instead, significant calibration errors apparently can occur at any institution at any time.

Approximately 230 new machines are installed each year at institutions participating in US NCI-funded clinical trials and may be subject to calibration errors as they are put into clinical service, with potentially serious results. These errors can occur as a result of changes in procedures and protocols, the recruiting of inexperienced personnel, operating new systems, etc. and are relatively frequently observed with the use of new treatment equipment.

The IAEA experience and that from the linked national groups in its developing national audit network projects is similar [7-9, 19, 20] in general terms. However this can be significantly modified overall, or in individual countries, by the much wider range of resource, staffing, training, equipment, infrastructure, prior audit experience, etc., as the centres audited are in many countries worldwide, with widely varying economic conditions.

3.3 On-site Visits for Dosimetry Audit and Dosimetry and QA Review

A dosimetry review audit has been recommended by several organizations, including the AAPM and the IAEA [41, 42]. In addition, some national audit systems, eg the UK Institute of Physics and Engineering in Medicine national audit network [6], use on-site visits, and not remote audits, as the basis of their regular routine audit. This has grown out of initial experience with ion chamber-based national dosimetry intercomparisons and audits [43, 44] and combines dosimetry audit and QA review at the same visit and enables greater flexibility in the implementation and development of routine audits [6, 45]. An independent on-site audit is especially important for solo practitioners but is a valuable exercise for all practicing clinical medical physicists. It need not be extensive, but should address key activities such as basic calibrations, the overall QA program and documentation

An independent dosimetry review visit consists of a review of the institution's QA procedures and documentation; a review of treatment records to ascertain the consistency of the procedures used for treatment planning and monitor unit calculations; and measurements of the radiation beams and radioactive sources [41, 42, 46]. The set of measurements to be carried out should be defined to align with the visit review remit and requirements. For example, depending on the scope of the visit, this should include mechanical alignment and accuracy of position readout devices, light versus radiation field congruency, calibration of treatment machine output, relative field-size dependence, percent depth dose, off-axis ratio, asymmetric jaw and irregular field parameters, and accessory transmission factors. Additional measurements could evaluate the basic data required for delivery of IMRT, including small-field output factors, and the performance of a multi-leaf collimator (MLC). A straight-forward spot-check of image-guided radiation therapy (IGRT) can be performed with a simple phantom [47]. Other modalities, such as brachytherapy source strength checks can also be included. Because of interest in treating protocol patients with protons, the IROC-H and other QA groups have more recently developed and implemented procedures for visits to proton-beam facilities [48].

The IROC-H has conducted on-site audits during its 45-year history, and has accumulated extensive measured data from several thousand photon beams which have been grouped into 96 combinations of manufacturer, model, and beam energy [49]. This database of "Standard Data" enables the IROC-H to provide assistance by comparing an institution's measured data with the Standard Data. Differences often point to measurement errors and help identify the source of calibration errors detected by a mailed audit.

3.4 Review of QA Programs

Some QA/audit systems, particularly those linked to clinical trial QA, formally review QA and dosimetry procedures and records from the participating institutions, to ascertain compliance with accepted published recommendations. These may change with time as recommendations evolve, but for example IROC-H currently compares to the guidelines in AAPM reports from Task Groups 40 and 142 on comprehensive quality assurance [50, 51]. Significant differences are found frequently, but in some cases these are justified by the institution's own procedures and measurements. More often, however, the institution has overlooked some component of recommended QA, or has allowed their program to lapse in some important aspect. A list of common failures or lapses in QA programs found by the IROC-H is given below [2]:

- QA records not available or maintained
- Annual calibrations or monthly checks not performed timely
- No record of comparison to clinical values on annual report
- No record of comparison of daily and monthly checks against annual "baseline" values
- Physicist review of daily checks not documented
- No record of corrective actions and repeat measurements
- Check of electron beam energy not performed as required by applicable regulations and recommendations
- Output or field flatness constancy with gantry angle not checked during annual calibration

3.5 Audits of advanced treatment methods

Whilst a number of advanced treatment audits have been linked to specific clinical trials (section 4) some have been carried out as more general audits of performance for planning and delivery of such treatments (ie not linked to a particular trial). Many of these have grown from earlier audits of 3DCRT, or of other lower level audits. For example, the IAEA is currently developing a TLD and radiochromic film-based IMRT audit, for single field delivery in a planar phantom and for combined field end-to-end tests in a head-and-neck-like phantom within its developing national audit group project [9], having previously developed increasingly complex audit levels involving non-reference and other complex and combined field audits [9]. In the early 2000s, an ESTRO project group developed and ran a general intercomparison of IMRT planning and delivery for a pelvic/prostate-like phantom across a range of centres with diverse TPS and delivery technology [52, 53] and other European advanced technology audits have been reported [54-58] which are not specifically linked to particular trials. Many of these have taken into account experience from earlier audits for 3DCRT [59-61]. The Australian ACDS has carried out some level III audits, building on other trial or general audits for IMRT [3, 62] and other Australian IMRT and advanced technology audits have been run [3, 63]. A general VMAT audit is currently underway [64] based on similar methodology to the UK rotational IMRT audit described below [65]; and pilot audits are currently being carried out in NSW for 4D lung treatment methods (3DCRT, IMRT and VMAT) using a locally-developed moving 'tumour' phantom and making dose measurements with both an ion chamber and radiochromic film [66, 67] and for FFF VMAT lung SBRT using an ion chamber and ArcCheck [68].

The UK audit group (RTTQA, www.uk-dan.co.uk) has carried out general IMRT audits [69] and rotational therapy audits [65, 70, 71] as well as national audits for brachytherapy [72] and is currently conducting an SBRT audit [73]. These are within a national interdepartmental audit structure that comprises regular routine audits within an audit network of 8 regional groups, linked by less frequent audits of beam calibration by the NPL and supplemented by one-off audits at the national level, as above. This structure grew out of the original national photon dose calibration and multi-field planning audits in the late 1980s [43] and electron audits in the early 90s [44], with the geographic regional network beginning to take shape in 1991/2 [6]. The latter used a range of simple geometric, semi-anatomic and pseudo-anatomic phantoms for flexibility for 3DCRT audit [5, 6, 43, 45, 74-76] some of which had originally been designed for clinical trial audits [77-80]. Almost all the audits have been via on-site visits and measurement methods have included ion chambers, TLDs, radiochromic film and latterly 2D arrays. Some audits have included in-vivo dosimetry using TLDs [81] or diodes [82] as one component.

Taking the UK IMRT audits as examples:

i) IMRT audit [69]: A mail-based IMRT audit was carried out between 2009-10. Measurements were made for each individual field in institution-selected current clinical IMRT plans. Each field was measured isocentrically in a flat water-equivalent phantom at a depth of 5 cm using film and alanine, processed centrally, with additional local ion chamber measurements also being made. 57/62 centres participated (78 plans submitted). For the film measurements, all 176 fields from less complex IMRT plans (mainly prostate and breast) achieved over a 95% gamma pass rate (using 3%/3mm for points above a 20% threshold). For more complex plans (mainly head-and-neck) 237/245 fields (98.7%) achieved more than a 95% gamma pass rate, using 4%/4mm criteria. 4/78 (5.1%) of the alanine measurements were > 5% different to the TPS predicted dose, three of which were large deviations (> 10%), but which were linked to procedural errors during the audit set-up and dose delivery and not to real clinical errors. All out of tolerance values were found to be within tolerance when repeated. Excluding those three results on that basis, the mean difference was 0.05% (sd 1.5%). The study concluded that the audit in 92% of UK centres supported confidence in implementation of IMRT in the country.

ii) rotational IMRT audit [65, 70, 71]: 34 UK cancer centres with 43 treatment delivery systems (25 Varian and 12 Elekta VMAT 6 helical tomotherapy) took part in the audits in 2013. A complex virtual phantom planning exercise with 5 PTVs and one OAR, designed to test limits of the delivery system [71], and a clinical trial planning exercise were carried out and independently measured in each institution using a PTW Octavius II phantom with a PTW seven29 2D-array [65]. Point dose differences and global gamma index were calculated in regions corresponding to PTVs and OARs. Point dose differences for the virtual phantom test gave a mean (±sd) of 0.1% (±2.6%) and for the clinical trial plans were 0.2% (±2.0%). For the virtual phantom test, 34/43 institutions achieved all measured planes with > 95% pixels passing gamma <1 at 3%/3mm. rising to 42/43 for the clinical trial plans. Again the study concluded accurate clinical implementation in the country.

4. Audits of advanced treatment dosimetry for clinical trials

4.1 Results of treatment planning benchmark tests

Some trial QA systems have developed techniques to credential institutions through the use of a specific treatment planning exercises called a "benchmark case" and this is an increasingly standard requirement, particularly for more advanced treatments, incorporating full 'dummy runs' of planning cases to meet trial protocol requirements. Its purpose is to test whether an institution can meet the trial-specified requirements, such as dose prescription, contouring, etc. To comply, an institution must download a standardized CT data set (or comparable imaging information) and generate a treatment plan that complies with the requirements of the relevant trial. This must then be submitted digitally for evaluation, generally consisting of review of target volume contours and DVHs [83, 84]. For IROC-H, it can also include an independent calculation of dose to the target. When benchmark cases fail to meet the criteria, the QA office can contact the institution, explain the discrepancies, and work with the institution to resolve them. The follow-up generally consists of irradiation of an anthropomorphic phantom as a more definitive end-to-end test of the treatment planning capability. The comprehensive EORTC trial QA procedures have emphasised dummy runs for some considerable time as summarised in a range of reviews of their QA activity [85-90], linking generic QA with specific planning QA on virtual phantoms and with practical measurement-based QA.

4.2 Results of anthropomorphic phantom reviews

For the more complex technologies and treatments, clinical trial QA systems are now more frequently requiring institutions to simulate, plan, and treat a geometric or anthropomorphic phantom in ways as close to those used for patients as possible (an "end-to-end test"). If the protocol requires a benchmark treatment plan (section 4.1), some QA offices such as the IROC-H review the institution's plan and re-calculate the doses at key locations to evaluate the accuracy of the planning system. This calculation is possible because the IROC-H has either measured data collected through visits to the institution, or is able to use its "standard data" [49]. When anthropomorphic phantoms are used, the delivered dose must be compared with the institution's plan to determine the agreement [21, 91-94] Similar approaches are used by other trial audit systems, eg a range of audits of institutions participating in trials run by the Trans-Tasman Radiation Oncology Group were reported or summarised by Kron et al. [3, 17] and Ebert et al [63]; and for the UK, for example, Clark et al reported detailed audit methods and results for the PARSPORT head-and-neck IMRT clinical trial [14, 95, 96]. The EORTC has supported a range of specific audits linked to trials including advanced treatments [85-90].

As one example, a review of the IROC-H results was recently reported by Molineu, et al [94]. During the time period 2001 to 2011 the IROC-H mailed head-and-neck phantoms to 763 distinct institutions. The phantom (figure 1) contains a primary and secondary PTV, along with an OAR and can hold point dosimeters and radiochromic film in two orientations. The institutions were instructed to perform imaging, develop a treatment plan using IMRT techniques, and then deliver the treatment to the phantom. A total of 1,139 irradiations were analyzed. Of these, 929 irradiations or 82% successfully met the irradiation criteria of 7% and 4 mm distance-to-agreement. (Had the dose criterion been reduced to 5%, only 69% of the irradiations would have passed.) The passing rate has increased steadily since the phantom's introduction in 2001 from approximately 69% to a current rate of 91% [94]. It has been noted that some institutions irradiated such phantoms multiple times to improve their initial irradiation results, test different treatment planning system algorithms or test different treatment delivery systems. As a comment on the usefulness of end-to-end measurement audits of this nature, a study can be considered which compared the results from the benchmark case used by the Quality Assurance Review Center as an IMRT treatment planning capability exercise, to the IROC-H IMRT H&N phantom irradiation results from the same group of 113 institutions. All of the institutions passed the IMRT benchmark case (planning study and self-reported IMRT QA measurement) while 20 of these same institutions did not pass the end-to-end independent phantom irradiation. The phantom audit is capable of detecting imaging, data transfer and delivery errors that cannot be detected by the institution's completion of a benchmark case. In addition, each institution's own IMRT QA results of their IMRT benchmark plans passed their own QA criteria. An independent end-to-end anthropomorphic QA audit is capable of detecting dosimetry errors that might otherwise go undetected.

Besides the head-and-neck phantom the IROC-H system has developed a range of other specific phantoms, including pelvis, spine, liver, thorax, SRS head [91-93]. Whilst many options for point dosimetry and for radiochromic film dosimetry are possible and routine in such phantoms, and in commercial anthropomorphic phantoms, the use of true 3D dosimeters is not yet routine [97]. The feasibility of using 3D gel [98] and PRESAGE® [99, 100] dosimetry in the IROC-H head phantom has been tested but is not yet easily available. Similarly, some commercial anthropomorphic can be supplied with the option of gel-holding inserts, but these dosimetry systems are still not robust enough for wider audit use.

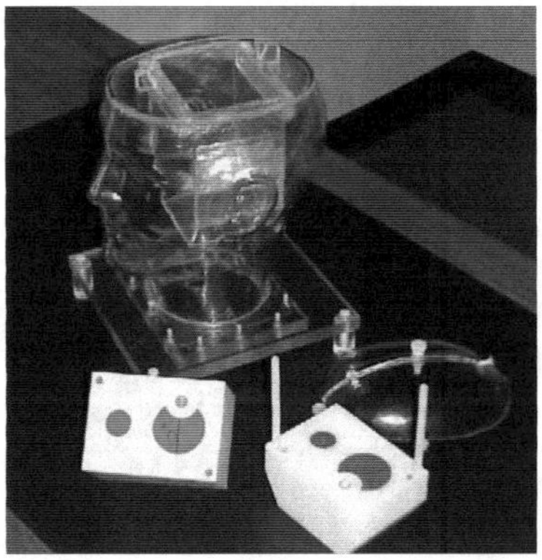

Figure 1. IROC-H head phantom

5. Final comments

Kron [3] identified some current challenges for audit dosimetry, which included: risk assessment approaches to audit design; scope and frequency for best use of limited resources; limited methods available for image-guided and adaptive radiotherapy, for motion management and for brachytherapy; and the need for access to treatment planning or trials review software [101] that can display dose distributions from a variety of different planning systems for comprehensive and flexible evaluation of dose distributions from measurements against those from TPS. Ebert et al [63] and the EORTC [10, 88, 89] have considered a range of organisational, administrative and financial issues relating to audit and practical aspects of audit measurements and challenges posed by the new technologies [88]. Many commentators and reviewers note that measurement audit cannot exist in isolation and increasingly must be part of a wider set of comprehensive audit of facility and systems performance [42].

6. Conclusions

Radiation therapy technology and techniques have been advancing rapidly over recent years and continue to do so. They have become increasingly complex through the widespread availability and implementation of IMRT, VMAT, other rotational and robotic delivery methods, etc. and all coupled with image-guidance. Increasingly too, other advanced methods are being used, including SBRT, FFF beams, adaptive radiotherapy, 4D radiotherapy, proton and other beams, etc. Institutional commissioning and QA has had to advance with these changes. Similarly dosimetry audit methods have had to evolve to meet the demands of advanced treatments. Many of these have been developed and piloted for clinical trial QA, as it is often in this context that advanced treatments begin to roll out more rapidly. It is critical for optimum trial outcome and confidence that participating centres are QA'd and audited carefully for these treatment methods, to provide comparison between centres of the level of performance consistency against defined external trial standards. A comprehensive trial quality assurance program can define the range of acceptable and unacceptable variations, detect and correct the causes of such variations, and document the frequency of variations. The ability of QA programs to reduce variations in treatment delivery and improve the quality of clinical trials and routine clinical practice in participating institutions has been demonstrated. Measurement and analysis methods for advanced treatment dosimetry QA and audit need to be accurate and precise, in line with demonstrating that the QA standards requirements are met. Ideally for complex 3D and 4D treatments, QA and audit measurement methods should be fully 3D and 4D, but practical systems that fully deliver this are still not available.

7. References

[1] CEBM. *Centre for Evidence-Based Medicine*. 2011; Available from: http://www.cebm.net.
[2] Ibbott G S 2010 *J. Phys: Conf. Ser.* **250** 1-7
[3] Kron T *et al* 2013 *J. Phys: Conf. Ser.* **444** 012014.
[4] Ibbott G S *et al* 2013 *Front. Oncol.* **3** 311
[5] Thwaites D I 2010 *IAEA SSDL News* **58** 7-9 (available from http://www-pub.iaea.org/books/IAEABooks/8501/SSDL-Newsletter-Issue-No-58-June-2010
[6] Thwaites D I *et al* 2003 *in Standards and Codes of Practice for Medical Radiation Dosimetry,* IAEA:Vienna, IAEA-STI-PUB-1153 **2** 183-90
[7] Izewska J and Thwaites D I 2003 *in Standards and Codes of Practice for Medical Radiation Dosimetry,* IAEA:Vienna, IAEA-STI-PUB-1153 **2** 249-68
[8] Izewska J *et al* 2007 *Radiother. Oncol.* **84** 67-74
[9] Izewska J *et al* 2011 *in International Symposium on Standards, Applications and Quality Assurance in Medical Radiation Dosimetry*, IAEA STI/PUB/1514; (IAEA: Vienna) **2** 443-51
[10] Melidis C *et al* 2014 *Radiother. Oncol.* **111** 327-9
[11] Bentzen S M *et al* 2000 *Eur J Cancer* **36** 615-20
[12] Peters L J *et al* 2010 *J. Clin. Onco.* **28** 2996-3001

[13] Followill D S *et al* 2012 *Front. Oncol.* **2** 198
[14] Clark C H *et al* 2009 *Br. J. Radiol.* **82** 585-94
[15] Mayles W P *et al* 2004 *Radiother. Oncol.* **73** 199-207
[16] Hurkmans C W *et al* 2011 *Int. J. Radiat. Oncol. Biol. Phys.* **80** 918-27
[17] Kron T *et al* 2002 *Int. J. Radiat. Oncol. Biol. Phys.* **52** 566-79
[18] Palta J R *et al* 2004 *Int. J. Radiat. Oncol. Biol. Phys.* **59** 1257-9; author reply 1259-61
[19] Izewska J and Andreo P 2000 *Radiother. Oncol.* **54** 65-72
[20] Izewska J *et al* 2011 *in International Symposium on Standards, Applications and Quality Assurance in Medical Radiation Dosimetry,* IAEA STI/PUB/1514; (IAEA: Vienna) **2** 335-48
[21] Ibbott G S *et al* 2008 *Int. J. Radiat. Oncol. Biol. Phys.* **71** (suppl 1) S71-5
[22] Kirby T H *et al* 1986 *Int. J. Radiat. Oncol. Biol. Phys.* **12** 261-5
[23] Homnick J *et al* 2008 *Med. Phys.* **35** 2994
[24] Dutreix A *et al* 1993 *Radiother. Oncol.* **29** 97-101
[25] Dutreix A *et al* 1994 *Radiother. Oncol. 32* 256-64
[26] Svensson *et al* 2000 *Radiother. Oncol.* **57** (suppl 1) S51-2
[27] Ferreira I H *et al* 2000 *Radiother. Oncol.* **55** 273-84
[28] Gomola I *et al* 2001 *Radiother. Oncol.* **58** 163-8
[29] Roue A *et al* 2006 *Radiother. Oncol.* **78** 78-83
[30] Mizuno H *et al* 2008 *Radiother. Oncol.* **86** 258-63
[31] IAEA 2011 *in International Symposium on Standards, Applications and Quality Assurance in Medical Radiation Dosimetry,* IAEA STI/PUB/1514 (IAEA: Vienna)
[32] Aguirre J *et al* 2011 *Med. Phys.* **38** 3515
[33] Rassiah P *et al* 2004 *Australas. Phys. Eng. Sci. Med.* **27** 25-9
[34] Rah J E *et al* 2009 *Appl. Radiat. Isot.* **67** 324-8
[35] Williams I *et al* 2012 *Australas. Phys. Eng. Sci. Med.* **35** 407-11
[36] Schaeken B *et al* 2011 *Radiother. Oncol.* **99** 94-6
[37] Olsson S *et al* 2011 *in International Symposium on Standards, Applications and Quality Assurance in Medical Radiation Dosimetry,* IAEA STI/PUB/1514; (IAEA: Vienna) **2** 423-31
[38] Ibbott G S 2009 *in Clinical Dosimetry Measurements in Radiotherapy* Madison, WI: Medical Physics Publishing.
[39] Pettersen M N *et al* 2008 *Radiother. Oncol.* **86** 195-9
[40] Ibbott G S 2011 *in Quality and Safety in Radiotherapy*. Boca Raton, FL: Taylor and Francis.
[41] Halvorsen P H *et al* 2005 *J. Appl. Clin. Med. Phys.* **6** 50-64
[42] IAEA *Comprehensive audits of radiotherapy practices: A tool for quality improvement - QUATRO2007*, 2007, International Atomic Energy Agency: Vienna.
[43] Thwaites D I *et al* 1992 *Phys. Med. Biol.* **37** 445-61
[44] Nisbet A and Thwaites D I 1997 *Phys. Med. Biol.* **42** 2393-409
[45] Thwaites D I 1996 *External audit in radiotherapy dosimetry, in 'Radiation Incidents',* (eds, Faulkner K and Harrison R M) BIR: London 21-8
[46] Chavaudra J *et al* 2007 *On-site visits to radiotherapy centres: medical physics procedures,* IAEA-TECDOC- 1543 , IAEA:Vienna
[47] Middleton M *et al* 2011 *Int. J. Radiat. Oncol. Biol. Phys.* **81** 1576-81
[48] Summers P *et al* 2012. *Med. Phys.* **39** 3866
[49] Lowenstein J *et al* 2012 *Med. Phys.* **39** 3754
[50] Kutcher G J *et al* 1994 *Med. Phys.* **21** 581-618
[51] Klein E E *et al* 2009 *Med. Phys.* **36** 4197-212
[52] Gillis S 2005 *Radiother. Oncol.***76** 340-53
[53] Bohsung J 2005 *Radiother. Oncol.* **76** 354-61
[54] Tomsej M 2005 *Radiother. Oncol.* **76** (suppl 1) S40
[55] Tomsej M 2007 *Radiother. Oncol.* **84** (suppl 1) S83

[56] Knoos T *et al* 2011 *in International Symposium on Standards, Applications and Quality Assurance in Medical Radiation Dosimetry,* IAEA STI/PUB/1514; (IAEA: Vienna) **2** 365-76
[57] Shiefer H *et al* 2010 *Med. Phys.* **37** 4424-31
[58] ESTRO 2008 *Guidelines for the verification of IMRT ESTRO booklet 9,* available at www.estro.be
[59] Johansson K-A 1987 *Radiother. Oncol.* **7** 269-79
[60] Wittkamper F W 1987 *Radiother. Oncol.* **9** 33-4
[61] van Bree N A M 1991 *Radiother. Oncol.* **22** 252-60
[62] Lehmann J *et al* 2014 *J. Phys.: Conf. Ser.* **489** 012094
[63] Ebert M A 2009 *J. Med. Imag. Radiat. Oncol.* **53** 119-31
[64] Haworth A 2014 private communication
[65] Hussein M *et al* 2013 *Radiother. Oncol.* **108** 78-85
[66] Kafrouni M *et al* 2014 *Australas. Phys. Eng. Sci. Med.* (in press)
[67] Caloz M *et al* 2014 *Proceedings of 8th International Conference on 3D Radiation Dosimetry (IC3DDose) 4-7 September 2014, Ystad, Sweden*
[68] Hansen C R *et al* 2014 *Proceedings of 8th International Conference on 3D Radiation Dosimetry (IC3DDose) 4-7 September 2014, Ystad, Sweden*
[69] Budgell G 2011 *Radiother. Oncol.* **99** 246-252
[70] Tsang Y *et al* 2013 *Br. J. Radiol.* **86** 1022-31
[71] Clark C *et al* 2014 *Radiother. Oncol.* (in press_
[72] Palmer A L 2013 *et al Phys. Med. Biol.* **58** 6623-40
[73] Clark C 2014 private communication
[74] Bonnett D *et al* 1994 *Br. J. Radiol.* **67** 275-82
[75] Blake S W 2002 *Br. J. Radiol.* **75** 754-81
[76] Palmer A 2011 *Br. J. Radiol.* **84** 733-41
[77] Aird E G A 1995 *Radiother. Oncol.* **36** 235-95
[78] Venables K *et al* 2001 *Phys. Med. Biol.* **46** 1937-48
[79] Moore A R *et al* 2003 *Clin. Onc.* **15** 58-9
[80] Moore A R *et al* 2006 *Radiother. Oncol.* **80** 82-85
[81] Venables K *et al* 2004 *Radiother. Oncol.* **71** 303-10
[82] Thwaites D I 2006 *et al Proc. Int. Conf. on quality assurance and new techniques in radiation medicine,* IAEA:Vienna, IAEA-CN-146, 402-03
[83] Mayo C S and Urie M M 2003 *Med. Dosim.* **28** 235-42
[84] Ulin K *et al* 2010 *Int. J. Radiat. Oncol. Biol. Phys.* **77** 1584-9
[85] Horiot J C *et al* 1993 *Radiother. Oncol.* **29** 81-4
[86] Kouloulios V E *et al* 2003 *J. Eur. Cancer* **39** 4303
[87] Poortmans P M *et al* 2005 *Eur. J. Surg. Oncol.* **31** 667-74
[88] Weber D C *et al* 2011 *Radiother. Oncol.* **100** 150-6
[89] Weber D C *et al* 2012 *Radiother. Oncol.* **105** 4-8
[90] Weber D C *et al* 2014 *Radiat. Oncol.* **9** 123
[91] Followill D S *et al* 2007 *Med. Phys.* **34** 2070-6
[92] Ibbott G S *et al* 2006 *Technol. Cancer Res. Treat.* **5** 481-7
[93] Molineu A *et al* 2005 *Int. J. Radiat. Oncol. Biol. Phys.* **63** 577-83
[94] Molineu A *et al* 2013 *Med. Phys.* **40** 022101
[95] Clark C *et al* 2009 *Radiother. Oncol.* **93** 102-108
[96] Nutting C *et al* 2011 *Lancet Oncol.* **12** 127-36
[97] Baldock C *et al* 2010 *Phys. Med. Biol.* **55** R1-63
[98] Babic S 2008 *Int. J. Radat. Oncol. Biol. Phys.* **70** 1281-91
[99] Sakhalkar H 2009 *Med. Phys.* **36** 3371-77
[100] Gorjiara T *et al* 2011 *Med. Phys.* **38** 2265-74
[101] Ebert M A *et al* 2008 *Radiother. Oncol.* **86** 200-10

Journal of Physics: Conference Series **573** (2015) 012003 doi:10.1088/1742-6596/573/1/012003

True 3D chemical dosimetry (gels, plastics): Development and clinical role

L J Schreiner
Department of Medical Physics, Cancer Centre of Southeastern Ontario at the Kingston General Hospital, Kingston, Ontario, Canada, K7L 5P9
Departments of Oncology and Physics, Queen's University, Kingston, Ontario, Canada K7L 5P9

E-mail: john.schreiner@krcc.on.ca

Abstract. Since the introduction of volumetric chemical dosimetry with Fricke gel dosimeters in the 1980s, three-dimensional (3D) dosimetry has been a promising technique for the clinic, since it provides a unique methodology for 3D dose measurement of the complex conformal dose distributions achieved by modern techniques such as Intensity Modulated and Volumetric Arc Radiation Therapy. In the last decade, the potential for improved clinical applicability has been advanced by the development of improved 3D dosimeters such as normoxic polymer gel systems, radiochromic plastics (such as PRESAGE) and, recently, newer radiochromic gel dosimeters. Some of these new 3D dosimetry systems were enabled by the availability of optical computed tomography imaging systems for fast dose readout. However, despite its promise, true 3D dosimetry is still not widely practiced in the community. Its use has been confined primarily to select centres of expertise and to specialised quality assurance or commissioning roles where other dosimetry techniques are difficult to implement. In this paper I review some of the current 3D chemical dosimeters available, discuss the requirements for their use and briefly review the roles that these systems can provide to complement the other dose delivery validation approaches available in the clinic. I conclude by describing two roles that may be uniquely served by 3D chemical dosimetry in end-to-end process testing and validation in the complex environment coming into play with the development of Image Guided Adaptive Radiation Therapy.

1. Introduction

Radiotherapy is a localized treatment used in the treatment of approximately 50% of all patients undergoing cancer care in North America. Treatment is personalized to the individual patient so that a sufficient (usually uniform) dose is delivered to the target to achieve tumour control, while the dose to adjacent normal tissue and organs at risk is limited in order to minimize unwanted complication (see figure 1). Over the years, radiation therapy has progressed considerably [1] by the development of three dimensional (3D) external beam conformal radiation therapy delivery such as provided by Intensity Modulated Radiation Therapy (IMRT), Volumetric Modulated Arc Therapy (VMAT), Stereotactic Ablative Radiation Therapy (SABR), etc. and through improved procedures for high-dose rate brachytherapy. The advances in treatment were made possible by enhanced three dimensional imaging for planning, by the superior treatment planning systems which more faithfully calculate radiation doses in the heterogeneous patient and better plan the treatment unit's delivery trajectory through inverse planning approaches, and by the sophisticated networking and control systems that enable the data to flow from the planning systems to the treatment unit [1]. The radiation delivery associated by these modern conformal techniques is complex and results in 3D shaped dose distributions structures that must be correctly registered to the patient's anatomy (see figure 2) to achieve the treatment intent. To this end treatment units have been enhanced with imaging to enable Image Guided Radiation Therapy (IGRT) for validation of the patient position immediately prior to treatment. Conventional IGRT systems use kV x-ray imaging with cone beam computed tomographic capabilities and novel

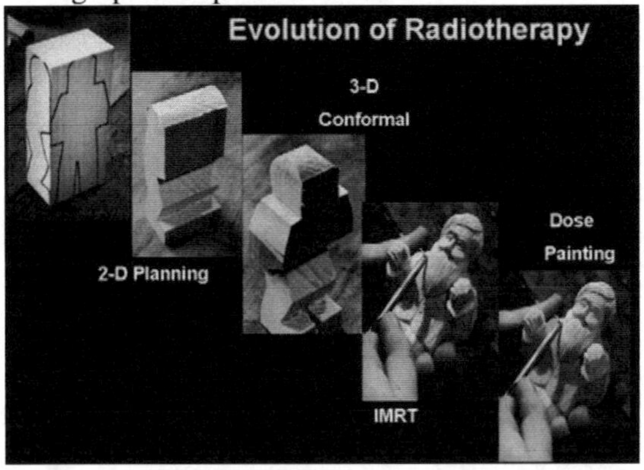

Figure 1. A simple analogy (from woodcarving) for modern radiation therapy [5]. The target treated by radiation often has a complex shape. The goal of the proposed research is to advance gel dosimetry to provide a 3D approach to delivery validation. The colours in the last panel suggest dose painting in which the dose to the target is intentionally not be uniform throughout (Rock Mackie, IC3D, 2010).

Figure 2. Dosimetric validation of modern radiation delivery is vital, since the radiation delivery has been made more precise. A miss that may have been acceptable in the past may in modern conformal delivery underdose target and overdose organs at risk. To account for displacement the target volumes are increased by some margin to allow for expected displacement over the course of treatment. (From [5])

units are under development incorporating magnetic resonance imaging systems [2-4]. The usual implementation of IGRT is to correct the observed displacements shown in figure 2 prior to the each treatment (see figure 3). The impact of such correction via IGRT has been a reduction of the margins used around targets during treatment, since these margins can now be patient specific rather than population based, again increasing the need for treatment validation. By using IGRT over the multiple fractions of the patient's treatment, there is a further potential to check for changes of anatomy in the region of the tumour and its surroundings [6]. Such changes, which may be seen over the weeks of treatment, have led to proposals for of off-line approaches in which the treatment plan is re-designed (daily or weekly) to maintain target coverage and healthy normal tissue sparing as the volume changes. Such Image Guided Adaptive Radiation Therapy (IGART) may incorporate the daily use of the on-line imaging (kV, MV or perhaps MRI) not only to ensure patient positioning or to monitor anatomy changes but also, perhaps, to flag re-planning (perhaps requiring new CT simulation, or maybe using the IGRT imaging itself [7]. In very simple terms, such *adaptive* radiation therapy approaches (ART) have defined a new class of radiation therapy in which the patient's treatment is modified during the course of care based on new information gained during the treatment [8, 9]. More sophisticated approaches for inter-fraction ART [10] are not limited to corrections for tissue geometry. The potential for on-line dosimetry (perhaps via exit beam dosimetry with electronic portal imaging devices, EPIDs, or tomotherapy CT detectors) during each treatment raises the feasibility of dose delivery monitoring. That is, with on-line dosimetry ART processes might be used to correct the under and overdosing in figure 2 in subsequent fractions of the treatment course by modifications of the future deliveries [11].

In all implementations, IGART processes consist of multiple steps for imaging, measurement, assessment, optimization, re-imaging, dose verification, etc. that may not be independent and further validation, perhaps with independent measurements, is necessary to establish the robustness of the adaptive treatment process. The challenge for delivery validation has been further increased [5] through the development of four dimensional treatment approaches in which the dose delivery is modified (through motion suppression techniques, increased margins, or beam gating) to account for the target motion inherent in the treatment of particular cancers.

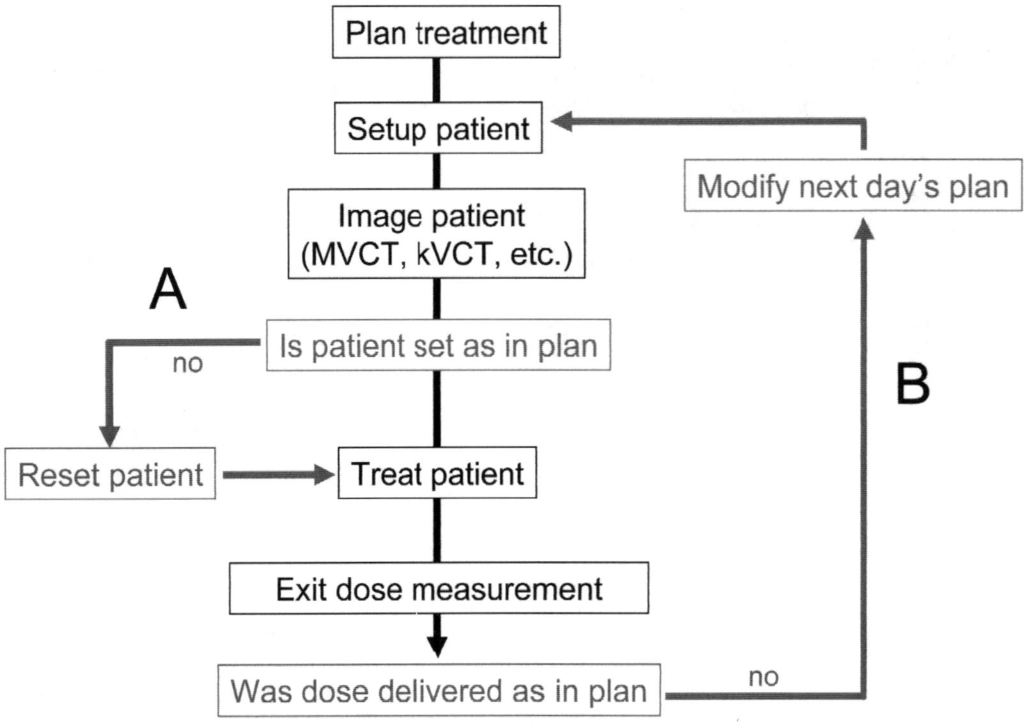

Figure 3. An example of two simple IGART processes: A) patient set-up verification at time of treatment (e.g., using kV CBCT image registration with a simulation CT reference image from the initial treatment planning) is already standard in the clinic. If a discrepancy is noted by the treating therapist the patient is moved to the correct position prior to irradiation. B) In a more sophisticated schema exit dose measurements could be used to determine the dose delivered in a given fraction. If the dose delivery was not as intended, one then could modify the next fraction to correct and bring the cumulative dose to the intended delivery over the multiple fractions. [from ref. 12].

All this is to say, modern radiation therapy has become very complex and as new approaches have advanced from 3DCT, through IMRT, VMAT, IGRT, ART to IGART - Image Guided ART, the quality assurance for technical components, the treatment unit and patient specific dose delivery validation, and the process assurance requirements have increased.

Numerous approaches have been established in the clinic to validate various points or steps in modern dose delivery [5, 13, 14]. In particular, there has been considerable development using 2D and 3D arrays of detectors to verify complex patient specific dose delivery [15-20]. And electronic portal imaging based assessment [21, 22] has been shown to provide convenient and sensitive patient specific delivery validation. These approaches have their place and are clinically very useful. But they typically provide sparse 3D data and only surrogate validation of 3D dose delivery. They do not fully fulfil the Resolution-Time-Accuracy-Precision (RTAP) performance criteria that were proposed by Mark Oldham *et al* [23, 24] as a useful benchmark for clinically useful true 3D dosimetry (see below). And there can be problems associated with their use if care is not taken; for example problems have been identified with the use of 2D data measurements of individual beams to validate patient specific IMRT delivery [25].

A novel class of scintillation and Cherenkov based radiation dosimeters has also been in development the last decades with recent advances to 3D measurement [26-28]. These detectors rely on the detection of immediate light emission from the irradiated media. Some of the systems under study do provide high resolution isotropic data; however, these systems are not dose integrating, and will also not be discussed further here.

This paper is a review of three dimensional chemical dosimeter systems that can provide, through some post irradiation imaging, continuous integrated dose measurement through an irradiated volume (see figure 4). The main dosimetry systems will be briefly described and the applicability of the dosimeters to the issues raised above addressed. This discussion is not new and many of the points addressed have been discussed previously [5, 12, 29]. My desire is to show that some of the considerations for the use of 3D dosimeters date back to the initial work motivating their development in the 1980's, while some are new. The description of these systems will be cursory; very comprehensive reviews of the systems, and of the requirements for good readout of the dose information, have been presented in past DosGel and IC3D proceedings articles freely available in the *Journal of Physics Conference Series*, and in the comprehensive review of polymer gel dosimeters by Baldock *et al* [30]. Additional fundamental descriptions of the systems are given in two comprehensive educational reviews by Oldham [24] and Schreiner and Olding [31].

This paper is intended to set the stage for remainder of the conference with the many contributions reviewing the fundamental science and technical challenges of 3D dosimetry and reporting the current status of 3D dosimetry in the clinic [14, 19, 20, 27, 28, 33, 40, 54, 56-59, 62, 64].

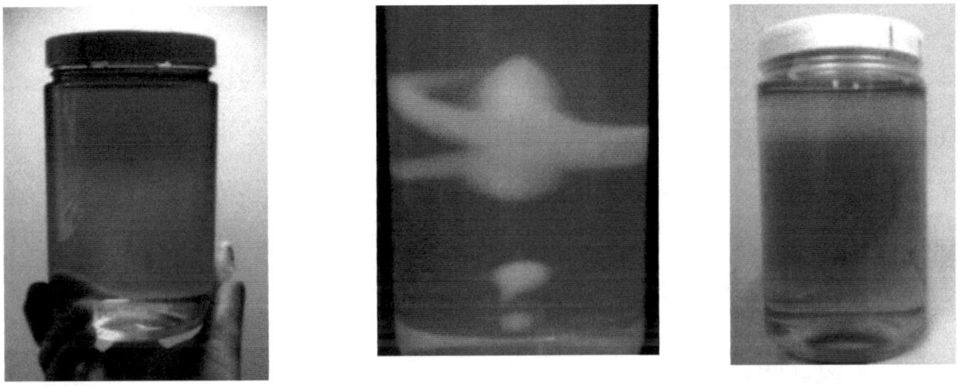

Figure 4. Illustration of the 3D dose information captured in three different gel dosimeters. (left) The Fricke-xylenol-orange-gelatin dosimeter shows a colour change in the volume irradiated with a12 MeV electron beams. (centre) The polyacrylamide polymer gel dosimeter shows increased scatter in the high dose areas radiated using a Cobalt-60 tomotherapy IMRT delivery. (right) A VMAT prostate irradiation of a Leuco-crystal-violet micelle gel dosimeter.

2. Chemical dosimeters for three dimensional measurement

As noted above, elements of the RTAP criteria [23, 24] are used in this work to set the definition of 'true' 3D dosimeters (with one additional feature). Under RTAP an ideal true 3D dosimetry system (dosimeter and associated readout) should be able to deliver dose measurements in a 3D volume with 1 mm isotropic spatial resolution in less than one hour with an accuracy of 3% and a precision of 1%. While the resolution, accuracy and precision criteria may be relaxed in various clinic practice, depending on the validation being performed; the criteria for high resolution isotropic measurement has limited true 3D radiation dosimeters (to date) to chemical radiation dosimetry based on quantifying the effects of radiation-induced chemical changes occurring within some volume of material [24, 29-31]. For clinical utility, the response of the true 3D dosimeter must also be reproducible, and stable [31-33]. The degree of the chemical change must be related to the absorbed dose and the changes must be able to be spatially localized in the irradiated volume by some imaging.

Historically the choice of appropriate chemical constituents for a 3D dosimeter has been determined a number of factors. The species that change under irradiation must be quantifiable by some approach. The species that changes under irradiation must be dispersible in a substrate that fixes radiation induced changes locally until they can be imaged [24, 29-31]. The initial 3D dosimeters primarily used gelatin or agarose as the localizing matrix, while some new dosimeters dissolve the radiation reporting leucodyes within a solid (when set) polyurethane matrix or in micelles dispersed through the aqueaous gel. There are distinct advantages to each system. Some of the properties of the various dosimeters with their appropriate readout systems are summarized in Table 1 above.

Table 1. A review of the main classes of 3D chemical dosimeters showing the basic mechanism for interaction, and conventional readout mechanism (with typical dose sensitivity). The dose sensitivities listed are rough ranges only as the sensitivity for a given dosimeter is highly dependent on the system preparation and readout details. More complete summaries detailing these characteristics are available in instructive reviews [24, 30, 31]. The acronyms for the polymer gel dosimeters follow common convention [31].

Class	Dosimeter		Readout		
	Basic mechanism	Usual Stabilizing substrate	MRI ($s^{-1}Gy^{-1}$)	Optical CT ($cm^{-1}Gy^{-1}$)	X-ray CT ($HUGy^{-1}$)
Fricke Gels			Δ relaxivity	Δ absorbance	N/A
Fricke	Fe^{2+} → Fe^{3+}	Gelatin or agarose	~0.04 (R1)	~0.01	
Fricke Xylenol			~0.009 (R2)	~0.1	---
Polymer Gels			Δ dynamics and structure	Δ scatter	Δ density
PAG, PAGAT, MAGIC, NIPAM, VIPAR, BANG etc.	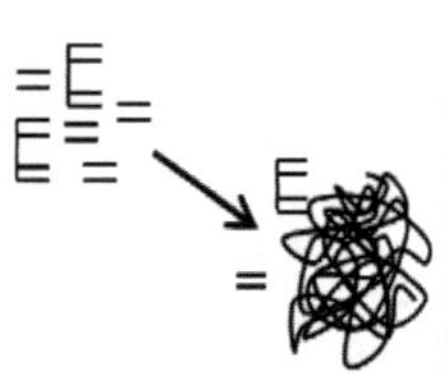	Gelatin or agarose	~0.1 – 1.0	~0.1	~0.25 – 0.85
Novel Radiochromic systems			N/A	Δ absorbance	N/A
Plastic	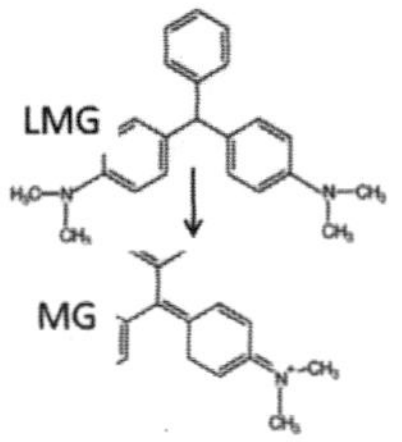	polyurethane	---	~0.01-0.05	---
Silicone		Poly-dimethyl-silocane	---	~0.01	---
Micelle gel	LMG or LVC → MG+ or CV+	Gelatin +micelles	---	~0.003-.007	---

Journal of Physics: Conference Series **573** (2015) 012003 doi:10.1088/1742-6596/573/1/012003

3. Readout methods for three dimensional measurement

The initial 3D dosimeter was the Fricke gel read out using magnetic resonance imaging MRI [34] although optical readout was soon proposed for polymer gels [35] and modified Fricke Xylenol dosimeters [36]. The motivation for optical readout was driven greatly by the desire to make 3D imaging readout more readily available since MRI access was often limited in the clinical setting of a radiation therapy centre. The desire for more accessible readout also initiated the development of x-ray computed tomography (CT) based 3D dosimetry with polymer gels [37, 38]. MRI, optical CT (optCT) and x-ray (CT) remain the main imaging modalities for 3D dose readout [24, 29-31].

The dose quantification using MRI results from the dose dependent change in the nuclear magnetic relaxation (NMR) properties of the dosimeter under irradiation. In Fricke gels the NMR spin-lattice relaxation rates (the usually measured parameter) depend directly on the concentration of the ferric (Fe^{2+}) and ferrous (Fe^{3+}) ion species since they have different relaxivity [31]. In polymer dosimeters the distinct monomer, polymer, gelatin and water protons environments of the protons providing the NMR signal change with dose. For the most part the relaxation of water protons in bulk and hydrating monomers remains unaffected as radiation induced polymerization progresses, however, the relaxation of the water molecules hydrating the growing polymer changes [30, 31]. The fraction of water hydrating these macromolecules increases with dose and the NMR spin-spin relaxation of this water is modified since it is mediated by chemical and physical interactions with the polymer (through mechanisms such as chemical exchange and magnetization transfer [39]).

While optical computed tomography readout was first proposed for polymer gels [35], the optCT technique has flourished mainly when coupled to radiochromic dosimeter (Fricke xylenol and the various leucodye) systems. In these dosimeters the radiation induced changes in colour or transparency at visual wavelengths enable optCT imaging based dose quantification [40]. In Fricke gels prepared with xylenol orange (a metal ion indicator) the change in Fe^{3+} concentration leads to the dose dependent optical changes [36, 40]. The radiation induced optical-contrast of leucodye radiochromic dosimeters is generated through the oxidation of a leucodye dispersed in stabilizing matrix that has been doped with a halogenated hydrocarbon free radical initiator. In the plastic PRESAGE dosimeter the dye is leuco-malachite-green (LMG), the dye which was also studied in early aqueous micelle gel dosimeters [40-42]. Newer radiochromic dye systems use leuco-crystal-violet (LCV) [43, 44]. These radiochromic dosimeters have the advantage that the optical response is primarily the result of light attenuation through absorption with minimal scattered light perturbation. This provides a major advantage over the optical readout of polymer gel dosimeters in which the primary mechanism for attenuation changes results from light scattering from particles which are formed as radiation induced polymer precipitates in the gel. This mechanism for change in polymer gels presents challenges to optical quantification which must be carefully considered for good dosimetry [24, 31, 45].

The relationship between the X-ray CT and material properties of irradiated polymer gels arises from x-ray CT quantifying and mapping linear attenuation coefficients [31, 37, 38]. For materials with dominant x-ray interactions similar to water, the CT number in Hounsfield units (HU) (a unit which normalizes a local attenuation coefficient to that of water) scales directly to a physical density [31, 46]. Thus, in X-ray CT based dosimetry, it is the change in density of the irradiated polymer gel that determines the change in CT number and enables dose quantification.

The availability of various imaging modalities has influenced the use of various dosimeters. For example, Fricke Xylenol gel dosimeters which had fallen out of favour because iron species diffusion limited the readout time to the order of an hour [47, 48], are now readily imaged within this constraint when optical scanning is available [49, 50]. It may be that the development of on-line imaging on treatment units may also widen the availability of dosimetric imaging. Recently, Bong *et al* investigated whether cone beam CT imaging on a treatment unit CBCT system could be used for dosimetric imaging [51]. The preliminary experiments confirmed the difficulty of image quantification with CBCT, but pointed to a potential development in the clinic as cobalt and x-ray linac treatment units are coupled to MRI on-line imaging systems.

4. Some practical considerations

The reproducible and accurate dose measurement with the 3D dosimeters reviewed above does require careful attention to particulars of set procedures [30, 31, 33, 38, 40, 52, 53]. However, such attention to detail is no different to the care required for good dosimetry with systems such as thermoluminescent, film and diode dosimeters which have been long, and regularly, used in the clinic. The perception in the community that the constraints for reliable 3D chemical dosimetry are more onerous is misplaced. What is required is a careful characterization and understanding of the conditions which need be faithfully maintained in the dosimetric procedures. Conditions such as temperature during preparation, storage in cool dark environments, and temperature control during irradiation may be of variable importance depending on the system used. But it would be best practice to maintain these conditions within established tolerances for the particular dosimeter being used in order to ensure reproducibility. Similarly, the timing of the stages in the dosimetry are important (see Table 2). In particular, when prepared in house, the dosimeters need some time to stabilize and set before irradiation. Because of auto-oxidation, it is usually good practice not to prepare the dosimeter too far in advance of irradiation. Afterwards the time between irradiation and readout should be controlled: for Fricke systems it cannot be too long or the spatial integrity of the dose information will degrade through diffusion; for polymer systems the time cannot be too short as the system needs time to develop as the radiation induced polymerization reactions proceed. The spatial integrity of the 3D dose distributions in polymer gels, some formulations of PRESAGE and the new leucodye micelle gels seems stable from months to years, indicating promise for dose delivery validation over multiple fractionations or for use in IMRT delivery validation by mailed phantoms from an external credentialing group such as the Imaging and Radiation Oncology Core-Houston (IROC-H) [52-54]. It should be noted that some of the leucodye radiochromic systems do change colour and darken over time after irradiation, but this behaviour is well behaved and can be characterized [55-57]. Furthermore, as the dose quantification typically proceeds with calibration samples or phantoms with the same time course, the dosimetry remains consistent and spatially stable [56].

Table 2. A broad overview of some typical time constraints in the use of the various 3D dosimeters. In some cases these are set by convention for convenient dosimetry (e.g., the maximum time from preparation to irradiation), in some cases (e.g. polymer gel stabilization after irradiation) the constraints are required for stable reproducible readout. It is assumed that the dosimeters are stored appropriately (in dark, cool storage) to maintain performance.

Dosimetry System	Time Needed /Constraint				
	Preparation	Wait between Preparation and Irradiation	Time for Irradiation	Wait Post irradiation to readout	Period for stable localization
Fricke Xylenol	45 min – 1 hr	12 hrs -1 week	< 20 min	30 min - 1.5 hrs	~ hrs or less
Polymer Gel	45 min – 1 hr	12 hrs -1 week	No limit	> 12- 24 hrs required for reactions to stabilize	years
PRESAGE	Commercially available	Can be weeks	No limit	Minutes (readings are f(t) but well behaved)	See text
Radiochromic Micelle Leucodye	45 min – 1 hr	24 hrs -1 week	No limit	> 20 min to hrs	See text
Silicone gels	1hr	12 hrs -1 week	No limit	> 2-3 hrs	See text

5. Clinical role (reprise)

The strength of 3D dosimetry as described in this review is particularly realized in cases where the dosimetry provides distinct benefits of isotropic high resolution dose measurement that may not be achieved with other dose measurement tools. The clinical role of 3D dosimetry has been long promoted and discussed [5, 14, 24, 29-31]. It has been noted that gel dosimeters mimic tissue extremely well and can be designed for insertion into cavities in anthropomorphic phantoms that can be used to to evaluate new techniques that are being brought on-line (see figure 5). Oldham [24] and Schreiner and

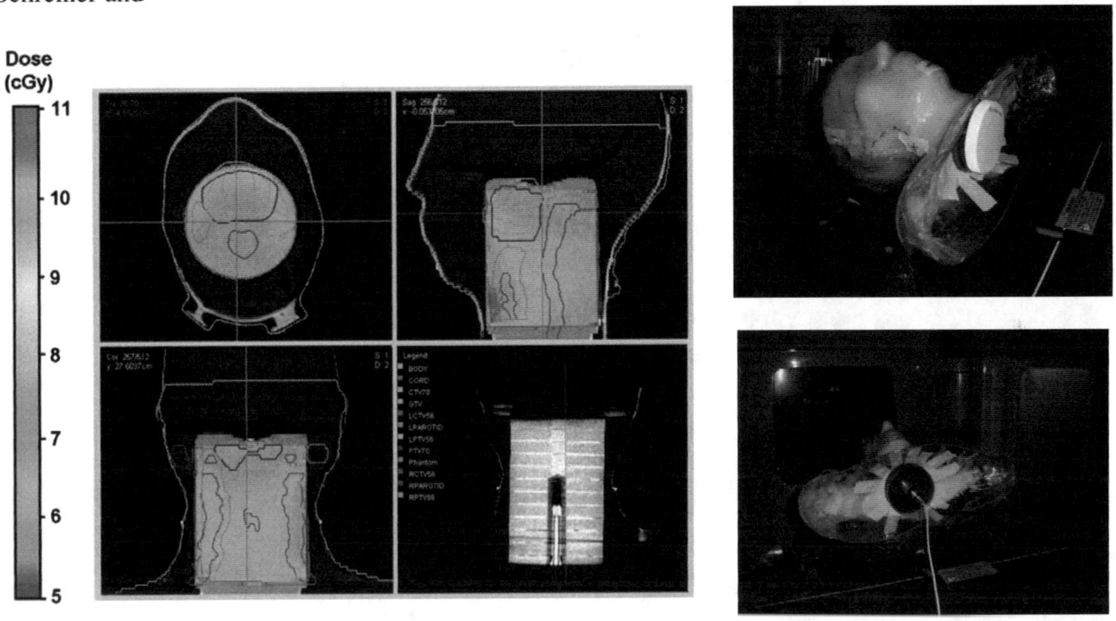

Figure 5. An illustration of a unique implementation of 3D dosimetry in Kingston. (left) A visualization of the radiation dose delivered to an in-house customized head-and-neck phantom containing an FXG gel dosimeter insert (top right) by an average of 5 full bowtie 100 kVp CBCT scans prior to an OBI upgrade. The dose is viewed in the CERR environment and is overlaid on the planning CT data. The bottom right corner of the CERR image frame shows a coronal CBCT slice of an acrylic insert in the wax-filled Rando reproduction (bottom right) containing an ionization chamber for dose readout over multiple CBCT scans. Roughly an order of magnitude reduction in the mean dose was observed following the upgrade of the Varian CBCT software (using gel dosimetry). This was verified at a single point using the acrylic-ion chamber measurement phantom.

Olding [31] have reviewed well multiple applications of 3D dosimeters in clinical validation. Their reviews include applications to small field dosimetry in radiosurgical dose delivery and in the characterization of brachytherapy seeds, to validation of IMAT and VMAT delivery prior to patient irradiation, to the assessment of SABR dose delivery under tumour motion, etc.. Much of this work has also been described in the proceedings of past and present IC3D conferences readily available in the Journal of Physics: Conference Series.

Two recent developments add further to the clinical potential for 3D dosimetry: the investigation of reusable dosimeters and of deformable systems. The results of such studies have not yet been reviewed as extensively since the investigations are recent, but results in both areas are presented in these proceedings [57-59]. The development of reusable dosimeters [54] may make commercial supply of dosimeters economically viable, which would likely extend clinical interest to groups that are currently reluctant to prepare dosimeters in-house. The development of commercial supply with appropriate manufacturing processes and quality control should also make the dosimeters themselves more consistent, and may help establish protocols that would increase the reliability and reproducibility of the dosimetry. The development of deformable dosimeters [57, 59-61] extends 3D dosimetry to areas currently limited to the domain of the treatment planning algorithms being implemented to account for anatomical changes as tissues respond to treatment as the Adaptive Radiation Therapy proceeds over multiple fractions. Initial reports by Juang, Oldham, *et al* [60, 61] suggest that there needs to be considerable additional validation of commercial deformable image registration packages before they are fully adopted in the clinic.

As noted in the introduction it is clear that the requirements for the validation of patient specific dose delivery at the treatment unit, and for the quality assurance of the radiation therapy processes, have increased over the years with the implementation of IGART. It has been suggested in the past, and in these proceedings, that clinical quality assurance should not be limited to specific steps in the process but that it should also include end-to-end process validation [12, 29, 54, 62]. Presently, this type of quality assurance is typically performed only when working with an external credentialing group such as IROC-H using phantoms they provide [54]. The task could be well executed in-house using 3D dosimeters inserted into an anthropomorphic phantoms which would then be passed on through the radiation delivery team to undergo the adaptive process to be tested [12, 29]. The phantom would be imaged by the CT therapists and contoured by the radiation oncologist. A treatment plan would be developed by dosimetrists and the treatment plan and phantom images transferred to the treatment unit. Setup verification would be completed through the use of on-board cone beam CT imaging on the linear accelerator, with adjustments made as necessary. Once positioning has been verified the planned dose delivery would be given by the treatment therapists. All the steps in the ART process under evaluation (say a dose correction on fraction 11) would be performed fully as specified in the adaptive protocol being tested. Once the protocol has been completed, the physics group would remove and image the 3D dosimeter, register the dose data with the planned deliveries, and evaluate the results. The ability to mimic a patient and measure dose throughout a full ART process, including the various adaptive steps, is a unique and key advantage provided by 3D dosimetry techniques.

We have been attempting to implement such an end-to-end QA in Kingston; a significant challenge has been to design and manufacture suitable opaque phantoms so that the interior is not easily observed without using the treatment localization tools that are part of the test. This work is still underway. One major step forward in the past year has been the development of 3D dosimetry tools in the open software suite Slicer-RT [63, 64]. This has reduced the analysis time for our work considerably. The development of process QA is a challenging task involving a number of health care personnel on the treatment team. But 3D dosimetry has matured sufficiently that the tools for process quality assurance and evaluation are no longer the limitation to the work.

Finally, 3D dosimeters may also have an important role in evaluating other dosimeter systems used in the clinic, especially in establishing their limitations. Most commercial dosimetry systems for clinical IMRT and VMAT delivery validation are complex [19, 20] and typically confined to specialized proprietary software to generate the dosimetric data and to analyse agreement with planned deliveries. For example, some pseudo 3D ion-chamber array systems interpolate or back project 3D dose distributions into regions outside the volume of the measurement points. 3D dosimeters can help in the evaluation of these systems through comparative dose measurements. It is important that the user understand the limitations of all delivery validation tools completely [19, 20, 66] and 3D dosimeters can help greatly in that assessment [67].

6. Authors note

The references are a selective list meant to point to select papers over a long history. The list forms an initial set of references on which the reader can build.

7. References

[1] Van Dyk J (ed) 1999, 2005, 2014 *The Modern Technology of Radiation Oncology: A Compendium for Medical Physicists and Radiation Oncologists* vol 1-3 (Madison, WI: Medical Physics Publishing)
[2] Jaffray D A *et al* 2002 *Int. J. Radiat. Oncol. Biol. Phys.* **53** 1337-49
[3] Yun J *et al* 2013 *Med. Phys.* **40** 051718
[4] Kron T *et al* 2006 *J. Med. Phys.* **31** 242–54
[5] Schreiner L J 2006 *J. Phys.: Conf. Ser.* **56** 1-13
[6] Woodford C *et al* 2007 *Int. J. Radiat. Oncol. Biol. Phys.* **69** 1316-22
[7] Ahunbay E E *et al* 2008 *Med. Phys.* **35** 3607-3615
[8] Yan D *et al* 1997 *Phys. Med. Biol.* **42** 123-32
[9] Martinez A A *et al* 2001 *Int. J. Radiat. Oncol. Biol. Phys.* **50** 1226-34
[10] Mageras G S *et al* 2007 Planning in the IGRT context: closing the loop *Sem. in Radiation Oncology* **17** 268-277
[11] Mackie T R *et al* 1995 *Int. J. Imaging Systems and Technology* **6** 43-55
[12] Schreiner L J *et al* 2011 Development of in-house process QA in modern radiation therapy, *Proc. Internat. Symp. Standards, Applications and Quality Assurance in Medical Radiation Dosimetry (IDOS) (Vienna, Austria 9-12 November 2010)* vol 2 (Vienna: IAEA) pp 197-205
[13] Low D A *et al* 1998, *Int. J. Radiat. Oncol. Biol. Phys.* **40**, 1231-1235
[14] Low D A 2014 Proceedings of 8th International Conference on 3D Radiation Dosimetry (IC3DDose) 4-7 September 2014, Ystad, Sweden
[15] Moran J *et al* 2005 *Med. Phys.* **32** 466-472
[16] Létourneau D *et al* 2009 *Med. Phys.* **36** 1813–1821
[17] Feygelman V *et al* 2011 *J. Appl. Clin. Med. Phys.* **12** 3346
[18] Neilson C *et al* 2013 *Medical Dosimetry* **38** 77-80
[19] Van Esch A *et al* 2014 Proceedings of 8th International Conference on 3D Radiation Dosimetry (IC3DDose) 4-7 September 2014, Ystad, Sweden
[20] Bäck A 2014 Proceedings of 8th International Conference on 3D Radiation Dosimetry (IC3DDose) 4-7 September 2014, Ystad, Sweden
[21] Mans A *et al* 2010 *Radiotherapy and Oncology* **94** 181–187
[22] Mijnheer B *et al* 2010 *J. Phys.: Conf. Ser.* **250** 012020.
[23] Oldham M *et al* 2001 *Med. Phys.* **28** 1436–45
[24] Oldham M 2014 *Advances in Medical Physics* ed Godfrey D *et al* (Madison, WI: Medical Physics Publishing)
[25] Nelms B E *et al* 2011 *Med. Phys.* **38** 1037–1044
[26] Beaulieu L *et al* 2013 *J. Phys.: Conf. Ser.* **444** 012013
[27] Beddar S 2014 Proceedings of 8th International Conference on 3D Radiation Dosimetry (IC3DDose) 4-7 September 2014, Ystad, Sweden
[28] Pogue B *et al* 2014 Proceedings of 8th International Conference on 3D Radiation Dosimetry (IC3DDose) 4-7 September 2014, Ystad, Sweden
[29] Schreiner L J 2009 *J. Phys.: Conf. Ser.* **164** 012001
[30] Baldock C *et al* 2010 *Phys. Med. Biol.* **55** R1-63
[31] Schreiner LJ *et al* 2009 Gel dosimetry *Clinical Dosimetry Measurements in Radiotherapy (AAPM Medical Physics Monograph No. 34)* Rogers D and Cygler J (ed.), 30, (Medical Physics Publishing, Madison WI, 2009) chapter 30 pp 979-1025
[32] De Deene Y 2004 *J. Phys.: Conf. Ser.* **3** 87–114
[33] De Deene Y *et al* 2014 Proceedings of 8th International Conference on 3D Radiation Dosimetry (IC3DDose) 4-7 September 2014, Ystad, Sweden
[34] Gore J C *et al* 1984 *Phys. Med. Biol* **29** 1189–1197
[35] Maryanski M J *et al* 1996 *Phys. Med. Biol.* **41** 2705–2717
[36] Kelly B G *et al* 1998 *Med. Phys.* **25** 1741–1750

[37] Hilts M *et al* 2000 *Phys. Med. Biol.* **45** 2559–2571
[38] Jirasek A 2013 *J. Phys.: Conf. Ser.* **444** 012005
[39] Lepage M *et al* 2002 *Phys. Med. Biol.* **47** 1881–1890
[40] Oldham M 2014 Proceedings of 8th International Conference on 3D Radiation Dosimetry (IC3DDose) 4-7 September 2014, Ystad, Sweden
[41] Guo P Y *et al* 2006 *Med. Phys.* **33** 1338-1345
[42] Jordan K *et al* 2009 *Phys. Med. Biol.* **54** 6773–6789
[43] Babic S *et al* 2009 *Phys. Med. Biol.* 54 6791–6808
[44] Vandecasteele J *et al* 2011 *Phys.Med.Biol.* **56** 627–651
[45] Olding T *et al* 2010 *Phys. Med. Biol.* **55** 2819–2840
[46] Rogers M *et al* 2005 *Proc. SPIE* **5745** 685-694
[47] Baldock C *et al* 2001 *Australas. Phys. Eng. Sci. Med.* **24** 19–30
[48] Schreiner L J 2004 *J. Phys.: Conf. Ser.* **3** 9-21
[49] Olding T *et al* 2010 *J. Phys.: Conf. Ser.* **250** 012003
[50] Olding T *et al* 2010 *J. Phys.: Conf. Ser.* **250** 012028
[51] Bong J H *et al* 2013 *J. Korean Phys. Soc.* **63** 1083-1087
[52] Sakhalkar H *et al* 2009 *Med. Phys.* **36** 3371-3377
[53] Juang T *et al* 2014 *Med. Phys.* **41** 071706
[54] Ibbott G *et al* 2014 Proceedings of 8th International Conference on 3D Radiation Dosimetry (IC3DDose) 4-7 September 2014, Ystad, Sweden
[55] Juang T *et al* 2013 *J. Phys.: Conf. Ser.* **444** 012029
[56] Nasr A *et al* 2014 Proceedings of 8th International Conference on 3D Radiation Dosimetry (IC3DDose) 4-7 September 2014, Ystad, Sweden
[57] Høye E *et al* 2014 Proceedings of 8th International Conference on 3D Radiation Dosimetry (IC3DDose) 4-7 September 2014, Ystad, Sweden
[58] Juang T *et al* 2014 Proceedings of 8th International Conference on 3D Radiation Dosimetry (IC3DDose) 4-7 September 2014, Ystad, Sweden
[59] De Deene Y *et al* 2014 Proceedings of 8th International Conference on 3D Radiation Dosimetry (IC3DDose) 4-7 September 2014, Ystad, Sweden
[60] Juang T *et al* 2013 *J. Phys.: Conf. Ser.* **444** 012080
[61] Juang T *et al* 2013 *Int. J. Radiat. Oncol. Biol. Phys.* **87** 414-21
[62] Knoos T 2014 Proceedings of 8th International Conference on 3D Radiation Dosimetry (IC3DDose) 4-7 September 2014, Ystad, Sweden
[63] Pinter C *et al* 2012 *Med. Phys.* **39** 6332-6337
[64] Alexander K M *et al* 2014 Proceedings of 8th International Conference on 3D Radiation Dosimetry (IC3DDose) 4-7 September 2014, Ystad, Sweden
[65] Juang T *et al* 2013 *J. Phys.: Conf. Ser.* **444** 012029
[66] Schreiner L J. 2011 *J. Med. Phys.* **36** 189-191
[67] Olding T *et al* 2013 *J. Phys.: Conf. Ser.* **444** 012052

Patient-specific QA using 4D Monte Carlo phase space predictions and EPID dosimetry

I A Popescu[1,2], P Atwal[1], J Lobo[2], J Lucido[3] and B M C McCurdy[4]
[1]British Columbia Cancer Agency, Vancouver, BC, Canada
[2]University of British Columbia, Vancouver, BC, Canada
[3]University of Colorado School of Medicine, Denver, CO, USA
[4]CancerCare Manitoba and University of Manitoba, Winnipeg, MB, Canada

E-mail: tpopescu@bccancer.bc.ca

Abstract. The goal of this review is to outline a solution for patient-specific QA of VMAT, IMRT, and other complex treatment delivery techniques. This solution has been developed in direct response to clinical needs, in order to allow our institution to offer VMAT to all patients who could potentially benefit from this advanced technique. To date, over 2500 VMAT patient plans and approximately 1000 IMRT patient plans have been verified by this method in Vancouver, while 40 other institutions worldwide have expressed interest in, or are already at various stages of implementing, this process. The addition of EPID *in vivo* dosimetry (i.e. data acquired during the patient treatment) and associated Monte Carlo predictions amounts to introducing a 'measurement component' in this QA process, which is currently mandated by the regulatory framework in some European countries, or for billing purposes in the USA. The fully automated, patient-specific, Monte Carlo based QA process described here is fast, maximally efficient in terms of departmental resources, and capable of simulating any plan in a single run, regardless of its complexity.

1. Introduction

From a quality-assurance and patient-safety perspective, the fundamental question is how to optimally use clinical resources in order to make available to patients the best radiation therapy techniques at our disposal, in a safe and accurate manner. A closely related question is whether our current QA practices can safeguard patients from radiotherapy incidents, such as the 'New York incident' [1], to give one example. Needless to say, this is a topic of great interest for the European (ESTRO), American (AAPM and ASTRO) and Canadian (COMP and CARO) organizations of radiotherapy practitioners, as well as for individual radiotherapy institutions worldwide. While it has to be recognized that pre-treatment VMAT QA is currently a clinical mainstay, we believe that an increasing focus on comprehensive critical-event and failure-mode analysis of the entire radiotherapy chain is more likely to prevent adverse clinical events.

During our more than a decade-long experience with the QA of IMRT and VMAT plans, our practice has gradually evolved from pre-treatment phantom-measurement QA for each patient plan towards a three-component procedure consisting of (1) independent 3D dose recalculation using Monte Carlo for each plan, with a full pre-treatment phantom measurement for every 50th plan (which currently occurs approximately once a week); (2) an increasingly comprehensive routine linac QA procedure, with emphasis on dynamic MLC, dynamic gantry, and variable dose rate tests; and (3) monitoring of dynalog/trajectory-log files acquired during patient treatment, with the option of rerunning the Monte Carlo dose calculation with these inputs. The fully automated Monte Carlo based QA process outlined in this paper has accumulated an extensive clinical track record, with over 2500 VMAT patient plans verified by this method in Vancouver and several hundred others world-wide. The core of this process is 'source 20' of DOSXYZnrc, developed by Lobo and Popescu [2, 3], for 4D Monte Carlo simulations of continuously-variable beam configurations, with arbitrary degrees of freedom. It would have been impossible to offer this advanced treatment to all patients who could have potentially benefited from it without a streamlined QA procedure that involved physics resources at a level commensurate with the clinical workload. One of the benefits of this process being automated is that the user is not required to be a specialist in Monte Carlo applications. The possibility of a user error is also eliminated by automation. This QA process has been applied to a variety of clinical cases, including jaw-tracking, non-coplanar, and multiple-arc VMAT, for both standard and flattening filter free (FFF) photon beams.

A quality-assurance protocol deserving of the 'patient-specific' title should be relevant to the actual patient treatment and be capable of providing *in vivo* dosimetry. Moreover, such a QA method should ideally be able to avert incidents such as the one quoted above. The major current commercially-available QA solutions, such as ArcCHECK, MapCheck, and EPIdose (Sun Nuclear, USA); Delta4 (ScandiDos, Sweden); Octavius (PTW Freiburg GmbH, Germany); or MatriXX (IBA Dosimetry GmbH, Germany), are valuable within the pre-treatment paradigm, but would not be able to catch an egregious error during treatment, such as patient miss-positioning, missing MLC, etc. In contrast, patient 3D dose calculation using real-time linac log information or transmission portal EPID dosimetry and cone-beam CT imaging to assess changes in patient anatomy, could give practitioners the data required for adaptive patient dose accumulation over the entire course of treatment and for inferring the associated radiobiological consequences. To this end, we have recently introduced the capability of simultaneously providing dose deposition in both the patient CT data set and any type of planar (entrance or exit) detector for VMAT Monte Carlo QA [4]. By enabling DOSXYZnrc to optionally output a 4D IAEA phase space during a source 20 simulation, it is now possible to collect the particles exiting the patient during a VMAT simulation. This phase space can be used for further synchronized simulations of EPID dose or images, either globally, incrementally, or per control point, allowing comparisons of the 4D phase space EPID prediction with the actual EPID data acquired with the patient on treatment.

2. Materials and methods

2.1 General considerations

Our Monte Carlo simulations use several recent developments of source 20, which were introduced in order to address specific challenges presented by TrueBeam (Varian Medical Systems, USA) simulations, namely the unavailability of a full linac model and the fact that input IAEA phase spaces may be non-planar. Some of these new features have been already included in the 2013 release (V4-2.4.0) of the BEAMnrc/DOSXYZnrc system (National Research Council, Canada) [5, 6], while others will be incorporated into the next release. The TrueBeam simulations used input phase spaces provided by the manufacturer and independently validated by several groups (for example, for the 6X FFF phase space, see the experimental validations by Gete et al [7] and – more extensively – by Belosi et al [8]). The number of histories for each simulation was selected such that the statistical uncertainty in the high dose voxels (PTV) was approximately 1%. The Monte Carlo simulation results were converted to absolute dose for the planned number of monitor units using the formalism of Popescu et al [9], with an amended methodology required for TrueBeam [10]. A detailed description of our Monte Carlo simulation parameters can be found in [3] and [9]. The patient treatment and verification plans were generated by the Eclipse (Varian Medical Systems, USA) treatment planning system (TPS), using AAA version 11.0.31.

2.2 Automated Monte Carlo workflow

Various components of the general process outlined here have been validated and presented in previous publications (in chronological order: [11, 9, 12, 2, 13, 3, 14, 15, 4, 16], as the system evolved over the last decade. At the pre-treatment QA stage, the input for the process consists of DICOM files exported from the TPS. In its current clinical version, the process is initiated by typing a command representing essentially the patient's hospital ID and no further user intervention is required. In an alternative version, which is used in a different institution, even this initial step is removed and replaced by a daemon-based monitoring of exported DICOM files. In that case, the process is automatically initiated once the presence of such files is detected. At treatment stage, we have developed the capability of taking into account linac log files (dynalog files for the Varian iX and related series and trajectory-log files for TrueBeam) as inputs for a secondary, post-delivery, Monte Carlo simulation. The flowchart of the process is shown in figure 1.

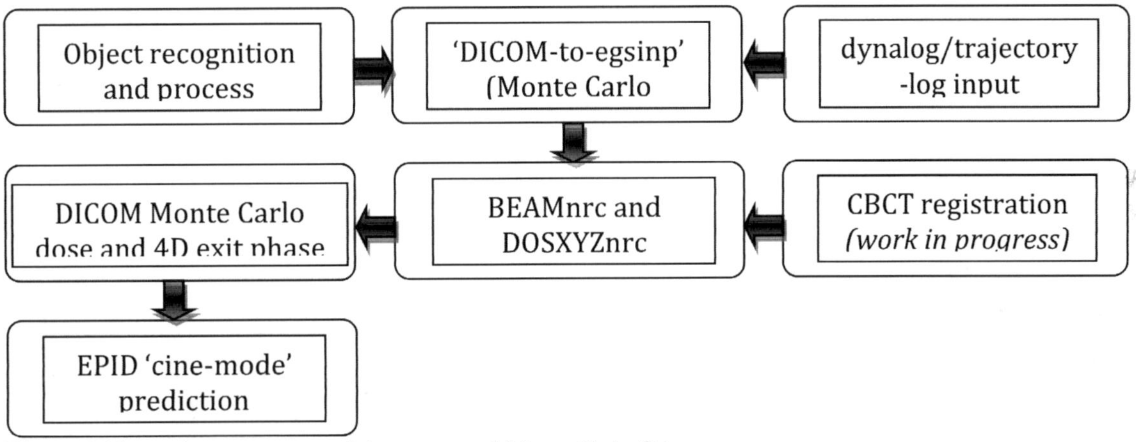

Figure 1. The main components of the automated Monte Carlo QA process.

2.2.1 Object recognition and process selection

The main input into the process is the DICOM RP plan file exported from the TPS. The scripts scans this file to determine whether the plan is for an iX (and related Varian Clinac series) or for a TrueBeam linac, the treatment unit identifier for the case of unmatched linacs, the type of MLC (Millennium or HDMLC), fluence type (standard or FFF beam), and any other characteristics that require specific input parameters. The export of the TPS DICOM RD total dose file is also expected, since it is overwritten in the end with the Monte Carlo dose. This allows the import of the Monte Carlo dose back into the TPS for in-depth comparisons with the planned dose, taking advantage of existing TPS tools, such as dose-volume histograms, plan comparison, dose subtraction, etc. If these are the only two files exported from a verification plan, the simulation will be performed by default on the corresponding verification phantom. Otherwise, if the set of CT DICOM files is also exported, the script will automatically detect that and use ctcreate to generate a patient-specific phantom for the simulation.

2.2.2 'DICOM-to-egsinp'

Once the process selection step is completed as described above, the script generates a single BEAMnrc egsinp file and a single DOSXYZnrc egsinp files, regardless of the number of fields in the plan. This step takes into account the transformation between the DICOM and DOSXYZnrc coordinate systems. Several other auxiliary files required for the post-processing step are generated at this time. We have used in the past a DICOM dictionary provided by the in-house code [17], but several other users have employed the open-source Python-based code [18]. We are also in the process of migrating towards the Python-based solution, since we believe that an open-source, publicly-available and supported code is preferable to an in-house one, for a variety of reasons, including the fact that other users could reproduce and independently verify the work carried out in our institution.

2.2.3 BEAMnrc and DOSXYZnrc

Our BEAMnrc simulations employ the SYNC* suite of component modules [5], allowing full flexibility of treating any plan as having jaw-tracking (whether it is actual dynamic jaw tracking or step-and-shoot style jaw setting specific to a multiple-field plan with different field sizes). Likewise, dynamic and/or step-and-shoot MLC for single or multiple-field plans is treated as a single dynamic MLC sequence. By employing the SYNC* modules, it is possible to use any BEAMnrc accelerator either as a shared library called by DOSXYZnrc at run time, or as a stand-alone executable. In the former case, no intermediate phase spaces are required. In the latter case, 4D IAEA phase spaces can be scored, preserving the full properties of beam particles, including the MU index, which allows for further synchronized simulations. For reasons partly historical and partly related to simulation speed, we collect a 4D IAEA phase space file under the SYNCJAWS module, which becomes the input for the downstream simulations. However, no further downstream phase spaces need to be collected, regardless of the plan complexity.

2.2.4 DICOM Monte Carlo dose and 4D exit phase space

Upon completion of the simulation, we apply a 3D locally adaptive Savitzky-Golay denoising filter [19] to the raw Monte Carlo dose, following the methodology introduced by Kawrakow [20]. A detailed description of the role of this filter in the context of our system can be found in [16]. The resulting dose distribution is then converted to absolute dose (in Gy), as mentioned in subsection 2.1, and cast in DICOM RD format, which allows import into the planning system. Optionally, a 4D IAEA phase space is output by DOSXYZnrc. The geometry of this phase space is that of the parallelepiped boundary of the patient CT phantom. The time-like dimension is provided by the MU index variable scored for each particle. This phase space can be scored in either BEAMnrc or DOSXYZnrc coordinate systems.

2.2.5 EPID 'cine-mode' prediction

When a 4D IAEA exit phase space is scored in the BEAMnrc coordinate system, it is straightforward to perform a synchronous simulation into an EPID phantom, since the EPID is static in this coordinate system, even for multiple-field non-coplanar VMAT plans. Such a simulation takes only a few minutes on our computer cluster, for one billion histories, since particles in the 4D exit phase space only have to travel a few centimetres in air before being incident on the EPID phantom. A single simulation is again required in all cases, without the need to build a phantom for each patient-EPID relative position, as we did in our earlier work [4]. As a by-product, simulations of cone-beam CT image acquisition and associated patient dose are also possible using the exit phase space. We have introduced a MU filter in source 20 that rejects particles outside a user-defined MU index interval. This filter is enabled by including two new optional variables defining the lower and upper limits of a MU index interval in the egsinp file. This feature can be used to calculate integral, incremental, or per control point, dose distributions either in the patient CT phantom, or in the EPID.

2.2.6 Dynalog/Trajectory-log input

We have developed the capability of generating Monte Carlo input files for post-delivery dose recalculation based on linac log files, namely 'dynalog files' [12, 13], for the Varian iX and Clinac series and 'trajectory-log' files for TrueBeam (unpublished). The information contained in these files is much more detailed than that in the DICOM RP file, resulting in Monte Carlo input files with up to approximately 3×10^4 control points. These files can be collected for every treatment fraction and reflect the performance of various dynamic linac components during the actual beam delivery.

2.2.7 Robustness and continuity checks
In the main script there are two instances (when running BEAMnrc and DOSXYZnrc) when the simulation is divided into a user-selected number of 'jobs' distributed to processors. This is done using the HTCondor (University of Wisconsin - Madison, USA) high-throughput distributed parallelization software, which allows for concurrent job allocation and dynamic resource management. In order to have a robust automated system, we implemented checks wherein HTCondor is queried to ensure that the jobs are running as expected. The first in the sequentially numbered set of distributed jobs creates a 'lock file', which all jobs in the task must access, one at a time, to determine how many particles are left to simulate and thus when the distributed simulation is complete. During the BEAMnrc simulation, each job creates an empty phase space file at the outset into which the resultant particle information is written. The script checks periodically if both the lock file and the phase space file for the first job exist. While the lock file exists, the script remains in a while-loop in which it periodically checks the status of the lock file. The lock file will be deleted by the last job to complete. Once the script determines that the lock file ceased to exist, it is assumed that the distributed task has completed successfully. If, however, either the lock file or the first job's phase space file do not exist, then the status of the first job is checked. There is a particular tag for this status, which must read 'R' for a running job. This status tag is checked periodically in a while-loop, until it reads 'R'. Within this loop the next thing to check is whether the status tag is set to any value at all or if it is empty. If it is not empty, then we assume that the job is in the process of being started, or it is idle while waiting in the HTCondor queue, and nothing further is done but to 'sleep' a few seconds and recheck the tag. On the other hand, if the status tag is found to be empty at this point, then the job can be assumed to be complete, but this fact must be verified. The script breaks out of the while-loop at this point and executes the code already mentioned regarding checks for the lock file. After determining that the BEAMnrc run has completed successfully, the phase spaces produced by each individual job are combined, an egslst file is generated (which contains the monitor chamber backscattered dose information required for the absolute dose conversion), and the script moves on to the DOSXYZnrc simulation. The DOSXYZnrc distributed simulation is checked in a similar fashion, except that each job creates a phase space file only if the option to generate a 4D IAEA exit phase space has been selected. In both instances, once it is determined that the task distribution has been completed successfully, any jobs for the given task which are still on the queue running (or to be run) are considered to be so-called 'orphaned' jobs and are deleted to ensure maximum availability of computing resources. After the end of the DOSXYZnrc simulation, the patient/phantom 3ddose is subjected to the post-processing steps described in 2.2.4, while the 4D exit phase space is used for the EPID simulation as mentioned in 2.2.5.

3. Results
Experimental validations of our Monte Carlo system, using ionization chambers, films, TLD in an anthropomorphic phantom, EPID, and the ArcCHECK system, have been previously published or reported at conferences (e.g. [4, 7, 11, 13, 16]). Of the independent validations performed at other institutions, the longest series has recently been reported by Cronholm et al [21] (unpublished). This group performed comparisons between Monte Carlo dose calculations using source 20 and the Delta4 system for a total of 100 clinical TrueBeam VMAT plans (30 plans with 6 MV beams and 70 plans with 10 MV beams). They concluded that the Delta4 measurements were generally in better agreement with the Monte Carlo than with the Eclipse computed dose distributions. The measurements were compared with MC and TPS results using a 3D gamma analysis based on a 3%/2mm criterion, with a lower cut-off of 20% of the maximum dose. The average and median pass-rates for MC were 97.7% and 98.1%, respectively.

In order to illustrate the versatility of our QA system, we present two clinical cases that have been encountered in our practice. Although they are different in terms of treatment techniques, they were both treated from the point of view of our MC QA system in the same way, namely as *single-field* treatments with multiple variable degrees of freedom.

The first case is of a patient who received treatment in both lungs simultaneously, for bilateral tumours. Each site was treated with two arcs, resulting in a four-arc VMAT plan with two isocentres, as shown in figure 2, with the field properties given in table 1.

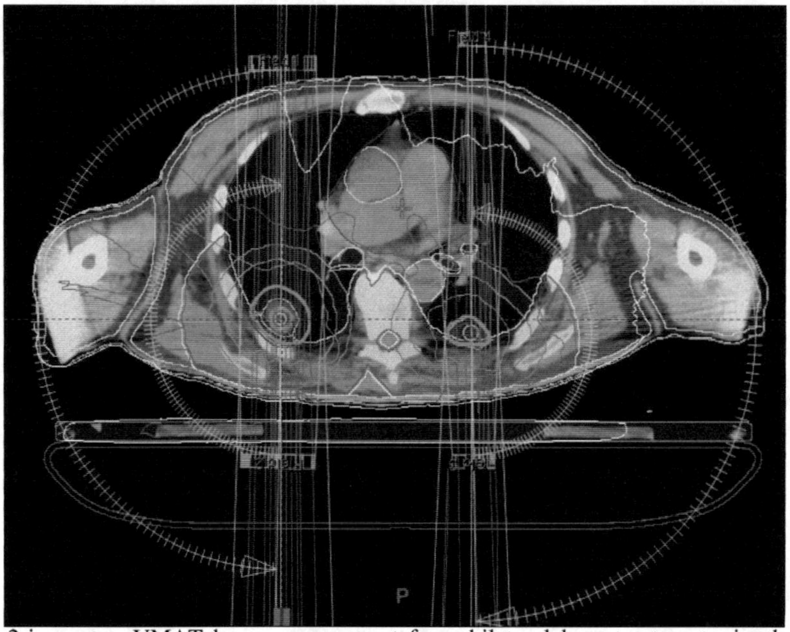

Figure 2. 4-arc, 2-isocentre, VMAT beam arrangement for a bilateral lung treatment, simulated as a 'single field' in source 20.

Table 1. Field set-up for the plan shown in figure 2.

Field ID	Machine/Energy	Gantry	Collimator	Couch	MU	X1 jaw	X2 jaw	Y1 jaw	Y2 jaw	x_iso	y_iso	z_iso
VMAT 1	TrueBeam 6 MV	0 ccw 181	45	0	2611	2.1	3.0	2.2	2.8	-7.00	0.00	0.00
VMAT 2	TrueBeam 6 MV	181 cw 0	315	0	2643	2.8	2.2	2.1	3.0	-7.00	0.00	0.00
VMAT 3	TrueBeam 6 MV	179 ccw 0	45	0	2068	3.7	3.2	3.7	3.1	7.00	1.00	-2.50
VMAT 4	TrueBeam 6 MV	0 cw 179	315	0	2052	3.1	3.7	3.7	3.2	7.00	1.00	-2.50

There were 9374.71 MU for the combined four fields. A single source 20 egsinp file was required for this simulation, containing 392 control points that defined the four partial arcs of this plan. Table 2 gives the beginning and the end of the control point sequence. On each line, the numbers represent the following variables: the (x, y, z) coordinates of the isocentre; the spherical coordinate angles defining the beam direction in the DOSXYZnrc coordinate system, θ and ϕ; the rotation angle about the beam axis, *phicol*; a variable with the role of *dsource*, which, in the case of a IAEA phase space input, is computed for each particle using the distance between the isocentre and the origin of the BEAMnrc coordinate system (typically, the linac SAD) and the z coordinate of the particle, as scored in the IAEA phase space - thus, it is this distance (SAD) that has to be specified in the egsinp file; and, finally, the fractional monitor units delivered up to this control point, *MU_index*.

Table 2. The first three and last three of the 392 control points, as written in the DOSXYZnrc egsinp file for the simulation of the plan shown in figure 2.

```
-7.24,   -28.27,   -87.887,   90,   270.00,   225,   100.0,   0.0000
-7.24,   -28.27,   -87.887,   90,   269.07,   225,   100.0,   0.0009
-7.24,   -28.27,   -87.887,   90,   267.20,   225,   100.0,   0.0030
[...............................................................]
 6.76,   -27.27,   -90.387,   90,    86.20,   315,   100.0,   0.9966
 6.76,   -27.27,   -90.387,   90,    88.07,   315,   100.0,   0.9989
 6.76,   -27.27,   -90.387,   90,    89.00,   315,   100.0,   1.0000
```

An axial slice through a cylindrical-phantom Monte Carlo *vs*. Eclipse dose comparison is shown in figure 3.

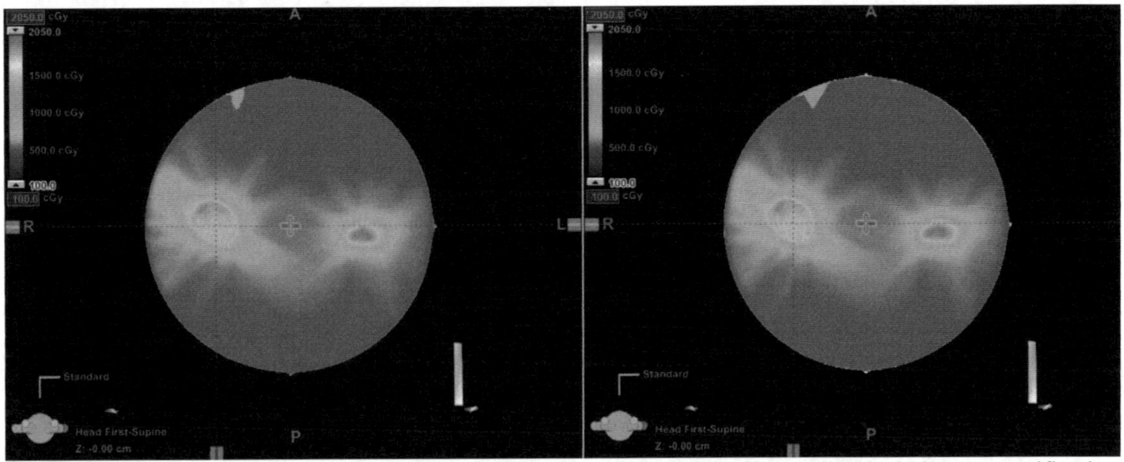

Figure 3. Monte Carlo dose (left panel) and Eclipse dose (right panel) for a cylindrical-phantom verification plan corresponding to the plan shown in figure 2. The 3D gamma agreement was 99.0%.

The second case is an 8-field, non-coplanar, 3D-conformal SBRT lung plan, with a combined 11877.76 MU over 4 fractions, corresponding to a SBRT prescription of 48 Gy. They were all small fields, with a maximum jaw setting of X1 = 1.4 cm, X2 = 1.3 cm, Y1 = 1.2 cm, and Y2 = 1.3 cm. The field arrangement is shown in figure 4.

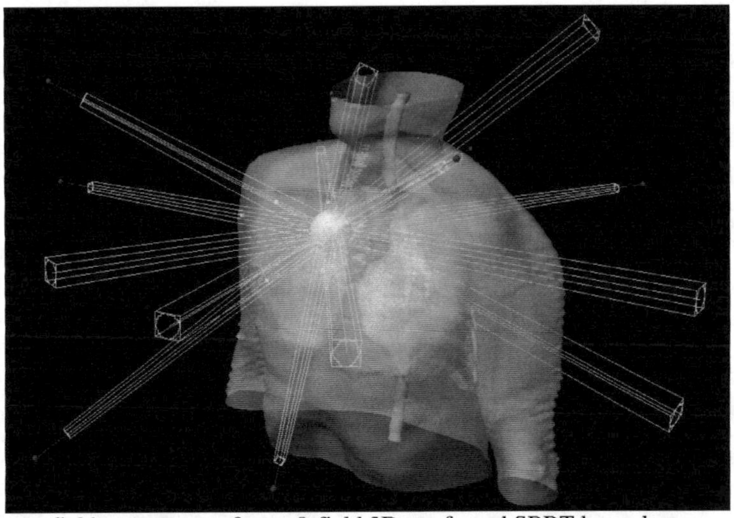

Figure 4. Non-coplanar field arrangement for an 8-field 3D-conformal SBRT lung plan.

Figure 5 shows an axial slice through the Monte Carlo *vs*. Eclipse dose comparison on the patient CT, while figure 6 shows a DVH comparison for the 4DCT-generated ITV and two organs-at-risk.

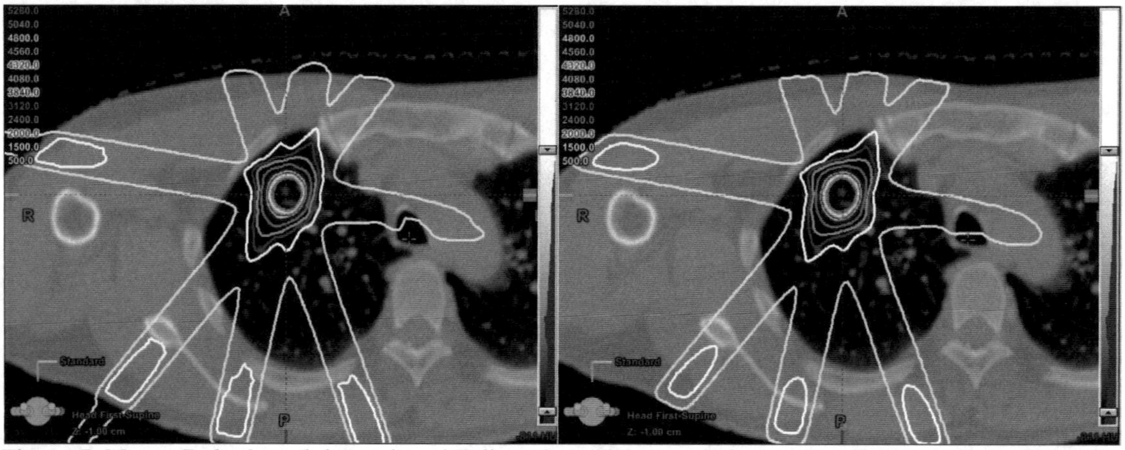

Figure 5. Monte Carlo dose (left panel) and Eclipse dose (right panel) for the plan shown in Figure 4, on the axial slice containing the isocentre.

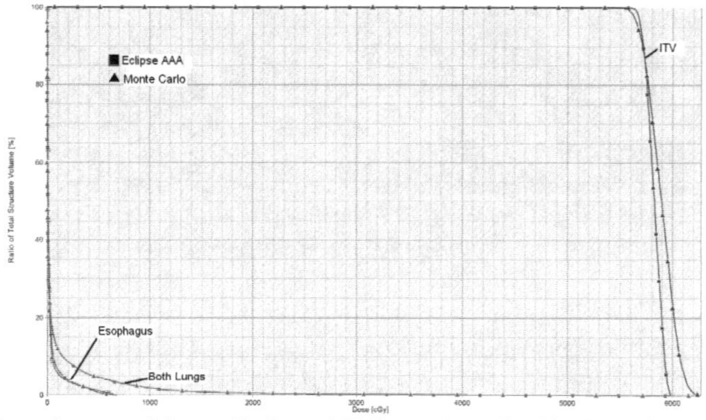

Figure 6. DVH comparison between Monte Carlo and Eclipse dose distributions, for selected structures, for the plan shown in figure 4.

Finally, figure 7 shows Monte Carlo predicted and EPID measured transmission doses for selected control points for a prostate VMAT plan. These images are formed by radiation transmitted through the patient body and our system has the capability of simultaneously providing the dose in the patient CT phantom (not shown here, but see [4] for details), as explained in subsections 2.2.4-5 above.

4. Conclusion and future directions

Our decade-long, thousand-case, pre-treatment QA measurement (ionization chamber and film) experience supports the idea that there is no additional patient-safety benefit from measurement based QA over Monte Carlo based QA, when a rigorous, failure-mode-analysis inspired, linac QA protocol is implemented along with dynalog or trajectory-log data monitoring and *in vivo* EPID dosimetry comparisons with the Monte Carlo 4D phase space predictions. By accumulating the patient dose delivered during the entire course of treatment (possibly in an adaptive manner, in conjunction with *in vivo* CBCT data), this process represents an authentic ‘patient-specific QA’ and could gradually replace the pre-treatment phantom-based QA process that an institution would typically use in the early stages of its VMAT program. We are also aware of the availability of at least one commercial solution that offers *in vivo* EPID dosimetry, EPIgray (DOSIsoft, France).

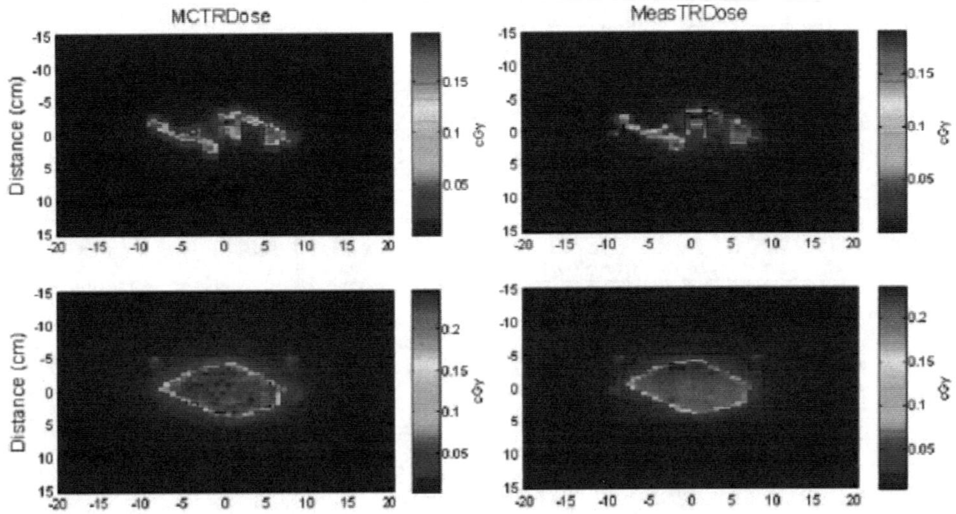

Figure 7. Monte Carlo calculated EPID transmission dose (left panel) and measured EPID transmission dose (right panel) for two randomly selected control points (68 and 93, of 178) of a VMAT prostate plan. (Image courtesy of Ganiyu Asuni).

Since the main source of potential variability of the dose distribution over the course of a fractionated treatment is the change in patient anatomy, a significant amount of work still has to be done on adaptive Monte Carlo dose accumulation, which will allow the monitoring of realization of planning constraints (on PTV and OAR) as dose accumulates over all fractions of the treatment. This requires recalculations on updated patient geometry, as provided by CT registration with CBCT imaging. A variety of commercial software tools have the potential of accomplishing this task, such as MIM (MIM Software Inc, USA), SmartAdapt (Varian Medical Systems, USA), or VelocityAI (Velocity Medical Solutions, USA). Eiland et al [22] analyzed the geometric and dosimetric performance of a deformable registration algorithm by comparing the CBCT-registered and deformed CT set with the actual rescan of the patient at some stage of the treatment. This group found that, in spite of certain variation in volume and dose between the two CT sets, the differences were within clinically acceptable limits for most of the patients.

The Monte Carlo based QA solution presented here is freely available upon request, in open-source format, and is distributed under a GNU General Public License. Source 20 of DOSXYZnrc and the SYNC* component modules of BEAMnrc are in the public domain, as part of the standard distribution of these codes. They are also freely available in open-source format and are distributed under a specific NRC license.

5. Acknowledgements

I would like to acknowledge the contributions of Ganiyu Asuni, Alanah Bergman, Karl Bush, Conor Shaw, Tony Teke, Ernest Tsang, and Sergei Zavgorodni, who have developed various elements of the Monte Carlo based QA process described here and with whom we have had many useful discussions. We also acknowledge the work of Fujio Araki, Rickard Cronholm, Jarkko Ojala, and Lixin Zhan, who have performed extensive independent validations of source 20 in their respective institutions. The first author would like to particularly acknowledge Wayne Beckham, who - many years ago, at a time of still insufficient computing resources - envisioned the main lines of this QA process and strongly supported the effort for its construction, for the benefit of patients and staff physicists alike.

6. References

[1] Bogdanich W 2010 Radiation Offers New Cures, and Ways to Do Harm *The New York Times*

(http://www.nytimes.com/2010/01/24/health/24radiation.html?pagewanted=all&_r=0)
[2] Popescu I A and Lobo J 2007 *Radiother. Oncol.* **84** S76
[3] Lobo J and Popescu I A 2010 *Phys. Med. Biol.* **55** 4431-43
[4] Asuni G *et al* 2013 A Monte Carlo tool for evaluating VMAT and DIMRT treatment deliveries including planar detectors *Phys. Med. Biol.* **58** 3535-50
[5] Rogers D W O *et al* 2013 BEAMnrc Users Manual *Report PIRS-0509(A)revL, National Research Council of Canada, Ottawa.*
[6] Walters B *et al* 2013 DOSXYZnrc Users Manual *Report PIRS-794, National Research Council of Canada, Ottawa.*
[7] Gete E *et al* 2013 *Med. Phys.* **40** 021707
[8] Belosi M F *et al* 2014 *Med. Phys.* **41** 051707
[9] Popescu I A *et al* 2005 *Phys. Med. Biol.* **50** 3375-92
[10] Popescu I A *et al* 2014. *56th Annual Meeting of AAPM, July 2014, Austin, Texas, USA*
[11] Popescu I A *et al* 2004 *Radiother. Oncol.* **73** (Suppl. 1) S87
[12] Popescu I A *et al* 2006 *IFMBE Proceedings* **14** 1801
[13] Teke T *et al* 2010 *Med. Phys.* **37** 116-23
[14] Teke T *et al* 2011 *Phys. Med. Biol.* **56** N295-305
[15] Moiseenko V *et al* 2013 *Phys. Med. Biol.* **58** 7107-16
[16] Bergman A *et al* 2014 *J. App. Clin. Med. Phys.* **15** 148-63
[17] Locke C and Zavgorodni S 2008 *Australas. Phys. Eng. Sci. Med.* **31** 290-99
[18] Mason D 2011 Pydicom: a pure python package for working with DICOM files (http://pypi.python.org/pypi/pydicom)
[19] Savitzky A and Golay M J E 1964 *Anal. Chem.* **36** 1627-39
[20] Kawrakow I 2002 *Phys. Med. Biol.* **47** 3087-103
[21] Cronholm R 2014 private communication
[22] Eiland R B et al 2014 *J. Rad. Res.* doi:10.1093/jrr/rru044 (advance access, June 6, 2014)

Real-time volumetric scintillation dosimetry

S Beddar

Department of Radiation Physics, The University of Texas MD Anderson Cancer Center, 1515 Holcombe Boulevard, Unit 94, Houston, Texas 77030

E-mail: abeddar@mdanderson.org

Abstract. The goal of this brief review is to review the current status of real-time 3D scintillation dosimetry and what has been done so far in this area. The basic concept is to use a large volume of a scintillator material (liquid or solid) to measure or image the dose distributions from external radiation therapy (RT) beams in three dimensions. In this configuration, the scintillator material fulfills the dual role of being the detector and the phantom material in which the measurements are being performed. In this case, dose perturbations caused by the introduction of a detector within a phantom will not be at issue. All the detector configurations that have been conceived to date used a Charge-Coupled Device (CCD) camera to measure the light produced within the scintillator. In order to accurately measure the scintillation light, one must correct for various optical artefacts that arise as the light propagates from the scintillating centers through the optical chain to the CCD chip. Quenching, defined in its simplest form as a nonlinear response to high-linear energy transfer (LET) charged particles, is one of the disadvantages when such systems are used to measure the absorbed dose from high-LET particles such protons. However, correction methods that restore the linear dose response through the whole proton range have been proven to be effective for both liquid and plastic scintillators. Volumetric scintillation dosimetry has the potential to provide fast, high-resolution and accurate 3D imaging of RT dose distributions. Further research is warranted to optimize the necessary image reconstruction methods and optical corrections needed to achieve its full potential.

1. Introduction

Volumetric scintillation is a promising new area of study aiming at making fast, high resolution and accurate measurements of absorbed dose distributions by imaging the light emitted by a scintillating medium in three dimensions. It was initially proposed by Kirov *et al* 2005 [1] for the dosimetry of brachytherapy Ru-106 eye plaque applicators using a liquid scintillator (LS) and then by Fukushima *et al* 2006 [2] for proton range measurement using a plastic scintillator block.

Our research group has been engaged over the last five years in the conception and development of real-time volumetric scintillation dosimetry using large-volume liquid or solid scintillators with the goal of rapidly measuring or imaging 3D dose distributions from photon or proton beams. Beddar *et al* 2009 [3] have investigated the feasibility of using a 3D LS detector system for the verification and characterization of proton beams in real time for intensity and energy-modulated proton therapy. They irradiated a plastic tank filled with LS with pristine proton Bragg peaks. Scintillation light produced during the irradiation was measured with a CCD camera. Acquisition rates of 20 and 10 frames per second (fps) were used to image consecutive frame sequences. These measurements were then compared to ion chamber measurements and Monte Carlo simulations. The light distribution measured from the images acquired at rates of 20 and 10 fps had standard deviations of 1.1% and

0.7%, respectively, in the plateau region of the Bragg curve. Differences were seen between the raw scintillation signal and the ion chamber due to the quenching effects of the scintillator and due to the optical properties of the imaging system. Beddar et al 2009 [3] have demonstrated the feasibility of using such detector system to measure the dose distribution produced by a proton beam in 2D and verified that the system is capable of acquiring multiple images within a single proton pulse with good signal -to-noise ratio and submillimeter image resolution. Although the size of the LS detector was limited (7 x 7 x 14 cm^3), they were able to conclude that a larger detector volume could be used.

Poenisch *et al* 2009 [4] investigated a larger LS system for application for high-energy photon beam dosimetry. They were able to acquire series of light signal images when irradiated with a 6 MV photon beam. The scintillation light data measured with the CCD camera were filtered to correct for scattering of the optical light inside the liquid scintillator. Depth-dose and lateral profiles as well as two-dimensional (2D) dose distributions were found to agree with results from the treatment planning system. Further, the corrected light output was found to be linear with dose and independent of dose rate. They concluded that the short time needed for image acquisition (10 s) and processing could make this system ideal for fast verification of the beam characteristics of the treatment machine and this new detector system shows potential usefulness for 2D Quality Assurance for 3D conformal RT or Intensity Modulated Radiation Therapy (IMRT) treatment verification. The limitation of this system is that it can only acquire images in two dimensions because there is no information as to the distance between the light source and the single camera. The addition of a second camera with an orthogonal view will lead to the eventual goal of measuring or imaging dose distributions in 3D.

Archambault *et al* 2012 [5] developed a similar LS detector system (20 x 20 x 20 cm^3) viewed by a CCD camera with a field view of ~ (26 x 20 cm^2) and a pixel size of 0.4 mm to measure the spatial location, intensity, and depth of penetration (energy) of the proton beamlets in near real-time. Pencil beams with nominal ranges in water between 9.5 and 17.6 cm were scanned to irradiate an area of 10 x 10 cm^2. Image frames were acquired at 50 ms per frame. Proton range measured from the light distribution produced in the LS was accurate to within 0.3 mm on average. The largest deviation seen between the nominal and measured range was 0.6 mm. The lateral position of the measured pencil beam was accurate to within 0.4 mm on average (see Section 3.3). The intensity of single proton spots was measured with precision ranging from 3% for the smallest spot intensity (0.005 MU) to 0.5% for the largest spot (0.04 MU).

Robertson *et al* 2013 [6] undertook the challenging task to study the quenching problem exhibited by scintillating materials when subjected to proton beams to be able to accurately convert the light signals to absorbed dose. The goal of their study was to develop an ionization quenching correction method to restore the linear dose response of scintillators when irradiated by proton beams. The quenching correction method they used was based on an empirical model that predicts quenching on the basis of the linear energy transfer (LET) of the proton beam developed by Birks 1951 [7,8]. The results of this study (see Section 5) demonstrated the effectiveness of a correction method that restores the linear dose response of a liquid scintillator throughout the proton beam range. The correction method was applied to a large-volume scintillation detector, addressing a major obstacle to fast 3D dosimetry of proton beams. This new correction method allows for closer agreement than a prior study by Wang *et al* 2012 [9] to correct for quenching in plastic scintillator detectors which showed an agreement within 5% between scintillator and ionization chamber measurements of proton beams.

Recently, a systematic study was performed by Robertson et al 2014 [10] to characterize the optical artefacts affecting measurement accuracy of such detector systems and to develop methods to correct for these artefacts. The optical artefacts addressed were photon scattering, refraction, camera perspective, vignetting, lens distortion, the lens point spread function, stray radiation, and noise in the camera. These artefacts were evaluated by theoretical and experimental means. The correction methods effectively mitigated the artefacts, increasing the average gamma analysis pass rate from 66% to 98% for gamma criteria of 2% dose difference and 2 mm distance to agreement.

Goulet *et al* 2014 [11] have recently developed a 10 x 10 x 10 cm^3 plastic scintillator embedded inside an acrylic phantom and imaged using a plenoptic camera. The 3D light distribution produced by the scintillator was reconstructed at a 2 mm resolution in all dimensions by back-projecting the light collected by each pixel of the light-field camera using an iterative reconstruction algorithm constrained by a beam's eye view projection of the incident dose acquired using the electronic portal imager integrated with the linac. The absolute dose difference between the reconstructed 3D dose and the expected dose calculated by the treatment planning software Pinnacle[3] was on average below 1.5 % of the maximum dose for both integrated IMRT and Volumetric Modulated Arc Therapy deliveries, and below 3% for each individual IMRT beam for a seven step-and-shoot IMRT plan of a brain tumor.

2. Materials and detector designs

All scintillation detectors that have been described in the literature have used either a liquid scintillator or a block of plastic scintillator, contained or embedded inside of an enclosure (tank or cell) and have at least one transparent side through which the light can be transmitted to a light imaging device. The scintillation light from the detector is collected and transmitted to a CCD or Complementary Metal Oxide Semiconductor (CMOS) chip using an appropriate optical lens that would capture the entire field of view onto the chip.

2.1. Material properties

Scintillators are designed to be transparent to their own light emissions. However, the scintillation photons produced within their own matrix material will experience Rayleigh scattering and additional scattering due to absorption and re-emission. The physical characteristics of the most widely used liquid scintillator (BC-531) [3,4,5,6,10] is compared to the most commonly known and used plastic scintillator (BC-400) [6] manufactured by Saint-Gobain Ceramics & Plastics, Valley Forge, PA and the recently used plastic scintillator EJ-260 [11] manufactured by Eljen Technology, Sweetwater, TX. These scintillating materials that do not contain any high Z materials are compared to polystyrene and water to ascertain there water-equivalence.

Table 1. Physical characteristics of the liquid scintillator BC-531 compared to the BC-400 plastic scintillator, EJ-260 plastic scintillator, polystyrene, and water.

	BC-531	BC-400	EJ-260	Polystyrene	Water
Emission (% of anthracene)	59	65	60	N/A	N/A
Peak wavelength (nm)	425	423	490	N/A	N/A
Electron density (10^{23} e/g)	2.930	3.272	3.350	3.238	3.343
Specific gravity (g/cm^3)	0.870	1.032	1.023	1.060	1
Composition	1:11.98	1:8.420	1:8.460	1:7.740	1:11.19
[Z: fraction by weight (%)]	6:88.02	6:91.58	6:91.54	6:92.26	8:88.81

Another organic liquid scintillator [12], OptiPhase HiSafe 3 (PerkinElmer, Waltham, MA) is being evaluated as a replacement for future scintillator detectors because its density is nearer to water than that of BC-531 (0.986 vs. 0.869 g/cm3) has not been included in the above table.

2.2. Detector design and construction

When designing any kind of detector system, one will always consider the primary clinical applications being sought, the type of ionizing radiation being detected and the required performance (dosimetric characteristics, spatial resolution, speed etc…). There will always be trade-offs in the selection and construction of all the components of such systems to make them useful for as many applications as possible, the most challenging one being clinically practical and cost effective.

2.2.1. Proton therapy beams. Figure 1 shows the LS detector systems used by Archambault *et al* 2012

[5] and Robertson *et al* 2013, 2014 [6,10]. The CCD camera used in both studies is a Luca^{EM} S 658M (Andor Technologies, Belfast, Northern Ireland). This is a 14 bits CCD with a resolution of 658 x 496 pixels and a physical pixel size of 10 μm, capable of measuring 37 full frames per second.

(a) (b)

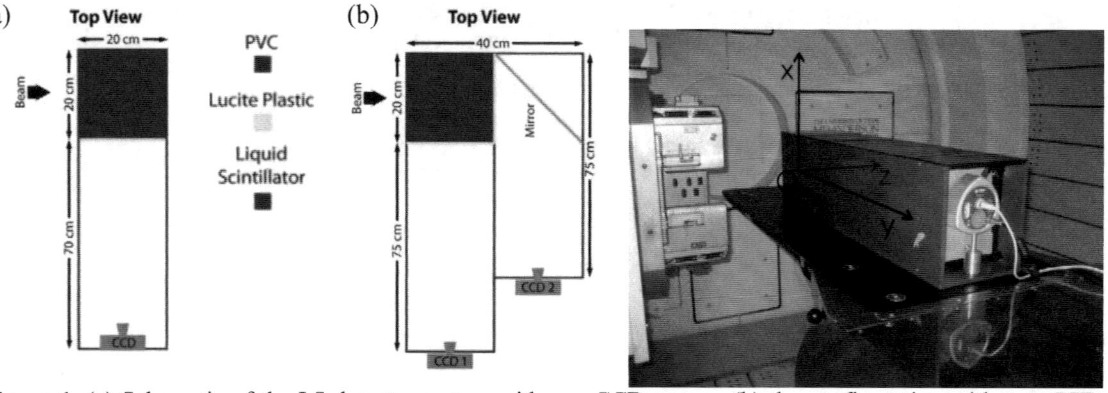

Figure 1. (a) Schematic of the LS detector system with one CCD camera (b) the configuration with two CCD cameras to resolve the 3rd dimension and (c) the actual LS detector system illustrated in Figure 1 (a) being set-up on a robotic couch at the scanning beam gantry at the Proton Therapy Center of the University of Texas MD Anderson Cancer Center in Houston TX.

2.2.2. Photon therapy beams. Although it is not evident to justify the need for developing additional 3D QA or *in vivo* dose verification of photon RT treatments, studies have looked at using volumetric scintillation dosimetry for these purposes and have proven to be an attractive and an efficient option.

(a) (b) (c) (d)

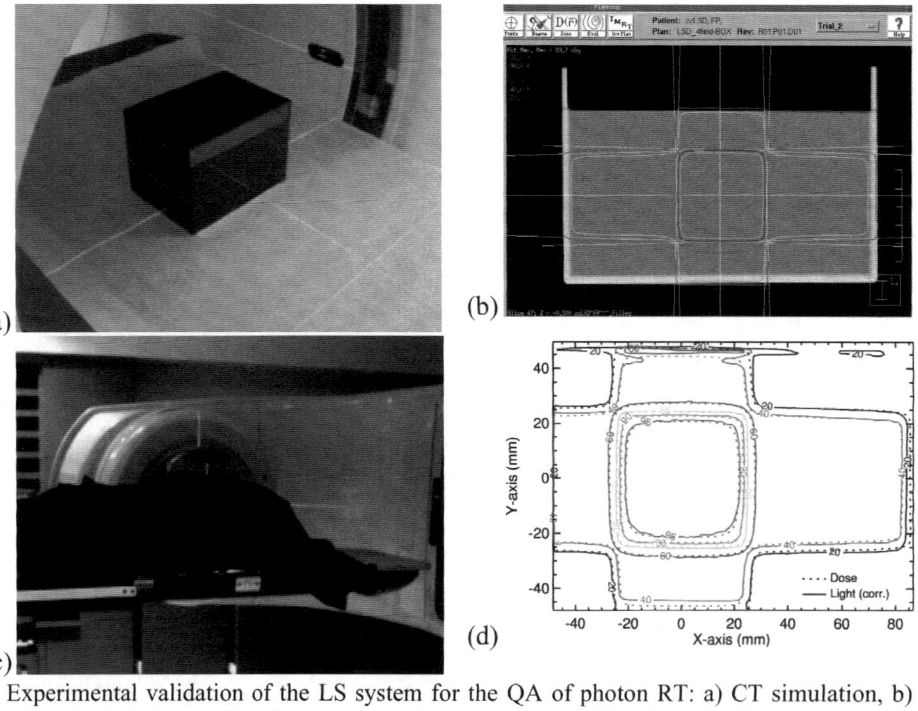

Figure 2. Experimental validation of the LS system for the QA of photon RT: a) CT simulation, b) treatment planning, c) treatment delivery, d) and image acquisition/processing of the light distribution produced within the LS detector compared to the expected dose distribution.

This approach would have to compete with EPID dosimetry, the ArcCHECK system, the ScandiDos Delta 4 system, the MatriXX system or the MapCheck system to establish itself in the marketplace. Despite these uncertainties, initial studies have been performed investigating the capabilities of these systems for photon RT. Fig. 2 shows an extension of the system that was described by Beddar *et al* [3] for IMPT and the steps undertaken by Poenisch *et al* 2009 [4] to verify the dose distribution of a simple four-field-box technique. Excellent agreement was found between the corrected scintillation light and the dose distribution calculated with the treatment planning system throughout both the high and low dose regions

The system used by Goulet *et al* 2014 [11] (*in press*) is shown in Fig. 3. They have developed and validate a novel type of high-resolution 3D dosimeter based on the real-time light acquisition of a plastic scintillator volume using a plenoptic camera. Using this, they were able to perform millimeter resolution, water-equivalent dosimetry of an IMRT and VMAT plan over a whole 3D volume.

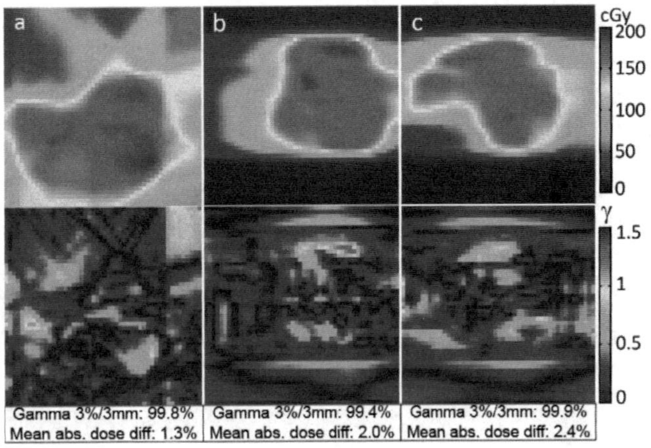

Figure 3. (Left): A plastic block scintillator system using a plenoptic camera in conjunction with an EPID for 3D photon QA. (Right) Axial (a), sagittal (b), and coronal (c) planes from a 3D reconstruction of an IMRT brain treatment measured with the plenoptic camera and the portal imaging device. The 2D gamma maps for the criteria (3% dose, 3 mm distance) are also shown.

3. Basic properties

3.1. Linearity with absorbed dose

An ideal detector system should respond linearly to the absorbed dose deposited within the detector. Poenisch et al [4] measured the light signal of the LS using a 6 MV photon beam, keeping the background noise constant by using a constant acquisition time of 25 s, a dose rate of 600 MU/min and a field size of 4x4 cm^2.

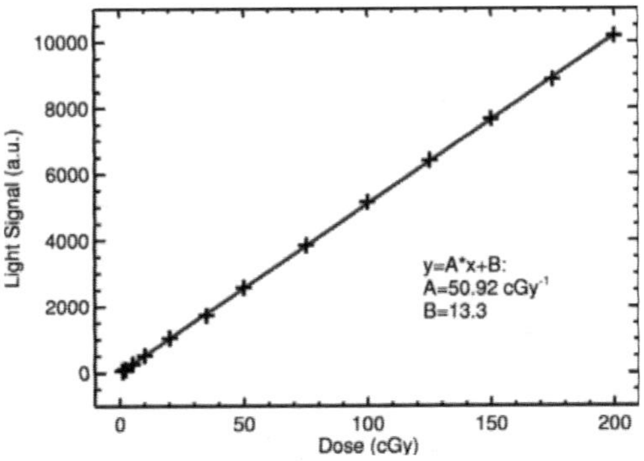

Figure 4. Measured light-dose response shown by symbols. The solid line depicts a linear fit with a correlation coefficient of R=0.99995.

Journal of Physics: Conference Series **573** (2015) 012005 doi:10.1088/1742-6596/573/1/012005

3.2. Spatial and temporal resolution
In general the spatial and temporal resolution is dictated by the physical characteristics of the CCD camera (pixel size and image acquisition time). However, these two properties should be experimentally determined for the particular LS system and application. Furthermore, the spatial resolution is determined by the blurring of the image due to light diffusion and the discrete nature of the pixels. The system described by Beddar *et al* [5] had a spatial resolution of 0.38 mm. In general all the systems that have been described here are capable of providing submillimeter spatial resolution.

The temporal resolution is limited by the maximum frame rate of the camera but also will be limited by the signal to noise ratio needed to make accurate measurements. This is one of the reasons that we selected cooled CCD cameras in our work. Another important consideration is the readout time of the camera, which determines the dead time of the detector between image frames.

3.3. Determination of particle range and lateral position for spot scanning
Fig. 5 (a) shows the difference between the nominal proton range in water and the range measured in the LS detector system as reported by Archambault *et al* 2012. They also determined the lateral position of single proton spots with great precision with an average difference between the measured and expected spot position of 0.11 ± 0.34 mm (1σ) and a maximum difference of 0.6 mm.

Fig. 5 (b) shows a sample of a proton spot measured with a similar LS detector system illustrating the intensity in color scale by Robertson (unpublished).

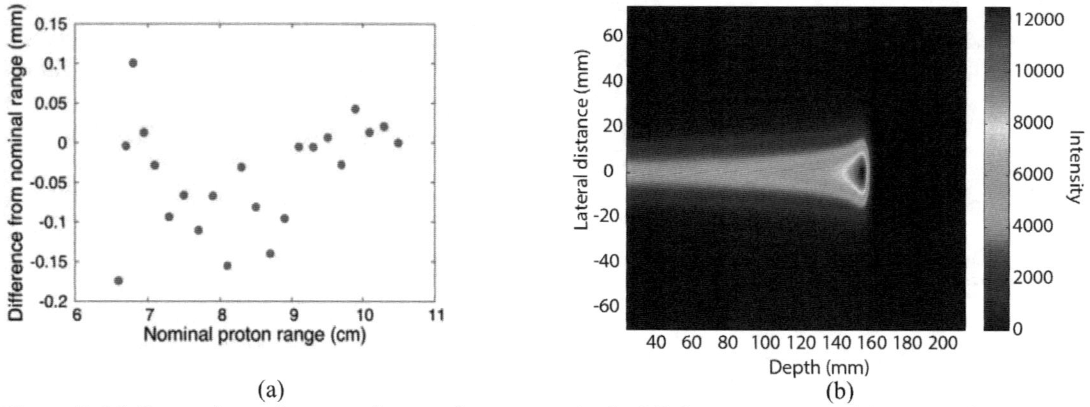

Figure 5. (a) Comparison of measured proton beam range in the LS detector system with expected range [5]. (b) A 144.9 MeV proton pencil beam measured with a liquid scintillator.

4. Optical artefact corrections

A number of artefacts need to be corrected in order to accurately measure the scintillation distribution produced inside the LS system. Table 2 lists their sources, physical phenomena and their effects [10].

Table 2. Optical artefacts in a volumetric scintillation dosimetry system.

Artefact source	Physical phenomenon	Effect
Light propagation in the scintillator and tank	Photon scattering	Blurring of light signal
	Refraction	Changes in effective pixel size and intensity
	Perspective	Changes in effective pixel size with depth
Optical train	Vignetting	Decreased brightness at image periphery
	Lens distortion	Radial variation in pixel size and location
	Lens point spread function	Blurring of light signal
CCD chip	Stray radiation	Hot pixels and streaks
	Background noise	Measurement uncertainty and pixel value offset

Robertson *et al* [10] developed correction methods for these artefacts and compared corrected light signals to calculated light distributions from a validated Monte Carlo model. The calculated and corrected light distributions were compared using gamma analysis. Table 3 gives the results, showing marked improvement in passing rates after applying the corrections to images of proton pencil beams.

Table 3. Gamma analysis pass rates for proton pencil beam light distributions as compared to projected dose distributions calculated using Monte Carlo methods.

Gamma Criteria	85.6 MeV		100.9 MeV		144.9 MeV		161.6 MeV	
	Orig	Corr	Orig	Corr	Orig	Corr	Orig	Corr
3%/3mm	80.7%	99.1%	81.9%	99.7%	84.5%	100.0%	95.3%	100.0%
3%/2mm	61.2%	95.3%	63.5%	98.4%	65.3%	99.9%	76.2%	99.9%
2%/2mm	60.6%	94.9%	62.9%	98.3%	64.6%	99.9%	75.5%	99.8%

5. Quenching correction

The LET-dependent response of organic scintillators, called ionization quenching, is problematic for proton beam dosimetry because of the steep increase in LET in the proton Bragg peak. The most common and simple model for ionization quenching of scintillators is the Birks model, which assumes that the degree of quenching is linearly proportional to the LET [7,8]. Using this model, Robertson *et al* [6] developed a voxel-by-voxel quenching correction method for use in volumetric scintillation dosimetry of proton beams. This model was tested on proton beams of four different energies. The corrected central-axis light signal matched the Monte Carlo-calculated dose within ±5% (Fig. 6a). Fig. 6b shows a comparison of the uncorrected light signal measured with the scintillator detector, the calculated dose, and the corrected light signal, showing a good qualitative agreement between the calculated dose and corrected light distributions.

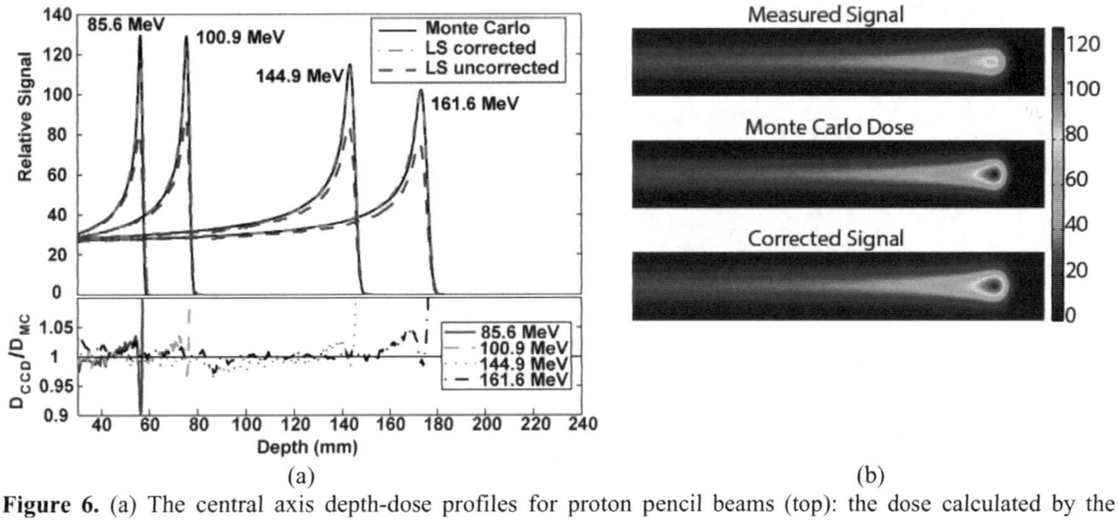

Figure 6. (a) The central axis depth-dose profiles for proton pencil beams (top): the dose calculated by the validated Monte Carlo model is shown in black, the uncorrected scintillation signal is shown in blue and the corrected scintillation signal is shown in red, and (below) the ratio of the corrected scintillation signal to the Monte Carlo dose. (b) The measured scintillation signal (top), the dose calculated using Monte Carlo methods (centre) and the corrected scintillation signal (bottom) for a 100.9 MeV proton pencil beam [6].

6. Conclusion and future directions

All the work developing liquid or plastic scintillator systems and the related theoretical studies that have been done so far have demonstrated the feasibility and utility of using such detector systems for 2D and 3D dosimetry for photon and proton RT. These systems have been shown to be capable of submillimeter spatial resolution and the ability to perform real-time measurements.

Studies have shown that optical artefacts introduce non negligible deviations into the scintillation light distribution measured by these volumetric detectors and indicated that it is essential to correct for them to accurately measure the intensity and the spatial distribution of the scintillation light emission.

When these systems are used in proton beams, one must also correct for scintillation quenching. Methods have been developed to do so with sufficient accuracy to fulfil dosimetric quality assurance and verification purposes. These methods require prior knowledge of the LET distribution of the beam and the Birks model parameters for the scintillator. Further research is still warranted in this area to get a better understanding of this phenomenon and the theory behind it and to improve the accuracy of quenching correction techniques.

Nevertheless sufficient work has been done over the past five years in real-time volumetric scintillation dosimetry that it is ready for technology transfer from the laboratory setting to commercial research and development (R&D) settings.

7. References

[1] Kirov A S *et al* 2005 *Phys. Med. Biol.* **50** 3063-81
[2] Fukushima Y *et al* 2006 *Phys. Med. Biol.* **51** 5927-36
[3] Beddar S *et al* 2009 *Med. Phys.* **36** 1736-1743
[4] Ponisch F *et al* 2009 *Med. Phys.* **36** 1478-85
[5] Archambault L *et al* 2012 *Med. Phys.* **39** 1239-46
[6] Robertson D *et al* 2013 *Phys. Med. Biol.* **58** 261-73
[7] Birks J B 1951 *Proc Phys Soc A* **64** 874-877
[8] Birks J B 1964 The Theory and Practice of Scintillation Counting (Oxford: Pergamon)
[9] Wang L L *et al* 2012 *Phys. Med. Biol.* **57** 7767-7781
[10] Robertson D *et al* 2014 *Phys. Med. Biol.* **59** 23-42
[11] Goulet M *et al* 2014 *Med Phys.* (In Press.)
[12] Ingram W S *et al* 2013 *Med. Phys.* **40** 227

Radiochromic 3D Detectors

Mark Oldham
Department of Radiation Oncology, Duke University Medical Center,
Durham, NC 27710, USA

E-mail: mark.oldham@duke.edu

Abstract. Radiochromic materials exhibit a colour change when exposed to ionising radiation. Radiochromic film has been used for clinical dosimetry for many years and increasingly so recently, as films of higher sensitivities have become available. The two principle advantages of radiochromic dosimetry include greater tissue equivalence (radiologically) and the lack of requirement for development of the colour change. In a radiochromic material, the colour change arises direct from ionising interactions affecting dye molecules, without requiring any latent chemical, optical or thermal development, with important implications for increased accuracy and convenience. It is only relatively recently however, that 3D radiochromic dosimetry has become possible. In this article we review recent developments and the current state-of-the-art of 3D radiochromic dosimetry, and the potential for a more comprehensive solution for the verification of complex radiation therapy treatments, and 3D dose measurement in general.

1. Introduction

End-to-end verification measurements of radiation dose distributions delivered to phantoms has long been a key component of comprehensive radiation therapy QA programs [1]. These measurements are typically performed during an intensive commissioning phase, when the accuracy of a new technique is established prior to implementation on patients in the clinic. Historically, in the absence of dosimetry tools that can measure the dose in 3-dimensions (3D), commissioning measurements were typically made at selected planes and points, utilizing film, TLD's, diodes, and ion chambers [2]. These low-sampling methods have important uses, but are incapable of providing comprehensive verification in 3D throughout the relevant volume at high resolution. The advent of sophisticated and complex radiation treatments has led researchers to consider whether 3D radiochromic dosimetry techniques can provide a more comprehensive solution to the problem of verification of complex radiation treatments. This review builds on work in Oldham [3], describing the development and current state-of-the-art of 3D radiochromic dosimetry.

The term 3D dosimetry was discussed in Oldham *et al* [4] who introduced the Resolution-Time-Accuracy-Precision (RTAP) criteria as a performance goal for clinical use. To meet RTAP, a dosimetry system should deliver a 3D dosimetric analysis of a treatment plan with 1mm isotropic spatial resolution, within 1hour, with an accuracy of within 3% of the true value, with 1% precision. The RTAP remains a useful benchmark, although others have noted that lower resolution is sufficient for some clinical situations, and a shorter time criteria would be ideal. Partial-3D dosimetry systems such as Delta4 and arc-check, have planes or surfaces of point detectors and interpolate to 3D dose [5-7]. These approaches represent momentum toward the ideal of a truly comprehensive 3D dosimetry system, but are not discussed further here.

Journal of Physics: Conference Series **573** (2015) 012006 doi:10.1088/1742-6596/573/1/012006

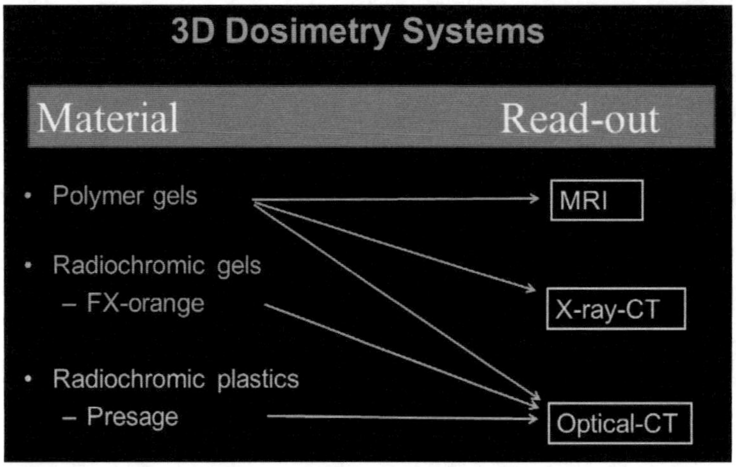

Figure 1. Summary of 3D dosimetry systems capable of directly measuring 3D dose isotropically with high resolution (2mm or better) and approaching the RTAP capability criteria [4].

3D dosimetry systems in current use include those illustrated in figure 1. These are chemical dosimetry systems, where a material exhibits a physical response to radiation that can be quantified and imaged by a readout system. Three categories of materials are shown: polymer gels, radiochromic gels, and radiochromic plastics. These three materials and associated readout (imaging) methods have been discussed recently in [3]. Polymer gel dosimetry has been much studied and while many applications and uses have been demonstrated [8, 9] the limitation to MRI or slower optical readout techniques have led some researchers to investigate the potential of radiochromic dosimetry. Radiochromic dosimeters are compatible with cheaper and potentially faster optical readout techniques, by virtue of relative insensitivity to stray-light artifacts (discussed below). Here we focus on developments in the radiochromic detectors – both materials and readout techniques. Alternative 3D dosimetry systems have also been recently proposed utilizing scintillation [10] and the Cerenkov effect [11], but are not discussed further here.

2. Methods of 3D Radiochromic Dosimetry

2.1. Radiochromic Gels

The ferrous sulphate (or Fricke dosimeter) has been used for many years as both as an absolute and relative dosimetry system [12, 13]. The Fricke dosimeter consists of a ferrous solution containing Fe2+, contained in small vials compatible with optical attenuation measurement in a spectrophotometer. Fricke dosimeters are tissue and water equivalent even for lower energy radiation fields [14]. When irradiated, Fe^{2+} is oxidized to Fe^{3+}, and the quantification of the proportion of Fe^{3+} (which is proportional to absorbed dose) is determined by spectrophotometry [15]. To achieve accurate measurements, the temperature of the reaction must be carefully controlled. Another challenge can be the limited sensitivity caused by the modest optical-density (OD) change of the Fe^{3+} ions at the probing light wavelength (224nm). Imaging at such a short wavelength is possible for small volumes of Fricke solution contained in optical cuvettes (with an optical path length of 1cm) but not feasible for large volumes due to strong light attenuation.

Gore *et al* 1984 [16] first proposed 3D Fricke based dosimetry by showing that ferric ions have a greater influence on proton relaxation times than ferrous, enabling 3D dose imaging by MRI [16, 17]. A linear relationship between spin-lattice relaxation rate R1 and dose was reported [18]. R1 relaxation measurement was preferred to spin-spin relaxation (R2) for Fricke gels, because of higher dynamic range (the unirradiated Fricke dosimeter exhibits low R1) [19]. A key limitation arises from diffusion of small Fe^{3+} ions in the gel immediately post irradiation, leading to degradation of the dose distribution. To mitigate this effect, the gel needs to be imaged within ~1 hour [20-22]. Attempts to stabilize the Fricke solution in gels were performed by a number of investigators ([14, 16, 23-26]. Early work reported low spatial resolution (thick slices) owing to the challenge of obtaining low noise with high spatial resolution using MRI. Recent work involving fast 3D MRI sequences has shown promising results: 1mm isotropic, in a 20 minute imaging session [27].

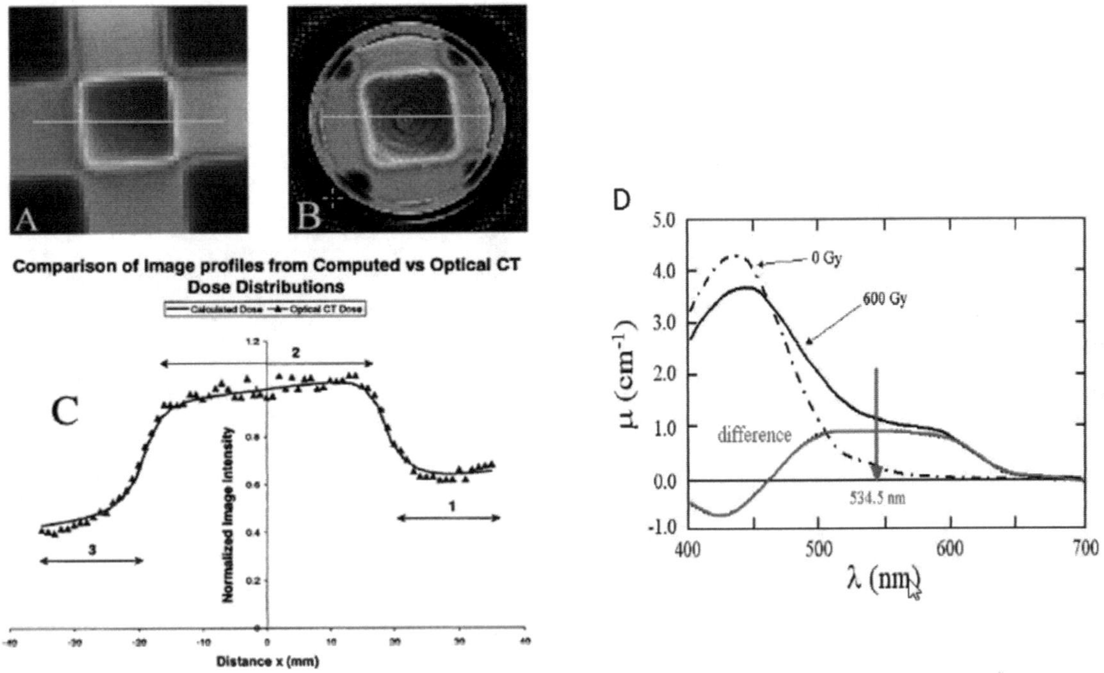

Figure 2. A) calculated dose distribution of a 4-field box irradiation (6MV and 4x4cm). B) Corresponding optical-CT scan through an irradiated FBX gel. C) Line profile comparison along the line indicated in A and B. D) The change in absorbance spectrum of an irradiated FBX dosimeter, indicating optimal readout wavelength of ~535nm. From Kelly *et al* [28].

The Fricke 3D dosimeter requires access to MRI, which is restrictive for many medical centres. Much effort has therefore been devoted to developing Fricke gels that could be imaged by optical-computed-tomography (optical-CT). The addition of a metal ion indicator (e.g. xylenol orange) leads to a visible colour change in the presence of ferric (Fe3+) ions [29]. This development opens up the facility for optically imaging larger Fricke dosimeters using longer wavelength light with greater penetrating power [17]. Kelly *et al* [28] demonstrated the feasibility of 3D optical-CT dosimetry of Fricke-benzoic-xylenol (FBX) gels (see figure 2). Their system used a scanning laser beam and was able to achieve sub mm spatial resolution and an accuracy of reconstructed attenuation values of within 2%, corresponding to dose measurement of within 5% in dose range 1-10Gy. A key advantage of the FBX dosimeter for optical-CT dosimetry, is the nature of the radiation induced optical contrast, which is absorbing in nature, incurring minimal stray light. This enables fast broad-beam scanning systems [30, 31]. The Fricke and FBX gels are relatively easy and convenient to make [32]. The main challenges relate to the need for an external casing to support the gel (largely a cost and convenience factor), and the diffusion limitation, requiring the gel to be imaged within 1 hour of irradiation.

2.2. Radiochromic Plastics

A radiochromic plastic material 'Presage' was first introduced in 2006 for 3D dosimetry by optical-CT [33]. Presage consists of a firm polyurethane matrix doped with a halogenated hydrocarbon free radical initiator and a leucodye (leuco-malachite-green). The polyurethane matrix is highly transparent: more so than gelatin or agarose. Radiation induced optical-contrast is generated through the oxidation of LMG to malachite green (figure 3). Presage thus maintains the advantage of radiochromic gels in the minimization of stray light through an absorptive radiation induced contrast mechanism. The hard polyurethane substrate removes the requirement of an external casing. Presage has proved highly versatile for moulding [34], and embedding cavities and channels (e.g. in brachytherapy applications [35, 36]). 3D printing of Presage dosimeters of various shapes, including anatomical shapes defined on CT scans, is an exciting new capability that has recently been demonstrated [37].

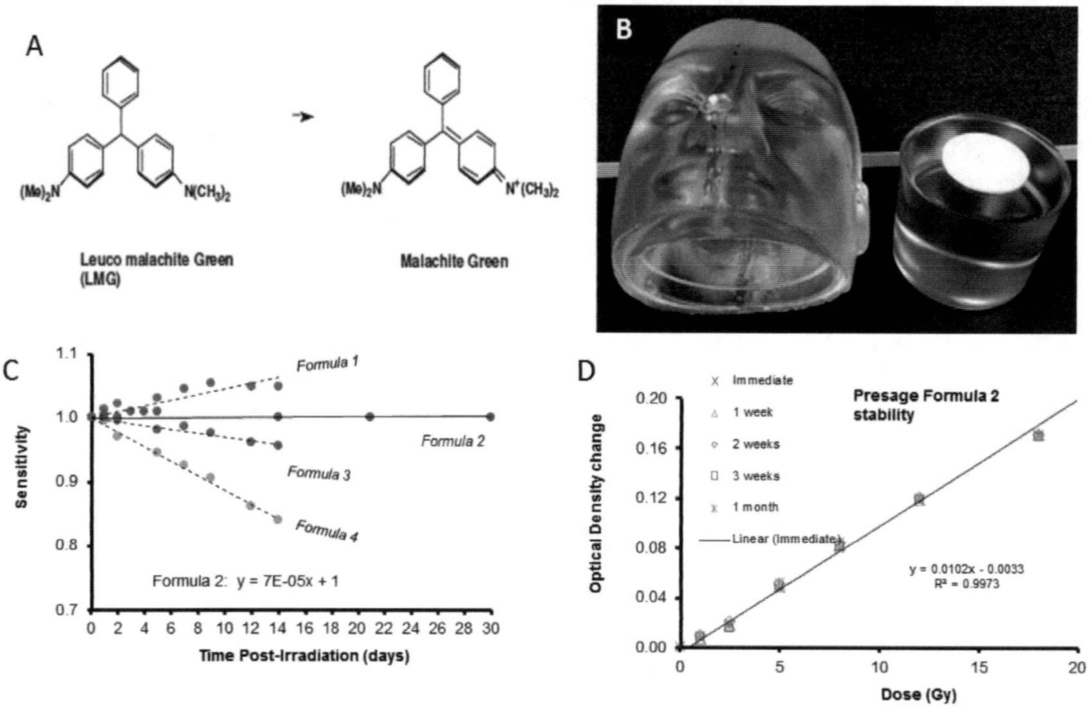

Figure 3. A) The transparent leuco-dye (LMG) and the absorbing oxidative product malachite green [33]. B) A cylindrical PRESAGE dosimeter (11cm diameter) which inserts into a poly-urethane head-phantom for radiosurgery verification [38]. C) post-radiation temporal stability of radiation sensitivity of several Presage variants, illustrating how minor changes in formulation can yield substantially different effects [34], D) linearity and stability of a presage formulation (unpublished data).

A linear optical dose response of Presage has been demonstrated up to doses as high as 100Gy [33, 39-41]. Some studies have reported negligible energy or dose-rate dependence in the region 145kVp to 18MV [33, 40-42]. However other studies have reported conflicting results as to the stability of the radiochromic response and the relative 3D distribution [43]. Substantial variations in sensitivities between formulations of the same batch, and even the same batch but different volumes have been reported [44]. These results suggest caution when evaluating and using different Presage formulations. The poly-urethane substrates part A and part B can be varied, in different proportions, to create dosimeters with markedly different hardness. Recent data from our lab (figure 3C) show wide variability in post irradiation stability associated with relatively minor changes in poly-urethane components. Like many materials (including PAGs, and EBT film) under-response has been reported in the Bragg-peak when using Presage in proton beams [45, 46].

2.3 MRI Dose Readout Techniques

MRI was first used for 3D gel dosimetry by imaging R1 (spin-lattice) relaxation of water molecules in Fricke gels. R1 imaging has higher dynamic range than R2 (spin-spin) relaxation. The mechanisms underlying the relaxation and paramagnetic molecules have been described previously [16, 18]. The main challenges for MRI 3D dosimetry of radiochromic Fricke gels include difficulty generating high resolution low noise images, and stabilizing the irradiated dose distribution. The diffusion of small Fe^{3+} ions within the gel degrades the dose distribution [13]. Given the challenges associated with accurate quantitative imaging with MRI, it is essential that the appropriate sequences and system characterization is implemented prior to 3D dosimetry studies. A number of limitations and challenges have been identified, leading to concern that expertise in MR imaging is a pre-requisite for accurate 3D dosimetry [47-49]. While some expertise is necessary, the prospect has been greatly simplified by recommendations in a recent report by De Deene *et al* [19].

2.4 Optical-CT readout of radiochromic gels

Optical-computed-tomography appears to have arisen independently in three specialty fields [50-54]. Optical-CT is analogous to x-ray CT, except visible light is used as the imaging modality. In both optical and x–ray CT, line integrals of attenuation are acquired at various views through the object to be imaged. The relatively small size of 3D dosimeters makes it practical to rotate the dosimeter, rather than the source and detector. The same reconstruction algorithms (e.g. filtered back-projection) can be used to reconstruct 3D maps of the local attenuation coefficients (x-ray or optical), and both modalities are susceptible to artefacts including scattered radiation, rings, beam-hardening, attenuation, motion, etc [55, 56]. Several different configurations of optical-CT systems for 3D dosimetry have been investigated as shown in figure 4. The primary advantages of optical-CT for 3D dosimetry, include reduced cost, increased accessibility, and potentially higher accuracy and precision in shorter imaging times [4].

The presence of scattered light (or stray light) poses challenges to obtaining accurate optical attenuation maps, in the same way that x-ray scatter can corrupt cone-beam-CT images [57, 58]. Accurate optical-CT is feasible if the stray light contribution to the line-integral measurements in projection images are minimal. The first optical-CT system for 3D dosimetry, the 'OCTOPUS' [52, 53], utilized a 1st generation scanning laser configuration, operating in a translate rotate manner. These systems are quite effective at reducing scatter, and can be used to image dose in dosimeters with high scattered light component, like PAGS [8, 59]. Several groups have investigated the performance of the OCTOPUS [42, 59-61]. Variations on the scanning laser system were also explored [4, 28].

The limitation of the laser scanning configuration is that long imaging times are needed to translate and rotate the dosimeter to acquire all required line profiles. Scanning times of several hours were reported for large volumes [4, 53, 61, 62]. Recent scanning laser systems have achieved faster speeds by up to a factor 4 [62, 63], and still faster scans are possible utilizing rotating mirrors [64]. For smaller samples good quality results are reported [65-67]. Despite these advances, it was not until the introduction of broad-beam scanning systems that 3D dosimetry became feasible in time frame within the RTAP criteria[4]. The first prototype broad beam optical-CT scanners were proposed at the turn of the millennium [68, 69]. The image quality from these systems was not ready for preclinical use, a result compounded by the lack of radiochromic (low-scatter) dosimeters at that time. Other groups also developed in-house broad beam systems with more sophisticated components and configuration [31, 69-71].

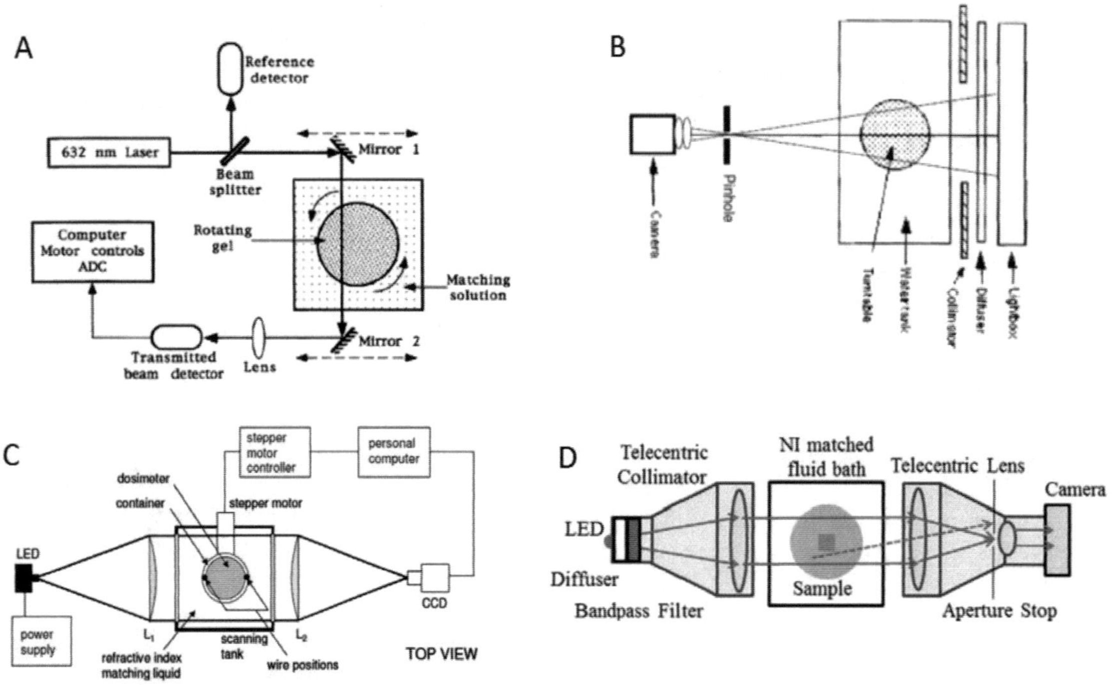

Figure 4. Optical-CT scanning configurations. A) first generation scanning laser (from Gore *et al* [53]), b) cone-beam-CT arrangement (from Wolodzko *et al* [68]), c) parallel beam arrangement (from Krstajic *et al* [72]), d) a fully bi-telecentric system (Thomas *et al* [73]).

Two categories of fast broad-beam optical-CT scanning configurations can be identified: parallel beam, and cone-beam, illustrated in figure 4. Cone-beam-CT systems are low cost, but have the potential disadvantage of reduced accuracy arising from stray-light artefacts [30, 31, 74-76]. Telecentric systems are more expensive (due to the inclusion of sophisticated lenses) but have the capacity for high accuracy through minimisation of stray-light [70, 77]. The precise limitations and balance of this trade-of is yet to be fully determined. Stray light artefacts have been reported when using cone-beam-CT scanners and high scatter polymer gels [30, 31]. A continued increase in sophistication of broad beam optical-CT scanners is evident in the literature [30, 31, 68-72, 77, 78].

Perhaps the current state-of-the-art broad-beam system was introduced by Thomas *et al* [73], consisting of a powerful large field-of-view matched bi-telecentric scanning system. This system was commissioned and benchmarked for clinical use, and set a new standard for 3D dosimetry capability [38, 73, 79]. Accompanying articles addressed minor corrections associated with stray-light reflections (mostly within the lenses) and spectral artefacts [80, 81]. These corrections are negligible except under extreme conditions (e.g. small field dosimetry).

3. Summary and Conclusion

3D dosimetry has proved an innovative field of research and clinical development for over two decades. The original motivation was to develop more comprehensive methods for verification of complex radiation therapy treatments. This goal is potentially even more relevant today with continued increase in sophistication of modern RT treatments. Current areas of innovation and further development are occurring in 3 areas: materials development, readout development, and novel applications. For radiochromic materials, key challenges include the development of more stable dosimeters for remote credentialing activities [82], and re-useable dosimeters for improved economics within an institution [34]. New approaches to dosimeters have emerged from micelle technology [83-85] to new radiochromic materials [86]. The challenge of a 3D dosimetry material without an LET dependence is also an active area of development [45, 87]. A recent development is the use of novel 3D printing technology to create anatomically accurate 3D dosimeters [37].

For 3D radiochromic dosimetry readout, improved optical-CT scanners and protocols are being developed, that are faster, more accurate (reduced scatter), cheaper, and with higher resolution [30, 73], as well as scanning configurations which require substantially less (or even no) fluid matching [88-90]. MRI sequences have been developed that minimize problems due to sample heating while preserving low noise and higher spatial resolution [47, 83]. For x-ray-CT readout, more sensitive formulations are being developed [91, 92].

A review of the current status of the field reveals that radiochromic approaches have contributed strong progress toward the long-sought goal of a reliable, practical, and cost-effective 3D dosimetry system that can match the RTAP criteria [4]. Several systems have demonstrated they can meet this goal, but are not yet realized in the commercial world. This will hopefully occur in the next few years, and will represent an important strengthening of core QA practices in radiation therapy, contributing to safer and more accurate treatment of our patients. A fairly comprehensive summary of the diversity of applications of 3D dosimetry in the clinical and pre-clinical worlds is given in Oldham [3].

4. Acknowledgments

I am very grateful to John Schreiner for proof reading parts of this article. I am also especially grateful to the many students, post-docs and colleagues who have worked with me on 3D dosimetry topics over the years.

5. References

[1] Moran J M *et al* 2011 *Med. Phys.* **38** 5067-72
[2] Ibbott G 2010 *J. Phys.: Conf. Ser.* **250** 0120001
[3] Oldham M 2014 *Advances in Medical Physics.* Devon Godfrey *et al* eds. Madison, WI: Medical Physics Publishing.
[4] Oldham M *et al* 2001 *Med. Phys.* **28** 1436-45
[5] Letourneau D *et al* 2004 *Radioth. Oncol.* **70** 199-206
[6] Letourneau D *et al* 2009 *Med. Phys.* **36** 1813-21
[7] Feygelman V *et al* 2011 *J. Appl. Clin. Med. Phys.* **12** 3346
[8] Baldock C *et al* 2010 *Phys. Med. Biol.* **55** R1-63
[9] McJury M *et al* 2000 *Brit. J. Radiol.* **73** 919-29
[10] Beddar S *et al* 2009 *Med. Phys.* **36** 1736-43
[11] Glaser A K *et al* 2013 *Phys. Med. Biol.* **58** 601-19
[12] Fricke H and Morse S 1927 *Am. J. Roent. Radium Ther. Nucl. Med* **18** 430-432
[13] Schreiner L J 2004 *J. Phys.: Conf. Ser.* **3** 9-21
[14] Kron T *et al* 1993 *Phys. Med. Biol.* o**38** 139-150
[15] Scharf K and Lee R M 1962 *Radiat. Res.* **16** 115
[16] Gore J C *et al* 1984 *Phys. Med. Biol.* **29** 1189-97
[17] Appleby A and Leghrouz A 1991 *Med. Phys.* **18** 309-12
[18] Podgorsak M B and Schreiner L J 1992 *Med. Phys.* **19** 87-95
[19] De Deene Y 2010 *J. Phys.: Conf. Ser.* **250** 1-10
[20] de Pasquale F *et al* 2006 *Radiat. Prot. Dosim.* **120** 151-4
[21] Tseng Y J *et al* 2002 *Mag. Res. Imag.* **20** 495-502
[22] Chu K C *et al* 2000 *Phys. Med. Biol.* **45** 955-69
[23] Hazle J D *et al* 1991 *Phys. Med. Biol.* **36** 1117-25
[24] Olsson L E *et al* 1992 *Radioth. Oncol.* **24** 82-6
[25] Olsson L E *et al* 1990 *Phys. Med. Biol.* **35** 1623-31
[26] Schreiner L J *et al* 1994 *Mag. Res. Imag.* **12** 901-7
[27] Cho N Y *et al* 2013 *Med. Phys.* **40** 052101
[28] Kelly R G *et al* 1998 *Med. Phys.* **25** 1741-50

[29] Gupta B L *et al* 1982 *Phys. Med. Biol.* **27** 235-45
[30] Olding T and Schreiner L J 2011 *Phys. Med. Biol.* **56** 1259-12
[31] Olding T *et al* 2010 *Phys. Med. Biol.* **55** 2819-40
[32] Jordan K 2010 *J. Phys.: Conf. Ser.* **250** 012043
[33] Adamovics J and Maryanski M J 2006 *Radiat. Prot. Dosim.* **120** 107-12
[34] Juang T *et al* 2013 *J. Phys.: Conf. Ser.* **444** 012029
[35] Adamson J *et al* 2013 *J. Phys.: Conf. Ser.* **444** 012100
[36] Adamson J *et al* 2012 *Med. Phys.* **39** 4515-23
[37] Bache S *et al* 2013 *Med. Phys.* **40** 490
[38] Thomas A *et al* 2013 *J. Phys.: Conf. Ser.* **444** 012050
[39] Wang Z *et al* 2010 *J. Phys.: Conf. Ser.* **250** 012058
[40] Sakhalkar H S *et al* 2009 *Med. Phys.* **36** 71-82
[41] Guo P Y *et al* 2006 *Med. Phys.* **33** 1338-45
[42] Guo P. *et al* 2006 *Med. Phys.* **33** 3962-72
[43] Yates E S *et al* 2011 *Acta oncologica* **50** 829-34
[44] Jackson J *et al* 2013 *Med. Phys.* **40** 215
[45] Zhao L *et al* 2012 *Phys. Med. Biol.* **57** N431-43
[46] Al-Nowais S *et al* 2009 *Appl. Radiat. Isotop.* **67** 415-8
[47] Vandecasteele J and De Deene Y 2013 *Phys. Med. Biol.* **58** 63-85
[48] Vandecasteele J and De Deene Y 2013 *Phys. Med. Biol.* **58** 43-61
[49] Vandecasteele J and De Deene Y 2013 *Phys. Med. Biol.* **58** 19-42
[50] Sharpe J 2004 *Annu. Rev. Biomed. Eng.* **6** 209-28
[51] Sharpe J *et al 2002 Science* **296** 541-5
[52] Maryanski M J *et al* 1996 *Phys. Med. Biol.* **41** 2705-17
[53] Gore J C *et al* 1996 *Phys. Med. Biol.* **41** 2695-704
[54] Winfree A T *et al* 1996 *Chaos* **6** 617-26
[55] Oldham M and Kim L 2004 *Med. Phys.* **31** 1093-1
[56] Oldham M *et al* 2003 *Med. Phys.* **30** 623-34
[57] Letourneau D *et al* 2005 *Radioth. Oncol.* **75** 279-86
[58] Oldham, M. *et al* 2005 *Radioth. Oncol.* **75**, 271-8
[59] Xu Y *et al* 2004 *Med. Phys.* **31** 3024-33
[60] Lopatiuk-Tirpak O *et al* 2008 *Med. Phys.* **35** 3847-59
[61] Islam K T *et al* 2003 *Med. Phys.* **30** 2159-68
[62] Campbell W G *et al* 2013 *Med. Phys.* **40** 061712
[63] Qian X *et al* 2013 *Phys. Med. Biol.* **58** N321-31
[64] Krstajic N and Doran S J 2007 *Phys. Med. Biol.* **52** N257-63
[65] Doran S J *et al* 2013 *Phys. Med. Biol.* **58** 6279-97
[66] Abdul Rahman A T *et al* 2011 *Phys. Med. Biol.* **56** 4177-99
[67] Doran S J *et al* 2010 *Phys. Med. Biol.* **55** 1531-47
[68] Wolodzko J G *et al* 1999 *Med. Phys.* **26** 2508-13
[69] Doran S J *et al* 2001 *Phys. Med. Biol.* **46** 3191-213
[70] Sakhalkar H S and Oldham M 2008 *Med. Phys.* **35** 101-11
[71] Babic S *et al* 2008 *Int. J. Radiat. Oncol.* **70** 1281-91
[72] Krstajic N and Doran S J 2006 *Phys. Med. Biol.* **51** 2055-75
[73] Thomas A *et al* 2011 *Med. Phys.* **38** 4846-57
[74] Oldham M 2006 *J. Phys.: Conf. Ser.* **56** 58-71
[75] Wuu C and Xu Y 2010 *J. Phys.: Conf. Ser.* **250** 200-9
[76] Babic S *et al* 2009 *Phys. Med. Biol.* **54** 2463-81
[77] Krstajic, N. & Doran, S. J. *Phys. Med. Biol.* **52**, 3693-3713, (2007).
[78] Thomas, A. & Oldham, M. *J. Phys.: Conf. Ser.* **250**, 1-5, (2010).
[79] Oldham M *et al* 2012 *Int. J. Radiat. Oncol.* **84** 540-6

[80] Thomas A *et al* 2011 *Phys. Med. Biol.* **56** 4433-51
[81] Thomas A *et al* 2011 *Phys. Med. Biol.* **56** 3403-16
[82] Juang T *et al* 2014 *Med. Phys* (in press).
[83] Vandecasteele J and De Deene Y 2013 *Phys. Med. Biol.* **58** 6241-62
[84] Nasr A T *et al* 2013 *Phys. Med. Biol.* **58** 787-805
[85] Jordan, K and Avvakumov N 2009 *Phys. Med. Biol.* **54** 6773-89
[86] Newton J R *et al* 2011 *J. Phys.: Conf. Ser.* **250** 49-53
[87] Lopatiuk-Tirpak O *et al* 2012 *Technol. Cancer Res. T.* **11** 441-5
[88] Rankine L and Oldham M 2013 *Med. Phys.* **40** 051701
[89] Doran S J and Yatigammana D N 2012 *Phys. Med. Biol.* **57** 665-83
[90] Ramm D *et al* 2012 *Phys. Med. Biol.* **57** 3853-68
[91] Johnston H *et al* 2012 *Phys. Med. Biol.* **57** 3155-75
[92] Jirasek A *et al* 2012 *Phys. Med. Biol.* **57** 3137-53

The quest for sensible data analysis in clinical routine: Study case on the new Octavius1500 array and its associated phantoms

Ann Van Esch[1,2], Dominique P Huyskens[1,2], K Basta[2], M Ghislain[2] and R Delvaux[2]
[1]7Sigma, QA-team in radiotherapy physics, Kasteeldreef 2, 3150 Tildonk, Belgium
[2]Radiotherapy Association, Centre Hospitalier Mouscron - Ste Elisabeth Namur, 7700 Mouscron - 5000 Namur, Belgium

E-mail: ann.vanesch@7sigma.be

Abstract. We have investigated the dosimetric characteristics of the new Octavius1500 array and its associated phantoms (Oct2D and Oct4D) and compared it to the previous systems. In line with the rising amount of publications advocating the retirement of the widely used but overly lenient 3%, 3mm (global dose) gamma analysis criteria, we have also focused on deriving appropriate evaluation criteria to be used with the new measurement systems in clinical routine. Using the Octavius1500 study as an example, we advocate the use of more selective, equipment specific gamma criteria, all starting from the same, very strict acceptance criteria but adding measurement uncertainty to these.

1. Introduction

The purpose of QA is simple: we want to make sure that the dose we have calculated for a radiotherapy treatment plan is the same as the dose we deliver. But how good does the agreement have to be? When looking into research and development, precision is the key and one can be as demanding as one chooses to be. When assessing clinical relevance, however, a more pragmatic approach is desirable and efficiency becomes a primary consideration in the QA process.

To perform QA in clinical routine, most physicists are dependent on the commercially available solutions. As rotational treatments are taking up an ever increasing share of all radiotherapy treatments, the most commonly used verification methods have gradually moved from field-by-field to composite verification through measurements. Composite plan measurement systems produce either planar (2D) or three dimensional (3D) dose information [1-23]. Planar data are obtained from the measurement in a straightforward manner, with data manipulation mostly limited to the application of energy (and sometimes angle) dependent calibration files and correction factors. Only the actually measured datapoints are subjected to gamma evaluation analysis. This is in contrast to most commercial 3D dosimetry systems, for which the 3D dose is not actually measured but reconstructed from a (series of) measurements. These QA systems are considerably more complex than their 2D predecessors and often suffer from a substantial black box-factor. And yet, for most of the radiotherapy centers, the purchased phantom is put to use with limited - if any - validation and the rather universal 3%G,3mm gamma criteria are applied, regardless of the specific dosimetric characteristics of the measurement system. We would therefore like to join Nelms et al. [24-29] in making a case for the revision of the way in which the gamma analysis is used in clinical practice. Rather than suggesting sufficiently large, universal acceptance criteria to ensure the continued use of the relatively insensitive, binary approach of the gamma evaluation (either pass or fail), we would advocate the use of more selective, equipment specific gamma criteria, all starting from the same, very strict acceptance criteria but adding measurement uncertainty to these. For now, as an initial starting point, we will assume that no clinically relevant changes are expected when the delivered ('real') and calculated doses in the homogeneous phantom agree within 2%L, 2mm for almost the entire irradiated volume (e.g. for 97% of the volume receiving at least 10% of the maximum dose). These starting criteria should be subjected to regular re-evaluation as radiotherapy systems further improve, but with a state of the art, properly commissioned, high quality radiotherapy system (comprised of a dose calculation and dose delivery system), this overall agreement between the calculated and the actual dose should be achievable in a homogeneous phantom for all but the lowest doses.

In the below study, we have used a prototype of the new Octavius1500 array (PTW, Freiburg, Germany) in combination with the Octavius4D phantom as a case-study to obtain the above mentioned, QA-system specific evaluation criteria. A more elaborate description of the below material was published in Medical Physics, September 2014 [30].

2. Methods, Materials and Results

2.1 The Octavius1500 system

The detector is shown in figure 1a. The 1405 ion chambers - of 0.44x0.44x0.30 cm^3 each - cover a 27x27 cm^2 active area and are arranged in a checkerboard pattern. The Octavius1500 arrray has twice the detector density of its 729 predecessor, resulting in an area coverage of almost 50%. If necessary, 100% coverage can be achieved by performing two measurements, shifting the setup over 5mm in between and merging both dataset within the software. As with every PTW ion chamber array, the array is already calibrated in the ^{60}Co beam at the PTW secondary standard dosimetry laboratory upon delivery. The detector matrix can be used in combination with solid water plates for orthogonal measurement setups as well as with phantoms of the Octavius® series, specifically designed for multidirectional irradiation.

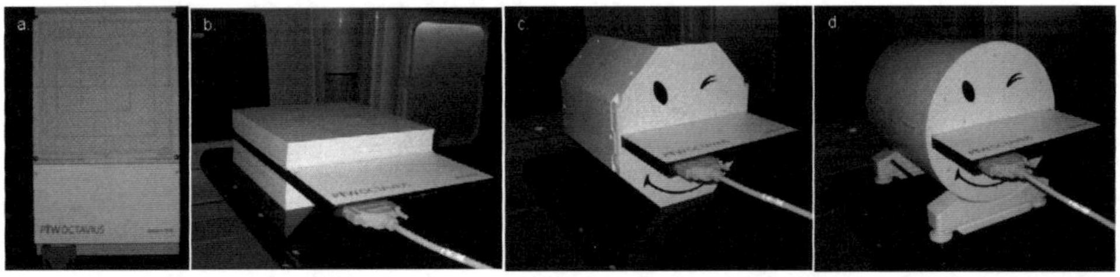

Figure1. (a) the Octavius1500 detector, (b) the orthogonal setup, (c) the Oct2D^{1500}/Oct1500 tandem (d) and the Oct4D/Oct1500 tandem.

In order to assess the uncertainties in the final 3D dose reconstruction, it is vital to understand the measurement process. The steps in the QA process by means of the Octavius4D rotational phantom are outlined in figure 2. The Octavius4D system (figure 1d) consists of a 2D array, a rotational phantom, a trolley for transport that also holds the electrometer and the control unit, a (wireless) inclinometer and the accompanying Verisoft software. The inclinometer is initialized at gantry zero. The responsibility of the control unit is to make sure that the phantom rotates along with the gantry angle, so the 2D array is always perpendicular to the beam axis and measures the full field output for every angle. To convert the detector signal to absolute dose, a simple energy dependent cross-calibration factor is sufficient. Every 200ms, the software stores the inclinometer readout and the corresponding 2D array readout in a simple ascii file (*.xcc format). Upon loading this file in the Verisoft software, the latte uses the series of planar measurements to perform the 3D dose reconstruction. In order to be able to do this independently from the TPS, it just needs depth dose information. So before the first use, the user has to measure the depth dose in the water tank for a small selection of open fields (from 3x3 to 27x27°) at SSD=85cm. Then, for every angular measurement point in the xcc file, an individual 3D dose reconstruction is performed based on the gantry angle readout, the 2D measurement orthogonal to the beam axis in mid-phantom and the stored depth dose data. The software deduces the field size from the actual measurement and selects the appropriate depth dose based on this. Then it sums up all individual contributions to obtain the total 3D dose. In order to do so, all data need to be sampled to the same grid, so data are interpolated to a user specified grid (typically the same as the TPS resolution, being 2.5mm) before they are all summed up to result in the final measurement reconstructed 3D dose.

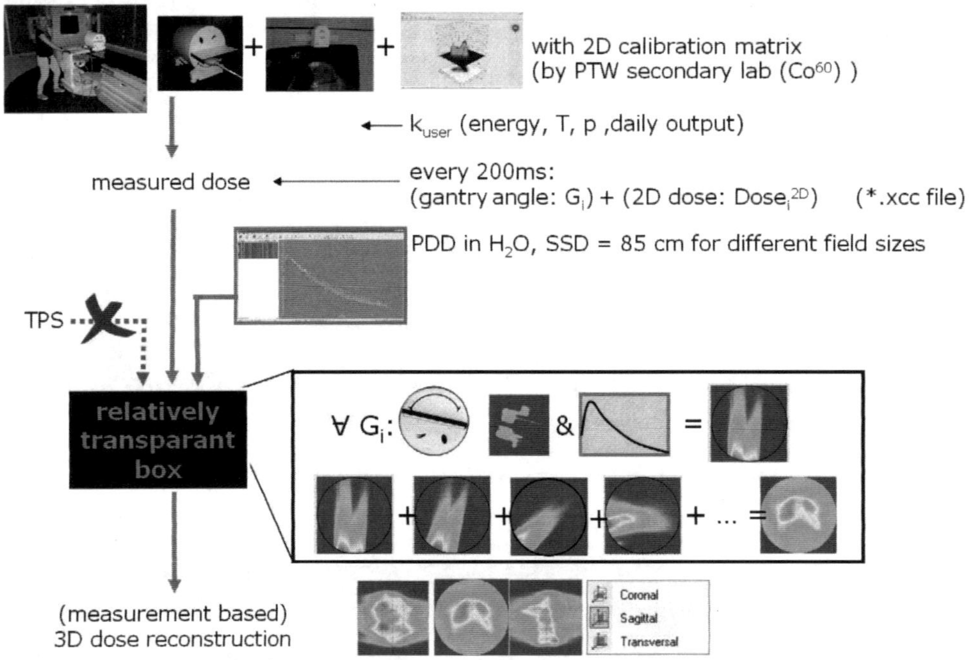

Figure 2. 3D dose measurement based reconstruction with the Octavius4D system.

2.2 Quantifying the measurement uncertainty of the Octavius1500 system

2.2.1 Basic characterisation of the Octavius1500 detector

For the basic characterisation of the detector itself we have simply used a series of open field measurements with the gantry at zero. The detector was first sandwiched between solid water, then slid into to upright rotational phantom. Detector stability, output factor behaviour (down to 3x3cm^2), dose rate dependence, linearity and energy dependence were all assessed and found to be most satisfying (all within 1%). The instant measurement stability is heaven, compared to the more cumbersome warm-up and cool-down behaviour of the previous Octavius729 model.

2.2.2 The Octavius1500/Octavius4D tandem

The assessment of the QA process with the rotational phantom is also performed one step at a time. Starting with simple open fields from different gantry angles (static or rotational), we subsequently move on to the validation of a reference sets of pre-validated IMRT & RA plans. This reference set contains 20 plans on artificial structures as well as clinical IMRT & RA plans (containing 6MV and 18MV treatment beams). We have used Eclipse for generating the plans and calculated the doses with AAA. All plans were delivered on Clinac21IX dual energy machines. All of these plans, we have tediously, elaborately, tirelessly validated before with all other means available to us. So for these reference plans, we are confident that the dose calculation really meets the true delivery within the 97% pass rate for the 2%L 2mm local gamma criteria in a homogeneous phantom. Subsequently, we have submitted 60 routine patient plans to the QA as they came along in clinic.

2.2.3 The gamma evaluation
To make sense of the vast amount of data, the availability of volumetric gamma analysis tools is most welcomed, if not mandatory. First, we did an in-depth evaluation for a variety of gamma criteria from the most tolerant 3 % 3mm global down to the most strict 2%,2mm local evaluation. And for these criteria, we looked at the corresponding pass rates for different threshold isodose levels from 10% up to 95% of the maximum dose. We have found the 10% and 50% dose thresholds useful to assure correct overall dose delivery in the organs at risk as well as in the lower target dose regions. The 95% dose level provides useful insights in the dose coverage of the (high dose) target. Table 1 presents a very compact summary of the whole dataset.

Table 1. Pass rates for different gamma analysis criteria on the pre-validated reference dataset and on the subsequent routine patient plans.

	$\gamma_{3D}(iso_{10\%})$	$\gamma_{3D}(iso_{50\%})$	$\gamma_{3D}(iso_{95\%})$
Reference dataset (20):			
3%G,3mm	99.99 ± 0.02	99.99 ± 0.04	99.92 ± 0.18
2%G,3mm	99.92 ± 0.10	99.87 ± 0.15	99.38 ± 1.13
2%G,2mm	99.31 ± 0.37	99.15 ± 0.48	97.84 ± 2.50
3%L,3mm	99.99 ± 0.02	99.99 ± 0.02	99.92 ± 0.18
2%L,3mm	99.72 ± 0.44	99.76 ± 0.31	99.31 ± 1.27
2%L,2mm	97.98 ± 1.26	98.48 ± 0.87	97.65 ± 2.71
RA patients (60)			
3%G,3mm	99.95 ± 0.11	99.92 ± 0.28	99.47 ± 1.95
2%G,3mm	99.77 ± 0.25	99.65 ± 0.58	97.62 ± 4.58
2%G,2mm	98.88 ± 0.74	98.59 ± 1.09	94.60 ± 6.67
3%L,3mm	99.89 ± 0.21	99.92 ± 0.28	99.47 ± 1.95
2%L,3mm	99.69 ± 0.32	99.64 ± 0.59	97.57 ± 4.62
2%L,2mm	97.40 ± 1.30	98.09 ± 1.44	94.03 ± 7.07

For the reference dataset, we are confident that the agreement without the measurement uncertainty should be at least 97% for the 2%L,2mm criteria. The fact that this is very close to what we obtain with the actual measurement indicates that, in theory, no considerable measurement uncertainties need to be attributed to the QA system. For the routine patients, we still get very good results, but we do notice that the spread around the mean values is larger and the average pass rates are lower. Reducing the DTA to 2mm makes almost half of the plans fail the 97% pass rate. The reference dataset was acquired in one or two nightly session, with the phantom set up very meticulously and without any real time pressure, apart from the eventual need for sleep. The routine patient dataset was acquired over numerous individual measurement sessions, accompanied by the well known pressure of clinical routine and limited machine availability. The phantom was positioned carefully, but quickly. As it turns out, a 1mm positioning uncertainty very easily slips into the measurement as such.

So rather than insisting on a more meticulous setup, we prefer to go for the pragmatic approach and include this small measurement setup error into the distance to agreement criterion. So in conclusion, we select the 2%L, 3mm gamma with 97% pass rates as a first filter.

Secondly, we did a detailed visual inspection of the data by going through the gamma evaluation for every slice of the 3D dose, and by looking at line profiles. We especially paid attention to the possible impact of the twofold detector density. An example of such an in-detail visual inspection of the data clearly showed the improvement in the 3D data reconstruction because of the twofold detector density. Figure 3 shows two gamma evaluation maps on the same treatment plan. Red and blue points fail the criteria, with red points indicating overdosage, blue points signaling underdosage:

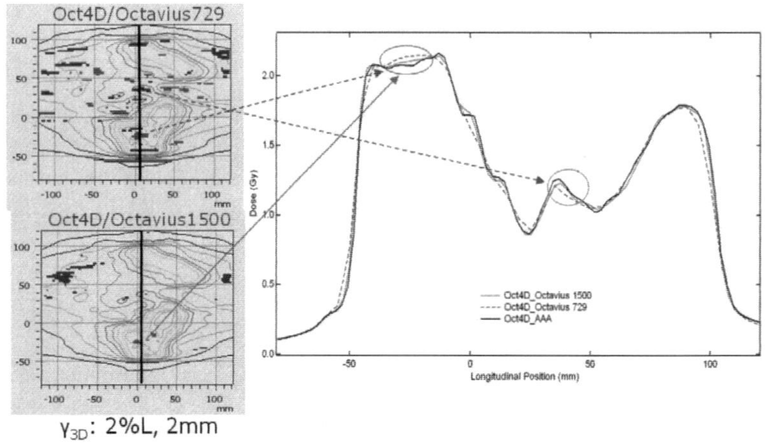

Figure 3. impact of the improved detector density of the Octavius1500 (50% coverage) compared to Oct729 (25%).

The upper gamma evaluation map was acquired with the Octavius729 array in the phantom whereas the lower one is with the new Octavius1500 detector. The necessary interpolation in between measurement points created substantially less falls alerts in the Octavius1500 system. The red dotted line is the Octavius729, the green line is the Octavius1500, the black line is the calculation. If desirable, one can even do two measurements with a 5mm shift in between, merge the data and have no more uncovered areas at all.

3. Conclusions

Having characterized the uncertainties we can expect in the measured 3D dose, we have made the following gamma evaluation flowchart (figure 4.) for use in clinical routine for the Octavius1500/Octavius4D 3D dose system.

We apply these strict gamma 3D criteria. If the 97% pass rate is met, all is well and there is no need to even look at the data any further. A green light is given for the treatment. If the 97%PR fails the data are submitted to further visual inspection. First the urgent question regarding the start of the treatment needs to be addressed. As no universal guidelines on the clinical impact of dose deviation are available, we apply a common sense filter, check by how much it failed and look for an explanation for the deviations. It is also wise to check in which dose areas or anatomical regions the deviations are situated (PTV? body dose? OAR?). When in doubt, we quickly run the more tolerant 3%,3mm check to proceed. If it passes, it can proceed to treatment. If the plan still fails this check, the plan is not be treated before the origin of the deviation is found.

Having addressed the imminent clinical importance of the QA, failed plans are investigated in more detail later on for research and development purposes.

As an overall conclusion, we would urge medical physicists to stop hiding behind sloppy criteria by default just because we aim for green lights in the patient files. We are missing out on a lot of possibly useful information if we do so. It pays off to be strict first as long as you are sensible later.

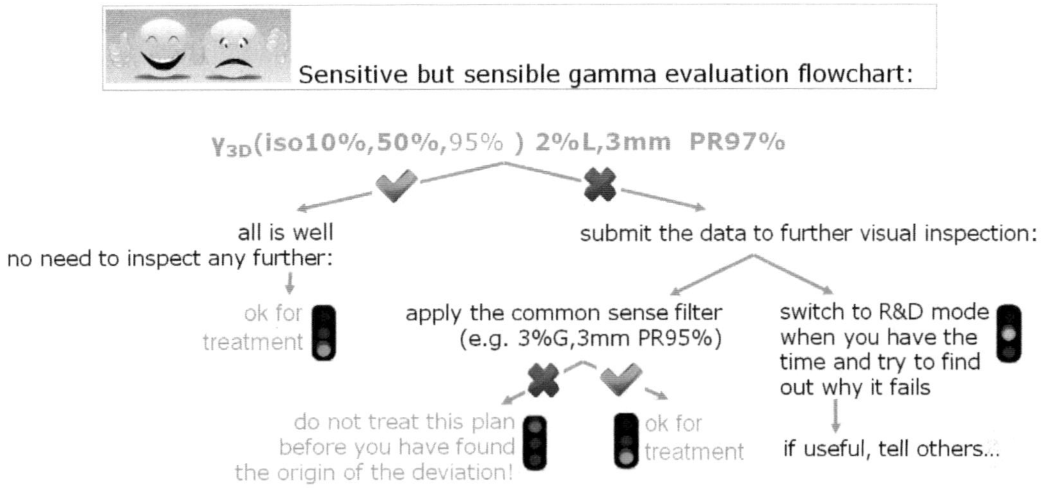

Figure 4. Proposal of flowchart for gamma evaluation with Octavius1500/Octavius4D QA-system.

4. References

[1] McGarry C K *et al* 2013 *Med. Phys.* **40** 091707
[2] Poppe B *et al* 2006 *Med. Phys.* **33** 1005-15
[3] Bakhtiari M *et al* 2011 *Med. Phys.* **38** 1366-73
[4] Spezi E *et al* 2005 *Phys. Med. Biol.* **50** 3361-73
[5] Poppe B *et al* 2013 *Med. Phys.* **40** 082106
[6] Poppe B *et al* 2007 *Phys. Med. Biol.* **52** 2921-29
[7] Korreman S *et al* 2009 *Acta Oncol.* **48** 185-91
[8] Bedford J L *et al* 2009 *Phys. Med. Biol.* **54** N167-76
[9] Van Esch A *et al* 2007 *Med. Phys.* **34** 3825-37
[10] Chandraraj V *et al* 2011 *J. Appl. Clin. Med. Phys.* **12** 338-49
[11] Feygelman V *et al* 2011 *J. Appl. Clin. Med. Phys.* **12** 3346
[12] Masi L *et al* 2011 *Med. Phys.* **38** 612-21
[13] Garcia-Vincente F *et al* 2012 *J. Appl. Clin. Med. Phys.* **13** 111-23
[14] Zhen H *et al* 2011 *Med. Phys.* **38** 5477-89
[15] Carrasco P *et al* 2012 *Med. Phys.* **39** 5040-9
[16] Stasi M *et al* 2012 *Med. Phys.* **39** 7626-34
[17] Olch A J *et al* 2012 *Med. Phys.* **39** 81-6
[18] Nelms B E *et al* 2012 *Med. Phys.* **39** 4228-38
[19] Feygelman V *et al* 2013 *Med. Phys.* **40** 021708
[20] Opp D *et al* 2013 *J. Appl. Clin. Med. Phys.* **14** 70-84
[21] Godart J *et al* 2011 *Phys. Med. Biol.* **56** 5029-43
[22] Boggula R *et al* 2010 *Phys. Med. Biol.* **55** 5619-33
[23] McCurdy B *et al* 2008 *Med. Phys.* **35** 2760
[24] Nelms B E *et al* 2013 *Med. Phys.* **40** 111722
[25] Cadman P F *et al* 2011 *Med. Phys.* **38** 4464
[26] Hussein M *et al* 2013 *J. Appl. Clin. Med. Phys.* **14** 274-92
[27] Low D A *et al* 1998 *Med. Phys.* **25** 656-61
[28] Depuydt T *et al* 2002 *Radiother. Oncol.* **62** 309-19

Journal of Physics: Conference Series **573** (2015) 012007 doi:10.1088/1742-6596/573/1/012007

[29] Ezzell G A *et al* 2009 *Med. Phys.* **36** 5359-73
[30] Van Esch A *et al* 2014 *Med. Phys.* **41** 091708-14

Uncertainty in 3D gel dosimetry

Yves De Deene[1,2] and Andrew Jirasek[3,4]
[1]Dept. of Engineering, Faculty of Science, Macquarie University, Sydney, Australia
[2]Institute of Medical Physics, School of Physics, University of Sydney
[3]Physics, I.K. Barber School of Arts and Science, University of British Columbia, Okanagan Campus, Canada
[4]Dept. Physics and Astronomy, University of Victoria, Canada

E-mail: yves.dedeene@mq.edu.au, andrew.jirasek@ubc.ca

Abstract. Three-dimensional (3D) gel dosimetry has a unique role to play in safeguarding conformal radiotherapy treatments as the technique can cover the full treatment chain and provides the radiation oncologist with the integrated dose distribution in 3D. It can also be applied to benchmark new treatment strategies such as image guided and tracking radiotherapy techniques. A major obstacle that has hindered the wider dissemination of gel dosimetry in radiotherapy centres is a lack of confidence in the reliability of the measured dose distribution. Uncertainties in 3D dosimeters are attributed to both dosimeter properties and scanning performance. In polymer gel dosimetry with MRI readout, discrepancies in dose response of large polymer gel dosimeters versus small calibration phantoms have been reported which can lead to significant inaccuracies in the dose maps. The sources of error in polymer gel dosimetry with MRI readout are well understood and it has been demonstrated that with a carefully designed scanning protocol, the overall uncertainty in absolute dose that can currently be obtained falls within 5% on an individual voxel basis, for a minimum voxel size of 5 mm^3. However, several research groups have chosen to use polymer gel dosimetry in a relative manner by normalizing the dose distribution towards an internal reference dose within the gel dosimeter phantom. 3D dosimetry with optical scanning has also been mostly applied in a relative way, although in principle absolute calibration is possible. As the optical absorption in 3D dosimeters is less dependent on temperature it can be expected that the achievable accuracy is higher with optical CT. The precision in optical scanning of 3D dosimeters depends to a large extend on the performance of the detector. 3D dosimetry with X-ray CT readout is a low contrast imaging modality for polymer gel dosimetry. Sources of error in x-ray CT polymer gel dosimetry (XCT) are currently under investigation and include inherent limitations in dosimeter homogeneity, imaging performance, and errors induced through post-acquisition processing. This overview highlights a number of aspects relating to uncertainties in polymer gel dosimetry.

1. Introduction

1.1. 3D dose verification with integrating dosimeters
3D radiation dosimetry provides a unique ability to display dose distributions occurring in clinical radiotherapy in three dimensions (3D) in humanoid shaped phantoms [1-3]. It has a unique role to play in safeguarding conformal radiotherapy treatments, as it is the only dosimeter that covers the whole treatment chain and provides the radiation oncologist with the integrated dose distribution in 3D.

Journal of Physics: Conference Series **573** (2015) 012008 doi:10.1088/1742-6596/573/1/012008

Moreover, 3D dosimeters can also be used to benchmark motion compensation radiotherapy technologies when mounted in a motion phantom. However, some doubts with respect to the overall reliability have overshadowed the wider dissemination of 3D radiation dosimetry in radiation centers.

1.2. 'Absolute' versus 'relative' 3D dosimetry

Although the terminology 'absolute' and 'relative' is somewhat ill defined in gel literature, we use the term 'absolute' to describe the approach where an independent external calibration of the 3D dosimeter is applied in contrast to the term 'relative' 3D dosimetry, where the 3D dose distribution is normalized towards some internal reference points (mostly the dose in the isocentre and the dose in a low or high dose region).

When utilizing polymer gels with MRI readout, significant differences in dose-R_2 response have been found between different 3D dosimeters that were poured in small calibration vials as compared to larger phantom recipients [4, 5]. This finding severely compromises the use of an externally derived calibration curve to convert the measured R_2 maps to dose maps. Several possible reasons have been proposed as explanation for the observed discrepancies such as differences in cooling rate [6], changes in temperature during irradiation as a result of the exothermal radiation induced polymerization reaction [7] and physicochemical changes related to an oxygen-antioxidant imbalance [8]. To circumvent the discrepancies in dose-R_2 response between small test tubes and larger volumetric anthropomorphic gel dosimeter phantoms, many researchers have chosen to apply an internal calibration (i.e. the R_2 is calibrated to dose by using two points within the phantom as 'known' doses), which is further referred to as 'relative' dosimetry. Regardless the fact that with an internal calibration, the ability of gel dosimetry to safeguard absolute dose information of a treatment is lost, it should also be emphasized that there is a potentially more severe danger with this strategy because the mentioned mechanisms may depend on radiation dose, dose rate, environmental conditions and size and shape of the phantom. Most often, a linear calibration curve is assumed with this approach. There is therefore no evidence that the same mechanism for the observed discrepancies would not cause any dose measurement errors within the volumetric polymer gel dosimeter phantom itself. Indeed, a major source of these discrepancies is attributed to temperature differences during scanning within the volumetric dosimeter phantom.

3D dosimetry using optical readout has mostly been performed with relative or internal calibration, although there doesn't seem to be a specific argument why absolute calibration could not be performed. The radiation absorptivity defined as absorption per unit of distance per dose should only depend on the dosimeter material and calibration samples can be exposed to known radiation doses and either scanned or readout using a benchtop optical spectrometer. Studies are underway to determine the uncertainty associated with external dose calibration of dye based plastic dosimeters.

Early investigations using x-ray CT polymer gel dosimetry (PGD) utilized linear calibration curves and "relative" dosimetry [9, 10] Calibrations were performed either in a purely relative sense (assuming a known dose at a given point) or via large phantom irradiation with small fields of known doses. However, calibration phantoms were not designed to match test (anthropomorphic) phantoms, thus introducing potential errors within the calibration procedure. Since those early experiments, refinements to the overall calibration procedure in x-ray CT PGD have been made. As in the case with MRI, it has been recognized that small vial irradiations can not be used to calibrate large test phantoms due to the differing response between small and large PGD samples. As mentioned above, this observation points to the differential response being due to properties of the gel reaction (exothermic etc) as opposed to an imaging modality-based explanation. Furthermore, it has been shown that a linear dose response can not be assumed when utilizing high dose-sensitivity polymer gel with x-ray CT imaging [11, 12]. Recent experiments have shown that when utilizing a non linear calibration curve obtained from the irradiation of a portion of the test phantom with a known dose distribution, absolute dosimetry can be performed which results in a dose dependent accuracy and precision, as described in further detail below.

1.3. Evaluating the uncertainty of 3D dosimetry

The evaluation methods that have been applied by many 3D dosimetry research groups to give evidence of the 'potential' of 3D dosimeters are often based on a percentage number of voxels that pass gamma criteria between a measured dose distribution and a treatment planning system (TPS) calculated dose distribution. This value of pass rates obviously depends on the number of slices and the size (field-of-view) of the dose maps taken into consideration. Some have suggested to use a dose threshold in which gamma is calculated but even in this case, care must be taken in using gamma pass rates that will still depend on the topological 'complexity' of the dose distribution. Although it is recognized that gamma maps are a valuable tool in benchmarking a radiotherapy treatment plan, the use of a percentage value of passing pixels in benchmarking a dosimetry technique against a TPS calculated dose distribution can be misleading as proof of the reliability of the dosimetry technique. Gamma map analysis should also be performed with much care as the measured dose distributions may contain stochastic noise. Gamma values may be significantly underestimated when the evaluated dose distribution contains noise [13]. Moreover, it is not beyond discussion to what extend a TPS calculated dose distribution should be used as 'golden standard' to benchmark a new dosimeter. Of course, it is recognized that the intrinsic problem with the benchmarking of a 3D dosimeter is that there does not exist another 3D dosimetric 'golden standard' to compare against. The most reasonable strategy is to compare doses obtained with gel dosimetry with doses obtained by the most reliable dosimetry techniques that apply to a certain spatial dimension. For example, dose profiles of a single field (photons and electrons) can be compared with dose profiles obtained with an ionization chamber or diamond detector [14, 15]. In two dimensions, 3D dosimetry can be compared with film dosimetry [3, 14, 16]. Dose distributions obtained with 3D dosimetry have been compared with those calculated with treatment planning software [3, 17-23]. Errors that compromise the accuracy may occur at different stages of the dosimetry procedure [24]. Given the uncertainties with respect to the calibration, we argue that a more stringent approach using other dosimeters and treatment planning should be applied to benchmark 3D radiation dosimetry than a gamma evaluation against a single treatment plan.

1.4. Accuracy and precision

Accuracy is defined as the degree of conformity of a measured or calculated quantity (a measurand) to its actual (true) value. *Precision* is defined as the ability of a measurement to be consistently reproduced. Accuracy and precision can be defined statistically in terms of the deviation of the mean value from the reference value and the standard deviation of the mean value of many measurements respectively (figure 1). The measurement uncertainty of a single experiment thus comprises both systematic and random errors. Uncertainties have further been classified in type A and type B uncertainties where type A standard uncertainty is obtained from a probability density function derived from an observed frequency distribution, while type B standard uncertainty is obtained from an assumed probability density function that is based on the degree of belief that an event will occur.

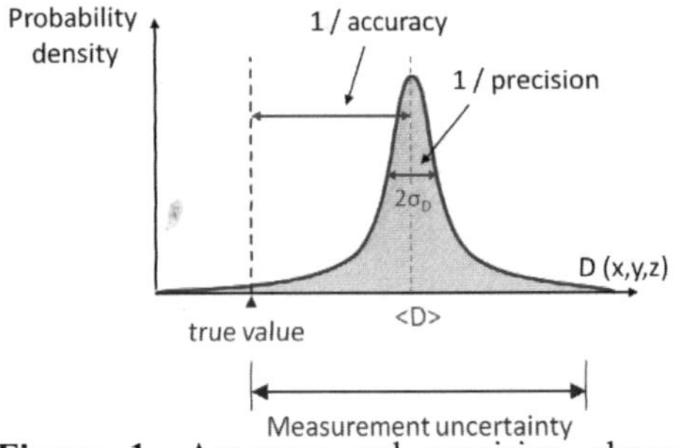

Figure 1. Accuracy and precision shown schematically in terms of probabilities of occurrence. Precision can be characterised in terms of the standard deviation of many measurements (experiments). The accuracy is characterized by the deviation of the mean of many measurements from the reference (known) value.

In the case of 3D radiation dosimetry, uncertainty applies to both dose and space. In a gel dosimetry dose verification experiment (and any radiation treatment), the spatial and dosimetric dimensions are interwoven. For large volume dosimeters it is theoretically impossible to extract both dosimetric and spatial errors from a measured spatial dose distribution (i.e. the result of a 3D gel dosimetry experiment). To comprise both spatial and dosimetric performance in one parameter, Low *et al* have introduced the gamma-index [25]. Other dose distribution comparisons are performed on the basis of maximum allowed dose differences [26].

The overall uncertainty of 3D radiation dosimetry can be achieved through a reproducibility study of the complete dosimetry experiment from gel fabrication to dose distribution analysis. An extensive reproducibility study, comprising the repetition of a gel dosimetry experiment has been carried out for polymer gel dosimeters with MRI readout [27]. It is imperative to realize that in a clinical verification with gel dosimetry, consisting of a single experiment, the uncertainty of the dose measured in each voxel comprises both the systematic inaccuracy and random errors, including stochastic image noise.

A typical clinical 3D dose verification experiment is performed in different steps and errors can occur at each stage (figure 2): (1) The polymer gel dosimeter is fabricated in a chemical laboratory and is stored until radiation. Any deviations in the chemical composition through inhomogeneous mixing, remnant impurities in the gel container or oxygen permeability within the gel containers can give rise to discrepancies between calibration and volumetric phantoms [8]. Moreover, a difference in temperature course during storage may also affect the dose-R2 response between calibration and test phantoms [6]. Most of these deviations are compensated for by using calibration phantoms that are (i) fabricated from the same batch of gel, (ii) of the same material composition, and (iii) of similar volumes. (2) The gel dosimeter phantom is then scanned with CT and a treatment plan is devised (3) on the CT images, resulting in a virtual planning target volume (PTV) and a set of critical organs. At this stage, just as with a patient, a reference coordinate system is allocated to the phantom by use of marker lines that are drawn on the phantom. In addition, for co-registration of the dose maps with the TPS, stereotactic fixtures or fiducial markers [28] can be placed on the phantom that will later be visible on the imaging modality of interest (MRI, optical CT, x-ray CT). Positional setup errors are likely to result in deviations between the planned and the measured dose distribution. These deviations are not intrinsic to the gel dosimeter but are indicative for errors that may occur during the actual patient treatment. (4) Upon irradiation, a complex set of radiation-induced chemical reactions take place. On a molecular level, these reactions are probabilistic in nature but on the spatial scale of clinical dose verification experiments, this radiochemical noise is insignificant.

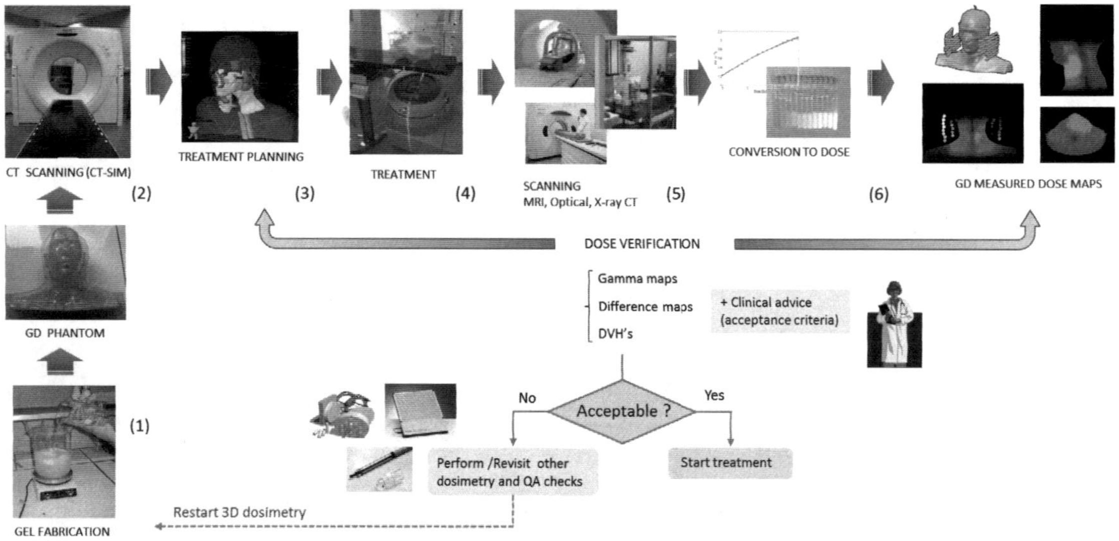

Figure 2. In a clinical dose verification experiment with 3D dosimetry, the dosimeter phantom is taken through the entire treatment chain. Possible dosimetric or setup errors can occur at the different stages. 3D dosimetry fulfills an important role in benchmarking the complete treatment chain.

Any positional set-up errors in treatment with respect to the planned treatment will affect the dose distribution. This error may also occur in patient treatment and is therefore not intrinsic to 3D dosimetry. In benchmarking 3D radiation dosimetry however, the experimental set-up error should be minimized as much as possible by adequate positioning aids in order to isolate the other error sources. Other sources of deviation in dose reading that may occur during irradiation of the gel dosimeter are related to the dependence of the irradiation response on the temperature of the 3D dosimeter during irradiation, on the dose rate and on the energy spectrum of the irradiation beam [29, 31]. It is important to realize that differences in dose rate (and energy) will occur in the irradiated 3D volume even if the photon fluence rate (in terms of MU/min) of the individual beams is kept constant. Another important characteristic of the dosimeter is its temporal and spatial integrity [32, 33]. These errors are inherent to the dosimeter material and should be minimized as they compromise the specificity of the dosimeter. After irradiation, the gel dosimeter is imaged (MRI, optical CT, x-ray CT) (5). During imaging, image noise will add to the measurements. In order to minimize the noise in the measured dose distribution the scanning parameters should be optimized [34-36]. The processing (fitting, filtering etc) of acquired data (6), may have a significant influence on the amplification of noise inherent in the data. For example, in quantitative R_2-NMR scanning it is found that a least-square fit will amplify the thermal noise in the base images to a larger extent than a chi-square (χ^2) based minimization. Imaging artifacts may result in systematic errors both dosimetric and spatially. A classification of the error sources in Type A and B uncertainties can be found elsewhere [37].

2. Uncertainties in polymer gel dosimetry with MRI readout

The overall dosimetric precision is compromised by random errors in the different operations (figure 2) carried out in the dosimetry experiment. The first step in a gel dosimetry experiment is weighting the chemicals. Stochastic variations in the weighting will result in variations in the measured dose-related value (R_1, R_2, MT) as the dose-response is determined by the chemical composition. It is found that other manufacturing conditions may also have an influence on the dose-response such as the temperature during fabrication and the oxygen concentration in the gel dosimeter. Stochastic variations in the controlled temperature between different experiments will therefore also lead to variations in the measured dose-related value. These variations only contribute to the overall uncertainty of the dosimeter if their contribution is different between the calibration phantom and the 3D dosimeter phantom, e.g. as a result of different cooling rates of small phantoms as compared to large phantoms. When gel dosimeters are handled under conditions of good practice [37], the physico-chemical mechanisms should not contribute to more than a total error of 2% in the dose measurement [38]. It is paramount that in some gel dosimeters (such as MAGAT) where a significant dose rate dependence is found [29], the figure of uncertainty may be exceeded.

Also during irradiation there are different sources of stochastic variable contributions that determine the overall dosimetric precision such as variations in the dose delivery, variations in the temperature during irradiation [29, 30] and stochastic variations in the positioning of the calibration phantoms. Similar factors of chemical and irradiation uncertainty are also present with optical and X-ray CT but their respective contribution may be different.

MR images of the gel dosimeter will contain noise in the images as a result of thermal detector and phantom noise. This noise component will propagate into the derived quantitative MR images. The noise contribution is also determined by some scan parameters that can be optimized to achieve a maximum precision [34, 35]. In gel dosimetry with MRI readout, a calibration of the scanned parametric map (e.g. R_2 map) is needed to obtain a dose map. The stochastic noise in the dose maps can be kept well below 0.8% standard deviation, resulting in a maximum deviation of 2.4% on an individual voxel basis with a confidence of 99% [24]. It is worth noting that the values mentioned here for thermal image noise are for imaging voxels with a volume of 3 mm^3 and can be proportionally reduced by scaling to a grid size similar to the treatment plan. Any random errors that operate on the calibration samples will also affect the uncertainty of the dose map. In addition, the stochastic variations in beam output during radiation of the calibration vials can also be assumed to be significantly less than 1% that is also supported by a reproducibility study where 20 (calibration) samples were irradiated with the same dose and under the same conditions. The resulting type A (random) uncertainty on the individual voxel level is therefore considered as predominantly originating from thermal noise during scanning.

Systematic errors (type B uncertainty) during MRI scanning of gel dosimeters are attributed to: B_0 and B_1-field non-uniformities; gradient non-uniformities; and temperature differences between calibration phantoms and volumetric dosimeter phantom [40-45]. Scanner related sources of uncertainty may be kept well below 3% by a proper scanning setup and/or by applying post-processing compensation methods such as for B_0- and B_1-field non-uniformity [2, 46]. A source of uncertainty that has a significant contribution to the overall uncertainty is related to temperature deviations and variations during scanning [39]. These effects can be minimized by use of an imaging sequence with centric k-space reordering [44] and/or an active temperature stabilization using thermal pads or a temperature stabilized water reservoir. An extensive set of guidelines on how to scan polymer gel dosimeters with MRI has been discussed in previous conference proceedings [37, 47].

3. Uncertainties in 3D dosimetry with optical readout

Different optical scanner prototypes [48-60] have been developed to scan 3D dosimeters. As with MRI scanning of 3D dosimeters, the uncertainties are determined by the combination of the dosimeter characteristics and the optical scanner, which are often coupled and can't be separated easily. For example, sensitivity to light scatter varies between scanner designs. As a result, the contribution of the uncertainty caused by light scattering in polymer gel dosimeters will depend on the scanner type.

For polymer gel dosimeters, the same physico-chemical sources of uncertainties, as discussed in section 2, apply. It has also been found that variations in the fabrication process and chemical composition in both micelle gel dosimeters [61, 62] and plastic dosimeters such as PRESAGE [63-65] and Flexydos3D [66] result in inter-batch differences in the optical dose response. Although it is expected that intra-batch variations are minimal, until now, their effect on the uncertainty as a result of physico-chemical differences between calibration samples and volumetric dosimeter phantoms has not been investigated. Temperature dependence during irradiation in the order of 2% per degree Celsius has been detected in both PRESAGE [63, 64] and micelle gel dosimeters [62]. The Flexydos3D dosimeters are still under investigation.

Detector noise will also contribute to pixel-by-pixel type A dose uncertainty. For CCD cameras, there is a square root relation between the noise and the signal intensity [67]. The propagation of noise from the detector noise to the signal-to-noise ratio (SNR) in the reconstructed optical density images can be easily simulated [67, 68]. It is found that the SNR in the reconstructed images also depends on the spatial frequency filter that is applied in the reconstruction algorithm. Computational simulations on synthetic phantoms can also be applied to detect the effect of the reconstruction algorithm on the modulation transfer function (MTF) and, hence, the spatial uncertainty introduced by the reconstruction [67, 68]. The spatial uncertainty can also be determined experimentally by using adequate resolution phantoms [54]. The SNR and the dynamic range of the detector are performance characteristics of the detector and are often associated with its cost. Just as with any other readout system, the SNR in the dose maps can be improved by averaging out several images at the cost of measurement time. The optimal dose range of the dosimeter is also determined by the SNR characteristics of the optical scanner, including both the detector noise performance and the reconstruction algorithm.

Systematic errors (type B uncertainty) during optical scanning may affect both the spatial and dose accuracy and are related to imaging artefacts that can be both scanner related or phantom related. Phantom related artefacts are caused by refractive index differences between phantom and refractive index matching fluid [55], light scattering [56, 69], refractive index variations in the phantom as a result of convection during the exothermal solidification process [67] and granular impurities such as dust particles in the phantom [72]. Phantom related systematic errors that affect calibration may also originate from temperature differences during irradiation [62-64] and physico-chemical sources of uncertainty such as temperature dependent auto-oxidation [70] of the leucodye and the effect of oxygen. Scanner related artefacts are scanner specific and can be caused by drift or fluctuations in the light source and/or detector [58, 71, 72], scratches on the optical components [71, 72], internal reflections in the optical components [71, 72], ambient stray light [71], desynchronization between galvanomirror and detector in laser scanners leading to misalignments in the sinogram [58, 72] and off-centric axis alignment with respect to the projections. Ray tracing [52] or Monte-Carlo simulation software [73] can be helpful in describing the performance of the scanner and in optimizing the design to minimize the instrumental uncertainties. Attempts to perform an independent calibration with calibration phantoms have been undertaken [71, 74, 75] but more studies are required to determine the achievable absolute dose accuracy with the various optically sensitive dosimeters. A comprehensive overview of optical scanning and tips-and-tricks can be found in previous conference proceedings [48, 76].

4. Uncertainties in 3D dosimetry with X-ray CT readout

All uncertainties and errors incurred during the PGD manufacture and irradiation steps of figure 2, and described above, are also present when imaging PGDs with x-ray CT. However, certain sensitivities, such as sensitivity to irradiation temperature, are less prevalent than when imaging with MRI. These observations speak to the low contrast resolution that is obtained when using x-ray CT for PGD imaging.

Imaging PGDs with x-ray CT is associated with a distinct host of uncertainties and errors that must be considered within any given experiment. We here subdivide these uncertainties and errors in to two categories: those induced through the imaging procedure; and those induced through post image acquisition processing. Within the imaging procedure, uncertainties within a resultant image are compounded by set-up errors and signal to noise ratio (SNR) considerations through imaging parameter optimization. An attractive feature of x-ray CT PGD is the fact that the CT process is already incorporated into the radiotherapy work-flow. Hence, physical set-up errors can be minimized through the use of standardized immobilization devices that are constant throughout the experiment (from treatment plan CT acquisition through to dose information extraction imaging). With rigorous attention to phantom set-up reproducibility, a ~1mm accuracy in set up can be achieved with an anthropomorphic head phantom and CT mask immobilization. In terms of imaging parameter optimization, work has been reported in the literature outlining the affects of parameter manipulation on resultant image SNR [36]. Image SNR can be optimized through judicious choice of CT parameter acquisition. While SNR can be improved through, essentially increasing the number of photons hitting any given detector, such a scheme comes at the expense of tube load and, therefore, image acquisition time, which is restricted by any required tube cooling.

Post CT imaging, data processing is required to convert raw CT images to high SNR dose maps. Figure 3 illustrates factors affecting systematic and random errors and uncertainties incurred through each processing step. Each step throughout the procedure outlined in figure 3 carries with it an inherent uncertainty which, ultimately, propagates through to the dose calibration function and hence also through to the calibrated dose maps. Recent work has incorporated errors throughout each step in order to obtain an estimation of the accuracy and precision in calibrated CT images. Results show that accuracy and precision throughout the calibration procedure are highly dose dependent and hence a single quantity for accuracy or precision can not be realistically quoted. Minimum errors in accuracy and precision occur in the "mid-dose" regions (~7 – 20 Gy) of the calibration and are quoted in the 5% range. It must also be noted that a non-linear calibration curve was essential in performing the "absolute" calibration in this case. Several non-linear functions were tested and a hyperbolic tangent function was found to provide the highest level of accuracy and precision in the resultant dose maps. This function was found empirically and does not necessarily offer insight in to reaction dynamics within PGDs, although this point has not been investigated in depth. In summary, a linear calibration would result in highly increased errors and is therefore not recommended when performing x-ray CT PGD.

Journal of Physics: Conference Series **573** (2015) 012008 doi:10.1088/1742-6596/573/1/012008

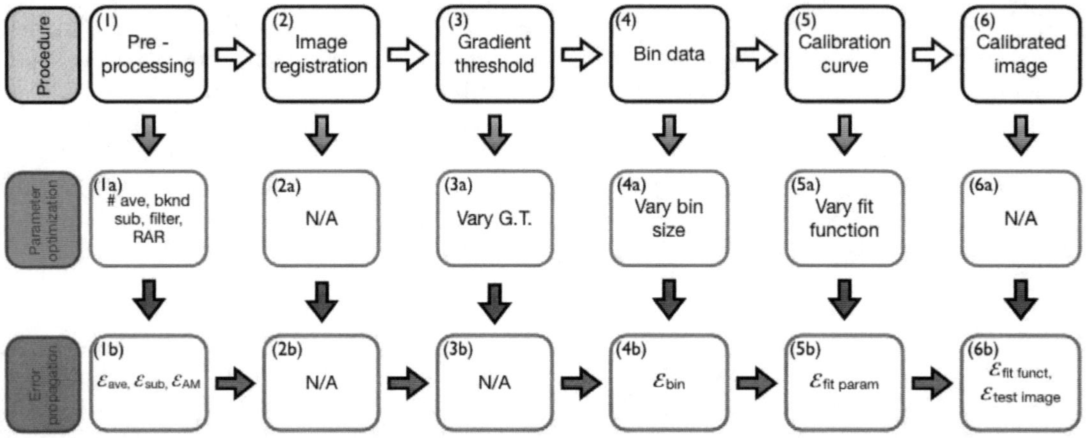

Figure 3. PGD CT image processing procedure with associated optimization parameters and errors. Red right-pointing arrows indicate an error propagation step. Taken from Jirasek *et al* 2014 with permission from Jirasek *et al* 2012.

5. Conclusions

Polymer gel dosimetry remains one of the most promising tools for measuring inherently complex 3D radiotherapy dose distributions. In fact, polymer gel dosimetry is currently feasible with MRI-based dose readout. However, the uncertainties within a given polymer gel system, irrespective of the imaging modality used to extract dose information, are non-trivial in nature and great care is required in understanding the magnitudes of errors and uncertainties within a full gel dosimetry protocol. With every dosimeter and imaging modality it is crucial to follow a careful optimized protocol and compensate for imaging artefacts to minimize the uncertainties. This review poses points for consideration when undertaking a benchmarking of accuracy and precision of polymer gel dosimetry imaged with MRI, optical or x-ray CT.

6. References

[1] Baldock C *et al* 2010 *Phys. Med. Biol.* **55** R1-63
[2] Vergote K *et al* 2004 *Phys. Med. Biol.* **49** 287-305
[3] De Deene Y *et al* 2000 *Magn. Reson. Med.* **43** 116-25
[4] Dumas E *et al* 2006 *J. Phys.: Conf. Ser.* **56** 239-41
[5] MacDougall N D *et al* 2008 *Br. J. Radiol.* **81** 46-50
[6] De Deene Y *et al* 2007 *Phys. Med. Biol.* **52** 2719-28
[7] Cosgrove V P *et al* 2000 *Phys. Med. Biol.* **45** 1195-210
[8] Sedaghat M *et al* 2011 *Phys. Med. Biol.* **56** 6083-107
[9] Hilts M *et al* 2000 *Phys. Med. Biol.* **45** 2559–71
[10] Audet C *et al* 2002 *J. Phys.: Conf. Ser.* **3** pp. 110-18
[11] Jirasek A and Hilts M *et al* 2014 *Phys. Med. Biol.* **59** 597-614
[12] Johnston H *et al* 2012 *Phys. Med. Biol.* **57** 3155-75
[13] Low D A 2010 *J. Phys.: Conf. Ser.* **250** 012071
[14] De Deene Y *et al* 1998 *Radiother. Oncol.* **48** 283-91
[15] Haraldsson P *et al* 2000 *Br. J. Radiol.* **73** 58-65
[16] Pappas E *et al* 2001 *Phys. Med. Biol.* **46** 783-97
[17] Ibbott G S *et al* 1997 *Int. J. Radiat. Oncol. Biol. Phys.* **38** 1097-103
[18] Oldham M *et al* 1998 *Phys. Med. Biol.* **43** 1113-32
[19] Meeks S L *et al* 1999 *Int. J. Radiat. Oncol. Biol. Phys.* **43** 1141-53

[20] Cosgrove V P *et al* 2000 *Phys. Med. Biol.* **45** 1195-210
[21] Ramm U *et al* 2000 *Phys. Med. Biol.* **45** N95-102
[22] Farajollahi A R *et al* 2000 *Phys. Med. Biol.* **45** N9-14
[23] Grebe G *et al* 2001 *Int. J. Radiat. Oncol. Biol. Phys.* **49** 1451-60
[24] De Deene Y 2006 *J. Phys.: Conf. Ser.* **56** 72-85
[25] Low D A *Med. Phys.* **25** 656-61
[26] Jiang S B *et al* 2006 *Phys. Med. Biol.* **51** 759-76
[27] Vandecasteele J and De Deene Y 2013 *Phys. Med. Biol.* **58** 19-42
[28] Meeks S L *et al* 1999 *Int. J. Radiat. Oncology Biol. Phys.* **43** 1135-41
[29] De Deene Y *et al* 2006 *Phys. Med. Biol.* **51** 653-73
[30] Salomons G J *et al* 2002 *Phys. Med. Biol.* **47** 1435-48
[31] Karlsson A *et al* 2007 *Phys. Med. Biol.* **52** 4697-706
[32] De Deene Y *et al* 2002 *Phys. Med. Biol.* **47** 2459-70
[33] Vergote K *et al* 2004 *Phys. Med. Biol.* **49** 4507-22
[34] De Deene Y *et al* 1998 *Signal Processing* **70** 85-101
[35] De Deene Y and Baldock C 2002 *Phys. Med. Biol.* **47** 3117-41
[36] Hilts M *et al* 2005 *Phys. Med. Biol.* **50** 1727-45
[37] De Deene Y and Vandecasteele J 2013 *J. Phys.: Conf. Ser.* **444** 012015
[38] Vandecasteele J and De Deene Y 2013 *Phys. Med. Biol.* **58** 43-61
[39] Vandecasteele J and De Deene Y 2013 *Phys. Med. Biol.* **58** 63-85
[40] De Deene Y 2004 *J. Phys.: Conf. Ser.* **3** 87-114
[41] De Deene Y 2009 *J. Phys.: Conf. Ser.* **164** 012033
[42] De Deene Y *et al* 2000 *Phys. Med. Biol.* **45** 1807-23
[43] De Deene Y *et al* 2000 *Phys. Med. Biol.* **45** 1825-39
[44] De Deene Y and De Wagter C 2001 *Phys. Med. Biol.* **46** 2697-711
[45] Doran S J *et al* 2005 *Phys. Med. Biol.* **50** 1343-61
[46] De Deene Y *et al* 2001 *Phys. Med. Biol.* **46** 2801-25
[47] De Deene Y 2010 *J. Phys.: Conf. Ser.* **250** 012015
[48] Doran S J and Krstajic N 2006 *J. Phys.: Conf. Ser.* **56** 45-57
[49] Gore J C *et al* 1996 *Phys. Med. Biol.* **41** 2695-704
[50] Cheng H-W *et al* 2011 *Radiat. Meas.* **46** 1932-5
[51] Campbell W G *et al Med. Phys.* **40** 061712-1-12
[52] Doran S J *et al* 2001 *Phys. Med. Biol.* **46** 3191-213
[53] Krstajic N and Doran S J 2007 *Phys. Med. Biol.* **52** N257-63
[54] Lopatiuk-Tirpak O *et al* 2008 *Med. Phys.* **35** 3847-59
[55] Maryanski M J and Ranade M K 2001 *Proc. SPIE* **4320** 764-74
[56] Olding T *et al* 2010 *Phys. Med. Biol.* **55** 2819-40
[57] Ramm D *et al* 2012 *Phys. Med. Biol.* **57** 3853-68
[58] Vandecasteele J and De Deene Y 2009 *J. Phys.: Conf. Ser.* **56** 45-57
[59] Xu Y and Wuu C-S 2013 *Phys Med Biol* **58** 479-95
[60] Sakhalkar H S and Oldham M 2008 *Med. Phys.* **35** 101-11
[61] Jordan K and Avvakumov N 2009 *Phys. Med. Biol.* **54** 6773-89
[62] Vandecasteele J and De Deene Y 2013 *Phys. Med. Biol.* **58** 6241-62
[63] Skyt P S *et al* 2012 *Med Phys* **39** 7232
[64] Guo P Y *et al* 2006 *Med. Phys.* **33** 1338-45
[65] Yates E S *et al* 2011 *Acta Oncologica* **50** 829-34
[66] De Deene Y et al 2014 Proceedings of 8th International Conference on 3D Radiation Dosimetry (IC3DDose) 4-7 September 2014, Ystad, Sweden
[67] Krstajic N and Doran S 2007 *Phys. Med. Biol.* **52** 3693-713
[68] De Deene Y 2014 Proceedings of 8th International Conference on 3D Radiation Dosimetry (IC3DDose) 4-7 September 2014, Ystad, Sweden

[69] Bosi S G *et al* 2009 *Phys. Med. Biol.* **54** 275-83.
[70] Skyt P S *et al* 2012 *Med Phys* **39** 7232-6
[71] Oldham M and Kim L 2004 *Med. Phys.* **31** 1093-104
[72] Xu Y *et al* 2004, *Med. Phys.* **31** 3024-33
[73] Oldham M *et al* 2003 *Med. Phys.* **30** 623-34
[74] Xu Y *et al* 2010 *Med. Phys.* **37** 861-8
[75] Alwan R *et al* 2008 *Nucl. Instr. Meth. Phys. Research B* **266** 834-40
[76] Doran S J 2010 *J. Phys.: Conf. Ser.* **250** 012086

The importance of 3D dosimetry

Daniel Low
UCLA Radiation Oncology, 200 Medical Plaza, Suite B265, Los Angeles, CA 90095 USA

E-mail: DLow@mednet.ucla.edu

Abstract. Radiation therapy has been getting progressively more complex for the past 20 years. Early radiation therapy techniques needed only basic dosimetry equipment; motorized water phantoms, ionization chambers, and basic radiographic film techniques. As intensity modulated radiation therapy and image guided therapy came into widespread practice, medical physicists were challenged with developing effective and efficient dose measurement techniques. The complex 3-dimensional (3D) nature of the dose distributions that were being delivered demanded the development of more quantitative and more thorough methods for dose measurement. The quality assurance vendors developed a wide array of multidetector arrays that have been enormously useful for measuring and characterizing dose distributions, and these have been made especially useful with the advent of 3D dose calculation systems based on the array measurements, as well as measurements made using film and portal imagers. Other vendors have been providing 3D calculations based on data from the linear accelerator or the record and verify system, providing thorough evaluation of the dose but lacking quality assurance (QA) of the dose delivery process, including machine calibration. The current state of 3D dosimetry is one of a state of flux. The vendors and professional associations are trying to determine the optimal balance between thorough QA, labor efficiency, and quantitation. This balance will take some time to reach, but a necessary component will be the 3D measurement and independent calculation of delivered radiation therapy dose distributions.

1. Introduction

Radiation therapy treatment complexity has continuously increased for the past 20 years. Prior to the advent of 3D radiation therapy treatment planning (figure 1), the medical physicist had to calibrate the linear accelerator by measuring the dose in a very controlled set of conditions and had to measure the relative doses for other clinically relevant conditions. Examples included varying field sizes, and depth and off-axis distance in medium. There were very few conditions under which a combination of the measured and therefore directly verified doses were not applicable or extrapolation was straightforward using the known behavior of the radiation beam, an example of which was moving the evaluation point towards and away from the source and invoking the inverse-square effect and Maynord F-Factor.

For clinical conditions that extended beyond the well-behaved and measured clinical parameter space, independent measurements were required. An example of this was total-body irradiation, which used source-to-patient distances of 4 to 5 meters.

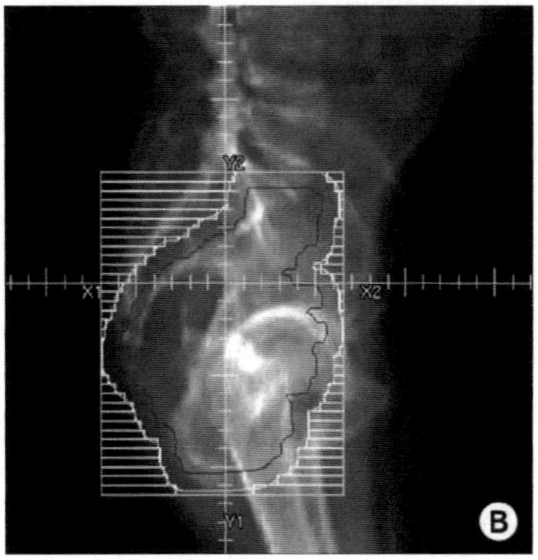

Figure 1. Beam's eye view of a conformal radiation therapy field. This type of treatment was pioneered in the 1980s and became the standard of care throughout the 1990s. One of the consequences of this treatment method was that the portals and dose distributions were essentially the same as the pre-conformal therapy era [1].

Dynamic treatments have been in common use for some time. For example, arc-based treatments were used in both stereotactic radiosurgery and prostate treatments (figure 2). In these cases, however, the field shapes were static and the total dose could be assumed to be a superposition of individual beams. Quality assurance was limited to hand calculating an approximate number of monitor units to provide the prescribed dose as well as the standard QA of the individual fixed beams. Dynamic arcs, consisting of a continuously varying portal shapes as a function of gantry angle, were employed in Japan but failed to gain much acceptance in the United States or elsewhere.

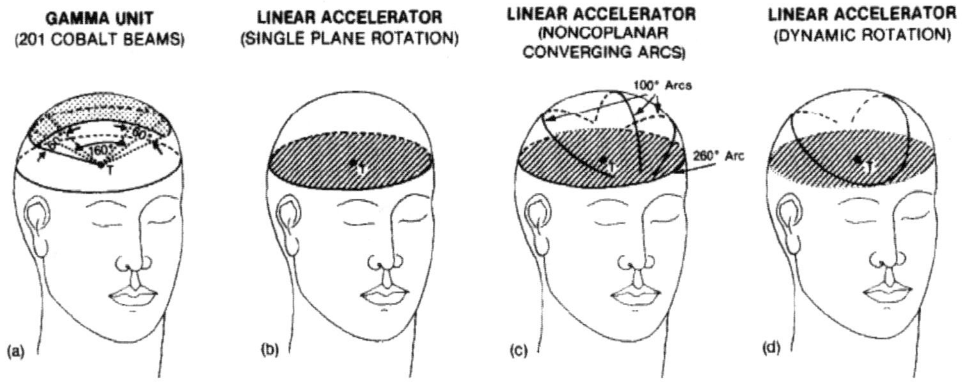

Figure 2. Points of beam entry into patient's skull for various radiosurgical techniques. (a) Gamma unit, (b) single plane rotation with a linear accelerator, (c) multiple non-coplanar arcs with a linear accelerator, and (d) dynamic rotation with a linear accelerator [2].

Quality assurance of the patient's treatment plan consisted of checking that the manually or treatment planning computer-generated monitor units were accurate and, the planned shapes of the beams were accurately reproduced on the patient, often by marking the patient immobilization system with the portal outline and using the field light to illuminate the radiation field and compare the two shapes, and portal images that showed both the portal shape and the intercepted anatomy [4].

The quality assurance of treatments began to get more challenging with the introduction of dynamically varying radiation fields. The first widely adopted example was dynamic wedge, a process by which one of the two upper collimators was moved while the beam was on (figure 3). A table of the relative monitor units versus the jaw position defined the wedge angle and the relative monitor units were multiplied by the total monitor units required to determine the relationship between jaw position and monitor unit. The dynamic wedge was designed to replace the physical wedges, which had the disadvantages of producing scattered radiation and of reducing clearance between the patient and linear accelerator.

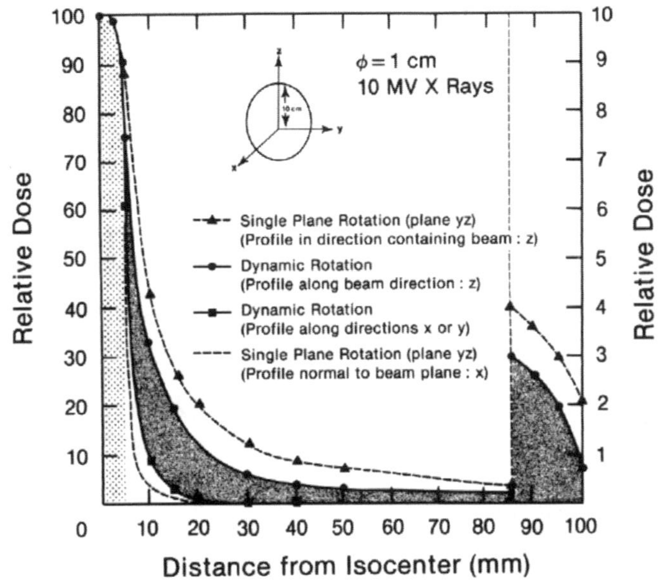

Figure 3. a) Example of stereotactic treament options including arc-based treatments [2]. b) Example profiles comparing single and multiple arc relative doses as a function of the distance from isocenter for a 1 cm diameter target. Single arc profiles have larger and smaller penumbra in and out of plane, respectively, and multiple arc approaches have more uniform penumbra [3].

The introduction of the dynamic wedge led to a rethinking of the quality assurance of wedged treatments. With physical wedges, one could use a standard water tank to measure the wedge profiles at different depths to provide input data for the planning system and to compare measurement against calculation. However, with a dynamic treatment, each position had to be measured throughout the irradiation, so scanning was impossible. Rather, ion chamber measurements were made at key points, at least at isocenter at a reference depth and source-to-surface distance, with relative measurements made at other positions and depths. These were typically made with radiographic film. Radiographic film had been used for quantitative measurements for stereotactic radiosurgery, which used radiation fields that were too small to be accurately scanned with the ionization chambers available then. However, the energy response of radiographic film meant that dose response varied with both field size and depth, so the physicist had to consider this when preparing dose response curves to convert optical density to dose.

2. Intensity Modulated Radiation Therapy

2.1 Challenges

Radiation therapy quality assurance needs remained stable until the advent of intensity modulated radiation therapy (IMRT). While IMRT provided unparalleled dose conformality, it also led to a vast array of quality assurance challenges. It is not an overstatement that the radiation field patterns we commonly use today would be considered impossible to accurately calculate before IMRT was developed (figure 4). Most physicists were of the opinion that secondary electronic equilibrium needed to be established in regions where the dose needed to be accurately known. Beam penumbrae, the regions where this feature was not true, were allowed to have looser tolerances because of the challenge of computing dose in regions without secondary electronic equilibrium. Stereotactic radiation beams had these challenges and many of its QA techniques were based on this issue. Many modern radiation beams are dynamic, with the dose rates, multileaf collimator geometry, and gantry angle continuously varying. The radiation portal often has a large portion that does not exhibit secondary electronic equilibrium, so most, if not all, of the patient is being irradiated by penumbra.

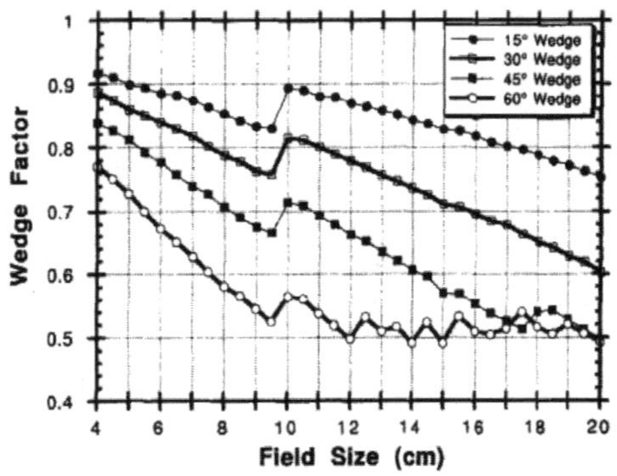

(a)

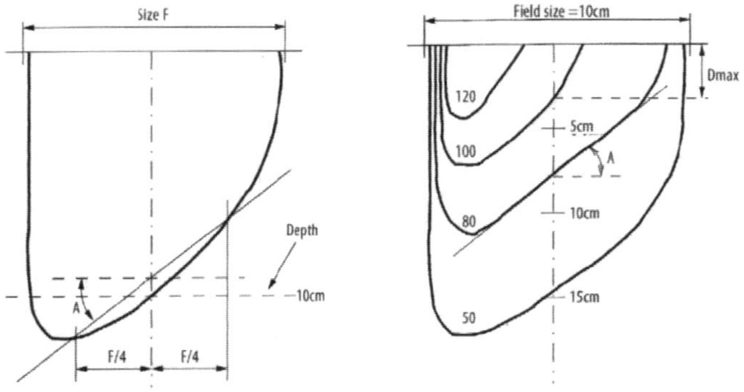

(b)

Figure 4. a) Dynamic wedge output factors showing more complex relationship between output factor and wedge angle than existed for physical wedges [5]. b) Wedge angle definition comparing dynamic and physical wedges [6].

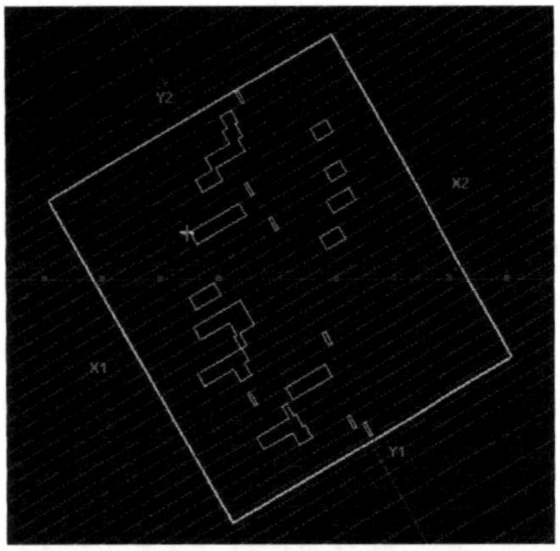

Figure 5. Example portal from a RapidArc treatment. The MLC leaf widths are 5 mm, so many of the radiation portals are as small as 5 mm x 2 mm. Secondary electronic equilibrium is not established in these small portals.

The increased hazards associated with the use of IMRT led to the development of the concept of patient-specific Quality Assurance. The use of manual or computer-aided verification of treatment planning system-calculated monitor units was insufficient given the hazards of IMRT. Therefore measurement-based approaches were adopted. These typically included making a planar measurement using film and a more quantitative measurement or measurements using an ionization chamber (figure 5). Absolute film dose measurements were difficult to do, so the ionization chamber measurement was used in essence to normalize the film-measured dose distribution so that the medical physicist could couple a relative two-dimensional dose distribution with a quantitatively measured single measurement point.

Figure 6. a) Example of film-measured dose distribution in early IMRT QA paper [7]. b) Example of an early commercial radiographic film phantom developed for IMRT [8]. c) The film phantom modified for other dosimeters (ion chambers in this example) [9].

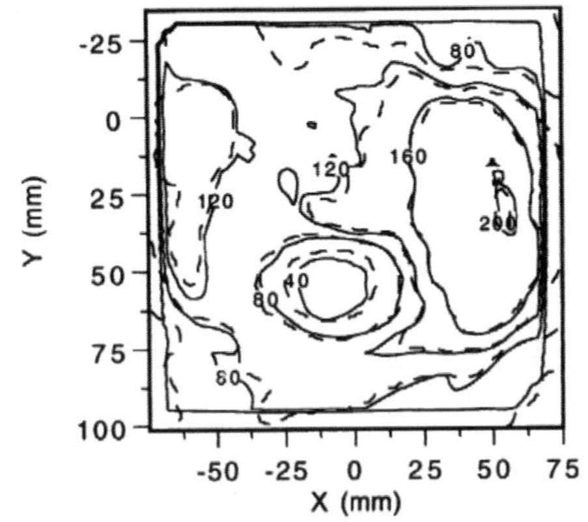

(a)

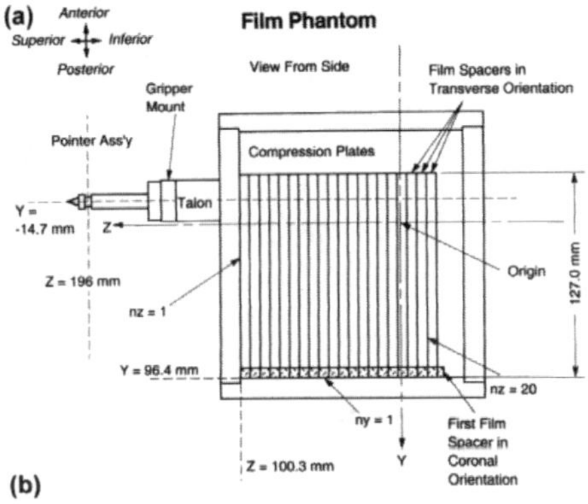

(b)

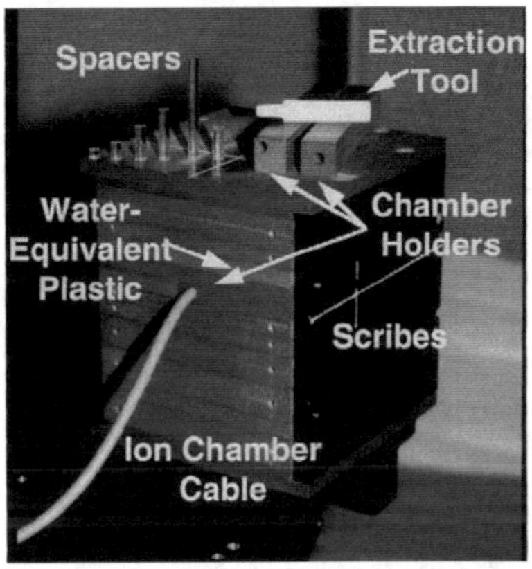

(c)

2.2 New Detectors

Eventually, dosimetry vendors responded to the need of physicists to make their quality assurance more efficient by producing novel detector arrays. Most often these were utilized diodes, but they also used ionization chambers. The detectors of most of these arrays were arranged in a plane, but some manufacturers created quasi-3D detection schemes by arranging the detectors in orthogonal planes or cylinders (figure 6).

The detector systems offered greater convenience and allowed the physicist to measure dose distributions without having to use film. They all had one thing in common. Their detectors were spaced a few mm apart. Therefore, high spatial resolution dosimetric features were impossible to characterize using these detectors (figure 7). The only options were film, but radiographic film was being phased out and many clinics were removing their film processors. Radiochromic film was a good alternative, but it still measured dose in only one plane. The complexity of IMRT dose distributions was such that a single planar dose measurement could not fully characterize the dose distribution.

Journal of Physics: Conference Series **573** (2015) 012009 doi:10.1088/1742-6596/573/1/012009

(a)

(b)
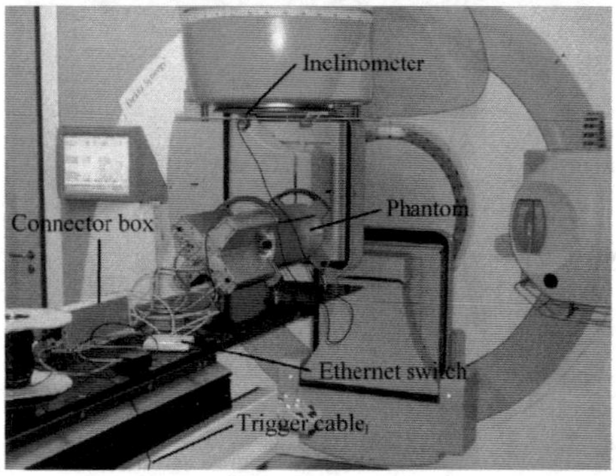

(c)

Figure 7. a) Planar diode array used for IMRT QA [10] b) Multiplane diode array used for IMRT QA. The diodes are distributed in two orthogonal planes [11]. c) Cylindrical diode array [12].

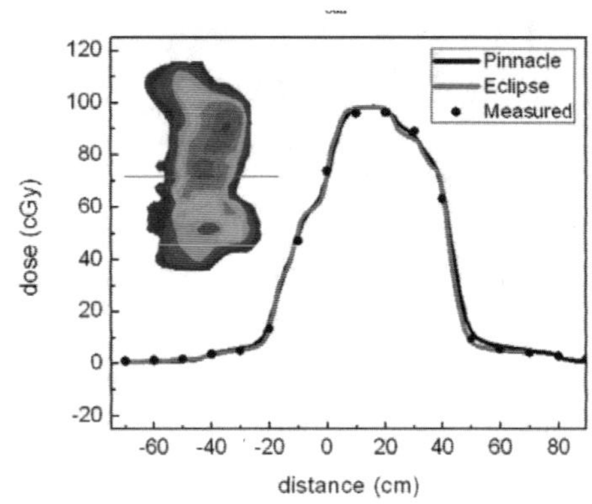

Figure 8. Example dose distribution calculations (lines) and diode array measurement (points) highlighting the relatively sparse measurement distributions provided by modern diode array systems [13].

For accurately characterizing complex 3D dose distributions, one needs a 3D dosimeter. It is tempting to use the 3D dosimeter for only end-to-end tests. Commission the planning system with a combination of scanning ion chambers, multidetector arrays, and film, and then at the end use a 3D dosimeter once to make sure that a full 3D dose distribution is accurately modeled and delivered. This would work well if the rest of the modeling was done sufficiently accurately and the planning system accurately computed the high spatial frequency features of the dose distribution, such as the tongue and groove and leaf end effects. Given that most planning systems do not, high spatial resolution 3D dosimetry should be employed to fully characterize the differences between the planning system and reality. This would allow the user to either a) tune the planning system to more accurately match the measured beams or b) provide an uncertainty estimate for the user when uncorrectable dose differences arise. In the future, one could imagine software that took these data and characterized the end to end test discrepancies and instructed the user whether the discrepancies were manifestations of the errors seen in the earlier measurements. 3D dosimeters would be used for all but the most straightforward of measurements, such as measuring the machine calibration. They would be fully integrated into the characterization of the treatment planning and radiation dose comparisons.

2.3 Dose Evaluation Tools

The use of 3D γ dose distributions, the analysis of γ passing rates Γ, and the analysis of differences of γ passing rates between different algorithms and/or experimental setups, $\Delta\Gamma$, would aid in the effectiveness of comparing the high spatial resolution dose distributions. Care would need to be taken to make sure that some of the high spatial frequency features of the dose distributions wouldn't be ignored by the γ, Γ, and $\Delta\Gamma$ analyses (due, for example, to use of a large distance-to-agreement criterion).

3. Summary and Conclusions

This need to accurately measure 3D dose distributions leads to the main current challenge of 3D dosimeters. None are mainstream. In each case, the 3D dosimeters are either manufactured in house, limiting use to select clinics, or sold by small companies that cannot generate sufficient scale to significantly reduce the price. Ideally, the 3D volume would match the size of the radiation beams being used for the treatment plans, but characterizing the treatment planning system and linear accelerator using 3D dosimeters would require many large-volume detectors, further limiting the feasibility of this approach. Finally, one advantage of dosimeters such as ionization chambers is that one gets a measurement while one is still at the machine. 3D dosimeters could provide results on similar timescales, say just after irradiation and before the setup has been torn down, but this means that the user would need to have their own scanner, further reducing the number of clinics that would take advantage of such a dosimeter.

In order to implement the model described above. We need convenient, relatively inexpensive, readily available, quantitative, water equivalent, and high spatial resolution detector systems to characterize the dose distributions we deliver to patients. End to end tests are useful for making sure we aren't going to cause catastrophic errors, but lack of a consistent, systematic use of 0D, 2D, and 3D detectors will keep the dose distribution accuracy below that which we enjoyed in the 1980s.

4. References

[1] Yu M *et al* 2013 *Int. J. Radiat. Oncol.* **31** 252-9
[2] Podgorsak E B *et al* 1989 *Int. J. Radiat. Oncol.* **16** 857-65
[3] Podgorsak E B *et al* 1988 *Int. J. Radiat. Oncol.* **14** 115-26
[4] AAPM 1994 Report of Task Group 40. *Med. Phys.* **21** 581-618
[5] Klein R J *et al* 1995 *Int. J. Radiat. Oncol* **31** 583-92
[6] Chelminski P R *et al* 2006 *Rep. Pract. Oncol. Radiother.* **11** 65-75
[7] Low D *et al Int. J. Radiat. Oncol.* 1998 **42** 681-92
[8] Low D *et al Int. J. Radiat. Oncol.* 1998 **49** 305-16
[9] Low D *et al Int. J. Radiat. Oncol.* 1998 **40** 1231-5
[10] Letourneau D *et al* 2004 *Radiother. Oncol.* **70** 199-206
[11] Bedford J L *et al* 2009 *Phys. Med. Biol.* **54** N167-76
[12] Petoukhova A *et al* 2011 *Phys. Med. Biol.* **56** 5411-28
[13] Sun B *et al* 2012 *J. Appl. Clin. Med. Phys.* **13** 140-54

Quantification of the absorbed dose in 3D by means of advanced optical diagnostics based on structured illumination

Elias Kristensson[1], Sofie Ceberg[2], Sven Bäck[2,3] and Kevin Jordan[4]
[1]Division of Combustion Physics, Lund University, Sweden
[2]Radiation Physics, Skåne University Hospital, Lund, Sweden
[3]Medical Radiation Physics, Lund University, Sweden
[4]London Regional Cancer Program, London, Ontario, Canada

E-mail: Elias.Kristensson@forbrf.lth.se

The purpose of this study was to present a novel optical diagnostic tool that corrects for undesired contribution of multiply scattered light, thus opening up for e.g. quantitative optical CT measurements of opaque samples. The approach is based on a technique called Structured Illumination (SI), which is commonly employed within microscopic imaging to enhance the depth-resolution. The concept of SI applies for many types of source-detector arrangements and the configuration employed in this paper relies on side-scattering detection. A nPAG polymer gel phantom was irradiated using 6 MV beam. Three-dimensional information was obtained by translating the sample perpendicular to the direction of light, thus sequentially probing different sections. These were then stacked together to form a 3D representation of the sample. By altering the polarization of the laser light during the data acquisition it was discovered that the aggregates responsible for the scattering of light followed Rayleigh scattering, implying that their individual sizes are smaller than, or in the order of, 500 nm.

1. Introduction

Multiple scatter of probe light with gel dosimeters limits the accuracy of transmission measurements. Development of more transparent 3D dosimeter [1] materials will minimize this problem. However, there exist several promising materials that would be more valuable if they could be optically scanned by a method that is less sensitive to multiple scatter. The geometry of the optical CT scanner is also important. Broad beam scanners have lower dynamic range than single ray scanners for scattering materials. Small acceptance angles are effective at stray light rejection. But can lead to relatively large optical CT scanners. Large diameter, telecentric optics is cost prohibitive for proposed clinical instruments. Scanned laser CT instruments are promising for scatter rejection. However, imaging through scattering media is a common problem and there may be many more optical approaches that should be investigated.

In this paper we present a novel optical diagnostic tool that corrects for this undesired contribution of multiply scattered light, thus opening up for e.g. quantitative optical CT measurements of opaque samples. The approach is based on a technique called Structured Illumination (SI), which is commonly employed within microscopic imaging to enhance the depth-resolution.

2. Material and methods

2.1 Structured Illumination

SI is an optical imaging technique [2], which can be combined with several different source-detector arrangements, such as back- and side scattering as well as transmission [3, 4]. The key feature with the technique is to add a recognizable structure to the illumination field, which will permit certain post-processing of the acquired data. Even though many different structures will suffice, it is mathematically convenient to utilize a sinusoidally varying fringe pattern, as shown in figure 1(a), as this simplifies the post-processing routines.

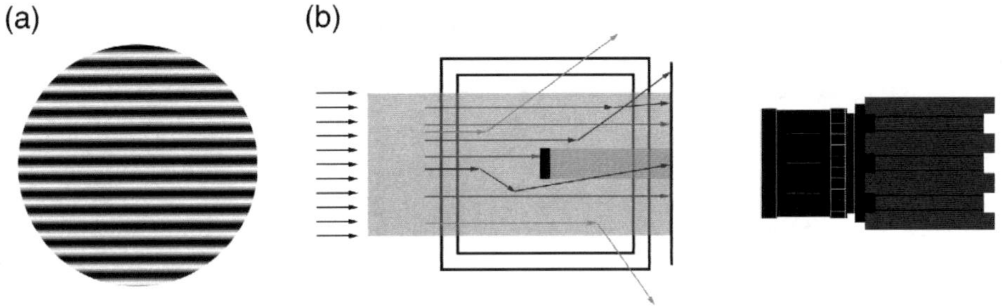

Figure 1. (a) Example of a laser beam with a sinusoidally modulated intensity field. (b) Transmission measurement of a turbid sample, the different possible outcomes for the photons are color-coded, red = unperturbed light, green = undetected scattered light and blue = scattered (singly and multiply) detected light.

To explain the technique, consider a non-absorbing opaque sample into which an obstruction is placed (see figure 1(b)), being illuminated with a wide laser beam that has an intensity profile according to that shown in figure 1(a). The transmitted light falls upon a screen, positioned behind the sample. A portion of the incident light will scatter away from the optical path, never to be detected. Other photons will propagate through the sample exit unperturbed, thus leaving a faithful shadow of the obstruction on the screen. Equally important, the spatial distribution of these undisturbed photons will follow the sinusoidal intensity pattern that was superimposed on the beam profile. A third group of photons will be scattered, either once or several times, yet still fall upon the screen. In contrast to the unperturbed light, the position where these photons end up on the screen is more random, meaning that (1) they will not cast a faithful shadow of the obstruction and (2) they will be independent on the sinusoidal intensity pattern. Removing this latter intensity contribution is key in order to probe a turbid sample accurately. Separating these two contributions can be achieved by acquiring three images, between which the incident line structure is shifted 1/3 of the sinus period. A filtered image (IS) is then attained by calculating

$$I_S = \sqrt{(I_0 - I_{120})^2 + (I_0 - I_{240})^2 + (I_{120} - I_{240})^2} \qquad (1)$$

where I_X denotes the three intensity modulated laser images (X is the spatial phase of the modulation). Equation (1) effectively removes all image features that are identical in the three images while all unique features are preserved. Since the end position (on the screen) for the scattered photons is random their intensity contribution will be nearly identical in all three images and thereby removed in equation (1). In contrast, by shifting the line structure the unperturbed light appears different in the three raw data images and is therefore not filtered out in equation (1). Figure 2 shows an example of a SI measurement performed for transmission imaging according to the setup given in figure 1(b). Here one of the three modulated beams is shown, together with a comparison between a non-filtered transmission image and an SI image. Notice how the contrast (between the obstruction an its surrounding) is considerably reduced in the non-filtered image. This reduced contrast is, as explained, a direct consequence of detecting scattered- and multiply scattered light and attempting to quantify the optical properties of the sample using the non-filtered image would render large errors.

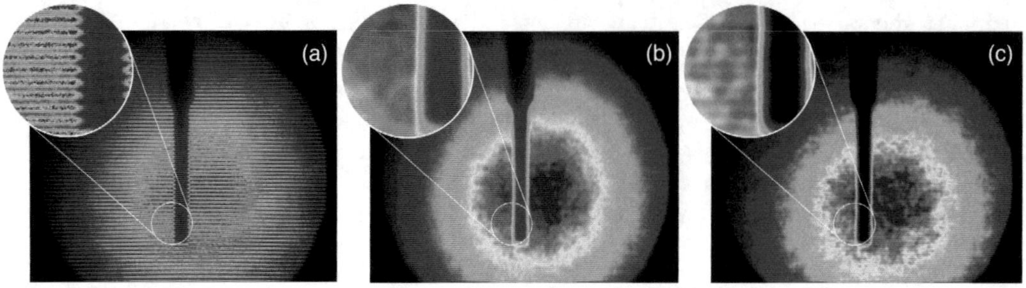

Figure 2. Example of an SI measurement on a turbid sample based on transmission imaging. (a) A sinusoidally intensity modulated laser light beam. (b) Transmission image without any filtering, with an estimated contrast of 49% (between obstruction and its surrounding). (c) SI transmission image where all scattered light has been filtered out, showing an estimated contrast of 98%.

2.2 Experimental set-up
As mentioned, the concept of structured illumination applies for many types of source-detector arrangements and the configuration employed in this paper relies on side-scattering detection. In the current configuration, laser light with a wavelength of 532 nm is formed into a thin (vertical) sheet of light and the sinusoidal modulation is applied along the vertical direction (see figure 3). The line structure is created by guiding the laser light through a transmission grating, comprising alternating opaque and transmissive stripes, which effectively casts shadows with a sinusoidal shape. Light that is scattered at a 90 degrees angle is collected by a CCD detector. Three-dimensional information is obtained by translating the sample perpendicular to the direction of light, thus sequentially probing different sections. These are then stacked together to form a 3D representation of the sample.

2.3 Preparing and irradiation of polymer gel phantoms
The normoxic polyacrylamide gel (nPAG) used in this experiment contained 89% w/w ultra-pure deionised water, 3% w/w acrylamide, 3% w/w N,N'-methylenebisacrylamide, 5% w/w gelatine and 10 mM tetrakis(hydroxymethyl)-phosphonium chloride (chemicals from Sigma Aldrich, Germany). The mixing procedures are described elsewhere [5]. Two gel phantoms from the same batch were prepared ($100x50x40mm^3$). One sample was used for background subtraction and the other sample received 3 Gy in dose maximum using a 6 MV 2x2 cm^2 beam entering the short end.

3. Results and discussion
Figure 3 shows an overview of the result for the performed measurement, the top row shows 2D sections while 3D views of the sample are given in the bottom row. The XY-view, in which the data is averaged over the Z-direction, shows a stronger signal where the x-ray beam enters the sample, as is expected due to attenuation. The same trend is seen in the YZ-view. The acquired data also suggests that there is a weak, yet detectable inhomogeneity along the Z-direction. This is most probably an image artifact arising because of the Z-scanning; as the sample is traversed along the Z-direction, the camera focus changes because of an increased optical path for the signal beams. The effect can be reduced by choosing a high f-number, which was the approach used during the current measurement, yet ultimately the effect must be corrected for.

The presented data is, at present, not quantitative. Two approaches should be considered in order to achieve this end. First, the Rayleigh scattering intensity can be quantified, given that a reference signal can be provided. Different reference samples could considered, such as a cuvette with a given amount of dye giving rise to a certain fluorescence signal or a sample with a well-defined number of scattering particles. Second, the incident and transmitted light intensity could be measured simultaneously and mathematically coupled to the scattered light intensity in each pixel to give an estimate of the local extinction coefficient, as demonstrated by Wellander *et al* [6].

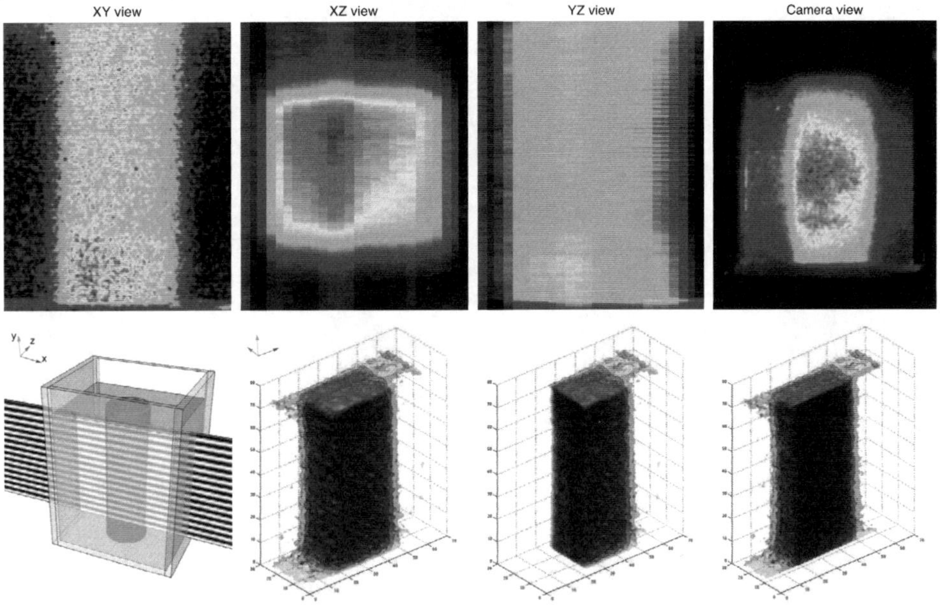

Figure 3. Overview of the results. Note that the 2D-sections are averaged along the third dimension.

4. Conclusion

In summary, the presentation describes the possibility of employing an imaging technique known as Structured Illumination to eventually use as means to quantify the radiation dose in 3D dosimetry. Relying purely on visible light, the approach has the prospect of reducing both cost and experimental complexity. In this first investigation, SI is combined with side-scattering detection, where 3D information was obtained by mounting the sample on a translational stage. Other experimental arrangements have been discussed and an example result from a transmission imaging setup has been shown.

By altering the polarization of the laser light during the data acquisition it was discovered that the aggregates responsible for the scattering of light followed Rayleigh scattering, implying that their individual sizes are smaller than, or in the order of, 500 nm. This further implies that it could be possible to eventually determine the size of the scattering particles by probing the sample at two (or more) different wavelengths. Whether this in turn could be coupled to the radiation dosage needs, however, to be investigated.

5. References

[1] Baldock C *et al* 2010 *Phys. Med. Biol.* **55** R1-63
[2] Neil M A A *et al* 1997 *Opt. Lett.* **22** 1905-7
[3] Kristensson E *et al* 2008 *Opt. Lett.* **33** 2752-54
[4] Kristensson E *et al* 2012 *Opt. Express* **20** 14437-50
[4] Ceberg S *et al* 2010 *Phys. Med. Biol.* **55** 4885-98
[5] Wellander R 2011 *Meas. Sci. Technol.* **22** 125303-15

Journal of Physics: Conference Series **573** (2015) 012011 doi:10.1088/1742-6596/573/1/012011

Clinical applications of 3-D dosimeters

Cheng-Shie Wuu
Department of Radiation Oncology, Columbia University, New York, NY, USA

E-mail: csw6@cumc.columbia.edu

Abstract. Both 3-D gels and radiochromic plastic dosimeters, in conjunction with dose image readout systems (MRI or optical-CT), have been employed to measure 3-D dose distributions in many clinical applications. The 3-D dose maps obtained from these systems can provide a useful tool for clinical dose verification for complex treatment techniques such as IMRT, SRS/SBRT, brachytherapy, and proton beam therapy. These complex treatments present high dose gradient regions in the boundaries between the target and surrounding critical organs. Dose accuracy in these areas can be critical, and may affect treatment outcome. In this review, applications of 3-D gels and PRESAGE dosimeter are reviewed and evaluated in terms of their performance in providing information on clinical dose verification as well as commissioning of various treatment modalities. Future interests and clinical needs on studies of 3-D dosimetry are also discussed.

1. Introduction

There have been many studies in the development of an accurate and reliable three-dimensional dosimetry system, prompted by the advances of technologies in radiation treatment of cancer patients. Advanced radiotherapy techniques such as intensity modulated radiation therapy (IMRT), stereotactic radiosurgery (SRS), high dose rate (HDR) brachytherapy, and proton beam therapy are aimed at localized dose deliveries within the target volumes. The dose distributions from these treatment methods are typically characterized with high dose gradients around the boundaries of the targets, which present certain challenges for one or two-dimensional dosimeters such as ion chamber, diode, and film.

The idea of gel-based 3-D radiation dosimetry was first proposed in 1950 [1]. More recently, polymer gel dosimeters and radiochromic plastic dosimeters (PRESAGE) have been studied intensively as candidates for next generation 3-D dosimeter [2-6]. When irradiated, polymer gels become opaque through polymerization and PRESAGE presents color change. Both MRI [7], optical CT [8] and x-ray CT [9] have been used to image polymer gel dosimeters irradiated with clinical beams [10], including conformal radiotherapy, IMRT [11-13], SRS [14, 15] brachytherapy [16] and proton beam therapy [17-19]. The fundamental science of polymer gels has been thoroughly reviewed by Baldock et al. [20].

This review will focus on clinical applications of 3-D gel dosimeters and radiochromic plastic dosimeters. 3-D dosimeters have been used in three clinical aspects: verification of dose distribution for complex radiotherapy techniques, investigation of dose characteristics unique to a specific treatment technique or treatment unit, and verification of geometric accuracy of clinical setup. A few examples in various clinical applications are presented in this review. Dosimetric issues and clinical implications in these applications are discussed. Improvements and needs for future studies are discussed. The purpose of this short review is to demonstrate how 3-D dosimeters can provide useful clinical information.

2. Verification of dose distribution for advanced treatment techniques

Currently, routine verification of dose distribution for complex treatment is predominantly done using 2-D array dosimeters. This can be partially attributed to the fact that all the reliable and fast QA tools are 2-D dosimeters. However, the spatial resolution of 2-D arrays of ion chambers or diodes was limited by the size and the distance between the centers of adjacent detectors.

There have been many reports on IMRT 3-D dose verification using gels or radiochromic plastic dosimeters. However, most of the dose distribution comparisons between measurements and treatment planning calculations were presented by 2-D dose maps or 1-D dose profiles. A few studies compared multiple 2-D dose maps in one or three orthogonal planes [21, 22]. A full 3-D gamma evaluation for dose distribution comparison can be performed [23], but the disadvantages are long computation time and difficulty in identifying the locations of large dose disagreement.

Volumetric modulated arc therapy (VMAT) is an advanced radiation therapy technique where the treatment is delivered during one or more gantry rotations. The radiation intensity from VMAT is modulated by simultaneously varying the MLC positions, the rotation speed of the gantry and the dose rate during the treatment. Dosimetry accuracy can be affected by plan parameters such as control point (CP) separation [24]. A finer CP separation could lead to a significant increase in dosimetric accuracy for plans with high leaf travel values. Nelm *et al* [25] reported that many forms of relevant systematic errors can go undetected when the currently prevalent metrics for IMRT/VMAT commissioning are used. Based on this study, alternative methods and metrics are needed instead of the conventional metrics so that these errors are more likely to be detected. These advanced diagnostic methods include dose profile examination, EPID-based measurements [26, 27], dose difference pattern analysis, 3-D measurement-guided dose reconstruction, and dose grid inspection and more sensitive metrics (2% local normalization/2 mm DTA and estimated DVH comparisons). These suggested methods are mainly based on 1-D and 2-D measurements. Gels/radiochromic plastic dosimeters can provide 3-D dosimetry information [28] and may be useful to the diagnostic analysis of these systematic errors in VMAT. 3-D dose verification of a RapidArc treatment plan and delivery was performed with polymer gels-magnetic resonance imaging and Monte Carlo simulation by Ceberg et al [29]. In this study, various methods were used for comparative dosimetry: 2-D dose maps in an axial plane, 1-D dose profile, cumulative dose volume histograms for the volume enclosed by 90% isodose surface (VOI_{90}), overlay of VOI_{90} from different techniques (as shown in figure 1), frequency distribution of voxel-by-voxel deviations within VOI_{90} (as shown in figure 2), 3-D gamma analysis within VOI_{90}.

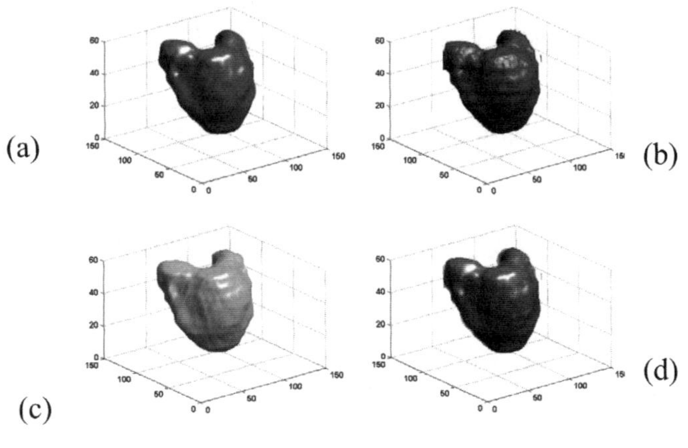

Figure 1. The 90% isodose surface projected into a 3-D view for (a) the RapidArc TPS, (b) the gel measurement and (c) the MC simulation. An overlay of all three volumes is presented in (d). Reproduced from [29].

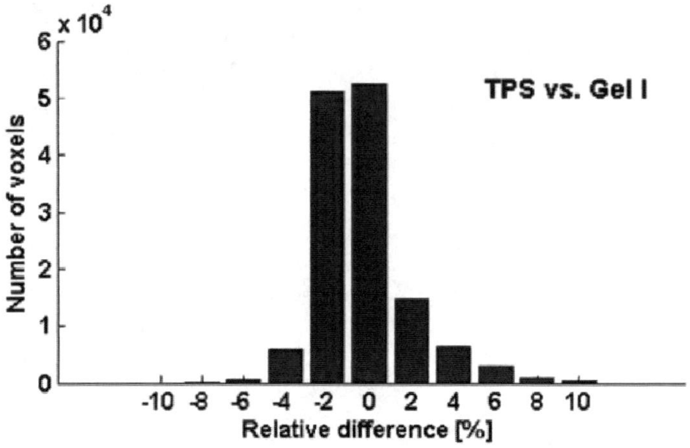

Figure 2. The distribution of voxel-by-voxel deviations between the different datasets within the volumes enclosed by the 90% isodose surface. Reproduced from [29].

These analysis techniques provide either detailed 1-D/2-D dose distributions, or part of overall 3-D dosimetry information.

3-D gel and radiochromic plastic dosimeters have been used for brachytherapy dosimetry in two categories: (1) TG-43 dosimetry characterization for new brachytherapy sources, (2) verification of dose distribution in clinical situations. One approach to validate gel measurement for brachytherapy is to present the dosimetry data according to the AAPM TG-43 dosimetric formalism. Polymer gel-MRI results concerning a new brachytherapy seed have been compared to corresponding, published dosimetric results using TLD, Monte Carlo simulation, and radiochromic film [30, 31]. TG43 parameters used in polymer gel studies include dose rate constant, Λ, radial dose function, $g(r)$, and anisotropy function $F(r,\theta)$. Petrokokkinos et al [32] verified the dosimetric accuracy of the TPS calculations for a treatment plan employing seven HDR source dwell positions and a partially shielded applicator, using Monte Carlo simulation and VIP polymer gel-magnetic resonance imaging. The dose responses of all the gel/radiochromic plastic dosimeters should be calibrated for each type of brachytherapy source with known air kerma strength and dose constant. The energy dependence at the energy range of brachytherapy sources has not been carefully investigated.

Proton radiotherapy is a good option for tumor in close proximity of critical structure because of its ability to deliver a sharp dose with practically no exit dose. The penumbra of a single proton field, although considerably sharper for shorter ranges compared to a photon field, steadily increases with range due to scattering in the patient. The sharper and complex dose distributions with proton therapy could benefit from 3-D dose verification with precision and high spatial resolution. However, the implementation of 3-D gels and radiochromic plasticdosimeters for proton therapy has not been as successful as that for photon therapy [17, 33] due to the fact that the gel/radiochromic plastic responses being LET dependent [34]. A typical under-dosage of 10-20 % (quenching) was observed near the Bragg peak region, as shown in figure 3.

Recently, a new polymer gel formulation, BANG3-Pro2, was developed by MGH Research to minimize the quenching effect [18]. This new gel formulation features a high viscosity gelatin matrix that limits the mobility of the growing polymer chains. Supposedly this feature also reduces the probability of premature chain termination in regions where high concentration of free-radical initiators prevail. In figure 4, left panel shows that this gel response under the Bragg peak and the distal end matches that of ion chamber with no observed dose quenching; right panel shows the comparison between the dose distribution from gel measurement and the calculated dose distribution from treatment planning system.

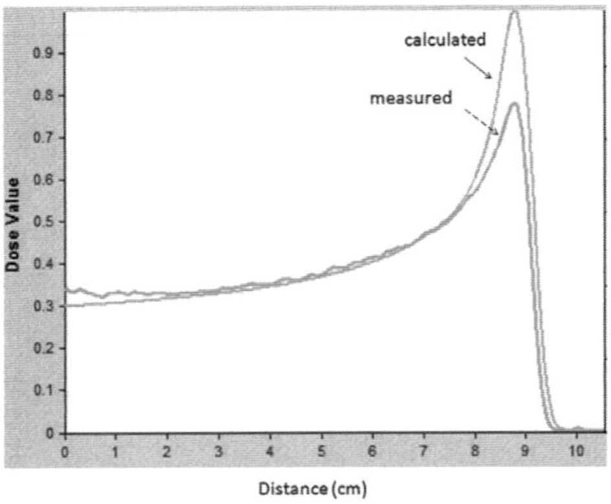

Figure 3. Percent depth dose curves comparison between the measured data from PRESAGE and the calculated from treatment planning system. Reproduced from [17].

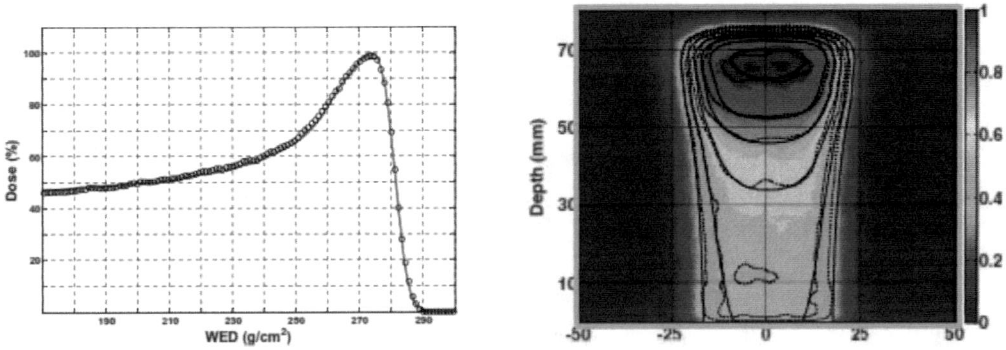

Figure 4. (Left) Comparisons of relative depth dose of the BANG3-Pro2 gel dosimeter (circles) to ion chamber scans in water (solid line); (Right): Comparison of dose distributions of irradiated gels by a non-modulated proton beam. The solid lines represents treatment planning system isodose lines and the dashed lines are measured gel data. Reproduced from [18].

3. Other clinical applications

3.1 Impact of MLC leaf width on SRS dose distribution

Wuu et al reported a study on the impact of MLC leaf width on SRS/SBRT clinical dosimetry, using a Varian Trilogy TX vs. a Varian 21 EX linear accelerator [28]. Trilogy TX has a 2.5 mm HD120® multileaf collimator system and 21EX has a 5 mm millennium multileaf collimator system. Dosimetric impact of MLC leaf width on an IMRT radiosurgery plan was evaluated by comparing the target coverage and the dose gradient around the target volume.

Figure 5 shows dose distribution at the central slice in an axial plane between the gel measurements and the treatment planning calculations. Both measured and calculated dose distributions are in good agreement for both machines. The difference in target coverage is small. However, as shown in figure 6, the dose falloff outside the target volume is sharper for 2.5 mm MLC compared to 5mm MLC.

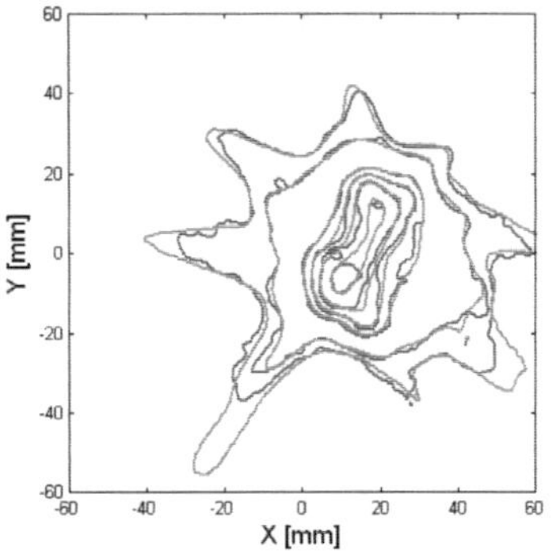

Figure 5. Comparison of the isodose lines (110, 100, 90, 80, 50, and 30 percent) from the Eclipse planning system (green lines) and the gel experiments for the central axil slice. Trilogy (2.5 mm MLC) experiment (left): red lines; 21EX (5 mm MLC) experiment (right): black lines.

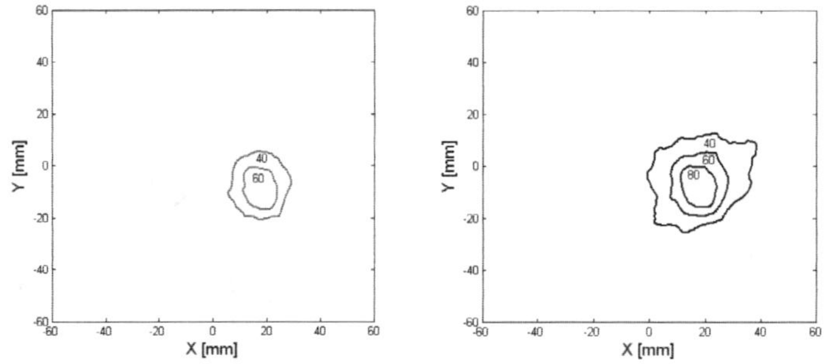

Figure 6. Isodose distribution (80, 60, and 40 percent) from the gel experiments for the axial slice that is 15 mm inferior to the central slice. Trilogy Tx (2.5 mm MLC): red lines; 21EX (5 mm MLC): black lines.

3.2 Verification of IGRT patient setup

Image-guided radiation therapy (IGRT) has been widely used to minimize the planning target volume (PTV) by imaging the patient target and repositioning the patient prior to treatment, accounting for the inter-fraction target movement. Nordstrom et all [35] investigated the feasibility of using a normoxic polyacrylamide gel (nPAG) dosimeter with implanted gold fiducial markers to evaluate the dosimetric impact of IGRT procedures using on-board imager (OBI) and Aria Record and Verify system (Varian medical systems). As shown in figure 7, the phantom consists of patient simulating volume, a bottle containing the nPAG gel, and the fiducial markers and fiducial support structures.

Treatment planning was performed on the CT images of the phantom. To test the validity of fiducial-OBI image guided process, the phantom was intentionally placed at an offset position from iso-center. Orthogonal imaging was performed with OBI, and setup corrections were made based on the orthogonal OBI images and integrated setup correction tools. Magnetic resonance imaging of the gels was performed after the irradiation. The dose distribution from gel measurement was normalized to the TPS calculated dose distribution. The 95% isosurface of the TPS calculated dose distribution was overlaid with 95% isosurface of the gel dose distribution for comparison and geometric verification. It was concluded that MRI based nPAG gel dosimetry can be used to verify the patient set up correction procedures using implanted gold fiducials.

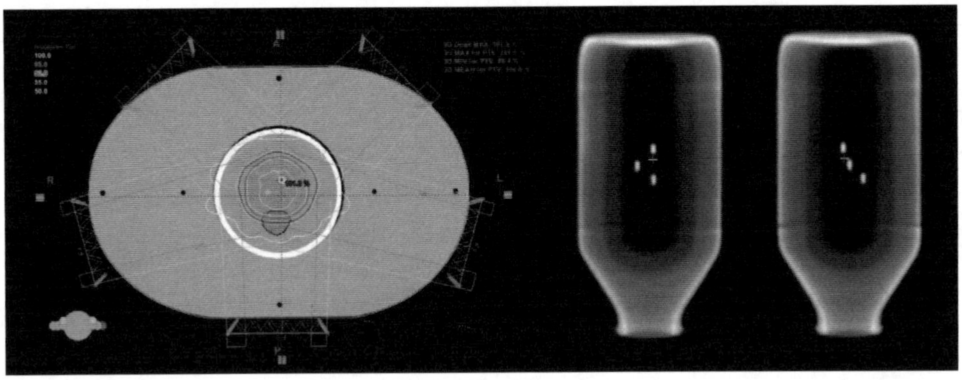

Figure 7. The IMRT treatment plan with structures from a dummy patient (left). The fiducials are clearly visible in the digital reconstructed radiographs (DRR) generated in 0° and 270°, respectively (middle and right).

4. Summary

3-D gel and radiochromic plastic dosimeters, in conjunction with MRI or optical CT scanner, have been demonstrated to be able to provide useful dosimetric information to many clinical applications. More efforts need be focused on the use of gel dosimeters in clinical situations where 3-D dosimetric information can be helpful for clinical decision.

5. References

[1] Day M J and Stein G 1950 *Nature* **166** 146-7
[2] Gore J C *et al* 1996 *Phys. Med. Biol.* **41** 2705-17
[4] Guo P Y *et al* 2006 *Med. Phys.* **33** 1338-45
[5] Brown S *et al* 2008 *Appl. Radiat. Isotopes* **66** 1970-4
[6] Gorjiara T *et al* 2011 *Med. Phys.* **38** 2265-74
[7] Murry P *et al* 2000 *Australas. Phys. Eng. Sci.* **23** 44-51
[8] Bosi S *et al* 2009 *Phys. Med. Biol.* **54** 275-83
[9] Baldock C *et al* 1994 *J. Roy. Soc. Med.* **87** 806-8
[10] Baldock C 2006 *J. Phys.: Conf. Ser.* **56** 14-22
[11] De Deene Y 2002 *Z Med. Phys.* **12** 77-88
[12] Oldham M 2006 J. Phys.: Conf. Ser. **56** 58-71
[13] Mather M L *et al* 2003 *Ultrasonics* **41** 551-9
[14] Grebe G *et al* 2001 *Int. J. Radiat. Oncol. Biol. Phys.* **49** 1451-60
[15] Pappas E *et al* 2001 *Phys. Med. Biol.* **46** 783-97
[16] Kipouros P *et al* 2003 *Med. Phys.* **30** 2031-9
[17] Zhao L *et al* 2012 *Phys. Med. Biol.* **57** N431-43
[18] Zaiden Q A *et al* 2010 *Med. Phys.* **37** 2145-52
[19] Gorjiara T *et al* 2012 *Med. Phys.* **39** 7071-9
[20] Baldock C *et al* 2010 *Phys. Med. Biol.* **55** R1-63
[21] De Deene Y *et al* 2000 *Magn. Reson. Med.* **43** 116-25
[22] Wuu C S and Xu Y 2006 *Med. Phys.* **33** 1412-9
[23] Wendling M *et al* 2007 *Med. Phys.* **34** 1647-54
[24] Masi L *et al* 2013 *Med. Phys.* **40** 071718
[25] Nelms B E *et al* 2013 *Med. Phys.* **40** 111722
[26] Greer P B *et al* 2007 *Med. Phys.* **34** 4389-98
[27] Vial P *et al* 2008 *Med. Phys.* **35** 1267-77
[28] Gustavsson H *et al* 2004 *Phys. Med. Biol.* **49** 227-41
[29] Ceberg S *et al* 2010 *Phys. Med. Biol.* **55** 4885-98
[30] Papagiannis P *et al* 2006 *Phys. Med. Biol.* **51** 2101-11

[31] Massillon-jl G *et al* 2012 *Phys. Med. Biol.* **57** 3407-18
[32] Petrokokkinos L *et al* 2011 *Med. Phys.* **38** 1981-92
[33] Gustavsson H *et al* 2004 *Phys. Med. Biol.* **49** 3847-55
[34] Wuu C S *et al* 2009 *J. Phys.: Conf. Ser.* **164** 012056
[35] Nordström F *et al* 2010 *J. Phys.: Conf. Ser.* **250** 012059.

Quasi 3D dosimetry (EPID, conventional 2D/3D detector matrices)

A Bäck
Department of Therapeutic Radiation Physics, Sahlgrenska University Hospital, Göteborg, Sweden

E-mail: anna.back@vgregion.se

Abstract. Patient specific pretreatment measurement for IMRT and VMAT QA should preferably give information with a high resolution in 3D. The ability to distinguish complex treatment plans, i.e. treatment plans with a difference between measured and calculated dose distributions that exceeds a specified tolerance, puts high demands on the dosimetry system used for the pretreatment measurements and the results of the measurement evaluation needs a clinical interpretation. There are a number of commercial dosimetry systems designed for pretreatment IMRT QA measurements. 2D arrays such as MapCHECK® (Sun Nuclear), MatriXX^{Evolution} (IBA Dosimetry) and OCTAVIOUS® 1500 (PTW), 3D phantoms such as OCTAVIUS® 4D (PTW), ArcCHECK® (Sun Nuclear) and Delta^4 (ScandiDos) and software for EPID dosimetry and 3D reconstruction of the dose in the patient geometry such as EPIDose™ (Sun Nuclear) and Dosimetry Check™ (Math Resolutions) are available. None of those dosimetry systems can measure the 3D dose distribution with a high resolution (full 3D dose distribution). Those systems can be called quasi 3D dosimetry systems. To be able to estimate the delivered dose in full 3D the user is dependent on a calculation algorithm in the software of the dosimetry system. All the vendors of the dosimetry systems mentioned above provide calculation algorithms to reconstruct a full 3D dose in the patient geometry. This enables analyzes of the difference between measured and calculated dose distributions in DVHs of the structures of clinical interest which facilitates the clinical interpretation and is a promising tool to be used for pretreatment IMRT QA measurements. However, independent validation studies on the accuracy of those algorithms are scarce. Pretreatment IMRT QA using the quasi 3D dosimetry systems mentioned above rely on both measurement uncertainty and accuracy of calculation algorithms. In this article, these quasi 3D dosimetry systems and their use in patient specific pretreatment IMRT/VMAT QA will be discussed.

1. Introduction

Verification measurements for patient specific pretreatment intensity modulated radiation therapy (IMRT) and volumetric modulated arc therapy (VMAT) quality assurance (QA) has been recommended [1, 2]. A pretreatment measurement can only verify a correct treatment delivery to a phantom at that particular point of time. It was shown in a study on the efficacy of common quality control (QC) checks that the pretreatment IMRT QA was the least effective check and detected only 1.4% of 4407 incidence error reports recorded in the course of clinical operations in 2 academic radiation oncology departments [3]. However, this result is not surprising since it is not the purpose of the pretreatment IMRT QA to find errors like for example wrong position of isocenter in the patient, wrong treatment plan selected for the treatment or wrong PTV expansion used for treatment planning.

The primary purpose of the pretreatment QA is to verify correct data transfer of the treatment plan parameters from the treatment planning system to the treatment machine and to verify that the treatment will deliver the planned dose distribution within a specified tolerance. IMRT treatment fields includes irregularly shaped multi-leave collimator (MLC) openings sometimes with small sub-field components. Advanced clinical dose calculation algorithms in treatment planning systems (TPS) have difficulties to estimate a correct dose distribution for small and irregular beam apertures [4]. Furthermore the delivery of treatment fields with small sub-field components is sensitive to small errors (within tolerance) in the machine, such as the MLC position [5]. Treatment plans including a large amount of small sub-field components will result in increased uncertainties with larger differences between planned and delivered dose distributions and are regarded as complex treatment plans that should be avoided to be used for treatment. Such treatment plans should be detected in the pretreatment measurement QA.

There are international recommendations to do 3D or at least quasi 3D measurements [1, 2] for pretreatment measurement QA but there are no specific recommendation on measurement and evaluation methods. The comparison of 3D dose distributions, i.e. the measured and the calculated dose distributions, require an evaluation method that include a large number of data points and preferably express the result in one or a few pass-fail criteria. A commonly used evaluation method is the gamma evaluation [6]. However, it has been shown that clinically relevant dose differences between planned and measured dose distribution are in some cases not detected based on common measurement procedures and gamma evaluation [7-10]. The international commission on radiation units and measurements (ICRU) have for IMRT/VMAT treatments replaced their previously recommended level of accepted uncertainty in delivered dose of 5% (1 standard deviation, SD) [11] by the recommendation that for low-gradient (< 20%/cm) regions the difference between the measured and the planned dose, normalized to the prescribed dose, should be no more than 3.5% (1 SD) and for high-gradient (> 20%/cm) regions the accuracy of distance to agreement should be within 3.5 mm (1 SD) [12]. During the implementation of pretreatment IMRT/VMAT QA in a clinic, the procedure, i.e. combination of dosimetry system, measurement and evaluation method, must be validated to be able to distinguish complex treatment plans. This must be done individually for treatment plans created at each clinic and therefore no general recommendation on measurement and evaluation criteria can be recommended [1, 2].

The ability to distinguish complex treatment plans puts high demands on the dosimetry system used for the pretreatment measurements. The results of the measurement evaluation needs a clinical interpretation. Besides phantoms for detector inserts, film and gel dosimetry (which will not be addressed in this paper) there are a number of commercial dosimetry systems designed for pretreatment IMRT QA measurements. 2D arrays such as MapCHECK® (Sun Nuclear), MatriXX^{Evolution} (IBA Dosimetry) and OCTAVIOUS® 1500 (PTW), 3D phantoms such as OCTAVIUS® 4D (PTW), ArcCHECK® (Sun Nuclear) and Delta^4 (ScandiDos) and software for electronic portal imaging device (EPID) dosimetry and 3D reconstruction of the dose in the patient geometry such as EPIDose™ (Sun Nuclear) and Dosimetry Check™ (Math Resolutions) are available. These quasi 3D dosimetry systems will be discussed in this article.

2. Quasi 3D dosimetry systems

The quality of pretreatment IMRT QA measurements are dependent on the properties and the characteristics of the chosen measurement equipment. The measurements should give information on the accuracy of the delivery in 3D and preferably with high resolution to cover both regions of target and organs at risk. The evaluation, i.e. comparison of the measured and the calculated dose distributions, must be done in multiple points in 3D and therefore traditional point dosimeters are not applicable. The general specifications of different common commercial dosimetry systems designed for IMRT QA purposes are summarized in table 1. The advantage of using ionization chambers in the measurement systems is the well established method to estimate absorbed dose. The advantage of measurement systems using diodes in front of ionization chambers is the small detector size which increase the resolution and the possibility to do more accurate measurements in regions with non-homogeneous dose distribution. A smaller detector will make it possible to decrease the distance between detectors. The detector distance and the number of detectors reveals the amount of information you can get from a measurement. A comparison of the measured and calculated dose distributions restricted to the measurement points is directly dependent on the distance between the detectors in the measurement. The 2D arrays MapCHECK®, MatriXX^{Evolution} and OCTAVIUS® 1500 have all the measurement points in one 2D plane. For the ArcCHECK® the diode detectors are placed in a helical grid in the outer parts around a cylinder shaped phantom. The diode detectors in the Delta^4 are placed in two orthogonal planes like a cross seen in a transversal slice of the cylinder shaped phantom. This means that for a comparison restricted to the measurement points there are volumes without measurement points. It is, for example, possible that an organ at risk might be situated in a volume that is not measured and that a dose difference between delivered and calculated dose will cause an undetected overdosage out of tolerance in that organ at risk. None of the dosimetry systems included in table 1 can measure the 3D dose distribution with a high resolution (full 3D dose distribution). Those systems can be called quasi 3D dosimetry systems. To be able to estimate the delivered dose in full 3D the user is dependent on a calculation algorithm provided in the software of the dosimetry systems.

Table 1. Summary of general specifications for different common commercial dosimetry systems designed for IMRT QA purposes.

	MapCHECK® 2 (Sun Nuclear)	MatriXX^{Evolution} (IBA Dosimetry)	OCTAVIUS® 1500 /OCTAVIUS® 4D (PTW)	ArcCHECK® (Sun Nuclear)	Delta^4 (ScandiDos)
Phantom shape	2D array	2D array	2D array/Cylinder	Cylinder	Cylinder
Detector	Diode	Ion chamber	Ion chamber	Diode	Diode
Detector area (mm^2)	0.64	15.9	19.4	0.64	0.78
Detector vol. (cm^3)	0.000019	0.08	0.06	0.000019	0.000039
Detector dist. (mm)	7.07	7.62	7.1	10	5/10
No. of detectors	1527	1020	1405	1386	1069
Detector pattern	Plane	Plane	Plane	Cylinder	2 orth. planes
Max field size (cm)	32 x 26	24.4 x 24.4	27 x 27	27	20
Weight (kg)	7.1		6/29	16	27

Measurements with the 2D arrays MapCHECK®, MatriXX^{Evolution} and OCTAVIOUS® 1500 as well as 2D measurements using the EPIDs are well suited to verify irradiation of a single beam direction, i.e. per beam verification of IMRT with fix beam angles. Such measurements can be used to identify systematic errors and can be an important part of the IMRT commissioning process. However, for pretreatment IMRT QA measurements, the composite dose distribution of all fields need to be measured to be able to do a clinical interpretation of the difference between the measured and the calculated dose distribution of the whole treatment plan. For a 2D array or an EPID, the composite dose can be measured as a sum of all beam directions directed perpendicular to the 2D array. This can be done either by delivering the whole treatment with the gantry angle at 0 degrees or to mount the 2D array, like the EPID, to the gantry to make the array follow the gantry rotation but always be positioned perpendicular to the beam direction. Such dose distribution cannot directly be related to the composite 3D dose of all beams but needs a calculation algorithm to reconstruct the measured dose to the dose in a 3D phantom or the patient geometry. The vendors of the all dosimetry systems in table 1 including those based on 3D phantoms, i.e. OCTAVIOUS® 4D, ArcCHECK® and Delta^4, all provide calculation algorithms to reconstruct the full 3D dose in the patient geometry. A full 3D dose reconstruction of the measurement in the patient geometry enables, for example, analyzes of the difference between measured and calculated dose distributions in dose volume histograms (DVH) of the structures of clinical interest. This makes the clinical interpretation clear and straight forward and is an efficient tool to be used for pretreatment IMRT QA measurements.

Different software for full 3D dose reconstruction in the patient geometry for 2D arrays, EPIDs and 3D phantoms are listed in table 2. In three of the software, COMPASS (IBA Dosimetry) described by Godart *et al* (2011) [13], Dosimetry Check™ (Math Resolutions) described by Renner (2007) [14] and the Delta^{4DVH} described by Gustafsson (2013) [15] reconstruction algorithm is based on an estimation of the delivered fluence, based on a measurement, that is used as input in a forward dose calculation algorithm to calculate the dose in the patient geometry. The estimation of the delivered fluence is done in different ways in the different software. Dosimetry Check™ and Delta^{4DVH} are stand-alone systems which estimates the delivered fluence without the need for information on the calculated dose distribution in the treatment planning system (TPS). The COMPASS system determines the delivered fluence by scaling the predicted response as calculated based on the calculated dose distribution from the TPS to the measurement. The estimated delivered fluence is used as input into a forward dose calculation algorithm to calculate the 3D dose distribution in the patient geometry. Those 3D dose calculation algorithms are different for the three software mentioned. The Dosimetry Check™ and Delta^{4DVH} software uses a generic pencil beam algorithm without the need for a full commissioning procedure while the COMPASS software are based on the Collapsed Cone algorithm [16] which require a full commissioning procedure the same as for any treatment planning system.

The algorithms in the software VeriSoft® (PTW) described by Allgaier *et al* (2013) [17] and 3DVH® (Sun Nuclear) described by Zhen *et al* (2011) [9] are based on scaling a known dose distribution along each beam ray through the phantom according to the measurement in one detector point along the ray. The relative dose distribution along each beam ray are in VeriSoft® based on depth dose curves that are measured and entered to the software during commissioning and for 3DVH® it is based on the dose distribution of the treatment plan in question as calculated in the TPS. The VeriSoft® is a stand-alone system since no dose information from the TPS is needed in the process of the 3D dose reconstruction. The cross section of the beam ray will be determined by the distance between the measured data points. The data points of the 2D arrays and the 3D phantom need to be interpolated to increase the number of data points for the 3D dose reconstruction. The VeriSoft® software uses a linear interpolation and 3DVH® is based on a more sophisticated interpolation called "smartinterpolation". Therefore the accuracy of the 3D dose reconstruction using VeriSoft® is directly dependent on the distance between the detectors in the measurement [17].

Table 2. Commercial software for full 3D dose reconstruction in the patient geometry.

Software	3DVH®/EPIDose™ (Sun Nuclear)	COMPASS (IBA Dosimetry)	Dosimetry Check™ (Math Resolutions)	VeriSoft® (PTW)	Delta^{4DVH} (ScandiDos)
Compatible with	MapCHECK®, ArcCHECK® and EPIDs	MatriXX^{Evolution}	EPIDs and other 2D arrays and 3D phantoms	OCTAVIUS® 4D	Delta^{4}

Efficient pretreatment IMRT QA measurements using the quasi 3D dosimetry systems mentioned above is not based on measurements only but rely heavily on calculation algorithms. Those algorithms need to be carefully validated before clinical use. Unfortunately the independent validation studies on the algorithms included in the software mentioned in table 2 are scarce and further studies are needed.

3. Summary and Conclusion

There are a number of commercial quasi 3D dosimetry systems designed for pretreatment IMRT QA measurements. Those systems does not measure the full 3D dose distribution. The full 3D dose needs to be reconstructed from the limited number of measurement points. To be able to evaluate the clinical relevance of the differences between measured and calculated dose distributions, algorithms for full 3D dose reconstruction in the patient geometry are available. This makes the QA rely not only on the measurement but also on a calculation algorithm in the software. Those algorithms need to be carefully validated before clinical use. Unfortunately the independent validation studies on those algorithms are scarce and there is a need for further studies.

4. Acknowledgments

I am very grateful to my close colleagues Anna Karlsson Hauer, Julia Götstedt and Magnus Gustafsson for fruitful discussions on the topics of this article and for proof reading. The Swedish Radiation Safety Authority are gratefully acknowledged for funding parts of my ongoing work related to the topic of this article.

5. References

[1] ESTRO 2008 *Guidelines for the verification of IMRT. ESTRO Booklet no. 9* Brussels, Belgium
[2] Moran J M, Dempsey M, Eisbruch A, Fraass BA, Galvin JM, Ibbott GS and Marks LB 2011 *Safety considerations for IMRT* (full report). Available from: http://www.practicalradonc.org
[3] Ford E C *et al* 2012 *Int. J. Radiat. Oncol. Biol. Phys.* **84** e263-9
[4] Das I J *et al* 2008 *Med. Phys.* **35** 206-15
[5] LoSasso T *et al* 1998 *Med. Phys.* **25** 1919-27
[6] Low D A *et al* 1998 *Med. Phys.* **25** 656-61
[7] Kruse J J 2010 *Med. Phys.* **37** 2516-24
[8] Nelms B E *et al* 2011 *Med. Phys.* **38** 1037-44
[9] Zhen H *et al* 2011 *Med. Phys.* **38** 5477-89
[10] Nilsson J *et al* 2013 *J. Phys.: Conf. Ser.* **444** 012048
[11] ICRU 1976 *Determination of absorbed dose in a patient irradiated by beams of X or gamma rays in radiotherapy procedures. ICRU REPORT 24* Oxford University Press
[12] ICRU 2010 *Prescribing, recording, and reporting photon-beam intensity-modulated radiation therapy (IMRT). ICRU REPORT 83* Oxford University Press
[13] Godart J *et al* 2011 *Phys. Med. Biol.* **56** 5029-43
[14] Renner W D 2007 *Med. Dosim* **32** 157-65
[15] Gustafsson A 2013 *Patient dose calculation based on ScandiDos Delta^{4PT} measurements.* ScandiDos White Paper
[16] Ahnesjö A 1989 *Med. Phys.* **16** 577-92
[17] Allgaier B *et al* 2013 *Dose reconstruction in the OCTAVIOUS 4D phantom and in the patient without using dose information from the TPS.* PTW White Paper

Cherenkov radiation dosimetry in water tanks – video rate imaging, tomography and IMRT & VMAT plan verification

Brian W Pogue[1,2], Adam K. Glaser[1], Rongxiao Zhang[2] and David J. Gladstone[1,3]
[1]Thayer School of Engineering, Dartmouth College, Hanover NH USA
[2]Department of Physics & Astronomy, Dartmouth College, Hanover NH USA
[3]Radiation Oncology, Norris Cotton Cancer Center, Dartmouth-Hitchcock Medical Center, Lebanon NH USA

E-mail: Brian.W.Pogue@dartmouth.edu

Abstract. This paper presents a survey of three types of imaging of radiation beams in water tanks for comparison to dose maps. The first was simple depth and lateral profile verification, showing excellent agreement between Cherenkov and planned dose, as predicted by the treatment planning system for a square 5cm beam. The second approach was 3D tomography of such beams, using a rotating water tank with camera attached, and using filtered backprojection for the recovery of the 3D volume. The final presentation was real time 2D imaging of IMRT or VMAT treatments in a water tank. In all cases the match to the treatment planning system was within what would be considered acceptable for clinical medical physics acceptance.

1. Introduction

Cherenkov light emission from gamma-ray and electron beam delivery to water tanks has recently been shown to be detectable and could be used in some applications for 3D dosimetry. The major potential benefit of Cherenkov imaging is that it is a way to image beams in real time with over 30 frames per second. As such, it is feasible to image IMRT and VMAT treatment beams in water tanks dynamically [1], and create composite visualizations of the treatment plans. This imaging can be used to verify new treatment plans prior to application to patients, or to quickly verify new machines, or testing in situations where access is limited. The strength of this optical imaging is that it is simple to implement, and provides immediate feedback. The drawbacks are related to minor differences between Cherenkov emission and dose, which make it not an exact measurement of dose.

The average emission intensity detected is in the range of 10's of nanoWatts/cm^2, and so typically would require the room lights to be off, but has been reliably imaged using standard CMOS cameras[2, 3]. However, since LINACs deliver radiation in 3microsecond pulses near 200 Hz repetition, the instantaneous emission intensity of Cherenkov is actually 10's of milliWatts/cm^2. Thus, if imaging is gated to this emission pattern, it is possible to image this with the room lights on using gated intensified cameras [4, 5]. The relationship between delivered dose and Cherenkov light emission is largely linear when the spectrum of electron energies is not changing significantly [6]. In most cases, the Cherenkov/Dose ratio does not vary by more than 20%, and under certain situations can vary as little as 1-2%. The situations where there is a high degree of linearity between Cherenkov emission and Dose deposited are when imaging broad surfaces, imaging lower energy beams and imaging symmetric treatment plans. Even in the cases where there is not good linearity, there is potential to calibrate the ratio based upon well-known estimates from Monte Carlo simulations or from measurements.

In this paper, a review of the approaches used to date is given. The use of a water tank doped with trace amounts of fluorophore was used, and the emission imaged laterally through the wall with a camera. The relationship between Cherenkov induced emission imaged and the expected radiation dose in water is shown. The steps attempted going from 2D planar beam imaging through to rotational 3D tomography are outlined. Finally, 2D temporal imaging of IMRT and VMAT treatment plans are examined to determine the value of exploiting the real-time value of Cherenkov imaging at near video rate.

2. Methods and Results

The setup and use of these systems is simply done with a cubic water tank, initially with 20cm length sides, and later increased to 30cm diameter sides. The lateral and back walls were blackened off with flat black paint, in order to minimize reflections. Addition of 1g/L quinine sulphate to the water is done to act as a fluorophore. This is done to absorb the large amount of UV-blue Cherenkov emission, which is unfortunately being emitted in a cone at a 41^0 emission angle from the direction of travel, and to re-emit it as blue fluorescence, which is isotropic in emission angle. An image of the emission from a 5cm diameter square beam are shown in figure 1(a), along with the linearity of emission versus dose in (b), and the expected square root dependence of Signal to Noise ratio (S/N) in (c). Measures of lateral and depth beam profiles along with the predicted dose lines from the Treatment Planning System (TPS) are shown in (d) and (e), respectively. Note that the lateral beam profile agreement between dose and measured signal is outstanding for either Cherenkov or the fluorescence signal. However for the depth dose profile agreement to signal, the Cherenkov signal decays faster than the dose. Yet the depth-fluorescence signal agrees quite well with depth dose curve.

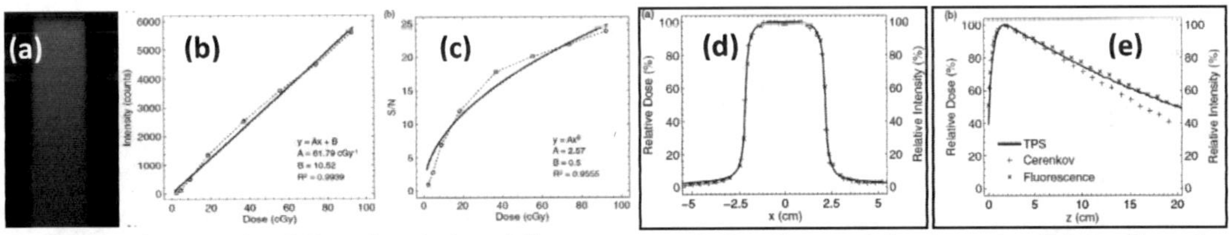

Figure 1. (a) Photograph of Cerenkov-induced fluorescent emission from quinine in water, (b) the linearity of image intensity with applied dose, (c) the signal to noise, following a shot noise limited square root function. In (d) a lateral beam profile is shown fluorescent beam (crosses) overlaid on the known beam profile (showing outstanding agreement), and (e) the depth dose curve is shown for the treatment planning system (TPS) along with Cerenkov and Cerenkov-fluorescence.

One of the more interesting opportunities is the ability to do 3 dimensional (3D) imaging of Cherenkov emission as a way to potentially image a surrogate of 3D dose distributions. The benefit is that the imaging is high resolution and rapid, so full 3D beam images could be obtained in a few minutes. The trickier aspects of this are that full recovery of the volume would require back projection reconstruction from the 2D projection images. Additionally, the geometry to avoid refraction trough the water tank walls is problematic, yet the easiest solution to this is to keep the camera and water tank wall in a fixed geometry and rotate the water tank. Rotational imaging around the water tank is possible but with refraction at the water tank surface, it is better to rotate the tank with the camera [3, 4]. An illustration of this is shown below in figure 2.

Journal of Physics: Conference Series **573** (2015) 012013 doi:10.1088/1742-6596/573/1/012013

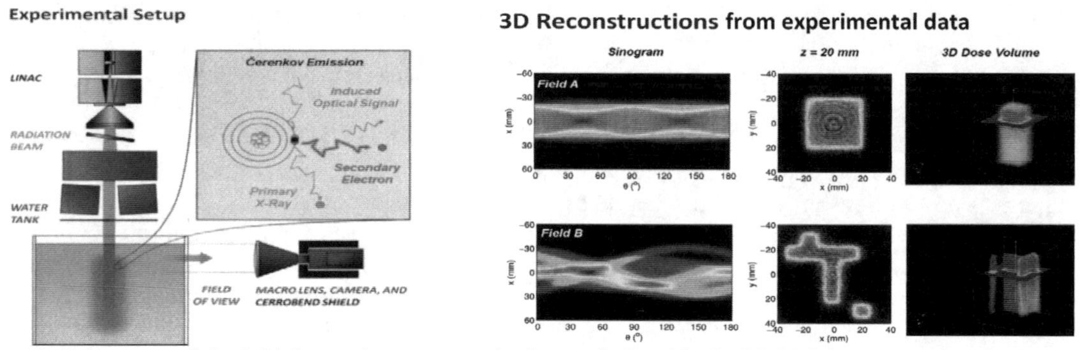

Figure 2. Schematic of the initial experiment set up is shown (left) with the LINAC beam entering the water tank. Čerenkov-induced fluorescence was imaged with the camera laterally, and repeated 1800 rotational angles. This lateral projection data was then combined in a sinogram (right) with tomographic reconstruction. This was completed experimentally on two test beams, rotating the camera, and imaging a simple square beam (right - top row) and a highly irregular beam (right bottom row). The full 3D visualizations of these beams are shown (far right) [7].

Perhaps the largest strengths of Cherenkov imaging as a surrogate for dose imaging, is that the emission capture is very fasts and high resolution, so while 3D imaging is interesting, the rotation around the beam is problematic for imaging of off-axis beams. However the most readily implementable use of this technology to medical physics QA is in single projection video imaging of beam dynamics through one side of a pre-existing water tank. As such, the ability to image the beam delivery dynamics of IMRT and VMAT is easily possible. Admittedly the single planar view through one surface does not provide a complete image of the deposited dose, rather is a projection through one direction view perspective of the beam. Yet it provides the fast feedback required for real-time verification, which could have distinct advantages in busy Medical Physics departments. An illustration of this is shown below in figure 3.

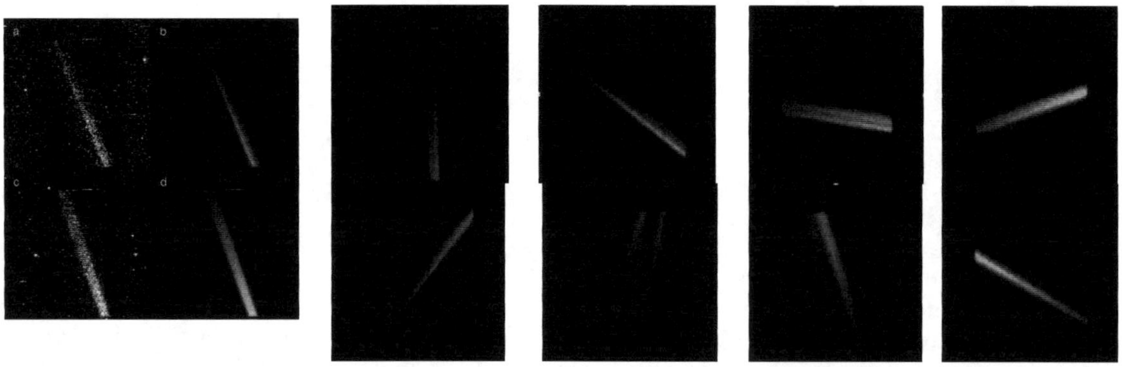

Figure 3. In raw images shown in gray scale, (a) shown after median filtering over 24 images (b), with VMAT on top and correlated IMRT images in (c) and (d). The blue images of the beam (right) are shown from a VMAT treatment for 8 positions of the dynamic beam, illustrating the fidelity of being able to capture images throughout the treatment verification process. These images can be integrated together to create deposited Cherenkov maps, which agree with dose maps to within 96% accuracy for gamma parameter of 3% and 3mm to agreement [1].

3. Discussions

This paper has presented a survey of three types of imaging of radiation beams in water tanks for comparison to dose maps. The first was simple depth and lateral profile verification, showing excellent agreement between Cherenkov and planned dose, as predicted by the treatment planning system for a square 5cm beam. The second approach was 3D tomography of such beams, using a rotating water tank with camera attached, and using filtered backprojection for the recovery of the 3D volume. The final presentation was real time 2D imaging of IMRT or VMAT treatments in a water tank. In all cases the match to the treatment planning system has been within what would be considered acceptable for clinical medical physics acceptance. Comparison to ionization chamber data has also been demonstrated, however because the situations of agreement and disagreement are so unique, better insight is provided by Monte Carlo simulations. This Monte Carlo study, previously published [6], showing that for broad beam profiles, and higher energy beams there is optimal match, and at the lower beam energies and smaller beam sizes the match will be less accurate. Still, the match in IMRT and VMAT treatment plans for a complex C-shaped treatment has been outstanding and so further experimental verification of different beam sizes and more complex treatment plans should be carried out to determine the limits of where there is good versus unacceptable match between Cherenkov and dose.

4. Acknowledgements

This work has been funded by NIH grant R21EB017559.

5. References

[1] Glaser A K *et al* 2014 *Med Phys* **41** 062102
[2] Glaser A K *et al* 2013 *Med Phys* **40** 012101
[3] Glaser A K *et al* 2013 *Phys. Med. Biol.* **58** 601-19
[4] Glaser A K *et al* 2012 *Opt. Lett.* **37** 1193-95
[5] Zhang R *et al* 2013 *Phys. Med. Biol.* **58** 5477-93
[6] Glaser A K *et al* 2014 *Phys. Med. Biol.* **59** 3789-811
[7] Glaser A K *et al* 2013 *Opt. Lett.* **38** 634-6

Current status of 3D EPID-based in vivo dosimetry in The Netherlands Cancer Institute

B Mijnheer, I Olaciregui-Ruiz, R Rozendaal, H Spreeuw, M van Herk and A Mans
Department of Radiation Oncology, The Netherlands Cancer Institute, Plesmanlaan 121, 1066 CX Amsterdam, The Netherlands

E-mail: b.mijnheer@nki.nl

Abstract. 3D in vivo dose verification using a-Si EPIDs is performed routinely in our institution for almost all RT treatments. The EPID-based 3D dose distribution is reconstructed using a back-projection algorithm and compared with the planned dose distribution using 3D gamma evaluation. Dose-reconstruction and gamma-evaluation software runs automatically, and deviations outside the alert criteria are immediately available and investigated, in combination with inspection of cone-beam CT scans. The implementation of our 3D EPID-based in vivo dosimetry approach was able to replace pre-treatment verification for more than 90% of the patient treatments. Clinically relevant deviations could be detected for approximately 1 out of 300 patient treatments (IMRT and VMAT). Most of these errors were patient related anatomical changes or deviations from the routine clinical procedure, and would not have been detected by pre-treatment verification. Moreover, 3D EPID-based in vivo dose verification is a fast and accurate tool to assure the safe delivery of RT treatments. It provides clinically more useful information and is less time consuming than pre-treatment verification measurements. Automated 3D in vivo dosimetry is therefore a prerequisite for large-scale implementation of patient-specific quality assurance of RT treatments.

1. Introduction

The use of *in vivo* dosimetry (IVD) in external beam radiotherapy (EBRT) has been addressed in many studies [e.g. 1-3]. In these papers a number of serious errors were identified that could not have been detected by other quality assurance (QA) checks, for instance by means of pre-treatment measurements. In addition to being able to detect major errors, the main application of IVD is to assess all clinically relevant differences between planned and delivered dose. In a recent paper the main characteristics of the most commonly applied IVD systems and their clinical use during EBRT have been summarized [4].

Despite the increasing use of EPIDs for dosimetry purposes [5, 6] only few groups have implemented *in vivo* EPID dosimetry in the clinic. Nijsten *et al* [2] presented an analysis of the results of routine EPID dose measurements obtained by correlating the dose measured with an EPID on the central beam axis with dose values at 5 cm depth. Piermattei *et al* [7] reported the first results of a national Italian *in vivo* dosimetry project in which a simple method was applied for the *in vivo* determination of the midplane dose using the central pixel values of an EPID. Recently Berry *et al* [8] applied an in-house developed algorithm to compare transit dosimetric EPID images, acquired during patient treatment, with predicted transit images. All these approaches verify the dose either at a point or in 2D.

In the Netherlands Cancer Institute (NKI) back-projection algorithms have been implemented for the 2D [9] and 3D dose verification [10] of intensity-modulated radiotherapy (IMRT) using amorphous silicon (a-Si) type of EPIDs, and more recently for the 3D verification of volumetric-modulated arc therapy (VMAT) [11]. In this presentation we will elucidate the current status of 3D EPID-based *in vivo* dosimetry as performed routinely in the NKI.

2. Materials and methods

Patient treatments are performed in our institution using SL20i linear accelerators (Elekta, Crawley, UK) having 6 and 10 MV photon beams. The linacs are equipped with MLCi (2x40 1cm leaves) or Agility (2x80 0.5cm leaves) MLCs and PerkinElmer RID 1680 AL5 amorphous silicon EPIDs (Elekta iViewGT).

Our EPID dosimetry method is a measurement-based approach applying a fast and simple 3D back-projection algorithm in combination with the planning CT data of a patient or phantom [9, 10]. For image acquisition during VMAT delivery, the in-house developed EPID acquisition software has been adjusted to save every detector frame separately (~ 3 frames/s), together with the accompanying gantry angle [11]. For dose verification of treatments of sites having tissue heterogeneities, a modification of our algorithm using the "*in aqua vivo*" approach is applied [12]. Phantom measurements showed that planned and measured dose values in the volume encompassed by the 50% isodose surface generally agree within 2% [10-12].

Treatment plans are generated with the clinical version of our treatment planning system (TPS) (Pinnacle v9.2, Philips Medical Systems, Eindhoven, The Netherlands), which includes the "SmartArc" module for VMAT plan generation. Reconstructed 3D dose distributions from acquired EPID images are compared with planned dose distributions using 3D γ-evaluation within the 50% isodose surface using global 3%/3mm criteria. All curative patients at NKI are irradiated either with an IMRT or VMAT technique; all others with a 3D conformal radiotherapy (3DCRT) technique. Almost all are verified by means of EPID-based *in vivo* dosimetry during three fractions in the first week of their treatment. Exceptions are single fraction irradiations, where both *in vivo* and pre-treatment verifications are performed, and when large field sizes are involved that will damage the EPID electronics.

New software tools were clinically introduced to allow automated image acquisition, to periodically inspect the record-and-verify database, and to automatically run the EPID dosimetry software [13]. The results of the analysis of the comparison of the EPID-reconstructed and planned dose distribution are available within a few minutes after delivery. Alerts are immediately raised without any human intervention when deviations are outside clinical criteria.

3. Results and discussion

The implementation of our 3D EPID-based *in vivo* dosimetry approach enabled the replacement of pre-treatment verification for more than 90% of the IMRT/VMAT treatments. The process has been integrated in our clinical workflow, and about 2,100 IMRT, 1,700 VMAT and 1,400 3DCRT treatments were verified in 2013. With our current set of clinical tolerance levels, approximately 20% of these verified plans showed deviations that required inspection by a medical physicist. After the inspection work, most of these deviations were found not to be clinically relevant. Only about one out of 300 verifications required consultation with the responsible radiation oncologist. These deviations were mostly patient-related (e.g. due to tumour shrinkage or disappearance of lung atelectesis) or due to changes in treatment parameters between planning and delivery (e.g. due to using wrong CT data or missing bolus during CT scanning), and could not have been traced with pre-treatment verification. The following example illustrates our procedure.

An EPID dosimetry report is automatically generated for each *in vivo* dose verification providing the mean γ (γ_{mean}), the 99th percentile (the near maximum γ value) ($\gamma_{1\%}$), and the difference in isocenter dose (ΔD_{isoc}) determined with the EPID and the TPS (see figure 1). A green dot in the EPID dosimetry report means that all alert criteria are within tolerance level ($\gamma_{mean}\leq 0.5$, $\gamma_{1\%}\leq 2$, and $\Delta D_{isoc}\leq 3\%$), a yellow dot indicates that at least one of the alert criteria is outside tolerance level but still within action level ($0.5<\gamma_{mean}\leq 1.0$, $2<\gamma_{1\%}\leq 4$, and $3\%<\Delta D_{isoc}\leq 5\%$), and a red dot indicates that at least for one of the evaluation criteria the error is outside the action level ($\gamma_{mean}>1$, $\gamma_{1\%}>4$, and $\Delta D_{isoc}>5\%$), i.e. it needs immediate action of a medical physicist. Note that a signed gamma analysis display is used; yellow and red mean that the EPID dose values are higher than in the plan, while green and blue display EPID doses lower than the corresponding data in the plan.

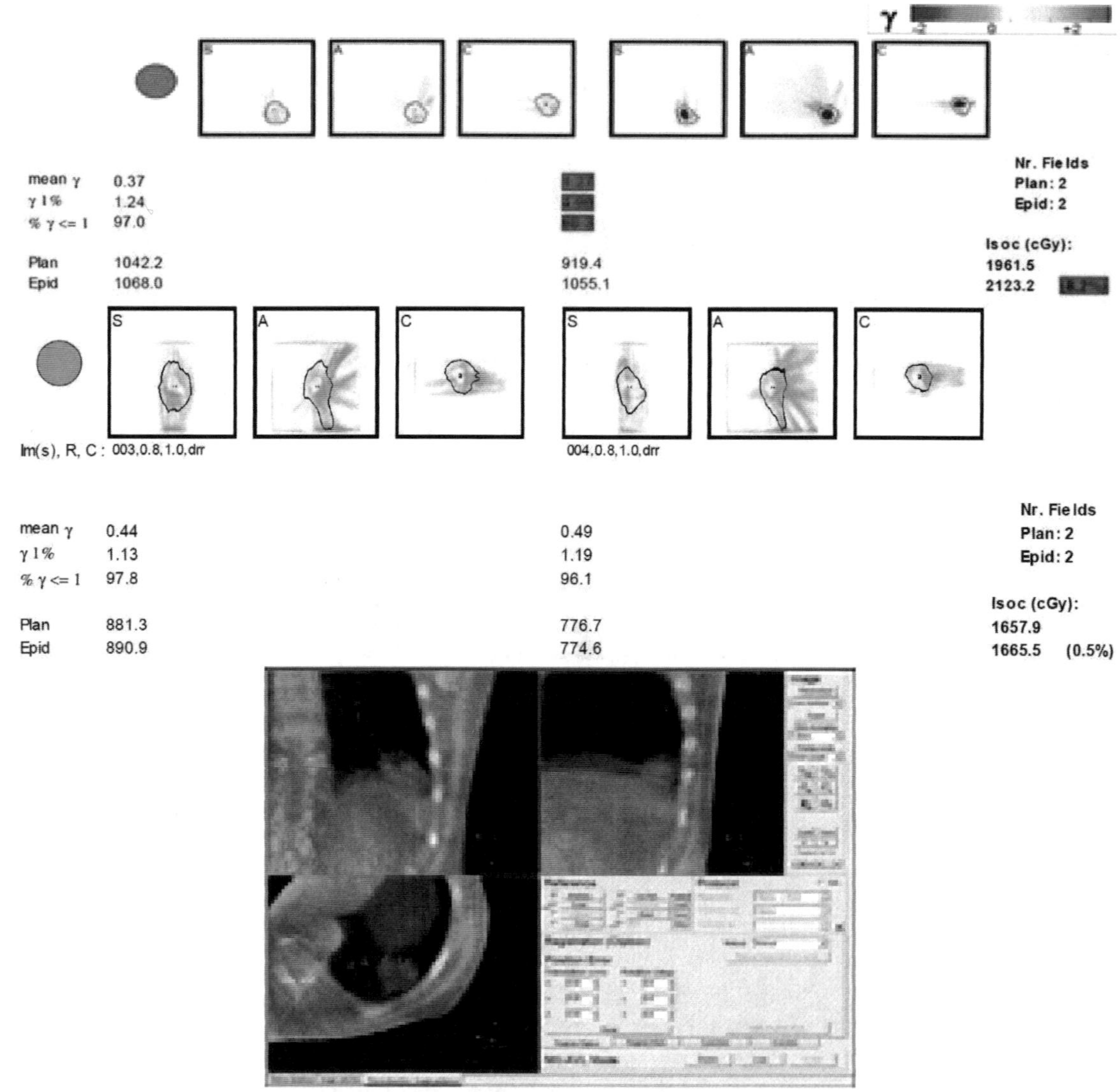

Figure 1. Results of a 3D EPID-based in vivo dose verification of a dual arc hypo-fractionated VMAT treatment of a lung cancer patient (top) and of a phantom measurement (middle) of the same irradiation. Indicated are the results of a 3D gamma evaluation in a sagittal, axial and coronal plane through the isocenter within the 50% isodose surface. The bottom part shows a comparison of a cone-beam CT scan (green) with the planning CT scan (purple) after the second arc in the three orthogonal planes.

The data presented in the upper part of figure 1 show that the results for the first arc were within acceptance criteria, but the data for the second arc were outside the action level. In order to verify if the plan was correctly calculated and delivered, a phantom measurement using the same irradiation

Journal of Physics: Conference Series **573** (2015) 012014 doi:10.1088/1742-6596/573/1/012014

conditions as used during the patient treatment was performed. The results shown in the middle part of the figure indicate that all evaluation statistics are within tolerance/action level. Inspection of the cone-beam CT scan made after the second arc as shown at the bottom part of the figure indicates a baseline shift of about 1.5 cm between the two arcs. As a result of such observations, in-line cone-beam CT scans [14] are now made routinely of both arcs of hypo-fractionated lung VMAT treatments.

4. Conclusions

Our EPID dosimetry approach is a fast and accurate method for 3D in vivo dose verification of RT treatments. It provides clinically more useful information and is less time consuming than patient-specific pre-treatment dose verification. Automated 3D in vivo dosimetry is therefore a prerequisite for large scale implementation of patient-specific QA of RT treatments.

5. References

[1] IAEA Human Health Report Nr 8 2012 *Development of Procedures for in vivo Dosimetry in Radiotherapy,* International Atomic Energy Agency (Vienna: IAEA)
[2] Nijsten S M J J G *et al* 2007 *Radiother. Oncol.* **83** 65-75
[3] Mans A *et al* 2010 *Med. Phys.* **37** 2638-2644
[4] Mijnheer B *et al* 2013 *Med. Phys.* **40** 070903-1-19
[5] van Elmpt W *et al* 2008 *Radiother. Oncol.* **88** 289-309
[6] Greer P B *et al* 2007 *Med. Phys.* **34** 4389-98
[7] Piermattei A *et al* 2012 *Nucl. Instr. Meth. Phys. Res. B* **274** 42–50
[8] Berry S L *et al* 2014 *Int. J. Radiat. Oncol. Biol. Phys.* **88** 204-9
[9] Wendling M *et al* 2006 *Med. Phys.* **33** 259-73
[10] Wendling M *et al* 2009 *Med. Phys.* **36** 3310-21
[11] Mans A *et al* 2010 *Radiother. Oncol.* **94** 181-7
[12] Wendling M *et al* 2012 *Med. Phys.* **39** 367-77
[13] Olaciregui-Ruiz I *et al* 2013 *Phys. Med. Biol.* **58** 8253-64
[14] van Herk M *et al* 2011 *Radiother. Oncol.* **100** 365-9

Strategies for quality assurance of intensity modulated radiation therapy

Hunor Benedek[1], Ulf Isacsson[2], Maria Olevik-Dunder[3], Mathias Westermark[3], Per Hållström[4], Jörgen Olofsson[5] and Magnus Gustafsson[6]
[1]Skåne University Hospital, Lund, Sweden
[2]Uppsala University Hospital, Uppsala, Sweden
[3]Karolinska University Hospital, Stockholm, Sweden
[4]Gävle Hospital, Gävle, Sweden
[5]Norrland's University Hospital, Umeå, Sweden
[6]Sahlgrenska University Hospital, Göteborg, Sweden

E-mail: hunor.benedek@med.lu.se

Abstract. In late 2011 The Swedish Society of Radiation Physics formed a working group to concentrate on the Quality Assurance of modern radiation therapy techniques. The given task was to identify and summarise the different QA strategies in Sweden and also the international recommendations. This was used to formulate recommendations for practical guidelines within Sweden. In this paper a brief summery of the group's work is presented. All the Swedish radiation therapy centres do a pre treatment verification measurement as QA for every new IMRT and VMAT plan. Physicists do it and they believe it to be time consuming. A general standpoint from all the centres was that new guidelines and legislation is needed to allow QA that does not require a measurement. Based on various international publications and recommendations the working group has presented two strategies, one where all new plans are checked through measurement and one where no measurement is needed. The measurement-based strategy is basically the same as the one used today with an extended machine QA part. The other presented strategy is process oriented where all the different parts of the treatment chain are checked separately. The final report can be found in Swedish on http://www.radiofysik.org.

1. Background and Purpose

The last few years have seen a rapid increase in the use of intensity modulated radiation therapy (IMRT) in Sweden. This trend is continuing with the implementation of rotational techniques such as VMAT (Volumetric Modulated Arc Therapy). Currently, the Quality Assurance (QA) of each treatment plan is regulated by the Swedish Radiation Safety Authority (SSM). The SSM's regulations say that each field needs to be assured by measurement during the first fraction of treatment. For IMRT techniques this is not feasible and a pre treatment verification measurement is done instead. This is very time-consuming, and considering the rapidly growing number of patients considered for IMRT a lot of resources are then required for the QA. The Swedish Society of Radiation Physics recognised a need for new guidelines and formed a national group to develop these. The group's tasks were:

i) To survey how IMRT QA is done at the different radiotherapy clinics in Sweden.

ii) Review existing international recommendations on IMRT QA.
iii) To consider quality assurance concepts used by the industry.
iv) To develop practical guidelines for IMRT and VMAT QA.

The new guidelines in the resulting report didn't specifically need to agree with the existing SSM regulations. The recommended guidelines can be found in the final report [11].

2. The survey

During the first quarter of 2012 a survey was sent out to the radiation therapy clinics in Sweden that were at that time performing IMRT/VMAT. Five clinics performed IMRT only, three only VMAT, while six clinics treated with both IMRT and VMAT. The average time spent on QA for two treatment plans was 65 min for IMRT and 43 min for VMAT including the preparations before measurement and evaluation afterwards.

One of the questions in the survey asked what changes were desired if there was no need to consider the legislation. All clinics wanted to shorten the time spent on QA. The four most common suggestions for time efficient QA were:

i) In –vivo EPID dosimetry, where there is no need for a phantom
ii) Independent dose calculation
iii) Measurement for the more complex cases only (complexity index)
iv) More focused and frequent machine QA to reduce patient QA

3. Published reports and guidelines

The task group chose to study publications by well-established organisations or internationally recognised institutions that have written guidelines for IMRT/VMAT or Tomotherapy. The studied publications are listed as references below.

All publications agree that some kind of patient specific QA is needed. The task group have chosen the following recommendations as support for the guidelines below.

i) A measurement is not necessary if an independent dose calculation is performed [1-3].
ii) The QA methods should be validated so they find the errors they are designed for [8, 9].
iii) The treatment plan integrity from approved status to treatment, must be verified [4, 9].
iv) Make sure the MLC, dose rate, collimator, and gantry angle parameters are well coordinated at all times [4].
v) One can use a complexity index to categorise treatment plans [5].

4. Suggested QA

The group have identified five parameters that need to be checked for every new treatment. MLC positions, Gantry angles, Dose/Monitor Unit, Data integrity and Dose calculation. These parameters is proposed to be assessed with two different strategies. The first, labelled measurement based QA , is the same as in use today where every plan is measured before treatment. The other strategy, labelled process oriented QA, is concerned with the control of each individual parameter affecting the treatment. The machine parameters such as MLC positions, Gantry angle and Dose/MU, are checked by an extended accelerator QA. The dose calculation and data integrity also needs to be checked for every plan. Process oriented QA is only recommended for standard (or 'routine') plans where the clinic has great experience and confidence in their delivery. The two strategies are illustrated in the flow chart below.

To implement the process oriented strategy a lot of work needs to be done on verification. All parameter controls must be verified and evaluated. Another challenge is the categorisation of the treatment plans. The non-standard plans needs to be identified so that they can be measured (see flow chart). This can be done using a complexity index. The index should be a measure of the plan's deliverability and every clinic should validate the index for their patient groups. Another way to regulate the plan complexity is to have strict treatment planning instructions and objectives.

Currently, using a purely process oriented strategy is not allowed in Sweden, because of the SSM legislation. In any case, the group once again wants to stress the importance of validation of the QA-methods no matter what method is chosen, so that your method of choice actually reveals the errors you aim to discover.

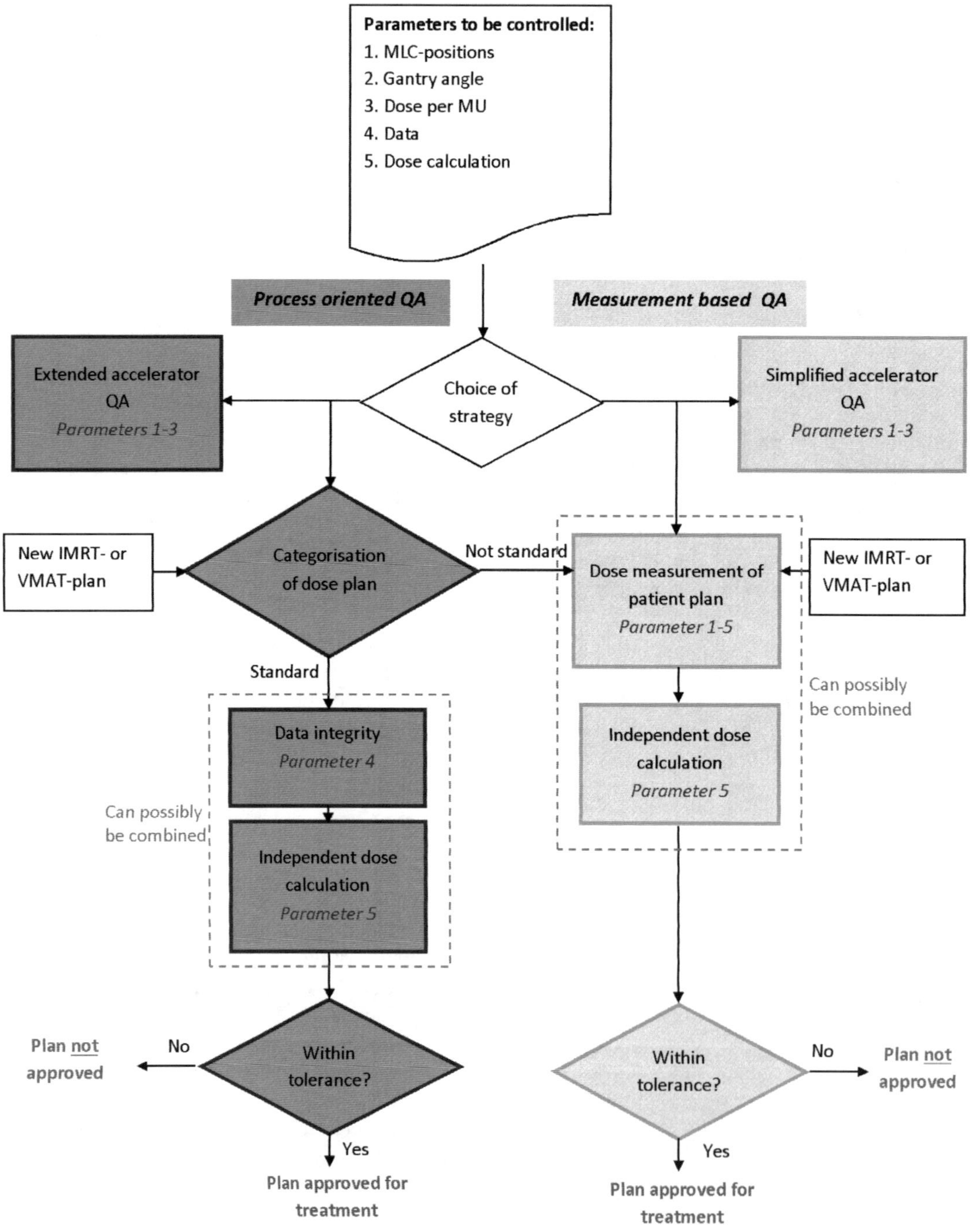

Journal of Physics: Conference Series **573** (2015) 012015 doi:10.1088/1742-6596/573/1/012015

5. References

[1] ESTRO Booklet no. 9: Guidelines for the verification of IMRT. Brussels, Belgium. ESTRO 2008
[2] ICRU REPORT 83: Prescribing, Recording, and Reporting Photon-Beam Intensity-Modulated Radiation Therapy (IMRT). Oxford University Press 2010
[3] Quality control for Intensity-modulated radiation therapy: Swiss Society for Radiobiology and Medical Physics 2007. (http://www.sgsmp.ch/)
[4] Hartford A C *et al* 2009 International journal of radiation oncology, biology, physics **73** 9
[5] Klein E E *et al* 2009 *Med. Phys.* **36** 4197-212
[6] Langen K M *et al.* 2010 *Med. Phys.* **37** 4817-53
[7] Low D A *et al* 2011. *Med Phys.* **38** 1313-38
[8] Moran J M *et al* 2011. *Med. Phys.* **38** 5067
[9] Moran J M *et al* 2011 Safety considerations for IMRT. (http://www.practicalradonc.org/)
[10] Thomadsen B R 2011. Draft of AAPM Task Group 100 report: Application of risk analysis methods to radiation therapy quality management
[11] Olofsson J, Gustafsson M *et al* 2014 Swedish Society of Radiation Physics REPORT 2014:1

Journal of Physics: Conference Series **573** (2015) 012016 doi:10.1088/1742-6596/573/1/012016

Complexity metric as a complement to measurement based IMRT/VMAT patient-specific QA

J Götstedt[1], A Karlsson Hauer[2] and A Bäck[2]
[1]Department of Radiation Physics, Sahlgrenska Academy, University of Gothenburg, Göteborg, Sweden
[2]Department of Therapeutic Radiation Physics, Sahlgrenska University Hospital, Göteborg, Sweden

E-mail: julia.gotstedt@vgregion.se

Abstract. IMRT/VMAT treatment plans contain treatment fields with MLC openings of various size and shape. Clinical dose calculation algorithms show limitations in calculating the correct dose in small and irregular parts of a MLC opening which leads to differences between the planned and delivered dose distributions. The patient-specific IMRT QA is often designed to compare planned and measured dose distributions and is therefore heavily dependent on the measurement equipment and the evaluation method. The purpose of this study is to develop a complexity metric based on shape and size of MLC openings that correlates to the dose differences between planned and delivered 3D dose distributions. Different MLC openings are measured and evaluated and used to determine a penalty function to steer the complexity metric and make the complexity scores correlate to dose difference pass rates. Results of this initial study show that a correlation was found between complexity scores and dose difference pass rates for static fields with varied complexity. Preliminary results also show that the complexity metric can distinguish clinical IMRT fields with higher complexity.

1. Introduction

Measurement based pre-treatment patient-specific quality assurance (QA) is a common approach to ensure the accuracy of the delivered absorbed 3D dose distribution of advanced radiotherapy treatments such as intensity modulated radiotherapy and volumetric modulated arc therapy treatments (IMRT/VMAT). Treatments with intensity modulated techniques are increasing in popularity due to the capability to create treatment plans of high conformity and to increase the target dose but still retain the dose level to the organs at risk [1]. The intensity modulation is mainly accomplished by varying the shape of the multi-leaf collimator (MLC) opening but also by varying the dose rate and sometimes the speed of the gantry rotation. Advanced dose optimization algorithms within the treatment planning systems (TPS) generate treatment plans that fulfill the user-selected constraints and objectives and contain treatment fields with MLC openings of various size and shape. Irregularly shaped MLC openings with smaller sub-field components are challenging from a dosimetric point of view [2]. Unfortunately, advanced clinical dose calculation algorithms have difficulties to estimate a correct dose distribution when it comes to small and irregular beam apertures where there is lack of charged particle equilibrium (CPE) [2]. This leads to a calculated dose distribution associated with increased uncertainties for treatment plans that consist of small and irregular MLC openings which in turn leads to dose differences between planned and delivered dose distributions for those plans.

Commonly used measurement based QA methods are designed to compare the calculated 3D dose distribution with a dose distribution measured using a semi-3D dosimetry system (e.g. Delta^{4PT}, ArcCHECK®, OCTAVIUS® etc.), which makes the method heavily dependent on the accuracy of the measurement equipment and the choice of evaluation method. Clinically relevant dose differences between planned and measured dose distributions in evaluated dose volume metrics of interest are in some cases not detected in common QA procedures [3,4]. The intention with the QA differs between departments and some departments might focus their QA on finding only the large errors that corresponds to something going wrong in the preparation procedure or the delivery. The purpose of this study, however, is to develop a complexity metric based on shape and size of MLC openings to identify all treatment plans with clinically relevant deviations between planned and delivered dose due to, for example inaccuracies in the absorbed dose calculation procedure. This initial study focuses on the limitations within the TPS to calculate a correct dose for small and irregular MLC openings. Delivery of treatment plans with similar dose distributions may differ very much in complexity due to different shapes of the MLC openings [5] and the use of a complexity metric to score different treatment plans will simplify the choice of plan to be used in order to deliver a robust treatment to the patient. This method will also make the patient-specific QA more effective by avoiding unnecessary time consuming phantom measurements that might limit the number of patients treated with advanced intensity modulated treatment techniques. The main idea is to develop a method that can be used to effectively distinguish treatment plans that need to be reconsidered or need extra attention during the quality assurance.

2. Material and method

An in-house developed MatLab® program (Mathworks, Natick, MA) is used to derive field specific complexity parameters. The MLC openings for each control point for both IMRT and VMAT plans are quantified as measured distances both parallel and opposed to the direction of the MLC leaves every 0.5 cm to cover the entire MLC opening (figure 1). Shorter distances are related to the complexity of irregularities and smallness of MLC openings and larger distances are related to non-complex large openings.

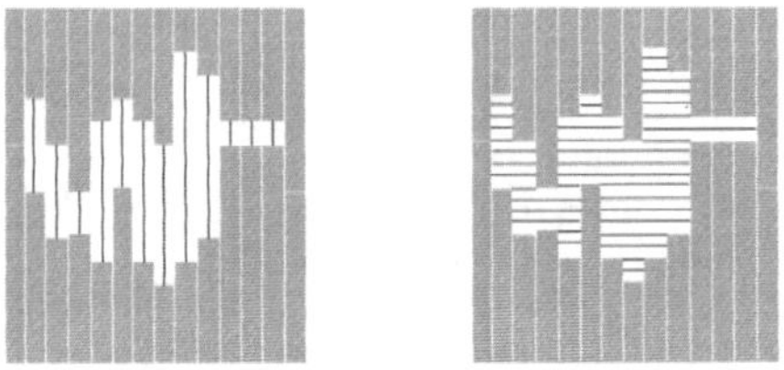

Figure 1. Measured distances both parallel and opposed to the direction of the MLC leaves.

The complexity metric is designed in two parts, $P = \overline{f(d_i)} \cdot f(a_{Eq})$, one part concerning the measured distances within the MLC opening and one part concerning the total field area. The distances d_i measured within the MLC opening and the equivalent field size a_{Eq} are penalized according to a penalty function f that is experimentally determined based on evaluated differences between planned and measured dose distributions. A control point can consist of more than one connected MLC opening and each MLC opening is treated separately in the complexity metric. The final complexity score for a control point is the mean value of all the individual MLC opening penalty values P. Static MLC openings of various complexities, each simulating one control point, are created in the Eclipse treatment planning system (Eclipse version 11, Varian Medical Systems) and grouped in seven series. The various shapes and sizes are determined based on the assumption that the TPS has difficulties to calculate a correct dose distribution for MLC openings smaller than 4x4 cm^2 [2]. Each series consists of MLC openings of similar shape, but modified to gradually become more complex regarding both the shape and size (figure 2). Series 4 and 5 have a constant area and series 3, 6 and 7 have a constant circumference.

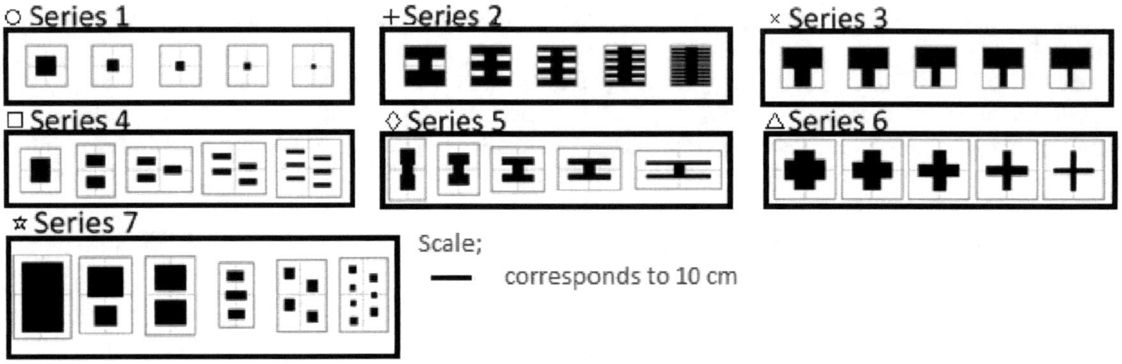

Figure 2. Static MLC openings simulating control points of various complexities.

All the 36 fields were delivered with a Clinac iX (Varian Medical Systems) linear accelerator and measured with an amorphous silicon electronic portal imaging device (EPID aSi 1000, Varian Medical Systems) in integrated image mode. The measurement session was started by calibrating the EPID for a 10 x 10 cm field shaped with the collimators. This field was repeatedly measured during the session to ensure that the measurements matched the calculated dose within 0.3 %. All measurements were carried out with the gantry at 0 degrees and a SSD of 100 cm. The number of MUs was adjusted to deliver 2 Gy centrally in the most open part of the field at a depth of 10 cm in a solid water phantom when calculated with the AAA algorithm in Eclipse. The measurements were analyzed within the Portal Dosimetry module in ARIA (Varian Medical Systems). The feature of auto align was used for 10 cm x 10 cm fields shaped with the MLC that also were repeatedly measured during the session, and the suggested correction was used to align all the static fields. The planned and measured doses were compared with different global dose difference criteria from 3-10% for each MLC opening. Pixels that received less than 10 % of the maximum dose were not considered in the evaluation. The relative number of pixels within the dose difference criterion was used as a measure of agreement between the planned and measured dose distributions. Dose difference pass rates of the static fields are used to derive the penalty function *f* in the complexity metric. The criteria of the penalty function was to give a penalty value between 0-1 and that a distance of 4-5 cm will be considered as non-complex and will be given the penalty value close to 1. The correlation between the calculated complexity score and the dose difference pass rate was evaluated with the Pearson's r-value. An initial test of the complexity metric was done for four clinical patient cases, two head and neck and two prostate IMRT treatment plans, both based on dynamic MLC with sliding window. All plans have been approved and used for treatment at Sahlgrenska University Hospital. These plans were measured field by field with EPID and evaluated in absolute dose mode using a 10 % dose cut off and the fields were auto aligned. The calculated complexity score for each field are compared to the global gamma (3%/2mm) pass rates, which is a commonly used evaluation method for patient specific QA.

3. Results and discussion

A penalty function *f* that showed a good correlation between calculated complexity scores and the dose differences pass rates for the static fields was an inverse exponential function, $f(x) = 1 - e^{-x}$. The correlation between the calculated complexity scores and the dose difference pass rates for the static fields are shown as a scatter plot in figure 3a.

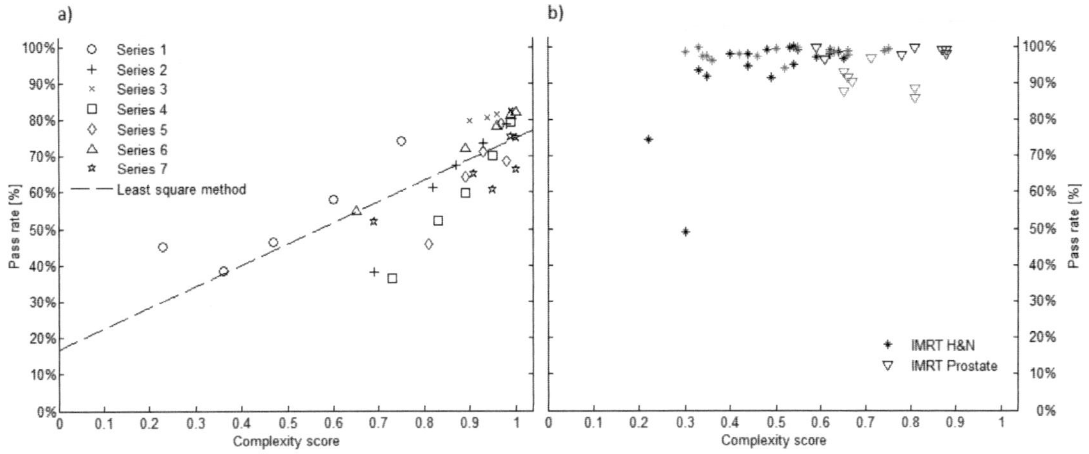

Figure 3. Scatter plots of the correlation between the calculated complexity scores and (a) the 5 % dose difference pass rates for the static fields and (b) the global gamma pass rates (3%/2mm) for the IMRT treatment plans.

The Pearson's r-value, which is a measure of the linear correlation, is 0.77 for the correlation of the static fields. Plans have been made to further develop the complexity metric in the near future.

The global gamma pass rates and the complexity score for the clinical IMRT treatment plans are shown in figure 3b. The complexity scores are generally lower for the H&N cases compared to the prostate cases as expected due to the more complex shape of the MLC openings in H&N treatment plans. Most of the fields of the clinical IMRT treatment plans that have been accepted and used for treatment have high gamma pass rates as expected. The two fields which have the lowest complexity scores are also the ones with the lowest pass rate and makes this initial test on clinical treatment plans promising. However, the complexity metric needs to be further validated to ensure that unacceptable treatment plans will be distinguished.

4. Conclusion

A complexity metric that is based on shape and size of MLC openings is developed. Initial results show that the extent of dose differences between planned and measured dose distributions correlates to calculated complexity scores. The design of the complexity metric needs further developments for better agreement between complexity scores and dose differences and further validation for clinical treatment plans.

5. References

[1] Yu C X *et al* 1995 *Phys. Med. Biol.* **40** 1435-49
[2] Das I J *et al* 2008 *Med. Phys.* **35** 206-215
[3] Nilsson J *et al* 2013 *J. Phys.: Conf. Ser.* **444** 012048
[4] Nelms B E *et al* 2011 *Med. Phys.* **38** 1037-44
[5] Petersson K 2014 Evaluation of plan quality and deliverability for automatically generated fallback radiotherapy treatment plans in PhD thesis: Optimising the clinical use of tomotherapy, Dept. of Medical Radiation Physics, Lund University

Journal of Physics: Conference Series **573** (2015) 012017 doi:10.1088/1742-6596/573/1/012017

Validation of high-resolution 3D patient QA for proton PBS and IMPT using laser CT of improved polymer gel dosimeters

A Cardin[1], X Ding[1], A Kassaee[1], L Lin[1], M J Maryanski[2] and S Avery[1]
[1]Radiation Oncology, University of Pennsylvania, Philadelphia, PA 19104, USA
[2]MGS Research, Inc., Madison, CT 06443, USA

E-mail: andrewj.cardin@gmail.com

Abstract. Laser CT scanning of LET-independent BANG3-Pro2® polymer gel dosimeters has recently shown potential in proton dosimetry. However, raw materials' impurities impart some variability. This study aimed to validate a new method of compensating for this variability, and to validate the suitability of the improved dosimeter for patient-specific QA in pencil beam scanning (PBS) and IMPT. Six modifications of the BANG3-Pro2® gel dosimeter were analysed for their sensitivity to proton dose and to LET. One formulation was selected for a clinical QA feasibility study, in which one composite IMPT plan, two single-field IMPT plans, and one SFUD plan were delivered to identical gel phantoms. New commercial VOLQA™ software (beta version) was used for data analysis. Both validations were successful.

1. Introduction

The need for a high resolution 3D dosimetry system has been reported since 2004 by The Council on Ionizing Radiation Measurements and Standards (CIRMS) [1]. This need is especially urgent in proton therapy pencil beam scanning (PBS) and intensity modulated proton therapy (IMPT), which deliver extremely complex dose distributions. Current quality assurance techniques are typically performed by multiple 2D measurements [2]. A high resolution 3D dosimeter, however, would improve proton therapy QA by providing more information while requiring much less beam-time.

1.1. BANG® Polymer Gel Dosimeters

A promising candidate commercial system for high resolution 3D dosimetry is the line of BANG® polymer gel dosimeters and OCTOPUS™ laser CT scanners made by MGS Research, Inc. (MGS, Madison, CT) [3-5]. This study intends to investigate the potential of this system as an effective dose verification and QA tool for the proton pencil beam scanning (PBS) and IMPT modalities.

1.2. LET Response

The BANG3-Pro2® polymer gel dosimeter formulation has removed a previously reported LET dependence [4]. A more recent investigation has demonstrated the feasibility of producing an LET-enhanced response in the BANG3-Pro2® which can potentially be used for mapping LET distributions in 3D [6].

As trace impurities in raw materials from which the gel is made can cause dose response variability, a new proprietary method was developed by MGS to control and compensate for this variability. Validation of this method was one of the objectives of the present research.

2. Materials and Methods

2.1. Materials and Equipment

All gel dosimeters (MGS's BANG3-Pro2® type, density 1.08 g cm^{-3}, proton stopping power ratio 1.085 [4]) were sealed inside identical thin-wall, photoprotective-coated Pyrex glass spheres, 166 mm in diameter. Four titanium wire markers were embedded in each gel for 3D image registration. A PBS nozzle developed by IBA was used for proton irradiation at the Roberts Proton Therapy Center, with nominal energies between 103 and 140 MeV. Gels were scanned by a SOMATOM® Sensation Open CT scanner, with 2 mm slice thickness, for planning with Varian's Eclipse™ TPS. During CT acquisition and irradiation, each gel was mounted in a special holder with the neck positioned horizontally. Gels designated for proton irradiation were always scanned and irradiated in a water tank as it was deemed necessary to have the protons incident upon a flat surface. A Varian Clinac was used to irradiate gels with 6 MeV photons. A new model of the commercial laser CT scanner, OCTOPUS-R&P™ (for Radiosurgery and Protons) was used for scanning the exposed gels with 1 mm slice thickness, 120 slices, and 1 mm in-plane pixel size (35 minutes scan time per gel). Calibration and 3D QA analysis was conducted using the beta version of the VOLQA™ software developed by MGS.

2.2. Method – Validation of Techniques for Stabilizing the LET Effect

Six pairs of six different formulations of the BANG3-Pro2® polymer gel dosimeter were made, and shipped to the University of Pennsylvania for irradiation. Two simple calibration plans were devised; a photon plan consisting of two lateral superior oblique fields and a proton plan consisting of a single vertex PBS field. The proton PBS plan specified four targets with planned mean doses between 30% and 106% of the prescribed dose. These two plans were delivered to all six pairs of gels, with formulation-specific scaling applied so that the prescribed dose was either 300 cGy or 500 cGy. One objective was to find the upper dose limit above which the LET effect would produce a departure of the proton calibration curve from the photon calibration measured for each gel formulation. Another was to examine the effect of the modifications on the dosimeter's sensitivity and dynamic range.

Each pair of exposed gels was shipped back to MGS for laser CT scanning and analysis, along with the DICOM RT-Dose and simulator CT scan files. A volume look-up table calibration method was used in which raw values of all voxels within the centrally positioned inner spherical volume of up to 60 mm radius are sorted in ascending order, for both OD/cm and RT-dose data. The raw data are plotted and fitted with a polynomial. In the LET-stabilization experiment, no polynomial fitting was employed, as quite complex departures from linearity were expected at the high-dose end of the proton calibration data where LET effects were likely to appear.

2.3. Method – Clinical Validation of Patient-Specific PBS and IMPT QA

Four gels were used for patient-specific QA validation, shipped and scanned in the same manner as described above. Two proton PBS plans-a two-field IMPT plan (a composite plan and individual fields) and one single field uniform dose (SFUD) plan-were evaluated.

OD/cm was calibrated to dose using the volume look-up table method described above, but the software can optionally import and apply a previously saved calibration polynomial. For QA analysis, the software generates, among others, 2D and 3D isodose overlays, orthogonal profiles, and 3D maps of distance to agreement, dose difference, and 3D delta index-an index equivalent to the ubiquitously used gamma index but more accurate and much faster (Maryanski MJ *et al* - unpublished result)-for which passing rates are computed at user-defined thresholds. The software also calculates the centers of mass of the planned and measured dose clouds, and recommends X/Y/Z shifts accordingly.

3. Results and Discussion

Raw data from the LET stabilization study are shown in figure 1(a). Q1 and Q2 series gels show the onset of the LET effect (i.e. significant and rapid departure of the dose response from the corresponding photon calibration curve) at about 400 cGy and 270 cGy respectively. The second formulations in each series shown in the figure (i.e. Q1d and Q2b) exhibit significantly greater sensitivity than the first (Q1c and Q2a). These combined results confirm the efficacy of the new method of compensating for the effects of raw materials' impurities. Figure 1(b) demonstrates excellent agreement for the PDD along an SOBP for one selected formulation.

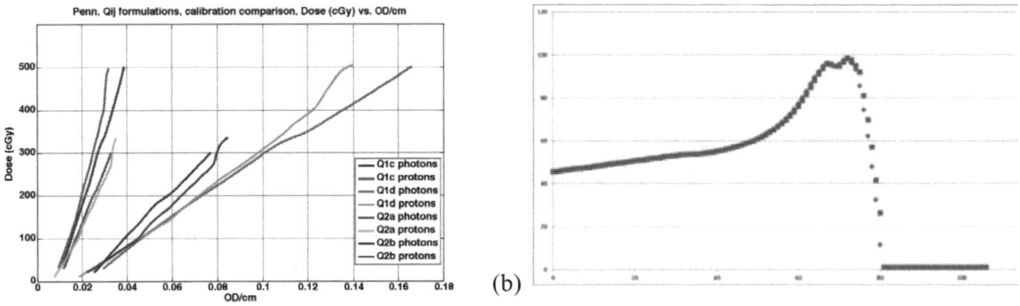

(a) (b)

Figure 1. (a) Calibration curves of select formulations. (b) Example proton SOBP PDD overlay.

Some proton calibration curves show inflections at lower doses (normally present in regions located at the periphery of the gel volume), where the slope of the curve rapidly changes. This is likely caused by insufficient temperature control at proton dose delivery. As stated, gels were submerged in a water tank immediately before proton irradiation. A temperature difference between the cold tap water and the gels (stored at room temperature) creates a temperature gradient within the gels, with outer layers being colder than the core. A colder gel responds more strongly-a dependency that is noticeable over a few degrees Celsius-hence low-dose irregularities in the proton calibration curves.

A smaller but similar effect appears in some photon calibration curves, where the core of the gel appears to have been slightly colder than the rest of the volume, as a result of incomplete thermal equilibration before exposure to photons (gels were shipped cold in insulated containers).

The results of the patient QA validation are shown in table 1. There was excellent agreement between the planned and measured dose distributions in the single IMPT fields and the SFUD field. A less than satisfactory agreement was seen in the composite IMPT gel, whose calibration curve indicated a significant temperature gradient inside the gel. For all other gels, calibration accuracy was found to be better than 2% (see figure 2a). Software-recommended shifts of the dose distributions were less than 4 mm. 3D isodose comparisons are shown in figure 3.

4. Conclusions

This study validates the new technique of compensating for raw-material impurity effects in BANG3-Pro2® polymer gel dosimeters and demonstrates that they can be used for accurate, high resolution 3D proton dosimetry and patient-specific QA. Mailed QA service provided by the manufacturer is a practical option when on-site laser CT scanning resources are unavailable. Temperature uniformity within the gel during dose delivery is critical for the accuracy of results.

Table 1. Results of clinical validation study

	IMPT Comp	IMPT Field 1	IMPT Field 2	SFUD
# Voxels Analyzed	219,499	41,370	163,392	259,255
Max Dose, cGy	184	102	165	214
Voxels with DTA < 3mm, %	94.2	99.7	97.2	94.8
3D Delta Index 3%/3mm Pass Rate, %	81.5	99.0	94.9	93.5
dX/dY/dZ Shifts, mm	0/3.9/2	-1/2.5/0	1/3.9/2	-1/0/1

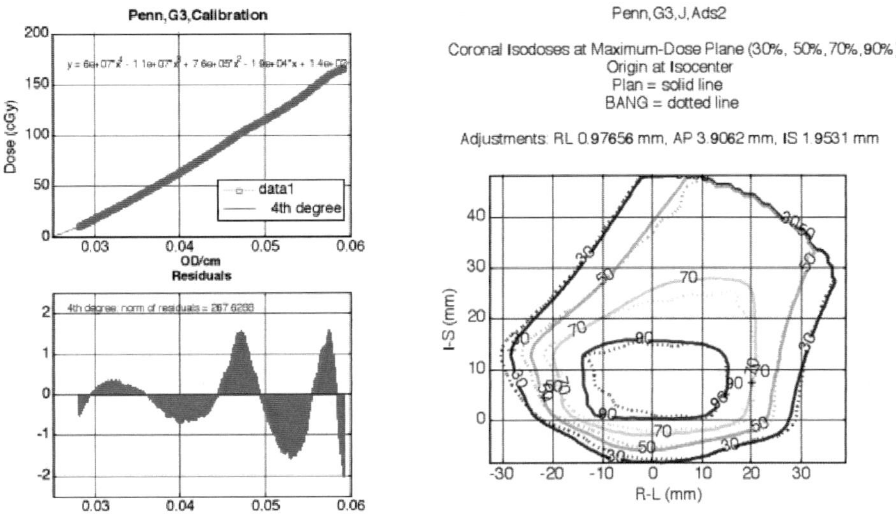

Figure 2. (Left) Auto-calibration for a single IMPT field, demonstrating a better than 2% accuracy, and (right) an example 2D isodose overlay.

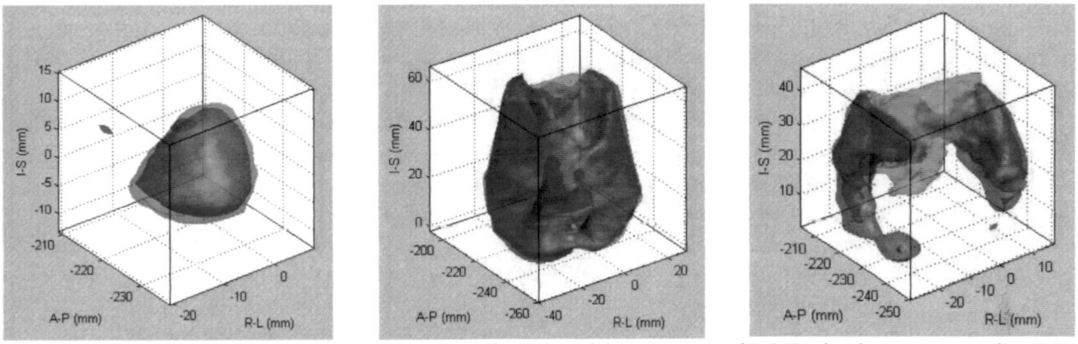

Figure 3. 3D isodose overlay comparisons. Plan is red, gel is green. (Left) 90% isodoses, composite IMPT, (middle) 70% isodoses, SFUD, (right) 90% isodoses, SFUD.

5. Conflicts of Interest

One co-author (M J Maryanski) is the founder and owner of MGS Research Inc. All other co-authors report no actual or potential conflicts of interest.

6. References

[1] 5th Report on Needs in Ionizing Radiation 2011. Prepared by the Science and Technology Committee of the Council on Ionizing Radiation Measurements and Standards
[2] Zhu XR *et al* 2011 *Int. J. Radiat. Oncol. Biol. Phys.* **81** 552-9
[3] Wuu C *et al* 2006 *Med. Phys.* **33** 1412-9
[4] Zeidan O A *et al* 2010 *Med. Phys.* **37** 2145-2152
[5] Low D *et al* 2011 *Med. Phys.* **38** 1313-38
[6] Lopatiuk-Tirpak O *et al* 2012 *Technology in Cancer Research and Treatment* **11** 441-5

Multicentre knowledge sharing and planning/dose audit on flattening filter free beams for SBRT lung

C R Hansen[1,9], J R Sykes[2,3,9], J Barber[3,9], K West[3], R Bromley[4], K Szymura[4], S Fisher[5], J Sim[5], M Bailey[6], D Chrystal[6], S Deshpande[7], I Franji[7], T B Nielsen[1], C Brink[1,8] and D I Thwaites[9]

[1]Laboratory of Radiation Physics, Odense University Hospital, Denmark
[2]St James's Institute of Oncology, Leeds, UK
[3]Department of Radiation Oncology, Nepean Cancer Care Centre, Sydney, Australia
[4]Northern Sydney Cancer Centre, Royal North Shore Hospital, Sydney, Australia
[5]Department of Radiation Oncology, Prince of Wales Hospital, Sydney, Australia
[6]Illawarra Cancer Care Centre, Wollongong, Australia
[7]Liverpool & Macarthur Cancer Therapy Centres, Sydney, Australia
[8]Institute of Clinical Research, University of Southern Denmark, Odense, Denmark
[9]Institute of Medical Physics, School of Physics, University of Sydney, Sydney, Australia

E-mail: christian.roenn@rsyd.dk

Abstract. When implementing new technology into clinical practice, there will always be a need for large knowledge gain. The aim of this study was twofold, (I) audit the treatment planning and dose delivery of Flattening Filter Free (FFF) beam technology for Stereotactic Body Radiation Therapy (SBRT) of lung tumours across a range of treatment planning systems compared to the conventional Flatting Filter (FF) beams, (II) investigate how sharing knowledge between centres of different experience can improve plan quality. All vendor/treatment planning system (TPS) combinations investigated were able to produce acceptable treatment plans and the dose accuracy was clinically acceptable for all plans. By sharing knowledge between the different centres, the minor protocol violations (MPV) could be significantly reduced, from an average of 1.9 MPV per plan to 0.6 after such sharing of treatment planning knowledge. In particular, for the centres with less SBRT and/or volumetric-modulated arc therapy (VMAT) experience the MPV average per plan improved. All vendor/TPS combinations were also able to successfully deliver the FF and FFF SBRT VMAT plans. The plan quality and dose accuracy were found to be clinically acceptable.

1. Introduction

Learning and understanding a new technology and then implementing it for clinical use is a core component of radiotherapy (RT). The rapid technological advances in RT over the last decade have changed cancer treatments such that what was considered cutting edge a few years ago is now routine treatment in most centres [1]. Currently flattening filter free (FFF) beams are becoming more widely available and are being increasingly used for Stereotactic Body Radiation Therapy (SBRT) treatments for lung cancer, often coupled with Volumetric-Modulated Arc Therapy (VMAT) delivery methods. FFF beams have a range of characteristics that have the potential to improve treatment plans in these situations, including higher dose rate, reduced lateral changes in beam hardening, reduced leakage and less out-of-field dose.

When implementing a new technology, many centres have collaborating institutions they can contact for good advice and published literature that can help and guide them to a safe implementation. This is often done by comparing the new technology to a well established one. For FFF technology, the natural comparison would be the conventional Flattening Filter (FF) technology. This has been done for FFF treatments for one specific vendor and treatment planning system (TPS) combination in several publications [2] and other vendor/TPS combinations are also now beginning to appear in the literature [3]. The next implementation step is often an internal and/or external dose audit; to validate the dose accuracy as sufficiently high for clinical use [4]. This has not been reported often for VMAT [5] and not at all for FFF.

The two purposes of the study were:

I) to audit the treatment planning quality and dose accuracy of SBRT VMAT treatments for lung tumours, for both FFF and FF beams, where the plans were created on different planning systems for delivery on different accelerators; and hence to assess the impact of different linacs, beam energies, TPSs and department experience for SBRT FFF VMAT lung treatments

II) to evaluate if sharing knowledge across the different centres with different TPS/linac combinations would improve plan quality.

2. Method and material

Three patients with a solitary lung tumour were selected, where one patient had a tumour close to the ribs, one close to the bronchial tree and one in the free part of the lung according to RTOG0915 [6]. The prescription doses were 48Gy/4fr., 50Gy/5fr., 54Gy/3fr. and the organ at risk (OAR) dose constraints were as specified by a combined protocol of RTOG0915 and Leeds Teaching Hospitals NHS Trust for an Australian SBRT phase II trial [6].

The three patient's CT-datasets and corresponding DICOM RT structures of targets and OAR were provided to six different cancer treatment centres for SBRT planning using FF and FFF VMAT beams. The centres were not allowed to change the provide structures, however they were allowed to add additional optimisation structures. All evaluation was performed on the provided structures. The different centre's linac/TPS combinations were: Elekta Agility/Monaco, Varian TrueBeam/Eclipse, Varian TrueBeam/Pinnacle, TomoTherapy/HiArt (TomoTherapy only with FFF plans) and two centres with Elekta Agility/Pinnacle. VMAT was used on all conventional linacs for all plans with a 200 degree arc avoiding the contra lateral lung. Mainly 6 MV FF and 6 MV FFF beams were used, however one centre used 10 MV FFF. The centres were encouraged to create the treatment plans in collaboration between a radiation therapist (dosimetrist) and a medical physicist.

The plan metrics evaluated were GTV and PTV mean doses, volume of PTV covered by 100% of the prescribed dose, mean and max organ at risk doses, integral dose (patient dose outside PTV), target conformity ($CI_{PTV}=V_{D100\%}/V_{PTV}$), and dose spillage conformity ($CI_{spillage}=V_{D50\%}/V_{PTV}$). Following the SBRT protocol, target coverage, organ sparing and conformity were evaluated as according to the protocol (per protocol). When the planning objectives were not met they were classified as minor or major protocol violations as defined in RTOG0915 [6].

To reduce the SBRT and/or VMAT experience level differences between centres, each centre was asked to plan the three patients according to the protocol and to upload the results to the principal investigator. A workshop was then arranged where each plan was discussed and debated by the radiation therapist and a medical physicist from each centre. The centres were given the opportunity to replan the patients before final submission. Three centres had clinical experience with both VMAT and SBRT and were classed as experienced. Three centres had no or limited clinical experience with VMAT and SBRT on the specific linac and were classed as having limited experienced.

The treatment plans were delivered on a specific linac in each centre and dose accuracy was measured on the Sun Nuclear ArcCHECK phantom. The clinically acceptable level was set to 95% gamma pass rate for 3% (of max dose) and 3mm.

Statistical significance was tested with Wilcoxon signed-rank non-parametric test, significant level was set to $p<0.05$.

3. Results

For all vendor combinations, differences in plan quality between FF and FFF plans were small and likely to be of no clinical relevance when evaluating target doses and OARs (table 1). After replanning, the dose spread was reduced for target mean doses and $CI_{spillage}$. The most significant difference between centres was the number of minor protocol violations (MPV) per plan, which seemed to correlate to the amount of clinical SBRT VMAT experience. The three most experienced centres had an average of 0.6 MPV per plan for both FF and FFF plans whereas the two centres with limited experience had 3.7 MPV per plan. The average MPV was significantly improved (p=0.002) after the sharing knowledge workshop and replanning.

Table 1. Bold number indicate statistically significant improvements after replanning

	Initial planning				Replanning			
	FF		FFF		FF		FFF	
Target dose	Mean	SD	Mean	SD	Mean	SD	Mean	SD
GTV mean dose [%]	114.7	8.9	115.0	7.4	119.2	7.3	118.4	6.3
PTV mean dose [%]	110.1	5.0	110.6	4.5	112.3	3.8	111.9	3.8
PTV $V_{100\%}$ [%]	98.6	0.6	98.9	0.8	98.8	0.7	99.0	0.6
OAR dose								
Cord max dose [Gy]	10.5	4.1	10.5	3.6	10.5	3.9	10.3	4.3
$V_{20\%}$ Lung-PTV [%]	5.1	2.0	5.3	2.0	4.9	1.8	4.8	1.8
Conformity								
Patient-PTV mean dose [Gy]	2.02	0.38	2.03	0.45	1.94	0.30	1.88	0.34
CI_{PTV} $V_{D100\%}/V_{PTV}$	1.14	0.10	1.18	0.18	1.10	0.06	1.11	0.06
$CI_{spillage}$ $V_{D50\%}/V_{PTV}$	6.14	1.74	6.26	1.84	5.58	0.70	5.51	0.71
Minor protocol violation pr plan								
MPV total group	1.7	2.0	2.1	2.0	0.5	0.7	**0.7**	1.0
MPV high experience	0.4	0.7	0.8	1.0	0.4	0.7	0.6	1.0
MPV limited experience	3.7	1.8	3.6	1.7	**0.5**	0.8	**0.8**	1.2

The dose measurements on the ArcCheck phantom were all but one within the pre-set clinically acceptable tolerance, with a 98.0±2.6% and 96.5±1.9% pass rate for FF and FFF plans, respectively.

4. Discussion

The VMAT optimisation algorithm is very different in the four TPSs [7-10], however the plan quality across the vendor/TPS combinations differed little on these three patients both for FF and FFF plans. Before the knowledge sharing workshop, the plans from the centres with limited experience showed signs of less optimal VMAT delivery. The number of MPVs were significantly reduced after the replanning; and also the general VMAT optimisation appeared to have improved indicating sub-optimal user inputs to the planning process before the knowledge sharing workshop.

At the knowledge sharing workshop, the joint discussion crossed several topics. For VMAT, a non-zero collimator angle is highly advisable, one TPS created better plans with collimator angles around 10 degrees, whereas the two other TPSs worked well around 25 to 35 degrees. Normalisation of the final dose to fit specific objectives, specifically target doses, were used in all TPSs; however one centre avoided this by finalising the calculation with 15 or less optimisation iterations. SBRT protocols often do not specify a maximum dose to the target thereby allowing for target over-dosage. This can reduce dose to the OAR, since demanding the homogenous target dose forces the TPS to modulate the beam more. Cancers in the lung have a range of OARs near the target and even though the treatment plan might meet all protocol objectives, there will still be benefits for the patients if it is possible to reduce OAR doses even further as long as it doesn't compromise target dose objectives. It is therefore important to make OAR prioritisation clear for the planner and a good consensus is often best reached by a multi disciplinary team discussion.

5. Conclusion

FFF can be used for SBRT planning, independent of the specific combination of linear accelerator vendor and treatment planning system. Centres with less experience in SBRT and VMAT did less well in the initial analysis but reduced MPVs occurred after knowledge sharing and replanning. The dose metric accuracy was only tested on the replanned treatment plans and was found to be good. All vendor/TPS combinations can produce and deliver FFF SBRT VMAT treatments to acceptable clinical requirements.

6. References

[1] Baldock C *et al* 2010 *Phys. Med. Biol.* **55** R1-63
[2] Lang D *et al* 2013 *Radioter. Oncol.* **106** 255-9
[3] Boda-Heggemann J *et al* 2013 *Radiother. Oncol.* **109** 505-9
[4] Miles E *et al* 2012 *Clin. Oncol.* **24** 710-2
[5] Hussein M *et al* 2013 *Radiother. and Oncol.* **108** 78–85
[6] Videtic G M M *et al* RTOG 0915: A Randomized Phase II Study Comparing 2 Stereotactic Body Radiation Therapy (SBRT) Schedules for Medically Inoperable Patients with Stage I Peripheral Non-Small Cell Lung Cancer. http://www.rtog.org/members/protocols/0915/0915.pdf
[7] Bortfeld T *et al* 1993 *Med. Phys.* **20** 311-8
[8] Otto K 2008 *Med. Phys.* **35** 310-7
[9] Cameron C 2005 *Phys. Med. Biol.* **50** 4317-36
[10] Sterzing F *et al* 2010 *Int. J. Radiation Oncology Biol. Phys.* **76** 1266-73

Delivery validation of VMAT stereotactic ablative body radiotherapy at commissioning

T Olding[1], K M Alexander[2], C Jechel[2], A T Nasr[3] and C Joshi[1,2]
[1]Cancer Centre of South Eastern Ontario at Kingston General Hospital, 25 King Street West, Kingston, ON, Canada, K7L5P9
[2]Department of Physics, Queen's University, Kingston, Ontario, Canada, K7L3N6
[3]Department of Chemical Engineering, Queen's University, Kingston, Ontario, Canada, K7L3N6

E-mail: tim.olding@krcc.on.ca

Abstract. Dosimetric validation of two volumetric modulated arc therapy (VMAT) stereotactic ablative radiotherapy (SABR) plans was completed as part of the commissioning process of this technique in our clinic. Static and dynamic ion chamber, EBT3 film and leuco crystal violet (LCV) micelle gel measurements were acquired using a motion phantom with appropriate inserts for each dosimeter. The results show good agreement between measured and calculated plan dose.

1. Introduction

Previous end-to-end delivery validation measurements of the non-coplanar 3D conformal lung SABR technique used at our centre have shown good agreement between Varian Eclipse plan dose and ion chamber, EBT3, EPID and Fricke-xylenol orange-gelatin (FXG) gel measurements [1]. This technique is employed as a treatment option for early stage (Stage I, T1/2, N0, < 5 cm), medically inoperable, non-small cell lung cancer. A process is now underway at our centre to transition to a VMAT SABR technique which offers the potential for reduced planning and delivery time. The purpose of this work is then to validate the VMAT SABR lung treatment delivery in a similar manner as before, using different dosimetric modalities. We have included a relatively new formulation of LCV gel dosimeter in this work which has an appropriate level of optical absorption-to-dose response for the high single fraction doses used in lung SABR, and minimal degradation of the optical response due to diffusion [2].

2. Materials and Methods

2.1. SABR Delivery Validation

The dosimetry tools used for SABR delivery validation in this study were: a) point single ion chamber (PR-05P 0.07 cc, Capintec, Ramsey, NJ), b) 2D Gafchromic EBT3 film (ISP, Wayne, NJ), c) 3D leuco crystal violet dye-micelle based gel dosimeter with optical CT readout [3] and d) 3D ArcCheck (Sun Nuclear, Melbourne, FL). Static and dynamic 4D computed tomography (4DCT) scans were acquired of a Quasar motion phantom (Modus Medical Devices Inc, London, ON, figure 1a) with a cedar insert containing a 3 cm diameter acrylic sphere 'tumour' and a hole for insertion of the ion chamber. A GE Lightspeed RT16 scanner (GE Healthcare, Maple Grove, MN) was used for CT data acquisition, with the addition of a Varian RPM motion tracking system for the dynamic scan. A similar cedar insert was used for the EBT3 film (figure 1b). The scans were then repeated with a custom gel insert in the Quasar phantom (figure 1c). A sinusoidal breathing cycle of 15 breaths per minute (BPM) and 20 mm amplitude peak-to-peak was used for all 4DCT scans.

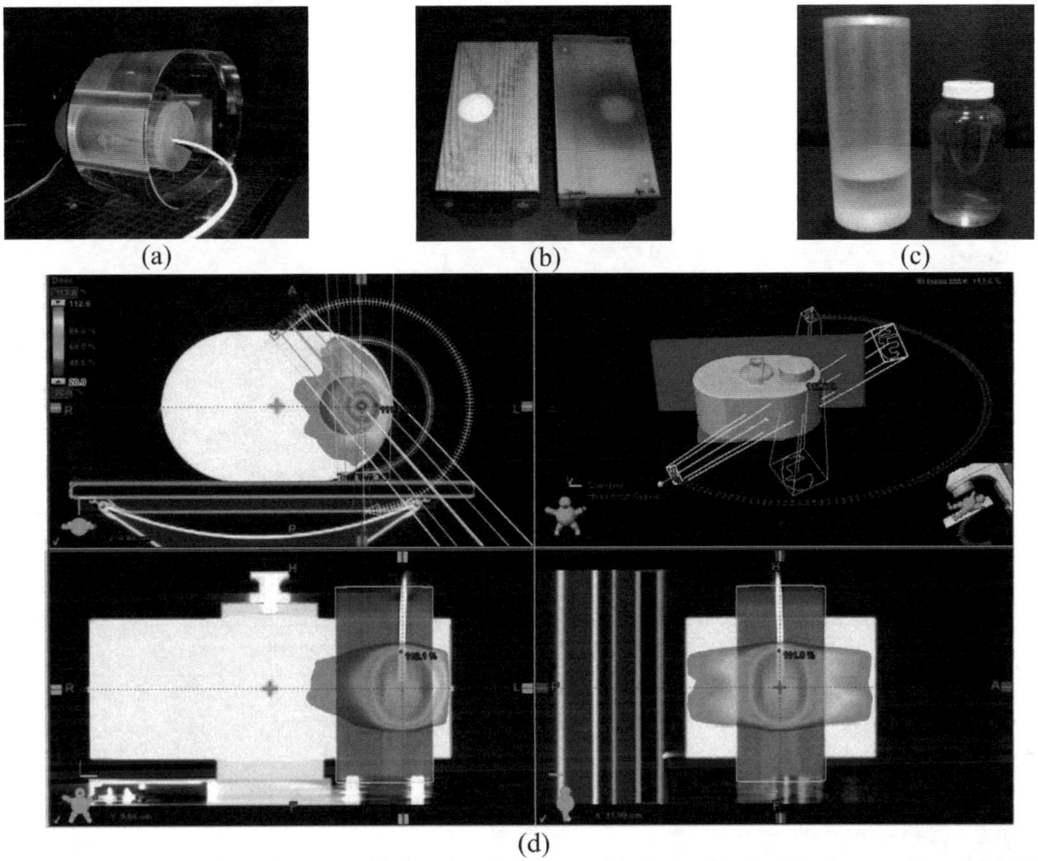

Figure 1. (a) Quasar motion phantom with ion chamber insert, (b) photo of cedar film insert with overlaid EBT3 film from dynamic delivery, (c) gel insert and dynamic delivery LCV gel, and (d) Eclipse screen capture of the high monitor unit dynamic ion chamber plan used in this study.

For VMAT SABR planning, dual coplanar clockwise/counterclockwise 6MV partial arcs were employed, extending from just outside the approximated edge of the contralateral lung anteriorly to 180 degrees posteriorly (figure 1d). The collimator angle on the two arcs was set to ±30 degrees to minimize discrepancies arising from the interplay effect[4]. High monitor unit (MU) and low monitor unit RapidArc^{TM} VMAT plans were optimized and calculated in Eclipse^{TM} v.10 (Varian Medical Systems, Palo Alto, CA) on the 4DCT 1.25mm slice thickness average scan of the ion chamber-Quasar phantom geometry. These plans were then re-calculated (but not re-optimized) on the rest of the ion chamber, film and gel static and dynamic CT phantom datasets. The high MU dynamic plan was optimized without any constraints on the total number of plan MUs. As a result, the high MU dynamic (HMUD) and static (HMUS) plans have reasonably complex multi-leaf collimator (MLC) leaf motions/apertures in the delivery. The low MU dynamic plan had a limit placed on the total number of MUs in the optimization stage of treatment planning; hence it has more open MLC apertures throughout the arc. A corresponding 30% decrease in the number of MUs was achieved in the low MU dynamic (LMUD) and static (LMUS) plans. The ion chamber and gel plans were normalized to the clinical single fraction prescription dose of 12 Gy, while the film plans were re-normalized (i.e. monitor units scaled down) to a prescription dose of 2 Gy to accommodate the dose

Journal of Physics: Conference Series **573** (2015) 012019 doi:10.1088/1742-6596/573/1/012019

range (up to 10 Gy) of EBT3 film. A screen capture of the high MU dynamic delivery ion chamber plan is shown in figure 1d.

All treatment plans were delivered on a Varian Trilogy 2100iX linear accelerator (Varian Medical Systems, Palo Alto, CA), with Advanced OBI™ imaging cone beam CT (CBCT) phantom setup verification. Each dose measurement was compared against calculated Eclipse treatment planning system (TPS) dose using software appropriate for that QA tool. EBT3 film measurements were acquired using an in-house built CCD-lightbox film scanner and analyzed in MATLAB (Mathworks, Newark, NJ). The reconstructed, calibrated gel dose data from the Vista™ optical CT scanner (Modus Medical Devices Inc., London, ON) was compared to Eclipse dose in 3D Slicer [5] (www.slicer.org).

3. Results

3.1. Dosimeter-Plan Dose Comparison

Dosimetric results for the dosimeters used in this work are summarized in Table 1. The near isocentre ion chamber point measurements are in good agreement with Eclipse. 2D Film and 3D gel results are consistent with this, although the point film measurements were lower than expected. Most of the dynamic film and gel disagreement (figures 2 and 3) is near the edges of the high dose region in the superior-inferior direction along the axis of phantom motion.

Table 1. Summary: measured versus Eclipse doses.

Dosimeter	Plan	Prescription (Gy)	Point Dose Difference (%)	% Gamma ≤1 (3%/3mm)
Ion Chamber	HMUS	12	-1.1	-
Ion Chamber	HMUD	12	-1.6	-
Ion Chamber	LMUS	12	-0.8	-
Ion Chamber	LMUD	12	-2.0	-
EBT3 Film	HMUS	2	-3.7	94.6
EBT3 Film	HMUD	2	-4.1	69.1
LCV Gel	HMUD	12	-1.3	86.1
ArcCheck	HMUS	12	-	100.0
ArcCheck	LMUS	12	-	99.4

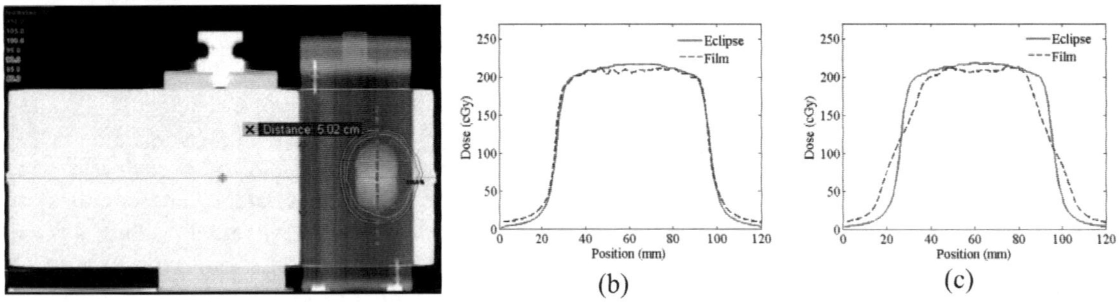

(a) (b) (c)

Figure 2. (a) Eclipse screen capture of the HMUD film plan with red dashed line along axis of motion showing location of profile measurements for the (b) static and (c) dynamic film deliveries.

4. Discussion & Conclusions

With reference to table 1, the slight increase in the point dose disagreement for the dynamic deliveries versus the static measurements is likely due to interplay effects [4, 6] which are not accounted for in Eclipse. The fact that all point dose measurements were less than that calculated by the TPS can be can also be partly attributed to planning system limitations in modelling dual electronic disequilibrium conditions (i.e. small field size and small target in inhomogeneous media). The disagreement near the edges of the high dose region in the dynamic delivery film and gel measurements is well-explained by motion during delivery, which is not accounted for by the TPS. The disagreement in the centre of the film for both deliveries (figure 2) leading to larger than expected central point dose differences (table 1) and gamma values (figure 3a and b) is believed to be due to multiple reflections off the camera lens in the custom CCD-lightbox scanner. This is expected to be resolved through the addition of an anti-reflection filter in front of the CCD camera. The ArcCheck results validate the delivery of both static plans, providing additional support to the assertion that the increase in point dose discrepancy of the dynamic plans near isocentre is largely related to interplay effects. Finally, gel dosimetry uniquely provides a 3D validation of the dynamic delivery. Good agreement is observed between plan and measurement over the full range of solid acrylic tumour motion (i.e. internal target volume (ITV), see figure 3c), suggesting that the 0.5 cm planning target volume (PTV) margin is adequate for the delivery.

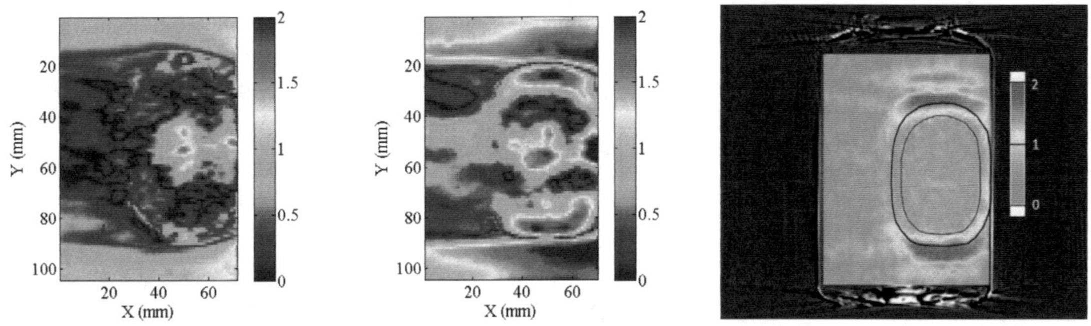

(a) (b) (c)

Figure 3. 3%, 3mm gamma comparison for the (left) static 2 Gy film delivery, (centre) dynamic 2 Gy film delivery, and (right) the central 9 cm high, 7 cm diameter cylindrical gel volume of the 12 Gy dynamic gel dose delivery, with ITV and PTV contours shown in red and blue respectively.

The 2 cm peak-to-peak motion chosen in this study represents an extreme clinical scenario. Tumour motion for most patients is typically less than 1 cm, and not entirely directed along the superior-inferior axis. With this in mind, it appears that both the LMUD and HMUD planning approaches are acceptable from the standpoint of delivery validation. Overall, good agreement is observed between the measured delivery and calculated Eclipse treatment planning system dose.

5. Acknowledgements

Research funding for this work has been provided by the Canadian Institutes of Health Research (CIHR) and the Cancer Centre of Southeastern Ontario.

6. References

[1] Olding T *et al* 2013 *J. Phys.: Conf. Ser.* **444** 012073
[2] Nasr A T *et al* 2014. Proceedings of 8th International Conference on 3D Radiation Dosimetry (IC3DDose) 4-7 September 2014, Ystad, Sweden
[3] Olding T *et al* 2010 *Phys. Med. Biol.* **56** 1259-79
[4] Ong C *et al* 2011 *Int. J. Radiation Oncology Biol. Phys.*, **79** 305-11
[5] Pinter C *et al* 2012 *Med. Phys.* **39** 6332-8
[6] Ceberg S *et al* 2013 *J. Phys.: Conf. Ser.* **444** 012098

3D dose verification with polymer gel detectors of brain-spine match line for proton pencil beam cranio-spinal: A preliminary study

S Avery[1], A Cardin[1], L Lin[1], M Kirk[1], A Kassaee[1] and M J Maryanski[2]
[1]Department of Radiation Oncology, University of Pennsylvania Medical Center, Philadelphia, PA, USA
[2]MGS Research, Inc., Madison, CT 06443, USA

E-mail: avery@uphs.upenn.edu

Abstract. This paper is intended as a preliminary study to demonstrate the quality assurance benefits from polymer gel detectors for proton pencil beam cranio-spinal treatments. A stable gel type was selected for protons to suppress the LET dependence at the end of the Bragg peak. The depth dose distributions in the gels were examined with regard of its dose dependences and compared to baseline measurements. The preliminary experimental results indicate polymer gel detectors may be able to verify dose in three dimensions along match line for proton therapy treatments.

1. Introduction

The purpose of our study is to examine the dose due to match lines using polymer gel detectors for quality assurance of proton pencil beam cranio-spinal treatment plans. An advantage of gels is that you have the ability to verify dose in three dimensions. The uses of polymer gel detectors as a tool for radiation dosimetry have been well investigated in the literature [1]. The Council on Ionizing Radiation Measurements and Standards reported in October 2011 that there's an urgent need for the introduction of high-resolution 3D methods of dosimetry, quality assurance and treatment verification [2]. Some of the advantages of gel dosimeter include potential applications of high linear energy transfer (LET) particles and proton therapy [3-5]. To improve quality assurance of particle therapy treatment of cancer and other diseases by the introduction of fast, accurate, high-resolution, three-dimensional dosimetry based on laser CT of soft-tissue-equivalent polymer dosimeters whose local optical density changes in proportion to local dose. This technology could be used in quality assurance (QA), and in patient-specific pre-treatment verification. Particle therapy is on the rise, but it lacks adequate dosimetry and QA tools. The promise of clinical benefits of particle therapy, which is characterized by superior tumor/normal tissue dose ratio, sharp range definition, enhanced relative biological effectiveness (RBE), and reduced oxygen sensitization effect in normal tissues, will materialize only to the extent that the accuracy of treatment delivery, both dosimetric and spatial, can be assured. The technology will become stagnate if new quality assurance methods are not developed.

In the study we generate a cranio-spinal plan and deliver it to a polymer gel phantom to interpret experimental results on the dose response of the gel from the irradiation of protons.

Journal of Physics: Conference Series **573** (2015) 012020 doi:10.1088/1742-6596/573/1/012020

2. Materials and Methods

If a patient has a tumor that has spread along the covering of the spinal cord (meninges) or into the surrounding cerebrospinal fluid, then radiation may be given to the whole brain and spinal cord. Tumors such as medulloblastomas are more likely to spread this way and often require cranio-spinal radiation [6]. We prepared a treatment plan using the Varian Eclipse treatment planning system. Field matching is achieved by matching the edges of the abutting fields. For example, the inferior edges of the field for the brain are aligned with the superior edge of the superior spine field. For the match line study using proton, all gels were irradiated with pencil beam line option at the Roberts Proton Therapy Center which is housed in the department of Radiation Oncology at the University of Pennsylvania Medical Center.

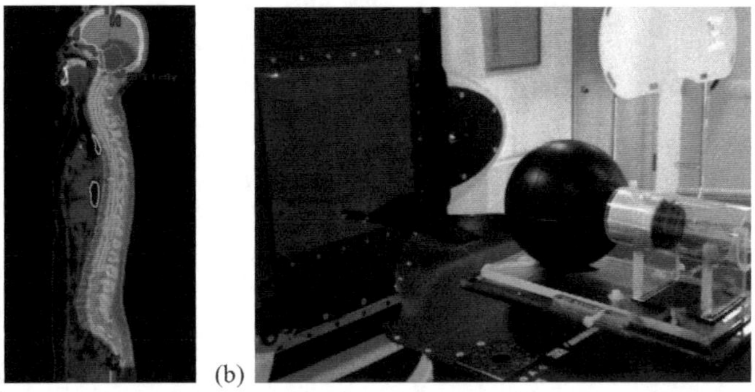

(a) (b)

Figure 1. (a) Colorwash of dose distribution for cranio-spinal plan with pencil beam protons. (b) Setup of polymer gel phantom for treatment position in proton room.

Figure 1 shows the patient plan and the polymer gel phantom setup in the treatment room. We used BBs the setup the phantom at isocenter. We first delivered the brain field then shifted the table to deliver the spine field.

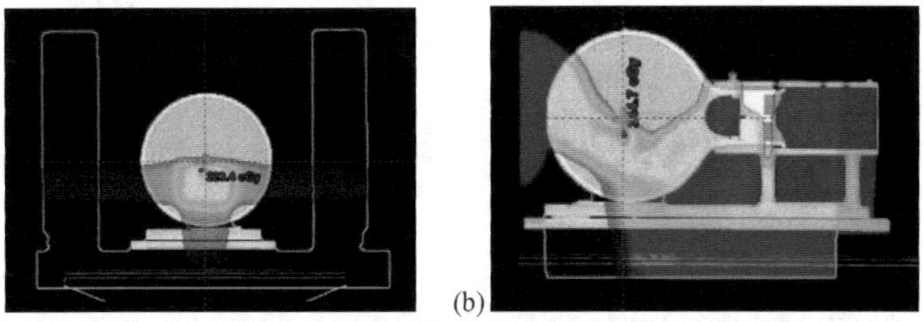

(a) (b)

Figure 2. (a) Axial view of dose plane with right and left lateral fields from brain plan Combined with PA field from spine plan. (b) Lateral view of brain-spine interface.

To determine the correct position of the matchline a verification plan was calculated using the CT study of polymer gel detector. Figure 2 shows the axial and lateral view of the dose distribution. The figure clearly shows the dose for the matchline is inside the polymer gel phantom.

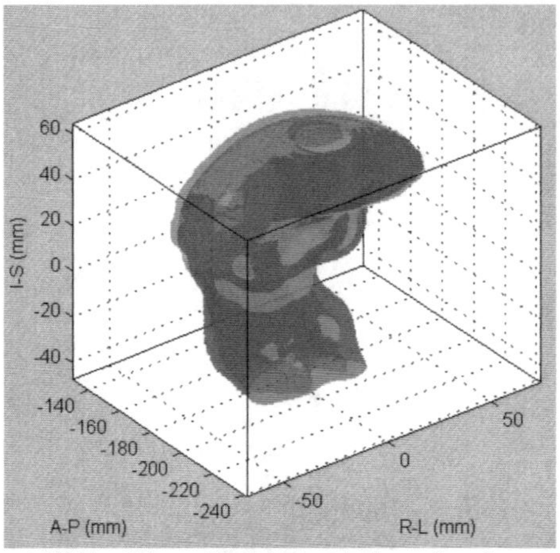

Figure 3. Isosurfaces at 50% of dose. Treatment is in red, gel measurement is in green.

Experimental measurements were scanned with OCTOPUS laser CT and compared to the treatment plan. Figure 3 shows the 3D overlay between the plan and measurements – initial results indicate good agreement of the gel dosimeter with the calculated plan.

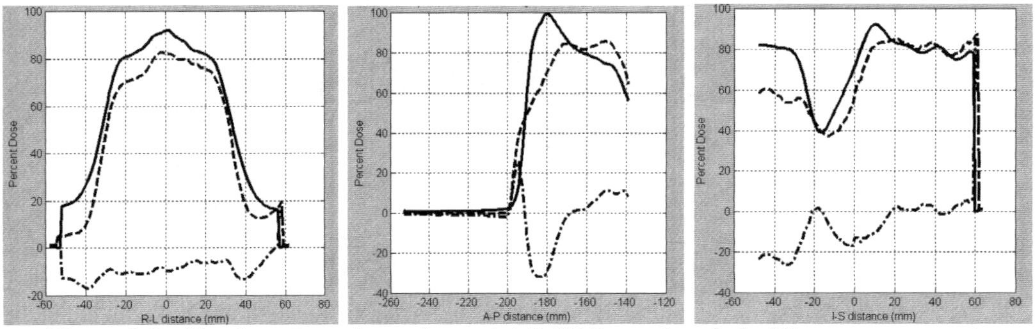

Figure 4. Profiles through maximum dose point. Solid line – plan; dashed line – measurement; dashed dot line – dose difference. Measurements were taken before stability tests.

Examining the profiles at the maximum dose point indicated the largest disparity in the anterior-posterior direction. This behaviour is expected due to the high LET dependence at the end of the Bragg peak. In figure 3 we see ~30% difference due to the end if range effects.

Stability studies of the polymer determine the best detector for measurements of proton beams. Figure 5 illustrates the improvement of the profiles – especially in the anterior-posterior direction where we saw the biggest effect due to LET dependence.

3. Results and Discussion

We determined a stable polymer gel detector could possibly be used for quality assurance of cranio-spinal fields from proton pencil beam plans Figure 6 shows the 30, 50 and 90% isodose planes for the coronal and axial views. From the analysis of the optical scans using our MATLAB code we were able to calculate a dose comparison from the proton irradiation. This study will allow serve as a baseline for dose to be calculated for future plans with proton beams.

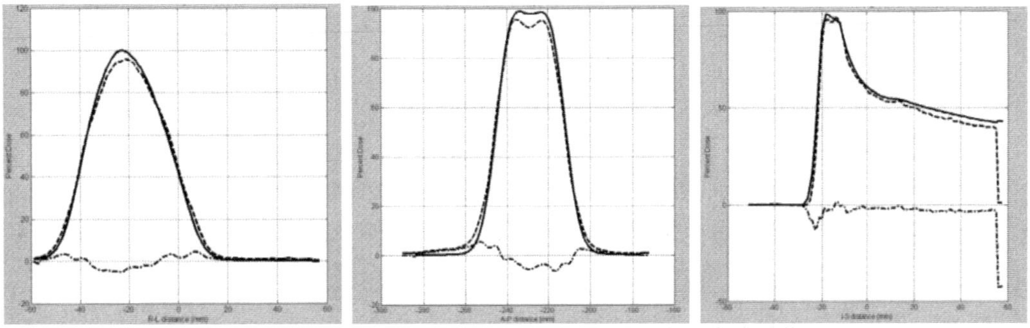

Figure 5. Profiles through maximum dose point. Solid line – plan; dashed line – measurement; dashed dot line – dose difference. Measurements were taken after stability tests.

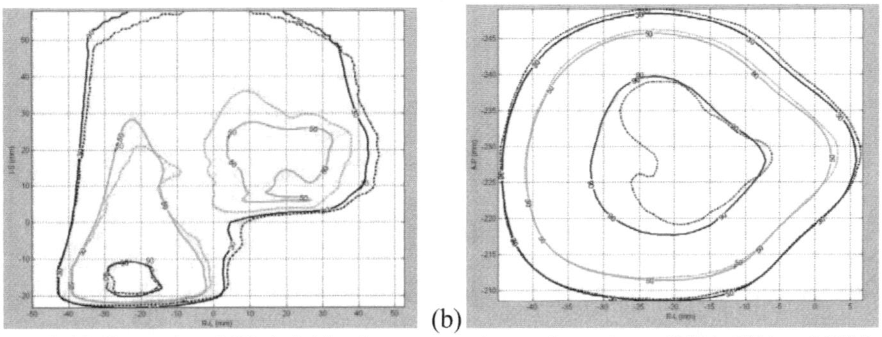

(a) (b)

Figure 6. (a) Coronal and (b) Axial isodoses at maximum dose planes (30%, 50% and 90%). Treatment plan – solid line and gel measurements – dotted line.

4. Conclusions

In this initial study, we demonstrated a method to verify the brain-spine matching using pencil beam proton beams. We developed a consistent method for preparing a stable gel and developing a procedure the compare with the treatment plan. The results indicate that the gels have the potential of measuring the dose response from proton beams. Further work is needed in this study to examine the characteristics of the dose response before we use them clinically; especially in match line for pencil beam proton plans. We also want to develop the system to be used routinely in the clinic with minimal workload on the part of the hospital physicist.

5. Conflicts of Interest

One co-author (M J Maryanski) is the founder and owner of MGS Research Inc. All other co-authors report no actual or potential conflicts of interest.

6. References

[1] Baldock C *et al* 2010 *Phys. Med. Biol.* **55** R1-63
[2] 5th Report on Needs in Ionizing Radiation 2011. Prepared by the Science and Technology Committee of the Council on Ionizing Radiation Measurements and Standards
[3] Heufelder J et al 2003 *Med. Phys.* **30** 1235-40
[4] Jirasek A *et al* 2002 *Med. Phys.* **29** 569-77
[5] Ramm U *et al* 2007 *Phys. Med. Biol.* **52** 2719-28
[6] Cheng C-W *et al* 2013 *Acta Oncologica* **52** 553-60

Sensitivity of collapsed arc QA method for delivery errors in Volumetric Modulated Arc Therapy (VMAT)

Tony Young[1], Aitang Xing[1], Philp Vial[1,2], David Thwaites[2], Lois Holloway[1,2,3,4] and Sankar Arumugam[1]

[1]Liverpool and Macarthur Cancer Therapy Centres and Ingham Institute, New South Wales, Australia

[2]Institute of Medical Physics, School of Physics, University of Sydney, New South Wales, Australia

[3]Centre for Medical Radiation Physics, University of Wollongong, New South Wales, Australia

[4]Uinversity of New South Wales, New South Wales, Australia.

E-mail: tony.young@sswahs.nsw.gov.au

Abstract. In this paper the sensitivity of an Electronic Portal Imaging Device (EPID) to detecting introduced Volumetric Arc Therapy (VMAT) treatment errors was studied using the Collapsed Arc method. Two clinical Head and Neck (H&N) and Prostate treatment plans had gantry dependent dose and MLC errors introduced to the plans. These plans were then delivered to an Elekta Synergy Linear Accelerator EPID and compared to the original treatment planning system Collapsed Arc dose matrix. With the Collapsed Arc technique the EPID was able to detect MLC errors down to 2mm and dose errors of down to 3% depending on the treatment plan complexity and gamma tolerance used.

1. Introduction

Modern radiotherapy techniques are increasingly complex in both planning and treatment delivery, with the Volumetric Modulated Arc Therapy (VMAT) technique following this trend [1-3]. The high complexity demands rigorous commissioning and also pre-treatment plan-specific dose verification using 3D dose measurements [4, 5]. All such radiotherapy techniques are dependent on strict machine tolerances to ensure consistent and accurate dose delivery [6, 7]. Errors due to both MLC [8] and dose differences [9] have been investigated and also different combinations of tools and commercial QA devices have been compared to varying extents [10, 11] for use in rotational therapies.

The collapsed arc technique involves the calculation of the dose matrix resulting from a VMAT arc in a plane perpendicular to the beam at isocentre. The algorithm used has been outlined elsewhere [12]. The technique allows the measurement of the VMAT arc using gantry mounted devices, or detectors which are always perpendicular to the treatment beam, but is not able to detect delivery errors from gantry angle position. Conventional linear accelerators have an EPID permanently attached with the imaging plate directly perpendicular to the treatment beam and so are well suited to capture fluence measurements in the treatment plane. Many different groups have utilized the EPID for QA purposes of different modality treatment plans, as well as the possibility of *in vivo* dosimetry [13-16]. In this study we investigate the sensitivity of the collapsed arc technique to detect deliberately-introduced errors to VMAT treatment plans measured using the EPID.

2. Material and Methods

2.1. VMAT Planning and Delivery

VMAT plans were generated using the Pinnacle, v 9.6, Treatment Planning System (TPS) for a 6 MV photon beam model for an Elekta Synergy (Elekta Ltd, Crawley, UK) linear accelerator (linac). Four clinical VMAT treatment plans were considered here – two prostate plans and two Head and Neck (H & N) plans. Each plan was collapsed to a single plane to produce a dose matrix. The linac was equipped with a MLCi head and Integrity v1.1, linac control software. The standard EPID panel for this linac was used for these measurements.

2.2. Studied delivery errors:

Gantry dependent MLC and dose error scenarios with varying levels of dosimetric impact [17] were considered. All errors were introduced to the treatment delivery file using an in-house computer program written in the Python language.

MLC error: MLC positional errors as a function of control point (CP) gantry angle were introduced to the VMAT arc. The error values were generated as a sine (SIN) function of gantry angle, with amplitudes of 1 to 7mm introduced to both X1 and X2 MLC banks (figure 1a).

Dose Error: CP weight errors as a function of gantry angle were introduced to all CPs of the arc. The error values were generated as a modulus of the sine (SIN) function of the CP gantry angle with amplitude ranging from 1% to 15% (figure 1b). The overall MU of the arc was kept constant.

All plans with introduced errors and the original error-free plan were measured using the EPID.

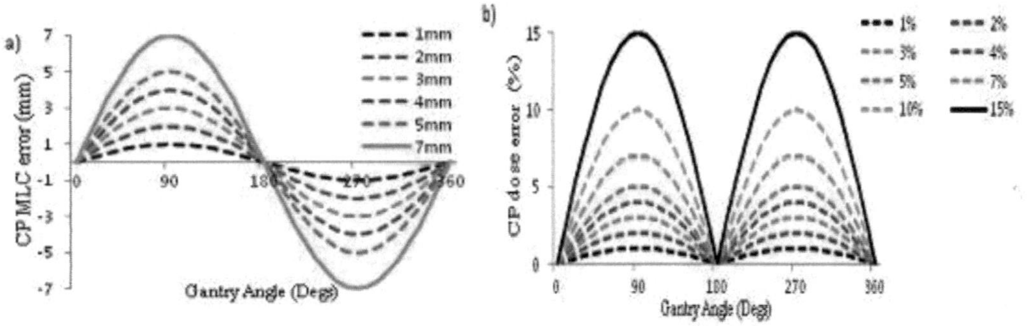

Figure 1. a) Magnitude of gantry angle dependent a. MLC error and b) weight error introduced to individual control points of studied VMAT arcs.

2.3. Treatment plan analysis:

The measured dose matrices were compared with the Pinnacle calculated error free dose matrices using gamma (γ) analysis [18] in OmniPro I'mRT (v1.6) software with a γ tolerance of 1. The γ analysis was performed with a tolerance of 2 % dose difference and 2 mm Distance-to-Agreement (DTA) (2%/2mm) and also 3%/3mm. Dose points receiving less than 10% of the maximum dose in the dose matrix were not considered in the analysis. The measured EPID images were converted into dose matrices using existing methodology and computer code outlined elsewhere [19].

3. Results

Both MLC and dose errors were detectable using the EPID. The magnitude of error detected varied depending on the complexity of the plan and the gamma tolerance criteria used.

3.1. MLC Errors

The gamma pass rates decreased as the magnitude of MLC error increased for both treatment sites. Considering a gamma pass rate above 90%, for the H&N plans a minimum 3mm MLC error was detected by the EPID with the tighter 2%/2mm gamma tolerance for HN1 and a minimum 4mm MLC error detected for HN2. For a 3%/3mm tolerance the MLC errors were not detected until beyond 5mm (HN1) or 7mm (HN2). For the prostate plans the MLC errors were detected by the EPID beyond 2mm for PRO1 using the 2%/2mm gamma tolerance, and beyond 3mm for PRO2. For a 3%/3mm tolerance the MLC errors were not detected until beyond 5mm (PRO1) or 7mm (PRO2).

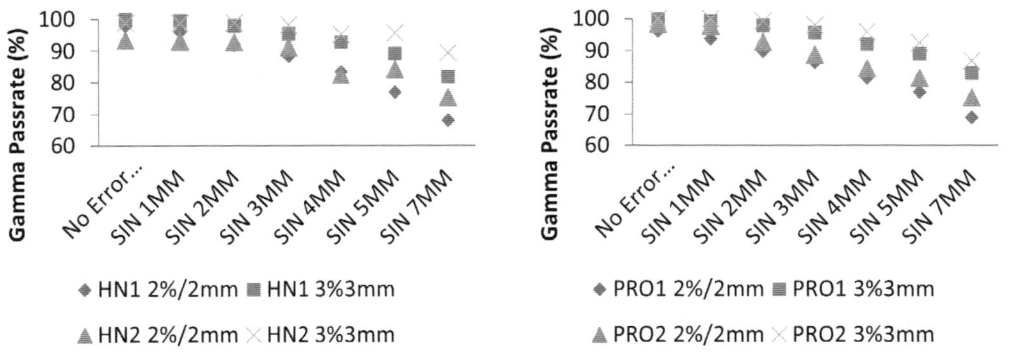

Figure 2. Results for plans with MLC errors introduced using a sine (SIN) function (figure 1)

3.2. Dose Errors

The gamma pass rates decreased as the magnitude of dose error was increased for both treatment sites. For the H&N plans the dose errors were detected by the EPID beyond 3%. For the prostate plans the dose errors were not detected by the EPID even at a 15% dose error, with the gamma pass rate above 90% even using the tighter gamma tolerance. For the H&N plans the dose error was detected at a 3% dose error for HN1 and at the 4% dose error for HN2 using the 2%/2mm gamma tolerance. Using the 3%/3mm gamma tolerance, the dose error was detected at 10% for HN1 and 15% for HN2. Only a limited subset of dose errors was able to be measured for HN1, leading to the possibility that the dose error may have been detectable between 4% and 10%.

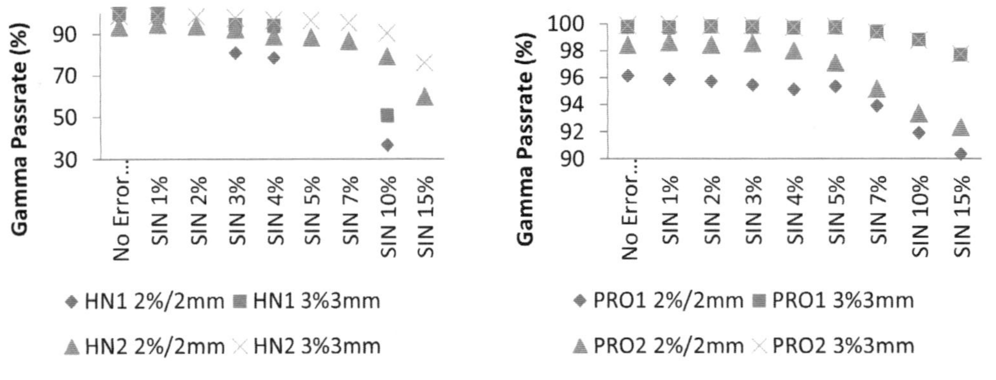

Figure 3. Results for plans with dose errors introduced using a sine (SIN) function (figure 1).

4. Discussion

The different complexity involved in treatment plans for the prostate and H&N resulted in different sensitivities to both MLC and dose errors dependent on gantry angle. The sensitivity in detecting MLC errors is dependent on the spatial resolution of the EPID. The ability of the EPID to resolve changes in fluence due to changes in MLC appears to be adequate for both prostate and H&N cases, with the gamma pass rates dropping with increased magnitude of errors in both cases. For detecting dose errors, the EPID is unable to quantify large dose differences for prostate plans, with gamma pass rates still above 90% even for introduced dose errors of up to 15%. The ability to resolve dose error is reduced due to collapsing the arc into a single dose matrix, which may potentially cancel out some of the introduced dose errors, depending on where the sinusoidal distribution is offset in relation to gantry angle. However the EPID was able to detect some dose errors for the much more complex H&N plans, though this was only when using the tighter gamma tolerance, suggesting that a tighter tolerance than 3%/3mm should be used to detect potential dose errors. For the simpler prostate plans, increasing the expected minimum gamma pass rate may enable detection of potential dose errors.

The Collapsed Arc technique enables 2D detectors to be used in conjunction with arc therapy. The algorithm outlined in [12] has been verified against a 3D dosimeter and a gantry mounted 2D array, with acceptable pass rates for 3%/3mm gamma tolerance. The main disadvantage of this Collapsed Arc single plane dose matrix is that it is unable to provide any information regarding possible gantry delivery errors, as the measurement plane is always perpendicular to the beam. Future work incorporating a gantry inclinometer feedback into the measurement may provide some ability to detect these potential errors. The method is suitable for pre-treatment patient QA of a radiotherapy system that has been verified by appropriate commissioning procedures and that is supported by an appropriate equipment QA program.

5. Conclusion

Using the collapsed arc technique, the EPID was able to detect the majority of MLC and dose errors of different magnitudes introduced to H&N and prostate plans, with sensitivity to errors depending on plan complexity. A gamma tolerance of 2%/2mm is more appropriate than 3%/3mm for use with the technique in detecting errors. The sensitivity of a clinical EPID is reasonable for pre treatment patient QA of VMAT plans at detecting potential delivery errors in MLC position and dose for complex plans. For simpler plans the expected minimum gamma pass rate may need to be increased to detect potential dose errors.

6. References

[1] Otto K 2008 *Med. Phys.* **35** 310
[2] Webb S *et al* 2009 *Phys. Med. Biol.* **54** 4345
[3] Baldock C *et al* 1994. *J. Roy. Soc. Med.* **87** 806-8
[4] Schreibmann E *et al* 2009 *Med. Phys.* **36** 4530-5
[5] Baldock C *et al* 2010 *Phys. Med. Biol.* **55** R1-63
[6] Greer P B *et al* 2007 *Med. Phys.* **34** 4389-98
[7] Vial P *et al* 2008 *Med. Phys.* **35** 1267-77
[8] Oliver M *et al* 2010 *Radiother. Oncol.* **97** 554-60
[9] Betzel G T *et al* 2012 *Med. Phys.* **39** 5882-90
[10] Fredh A *et al* 2013 *Med. Phys.* **40** 031716
[11] Masi L *et al* 2011 *Med. Phys.* **38** 612-21
[12] Arumugam S *et al* 2013 *Med. Phys.* **40** 071724
[13] Iori M *et al* 2010 *Med. Phys.* **37** 377-90
[14] Bailey D W *et al* 2012 *J. Appl. Clin. Med. Phys.* **13** 82
[15] Mijnheer B *et al* 2010 *J. Phys.: Conf. Ser.* **250** 012020
[16] Mans A *et al* 2010 *Radiotherapy and Oncology* **94** 181-7
[17] Arumugam S *et al* 2013 *J. Phys.: Conf. Ser.* **444** 012019
[18] Low D A *et al* 2003 *Med. Phys.* **30** 2455
[19] Xing A *et al* 2014 *J. Phys.: Conf. Ser.* **489** 012084

Journal of Physics: Conference Series **573** (2015) 012022 doi:10.1088/1742-6596/573/1/012022

Dose verification of radiotherapy for lung cancer by using plastic scintillator dosimetry and a heterogeneous phantom

W Ottosson[1,2], C F Behrens[2] and C E Andersen[1]
[1]Center for Nuclear Technologies, Technical University of Denmark, DTU Risø Campus, DK-4000 Roskilde, Denmark
[2]Department of Oncology, Radiotherapy Research Unit, Herlev Hospital, University of Copenhagen, DK-2730 Herlev, Denmark

E-mail: wiot@dtu.dk

Abstract Bone, air passages, cavities, and lung are elements present in patients, but challenging to properly correct for in treatment planning dose calculations. Plastic scintillator detectors (PSDs) have proven to be well suited for dosimetry in non-reference conditions such as small fields. The objective of this study was to investigate the performance of a commercial treatment planning system (TPS) using a PSD and a specially designed thorax phantom with lung tumor inserts. 10 treatment plans of different complexity and phantom configurations were evaluated. Although the TPS agreed well with the measurements for the least complex tests, deviations of tumor dose > 4% were observed for some cases. This study underpins the dosimetric challenge in TPS calculations for clinically relevant heterogeneous geometries. The scintillator system, together with the special phantom, provides a promising tool for evaluation of complex radiotherapy dose calculations and delivery.

1. Introduction

Fiber-coupled organic plastic scintillator detectors (PSDs) feature advantages suitable for complex and dynamic radiation dosimetry in megavoltage photon beams [1-3]. When it comes to heterogeneous setups with lack of charged particle equilibrium (CPE), there are recognized calculation challenges for most commercial treatment planning systems (TPSs). Thus, volumes containing bone, air passages, cavities and lung may deteriorate the TPS dose calculation accuracy [4, 5].

The objective of this study was to investigate the performance of a TPS dose calculation algorithm by using a PSD in a heterogeneous setup, analogous to the geometry of a lung cancer patient, while delivering clinical relevant treatment plans of varying complexity.

2. Material and methods

2.1. Phantom design

A thorax phantom, analogous to a lung cancer patient, was constructed to perform PSD dosimetry in a well-defined heterogeneous geometry. The body of the phantom is made of PMMA, 34 cm in width (W), 23 cm in height (H) and 40 cm in length (L) (figure 1). It contains three hollow cylinders of L:50 cm, and a diameter (Ø) of 10 cm. These cylinders can be filled with several inserts of various materials to simulate different homo- and heterogeneous geometries. The various inserts are made of the copolymer polyoxymethylene (POM-C), balsa wood, and PMMA representing bone, lung and soft tissue, respectively (figure 1, table 1-2). The lung inserts were 15 cm long with a Ø of 9 cm, mimicking a human lung in size. PMMA spheres of various sizes (1-8 cm in diameter) embedded in balsa wood are available to simulate tumors in lung (figure 1 (d)). In the lower part of the body, two smaller cylindrical holes of Ø:2 cm and Ø:3 cm (which also can be altered to Ø:2 cm) are positioned at different distances from the phantom center, i.e. 6.5 cm and 9.5 cm (table 2). These holes can, one at a time, be filled with a POM-C rod to simulate the spinal column at different diameters and position from the center of the phantom (table 2).

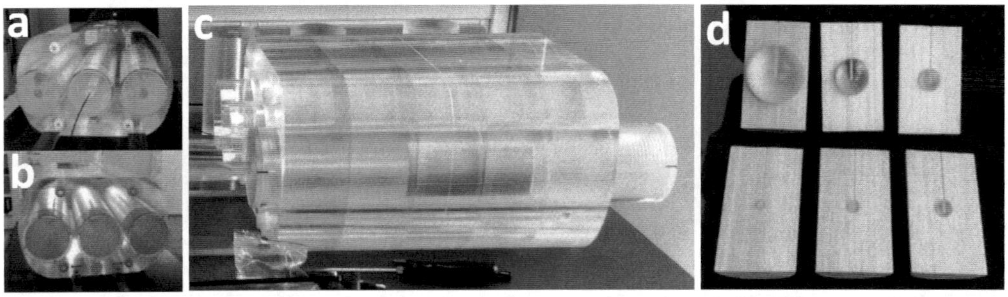

Figure 1. (a) A heterogeneous setup where the two lateral body cylinders are filled with balsa wood inserts. (b) A homogeneous setup, where the whole phantom are filled with PMMA inserts. (c) The heterogeneous setup described in (a) viewed from the side, where the lateral body cylinder containing the lung insert is longitudinal shifted from the central position in the phantom. (d) Balsa wood lung insert with associated tumors, ranging from 1-8 cm in diameter. (Color version of figure is available online.)

2.2. Image acquisition and target definition

Four phantom configurations ('Homo.', 'Hetero.', '3 cm tumor' and '5 cm tumor') scanned in a 16 slice Philips Brilliance CT Big Bore, version 3.5.17001 (Philips Medical Systems, Cleveland, OH) using a standard thorax scanning protocol were used in this study (illustrated in table 2). Table 1 presents the CT image characteristics of the phantom materials of 10 CT series of the phantom compared to corresponding human tissue data for 10 randomly picked lung cancer patients.

Table 1. CT image characteristics of the phantom materials compared to human tissue. Mean HU values and (range) for 10 CT series of the phantom and corresponding tissue data for 10 randomly picked lung cancer patients. Paired t-tests were performed for each tissue type, to check for differences in the mean HU value between the phantom material and the patient tissue data. No significances were found, using $p < 0.05$, i.e. there are good agreement between human tissue and the phantom materials.

Tissue	Phantom Material		HU[b]	
		Density / [g/cm^3]	Phantom	Patients
Bone	POM-C	1.40	319 (309;327)	313 (210;413)
Lung	Balsa wood	0.10	-913 (-917;-888)	-901 (-977;-770)
Soft tissue	PMMA[a]	1.18	116 (103;123)	118 (84;143)

[a.] Poly(methyl methacrylate)
[b.] Using a standard thorax CT scanning protocol by Philips.

Delineations of the anatomical structures GTV, CTV, PTV, medulla, lung and body were performed on all image sets in the treatment planning system (TPS) Eclipse v. 10 (Varian Medical Systems, Palo Alto, CA, USA). CTVs and PTVs were defined as a 5 mm and 10 mm symmetrical expansion of GTV, respectively.

2.3. Experimental setup and calibration conditions

The scintillator used was the BCF-60 (Saint-Gobain Ceramics & Plastics Inc.) with Ø:1 mm and L:2 mm, described by Beierholm *et al* [1]. The PSD was calibrated according to the procedure (method C) described by Guillot *et al* [6] in a solid water calibration phantom. Measurements were carried out using the ME40 scintillator dosimetry system (DTU Nutech) [2]. The reference dose (100 MU, 10x10 cm^2 field) was measured by a Farmer ionization chamber, type 30011 (PTW, Freiburg, Germany) at a depth of 10 cm in the PMMA phantom QUASAR Multi-Purpose Body Phantom (Modus Medical Devices Inc., London, ON, Canada). Irradiation was delivered by a Varian Clinac iX 2300 linear accelerator (Varian Medical Systems), with a beam energy of 6 MV at a dose rate of 600 MU/min [7]. The accumulated PSD dose for the various treatment plans were compared with corresponding TPS calculated point doses (table 2).

Table 2. TPS calculated isocentric point doses compared with corresponding PSD measured dose. (Color version of figure is available online.)

Isocenter is positioned in the center of the phantom (Homo. setup / Hetero. setup)		Fiber dose [Gy] Homo.	Fiber dose [Gy] Hetero.	TPS dose [Gy] Homo.	TPS dose [Gy] Hetero.	Dose deviation [%] Homo.	Dose deviation [%] Hetero.
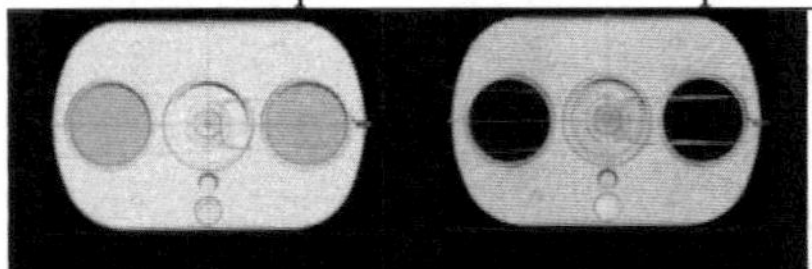	Conv.	2.001	2.005	2.000	2.000	-0.1	-0.3
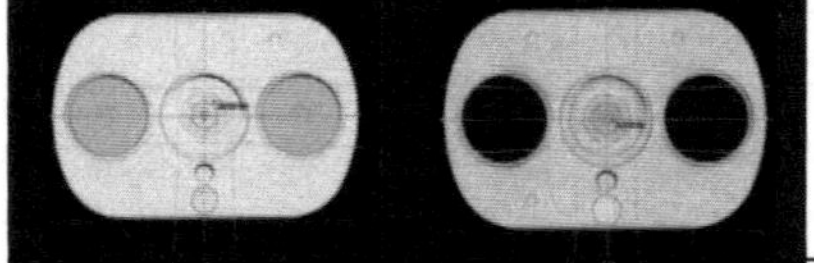		2.005	2.006	2.000	2.000	-0.3	-0.3
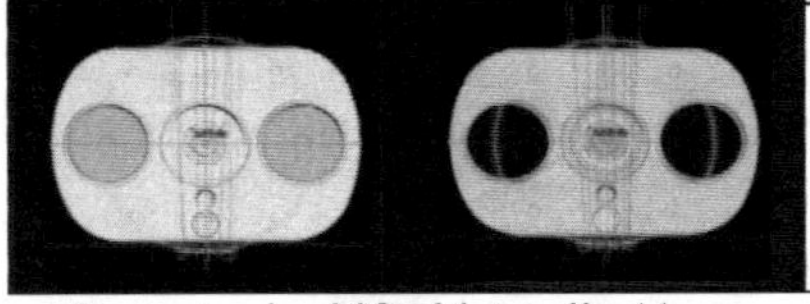	RA	1.976	1.953	1.984	1.965	0.4	0.6
Isocenter is shifted laterally 11 cm (3 cm tumor / 5 cm tumor)		**3 cm tumor**	**5 cm tumor**	**3 cm tumor**	**5 cm tumor**	**3 cm tumor**	**5 cm tumor**
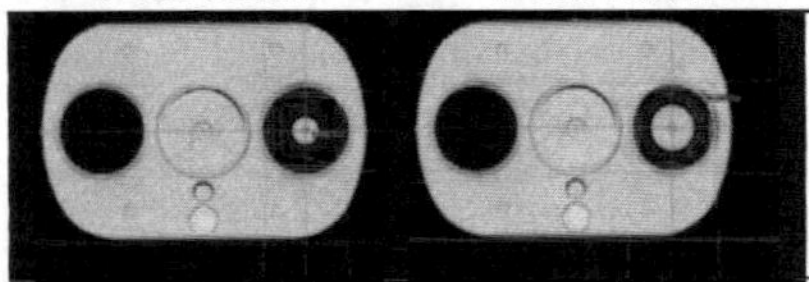	Conv.	2.114	2.083	2.000	2.000	-5.4	-4.0
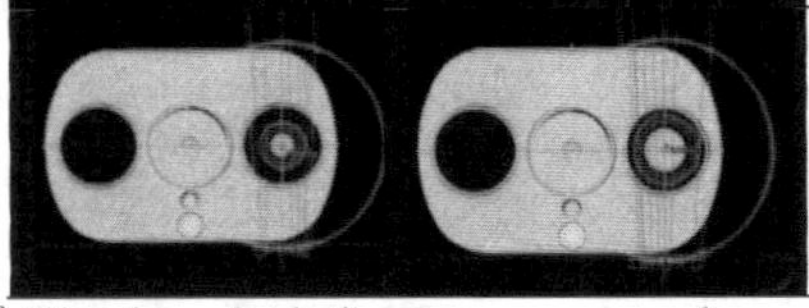	RA	2.323	2.203	2.268	2.175	-2.4[a]	-1.3

[a] The dose deviation per arc was as large as -7.6%.

2.4. Treatment plans and delivery

In total, 10 treatment plans of different phantom configurations and isocentric field techniques (single field, 4-field conventional and two-arc RapidArc (RA) plans) were created. For each treatment plan the isocenter was positioned in the center of the GTV. For six treatment plans the isocenter was positioned centrally in the phantom. Three of these had a homogenous setup (figure 1 (b), table 2), and three of them had a heterogeneous setup (figure 1 (a), table 2). For the remaining four treatment plans the isocenter were shifted 11 cm laterally, because the GTVs were situated in the left lung of the phantom (table 2). Doses were calculated using the AAA algorithm, with a prescribed dose of 2 Gy to the PTV. The RA plans were normalized to the mean dose of the PTV, while the other plans were normalized to the isocenter. All treatment plans satisfied the clinical dose coverage criteria. For the tumors situated in the lung 98% of the PTV volume was covered by minimum 90% of the prescribed dose. For the treatment plans not situated in the lung the corresponding dose coverage was 95%. The beam energy and dose rate used were 6 MV and 600 MU/min, respectively.

3. Results/Discussion

Under calibration conditions the PSDs agreed with the TPS calculations to 0.1%.

Deviations less than 1% were observed between calculated and measured doses when the isocenter was located in the middle of the phantom. For the homogeneous configuration, deviations were in the range of (-0.1%;0.4%) and for the heterogeneous configuration, deviations were in the range of (-0.3%; 0.6%) (table 2). The RA plans generally resulted in larger total dose deviation (0.4%;0.6%) compared with the simple conventional techniques (-0.3%;-0.1%). These low discrepancies, for the centrally positioned point in the phantom, probably illustrate that there is enough distance to adjacent heterogeneities in the phantom to be able to establish CPE. Larger TPS dose deviations (-5.4%;-1.3%) were observed when the isocenter was shifted laterally, since the GTV was situated in the left lung of the phantom. These substantial deviations could potentially be due to lack of sufficient spread of lateral radiation to obtain CPE. Even larger dose deviations (-5.4%;-2.4%) were observed for the smallest tumor size investigated (3 cm in diameter). This small size of tumor is not large enough to re-establish the CPE condition, and this is most likely the reason why the smallest tumor size results in the highest dose deviation. For the laterally shifted phantom configuration, the simple conventional technique resulted in a higher total dose deviation (-5.4%;-4.0%) compared to the more complex RA (-2.4%;-1.3%). The lower dose deviation, when using RA, might be due to the spread of incident radiation over the whole phantom compared to limited incident angles through heterogeneous medium, when using conventional technique.

4. Conclusion

Dose deviations of $< 1\%$ were observed for isocentric field techniques centered in the middle of the phantom, whereas dose deviations $> 4\%$ were observed for some laterally shifted treatment plans. The study confirmed that the smallest tumor size results in the highest dose deviation. The scintillator system and the heterogeneous phantom provide a promising tool for critical evaluations of complex radiotherapy calculations and dose delivery.

5. Acknowledgment

Søren Dalsgaard and the DTU Nutech workshop are thanked for making the phantom. Anders Beierholm is acknowledged for his valuable input to this study. The work was partly carried out within the EMRP: "Metrology for radiotherapy using complex radiation fields" (HLT09). The EMRP is jointly funded by the EMRP participating countries within EURAMET and the European Union.

6. References

[1] Beierholm A R *et al* 2011 *Phys. Med. Biol.* **56** 3033-45
[2] Beierholm A R *et al* 2013 *Radiat. Meas.* **56** 290-3
[3] Beddar A S *et al* 1992 *Phys. Med. Biol.* **37** 1901-13
[4] Papanikolaou N *et al* 2004 *AAPM Report no. 85: Tissue inhomogeneity corrections for mega voltage photon beams* (Madison, WI, USA)
[5] Knöös T *et al* 2006 *Phys. Med. Biol.* **51** 5785807
[6] Guillot M *et al* 2011 *Med. Phys.* **38** 2140-50
[7] Sjöström D *et al* 2009 *Acta Oncol.* **48** 192-200

Development of phantom and methodology for 3D and 4D dose intercomparisons for advanced lung radiotherapy

Misael Caloz[1,2], Marilyne Kafrouni[1,2], Quentin Leturgie[1,2], Stéphanie Corde[2], Simon Downes[2], Joerg Lehmann[1] and David Thwaites[1]
[1]Institute of Medical Physics, School of Physics, University of Sydney, NSW, Australia
[2]Radiation Oncology, Prince of Wales Hospital, Randwick, NSW, Australia

E-mail: david.thwaites@sydney.edu.au

Abstract. There are few reported intercomparisons or audits of combinations of advanced radiotherapy methods, particularly for 4D treatments. As part of an evaluation of the implementation of advanced radiotherapy technology, a phantom and associated methods, initially developed for in-house commissioning and QA of 4D lung treatments, has been developed further with the aim of using it for end-to-end dose intercomparison of 4D treatment planning and delivery. The respiratory thorax phantom can house moving inserts with variable speed (breathing rate) and motion amplitude. In one set-up mode it contains a small ion chamber for point dose measurements, or alternatively it can hold strips of radiochromic film to measure dose distributions. Initial pilot and feasibility measurements have been carried out in one hospital to thoroughly test the methods and procedures before using it more widely across a range of hospitals and treatment systems. Overall, the results show good agreement between measured and calculated doses and distributions, supporting the use of the phantom and methodology for multi-centre intercomparisons. However, before wider use, refinements of the method and analysis are currently underway particularly for the film measurements.

1. Introduction

As part of an evaluation of the consistency and accuracy of the implementation of advanced radiotherapy (RT) technology for the NSW Health Department, the Institute of Medical Physics (IMP) and a clinical partner, the Prince of Wales Hospital (PoWH), have developed and tested a phantom for end-to-end dose intercomparison of planning and 'treatment' delivery for lung radiotherapy, aimed at including IMRT, VMAT, hypo-fractionated SBRT, 4D RT and the use of FFF beams, across a range of treatment planning systems (TPSs), delivery systems and techniques. There are few, or no, reported audits/intercomparisons of many of these combinations [1-3], thus this work is aimed at providing initial testing of potential audit methods, with the developments linked with the Australian RT clinical trials (TROG) QA group [3] and the Australian national RT dosimetry audit system (ACDS:ARPANSA) [3, 4].

2. Materials and Methods

2.1 The phantom and measuring systems

The respiratory phantom was originally designed in-house at PoWH as a QA and verification tool for local 4D treatment commissioning and QC, based on single point ion chamber measurements and to fulfill the requirement for strict QA for 4D methods [5]. It was then further developed for wider use with different systems in different centres and to contain radiochromic film for dose distribution measurements. The phantom (figure 1) was to be representative of the thorax and was constructed of water-substitute plastic, with lung-substitute plastic to form two lungs. In the original version, one of the lungs contained a small cylindrical water-equivalent 'lung tumour', 1.4 cm in diameter and 2.5 cm high. This was positioned in a 2.5 cm x 2.5 cm x 25 cm movable insert, which could move in the cranio-caudal direction with a range of 'breathing rates' and amplitudes. These could be set to variable values: between 9 and 24 breathing cycles per minute (bpm); and four amplitudes of 0.5, 0.9, 1.9 and 2.8 cm. Initially, point doses were measured in the centre of the tumour using a small volume CC04 ion chamber (IBA). The pilot use of this in a multi-centre intercomparison has been reported [6, 7], where a pre-set tolerance of 5% was used for moving target situations [8, 9].

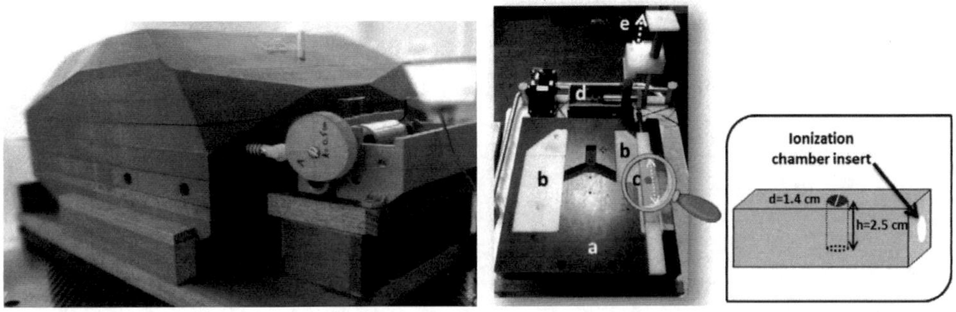

Figure 1. 'Respiratory' phantom in original ion chamber measurement mode, showing: overall appearance; phantom split to reveal internal construction; and schematic of the moving insert. In the central figure: (a) 'Thorax' (water-equivalent slabs); (b) 'lungs' (tissue-equivalent foam); (c) 'Tumour' (Solid Water™); (d) Motorized drive system, with variable speed selection and interchangeable eccentric 'wheels' to control amplitude; (e) Vertically moving platform added for Varian RPM™ system.

The phantom has been further developed to also house radiochromic film for dose distribution intercomparisons. In this version, the moving insert has been increased in size to 6 cm x 6 cm x 25 cm, constructed from cedar (relative electron density ~ 0.3) and containing a cylindrical Solid Water 'tumour' oriented along the direction of motion, of 2 cm diameter and 2 cm length. The insert can be split to house and reproducibly position strips of EBT3 Gafchromic film of 5.5 cm width and 16 cm length (figure 2). The whole insert can be positioned in one of two orientations, so that film measurements can be performed in either sagittal or coronal planes. Films were read out on an Epson Expression 10000XL scanner, processed with a net OD method using locally written Matlab procedures (calibration curves, film positioning, conversion to dose on the red channel). Further analysis, including gamma analysis and dose convolution, was performed using RIT113 v6.2 and OmniPro I'mRT v1.7 software.

It may be noted that the changes to the phantom to house a larger insert for film measurements necessitated a design change for ion chamber measurements. The original smaller ion chamber insert has now been replaced to fit the enlarged void designed for the film insert. The larger ion chamber insert now has a 'tumour volume' matching that in the film insert for consistency and is machined to house the CC04 chamber in the centre of that volume.

Figure 2. The 'respiratory' phantom in film measurement mode, showing: moving insert seen from the rear of the phantom and extra side-scatter attached to increase the 'chest-wall thickness'; and the cedar film holder split to show the Solid Water 'tumour' along its cylindrical axis and the pegs to reproducibly position the film strips.

2.2 Pilot and feasibility measurements

The intention is to use the phantom and measurement systems to intercompare performance in different centres, having a wide range of approaches to 4D lung radiotherapy, and to carry out end-to-end testing to include imaging, delineation, planning and treatment dose delivery according to local 4D protocols. Detailed pilot and feasibility measurements were carried out initially in PoWH to thoroughly test the methods and procedures before such wider use. Thus the phantom underwent the current PoWH 3D and 4D imaging, planning and treatment delivery procedures for the static situation and for a range of combinations of motion amplitude and breathing rate (up to 12 scans, plans and deliveries to encompass up to 4 amplitudes and 3 breathing rates). Planning scans were obtained using a Toshiba Aquilion LB scanner, initially for exhale and inhale breath hold scans and also for a free-breathing scan and later using the Varian RPM system, fusing scans to delineate the ITV. Any ion chamber air cavities were over-written with surrounding 'tissue' density. Plans were created on the exhale phase (for comparison to static measurements) and also on average scans (for motion) with either CMS XiO 4.6 or Monaco 3.3, using 8 conformal beams to encompass the tumour motion. They were calculated for a fraction dose of 8 Gy for 6 and 10 MV beams on an Elekta Axesse linac. Plans were then delivered to the phantom using those beams and the measured doses (ion chamber) or distributions (film) were compared to planned values. Planned 'point' doses were calculated as a mean of relevant interest points placed in the plan along the range of tumour (chamber) movement. Dose deliveries were repeated three times per situation to test the reproducibility. For the film measurements, a more restricted set of conditions has been investigated (4 amplitudes, but for only one breathing rate of 14 bpm), but repeated three times for each film orientation. Gamma analysis was carried out using a 10% threshold, global normalization and analysis criteria of 3%/3mm and 5%/3mm, where the first tolerance criteria were selected to comply with general clinical practice tolerances and the second wider set to be consistent with the ion chamber intercomparison tolerance.

3. Results and Discussion

3.1 Ion chamber point dose measurements

Agreement between measured doses and mean planned doses in moving conditions at 14 bpm was observed to be within 1.4%, with a minor increase in the difference as motion amplitude increased. Specifically, average (and standard deviation) dose differences between measured and planned doses were observed of -0.32 (± 0.04)%, 0.44 (± 0.01) %, 0.83 (± 0.02)% and 1.40 (± 0.01)% for motion amplitudes of 0.5, 0.9, 1.9 and 2.8 cm respectively. Negative dose differences indicate that measured doses were less than planned and vice versa. In each case, the average and standard deviation values are from the three repeated measurements taken in each situation. The measurements were generally reproducible, as represented by the standard deviations. The influence of breathing rate was also assessed and in the range used (9-24 bpm) was found not to have any significant effect, ie within experimental uncertainty the dose differences were the same for a given motion amplitude, whatever the breathing rate. These results are for 3D conformal treatments and hence do not involve dynamic deliveries; however the preliminary results from pilot studies involving IMRT and VMAT indicate that this finding is also supported for dynamic treatments, for this range of breathing rates [6,7].

3.2 Film measurements

Using repeated scans for the films from the static target case (no motion applied), the reproducibility of the scanner and the scanning methodology and process was high, with variations of no more than 0.5% in gamma evaluation pass rates. The preliminary results using Gafchromic film in the moving target phantom to measure dose distributions delivered in static mode agreed well with the planned values calculated on the exhale scan, with more than 95% of points passing a 3%/3mm gamma criteria and more than 99% of points passing a 5%/3mm gamma criteria. Under moving conditions, direct comparison of dose distributions calculated on the average scan and measured film values lead to gamma analysis results ranging from 98 to 32% of points passing the criteria from the smallest to the largest amplitudes for 3%/3mm (99 to 38% for 5%/3mm). The increasingly poorer agreement as motion amplitude increases is due to the average plans used not reflecting the whole effects of motion. The use of the ITV concept ensures that the GTV receives the expected dose, but the planned distribution is an average planned value whereas the measured distribution is a true smeared distribution as delivered. By convolving the dose distributions calculated according to the motion pattern, agreement better than 96% and 99% respectively were obtained, indicating that the methods are satisfactory, but that the region of interest for comparison needs further consideration.

4. Conclusion

The results indicate that the in-house respiratory phantom and associated methods can be used successfully to perform 4D commissioning and end-to-end QA, using different inserts to measure ion chamber point doses or film-measured dose distributions. Various issues observed in the feasibility studies, particularly for the film measurements, are being resolved by refinements of the method and analysis before beginning to use the system more widely for pilot multi-centre inter-comparisons over a range of scanners, TPSs, linacs, beam energies and modalities (FF and FFF) and delivery methods (3DCRT, IMRT, VMAT). These initial measurements are also being used to inform the decisions on tolerances for wider audit use. The phantom is versatile, modular and cost-effective for the intended purposes and besides being used for multi-centre intercomparisons has the potential also to be used for patient-specific validation of 4D hypo-fractionated lung treatment plans in a given centre.

5. References

[1] Thwaites D 2012 *J.Phys.:Conf Ser.* **444** 012006
[2] Ibbott G and Thwaites D2014 *J.Phys.:Conf. Ser.* This conference proceedings
[3] Kron T *et al* 2013 *J. Phys.: Conf. Ser.* **444**, 012014
[4] Williams I *et al* 2012 *Australas Phys. Eng Sci Med.* **35** 407-11
[5] AAPM 2006 American Association of Physicists in Medicine Report 91 http://www.aapm.org/pubs/reports/RPT_91.pdf
[6] Kafrouni M *et al* 2014 *Australas. Phys. Eng. Sci. Med.* **37** 222
[7] Corde S *et al* 2014 *Radiother. Oncol.* 111 705-6
[8] AAPM 1998 American Association of Physicists in Medicine Report 62 (TG53) https://www.aapm.org/pubs/reports /rpt_62.pdf
[9] IAEA 2004 International Atomic Energy Agency Technical Reports Series no. 430 http://www-pub.iaea.org/mtcd/publications/pdf/trs430_web.pdf

Journal of Physics: Conference Series **573** (2015) 012024 doi:10.1088/1742-6596/573/1/012024

Reproducibility assessment of dynamically deforming DEFGEL in a respiratory motion phantom

R D Franich[1], J R Supple[1], B Lindsay[1], U J Yeo[1], P Lonski[1,2], R L Smith[1,3], M L Taylor[1], L Dunn[1] and T Kron[1,2]
[1]School of Applied Sciences and Health Innovations Research Institute, RMIT University, Melbourne, VIC, Australia
[2]Physical Sciences, Peter MacCallum Cancer Centre, Melbourne, VIC, Australia
[3]William Buckland Radiotherapy Centre, The Alfred Hospital, Melbourne, VIC, Australia

E-mail: rick.franich@rmit.edu.au

Abstract. Conducting four-dimensional deforming dosimetry with a deformable gel dosimeter presents several challenges for implementation and interpretation of results. In this work, we describe a modification to a respiratory motion phantom to accommodate the DEFGEL dosimeter, and an evaluation of the reproducibility of the setup and deformation cycle. Setup errors are sub-mm while deformation reproducibility is typically less than half the CT voxel size (1 mm AP and LR; 3 mm SI). 4DCT confirms the absence of hysteresis.

1. Introduction

We have previously introduced the deformable gel dosimeter, DEFGEL, and demonstrated its use for direct measurement of dose accumulation in different states of deformation [1]. This has been shown to be an invaluable tool for the experimental validation of mathematical deformable dose accumulation approaches such as DIR–based dose warping [2].

To extend this to a range of situations involving more complex geometries and motions, we have adapted a QUASAR respiratory motion phantom (Modus Medical Devices Inc.) to accommodate the DEFGEL dosimetry element. The phantom was originally designed to sinusoidally oscillate a wooden lung-like component in a torso analogue. Our group has designed and implemented an upgrade to the phantom to enable more complicated motion profiles including actual patient breathing profiles recorded with Varian's RPM system at the time of 4DCT [3].

In this work, we have created a custom module to allow the phantom to accommodate a DEFGEL in a piston-driven deformation component. We describe the modifications to the phantom and its drive motor, and the evaluation of its suitability for performing dynamically deforming dose integration.

The application of dynamic deformation to a 3D dosimeter [4] presents several complications for both implementation and interpretation of measurements. In common with, for example, *in vivo* dosimetry in the presence of movement, the measurement is critically dependent on the reproducibility of the geometry with respect to the dosimeter, source and phantom/patient. In the special case of a 3D dosimeter used to experimentally validate dose accumulation over multiple motion cycles, two factors must be verifiably reproducible:

(i) the position of the dosimeter in the phantom, and
(ii) the location and shape of the dosimeter as it is deformed over multiple cycles.

In this study, we describe the use of a DEFGEL phantom with embedded fiducial markers, combined with CT scanning, to quantify the reproducibility of sample positioning and cyclic deformation.

2. Method

2.1. Phantom modification

The QUASAR normally drives a wooden cylinder back and forth through a cylindrical hollow through the torso phantom. To deform a DEFGEL, a cylindrical insert, closed at one end, was locked in place with a clamp and a piston was made to be driven by the stepper motor to apply pressure to the DEFGEL. The closed-end cylinder liner, piston and clamp are shown in figure 1(a).

A DEFGEL, approximately cylindrical as shown in figure 1(c), is supported in the deformation chamber by two end-cups, indented to support the gel in place (see figure 1) and featuring air escape holes. The assembled QUASAR with the deformation chamber installed, and the cylinder liner clamped in place, is shown in figure 1(b).

Initial trials of the system identified a weakness of the QUASAR drive system that was not able to exert enough pressure on the DEFGEL to deform it significantly without the adjustable amplitude drive screw slipping. Thus a new drive screw coupling was fashioned (also included in figure 1(a)).

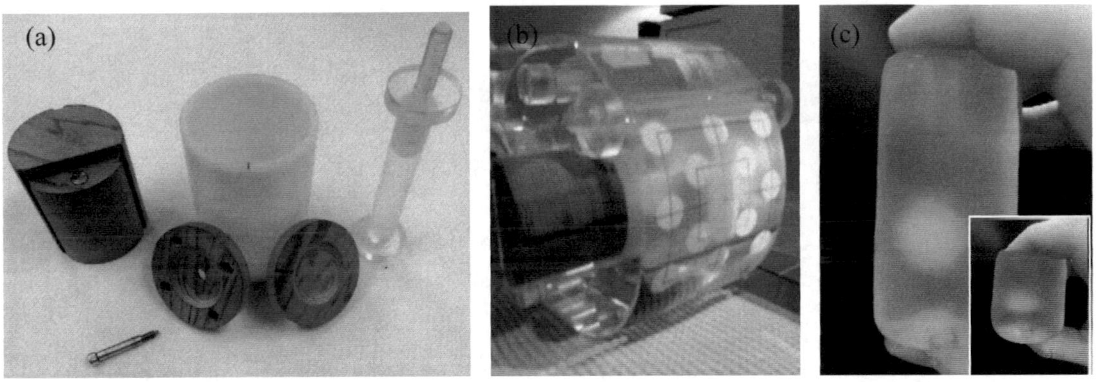

Figure 1. (a) Components used in the modification of the QASAR phantom, (b) the assembly with cylinder liner clamp in place, and (c) a deformable DEFGEL phantom.

2.2. DEFGEL with markers

To assess the position of the DEFGEL when installed in the chamber inside the QUASAR, and also later when being driven, a DEFGEL was made with 13 small aluminium fiducial markers (FM) embedded at the time of gel pouring. These markers, visible in CT imaging, have previously been used for motion and deformation assessment for the evaluation of deformable image registration algorithms applied to DEFGEL deformations [5]. The DEFGEL with fiducial markers is shown in figure 2.

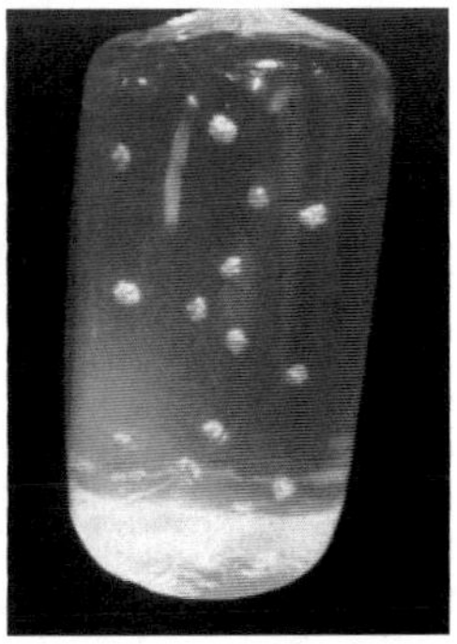

Figure 2. DEFGEL phantom with aluminium fiducial markers embedded.

2.3. Reproducibility of sample positioning

The ability to reproducibly locate the sample inside the phantom is important for consistency especially in studies where pre-irradiation CT imaging of the phantom will be performed for the purpose of treatment planning system calculations to be compared to the measurements.

To assess this reproducibility and effectively assess any setup errors, the DEFGEL with markers was installed in the phantom and then CT scanned. The whole assembly was dismantled, re-assembled and re-imaged 4 times (for a total of 5 scans). The position coordinates of the markers were extracted using ImageJ and the mean error and standard deviation were recorded.

2.4. Reproducibility of deformation

The DEFGEL was deformed with driving amplitude of 20 mm, giving the individual markers various movement amplitudes between 0 and 20 mm (as the distal end of the DEFGEL from the piston is a stationary point). Computed tomography imaging was performed before and after 30 cycles. The marker coordinates were determined at the extrema of the motion and at the mid-point, with the phantom stationary to avoid motion artefacts confusing the marker positions.

2.5. Analysis of hysteresis of dynamic deformation

A 4DCT image set was acquired while the QUASAR dynamically deformed a DEFGEL by applying a 20 mm peak-to-peak sinusoidal piston motion. Marker coordinates were compared in complementary phases of the deformation to ascertain whether any hysteresis was present (i.e. compare the 10 % phase image to the 90 % phase image, 20 % to 80 %, etc.).

3. Results and Discussion

3.1. Reproducibility of sample positioning

The distribution of errors of the FM positions relative to the mean position is shown in Table 1, for each of the 3 coordinate directions. The sample could usually be positioned with sub-mm reproducibility with few exceptions. Coarse resolution of the CT imaging in the SI direction (slice thickness of 3 mm) resulted in a one-voxel error occurring.

Table 1. Setup errors associated with positioning the DEFGEL sample in the phantom.

	AP	**LR**	**SI**
Mean absolute error (mm)	0.53	0.44	0.15
Standard deviation (mm)	0.23	0.36	0.89
Min. (mm)	0.02	0.01	0.00
Max. (mm)	2.21	1.69	3.00
Voxel dimension (mm)	1.17	1.17	3.00

3.2. Reproducibility of deformation

Table 2 describes the displacement of all markers after 30 cycles of sinusoidal deformation in each of 3 positions: at each extremum of the motion (piston position +/- 10 mm) and half way between (0 mm). The reproducibility is consistent with the setup accuracy. The larger of the errors observed are a consequence of coarse resolution in the SI direction (slice thickness = 3 mm) and partial volume averaging of the markers in the CT images.

Table 2. Reproducibility errors associated with the cycle of deformation of the DEFGEL within the phantom.

	AP			LR			SI		
Piston position (mm)	-10	0	+10	-10	0	+10	-10	0	+10
Absolute mean (mm)	0.27	0.25	0.28	0.35	0.38	0.39	2.10	1.80	1.50
St. dev. (mm)	0.29	0.22	0.40	0.32	0.25	0.35	1.45	1.51	1.58
Min. (mm)	0.01	0.01	0.00	0.05	0.03	0.12	0.00	0.00	0.00
Max. (mm)	0.89	0.74	0.93	0.93	1.10	1.12	3.00	3.00	3.00
Voxel dimension (mm)		1.17			1.17			3.00	

3.3. Analysis of hysteresis of dynamic deformation
Table 3 describes the displacement of all markers between phase images acquired at the same piston position but with opposite piston trajectories. No obvious hysteresis effects can be observed as the differences are consistent with setup accuracy and CT resolution.

Table 3. Marker displacements associated with hysteresis during deformation of the DEFGEL within the phantom. All measurements are in millimetres.

Image phases	AP		LR		SI	
compared (%)	Abs. mean	St. Dev.	Abs. mean	St. Dev.	Abs. mean	St. Dev.
10 - 90	0.29	0.42	0.25	0.18	0.21	0.20
20 - 80	0.03	0.24	0.10	0.14	0.13	0.48
30 - 70	0.42	0.04	0.24	0.11	0.22	0.27
40 - 60	0.35	0.22	0.32	0.29	0.23	0.23
Voxel dimension	1.17		1.17		3.00	

4. Conclusions
It is imperative to quantify setup uncertainties and reproducibility of dosimeter position when using 3D dosimetry to evaluate dose redistribution arising from deformation. We have demonstrated a respiratory phantom modification that establishes sub-mm sample positioning, and deformation reproducibility typically within +/- 0.5 mm (AP & LR) and 2 mm (SI) over multiple cycles for measuring dose redistribution with DEFGEL.

5. References
[1] Yeo U J *et al* 2012 *Med. Phys.* **39** 2203-13
[2] Yeo U J *et al Med. Phys.* **39** 5065-72
[3] Dunn L *et al* 2012 *Australas. Phys. Eng. Sci. Med.* **35** 93-100
[4] Baldock C *et al* 2010 *Phys. Med. Biol.* **55** R1-63
[5] Yeo U J *et al* 2013 *Med. Phys.* **40** 101701

Flexydos3D: A new deformable anthropomorphic 3D dosimeter readout with optical CT scanning

Yves De Deene[1,2], Robin Hill[3], Peter S Skyt[2,4] and Jeremy Booth[5]
[1]Dept. of Engineering, Faculty of Science, Macquarie University, Sydney, Australia
[2]Institute of Medical Physics, University of Sydney, Sydney, Australia
[3]Chris O'Brien Lifehouse, Sydney, Australia
[4]Department of Medical Physics, Aarhus University Hospital, Aarhus, Denmark
[5]Department of Radiotherapy, Royal North Shore Hospital, Sydney, Australia

E-mail: Yves.DeDeene@mq.edu.au

Abstract. A new deformable polydimethylsiloxane (PDMS) based dosimeter is proposed that can be cast in an anthropomorphic shape and that can be used for 3D radiation dosimetry of deformable targets. The new material has additional favorable characteristics as it is tissue equivalent for high-energy photons, easy to make and is non-toxic. In combination with dual wavelength optical scanning, it is a powerful dosimeter for dose verification of image gated or organ tracked radiotherapy with moving and deforming targets.

1. Introduction

Three-dimensional radiation dosimetry has received growing interest with the implementation of highly conformal radiotherapy treatments [1]. Much research has been conducted towards both the development of different radiation sensitive three-dimensional dosimeters and adequate scanning techniques. Although an acceptable precision and accuracy has been achieved with polymer gel dosimeters and MR scanning [2-4], the procedure is relatively labor intensive and some expertise is needed to acquire reliable quantitative MR images. As an alternative to MR scanning, optical CT scanning has been proposed as an alternative [5]. Different staining dosimeters have been proposed as a better alternative to polymer gel dosimeters including polyurethane based dosimeters, PRESAGE™ [6] and micelle gel dosimeters [7, 8]. We here propose a new 3D dosimeter with favorable properties. In contrast to polyurethane dosimeters such as PRESAGE™, a non-toxic matrix is used and it doesn't require a specialized chemistry laboratory to construct phantoms. Moreover, as a result of the slow curing time and therefore reduced heat dissipation, it does not display non-uniform densities in the material that in PRESAGE™ dosimeters may give rise to Schlieren artifacts. As a proof-of-principle, we have created some anthropomorphic and highly flexible dosimeter phantoms.

2. Methods and Materials

2.1. Dosimeter fabrication

The dosimeter consists of an optically transparent silicone elastomer, a curing agent, chloroform as a source of radiation induced radicals and the leucodye, leucomalachite green (LMG).

2.2. Dosimeter properties

The concentration of chloroform and LMG was varied in order to optimize the sensitivity. The dosimeter was also found to be sensitive to UV light. To facilitate the optimization of the chemical composition of the dosimeter, a UV light source was used to expose the dosimeter to a source of ionizing radiation for various time periods. Once a preliminary idea on sensitivity was obtained, the dosimeter was further analyzed by irradiation with X-ray photon beams from a clinical linear accelerator. We assumed that the dose response upon UVC exposure correlates well with the dose response of high energetic X-rays, which has been confirmed through several additional experiments.

2.3. Deformable cardiac phantom

Several anthropomorphic shaped dosimeter phantoms have been constructed. As an example, we demonstrate a cardiac dosimeter phantom with a single cavity that can be pumped in order to simulate the deformation of the myocardium.

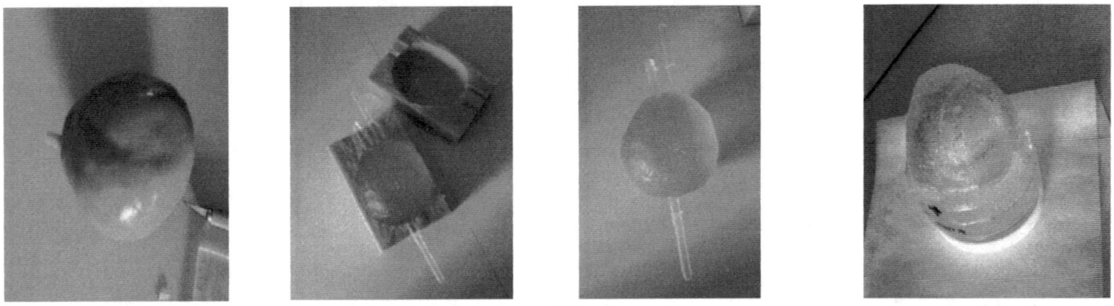

(a) **(b)** **(c)**

Figure 1. Construction of a heart shaped deformable radiation dosimeter using handmade casts and first dose maps after exposure with a square beam of high energetic photons. Handmade positive mold of the heart (a), inner cavity mold made from a hydrogel (b), a first prototype of a heart shaped phantom (c).

3. Results and Discussion

3.1. Optimization of the dose sensitivity

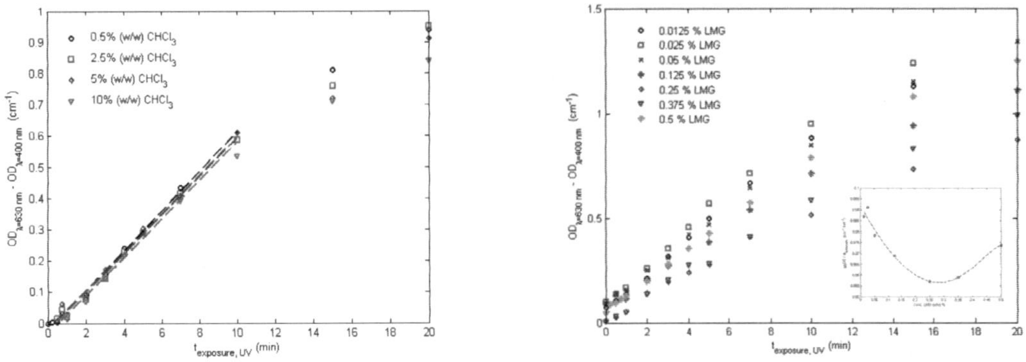

(a) **(b)**

Figure 2. Dose-response curves for different exposure times to a UV light source for different concentrations of chloroform (a) and different concentrations of LMG (b). The inset shows the change of slope (sensitivity) with increasing LMG concentration.

Varying the chloroform (CHCl3) concentration between 0.5% (w/w) and 10% (w/w) did not have a significant effect on the dose sensitivity of the dosimeter while keeping the amount of LMG fixed at 0.5% (w/w).

The optimum concentration of LMG is found to be 0.025% (w/w). For LMG concentrations higher than 0.5% (w/w) the dosimeter became turbid. For the optimum composition (0.025% (w/w) LMG), the dose sensitivity for high energetic photon beams (6 MV) was found to be 0.01 $cm^{-1}.Gy^{-1}$.

3.2. Tissue equivalence

The basic composition of the dosimeter is polydimethylsiloxane. Within the range of 100 kV to 20 MV, the dosimeter is highly tissue equivalent. The relative electron density of the dosimeter as compared to the electron density of water is in the order of 1.0059 (for a chain length of 10). For higher chain lengths the electron density converges to that of water.

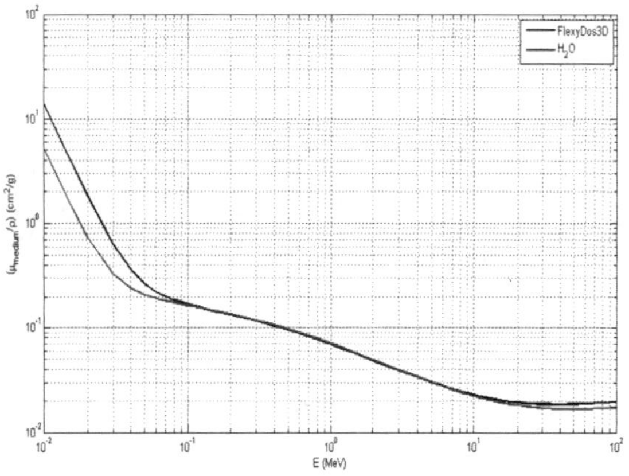

(a)

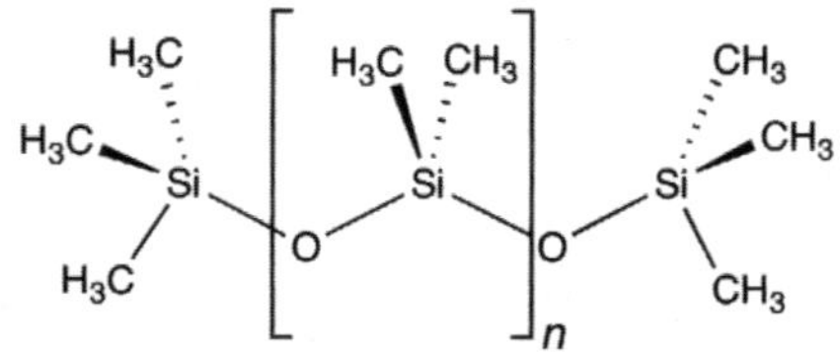

	n = 10	n = 20
H	8.7 %	8.4 %
C	33.6 %	33.6 %
O	20.4 %	20.4 %
Si	37.3 %	37.6 %
(e/e_{water})	1.0059	1.0036

(b)

Figure 3. Mass attenuation coefficient as a function of photon energies for FlexyDos3D (black) and water (blue) calculated using the XCOM NIST online database [9]. (b) Polydimethylsiloxone (PDMS) is the main component in FlexyDos3D. The weight fraction of atoms is listed for different chain lengths.

3.3. Dose rate dependence

The dose rate dependence of the Flexydos3D dosimeter is shown in figure 4 for dose rates between 200 cGy/min and 400 cGy/min. Remarkable is that the dose rate dependence seems to be more pronounced at higher dose levels and that the dose response is non-linear. The non-linearity in dose response increases with decreasing dose rate. For dose values between [0 Gy, 15 Gy], the dose rate dependence of the dose sensitivity amounts to 12%, while in the dose range [15 Gy, 40 Gy] the dose sensitivity varies to up to 32% for dose rates between 200 cGy/min and 600 cGy/min.

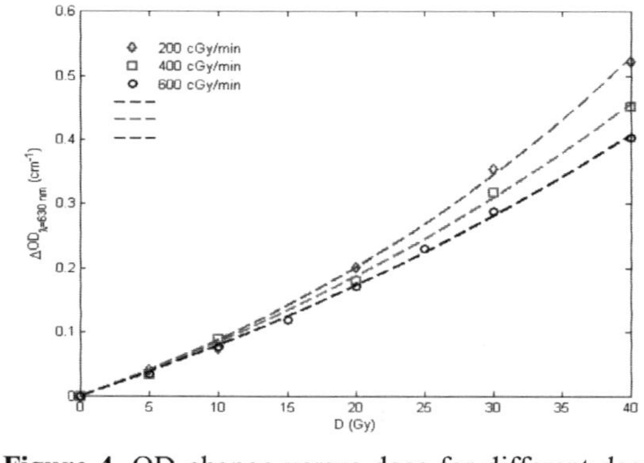

Figure 4. OD change versus dose for different dose rates (200 cGy/min, 400 cGy/min, 600 cGy/min).

The dose rate dependence was only studied for a single composition of gel type. Future experiments can be conducted for different dosimeter formulations in order to decrease the dose rate dependence of the FlexyDos3D dosimeter.

3.4. Energy dependence

The difference in dose sensitivity for FlexyDos3D dosimeters irradiated with photon beams of 6 MV and photon beams of 18 MV amounted to less than 5% which was within the measurement uncertainty for these measurements.

3.5. Deformable cardiac dosimeter phantom
By way of illustration, a deformable dosimeter phantom in the shape of a human heart with inner cavity was constructed. The cardiac phantom was connected to a pulsating hydraulic actuator that mimics the pulsation of the heart. The pump is controlled by use of an in house built controller that uses pulse-width modulation (PWM) to vary the pumping speed. This dosimeter phantom can be a valuable tool to verify image guided tracking treatment of stereotactic radiosurgery of cardiac ablation [10]. The cardiac phantom dosimeter was exposed to a static square photon beam and was scanned with a dual wavelength optical cone beam CT scanner (figure 5c).

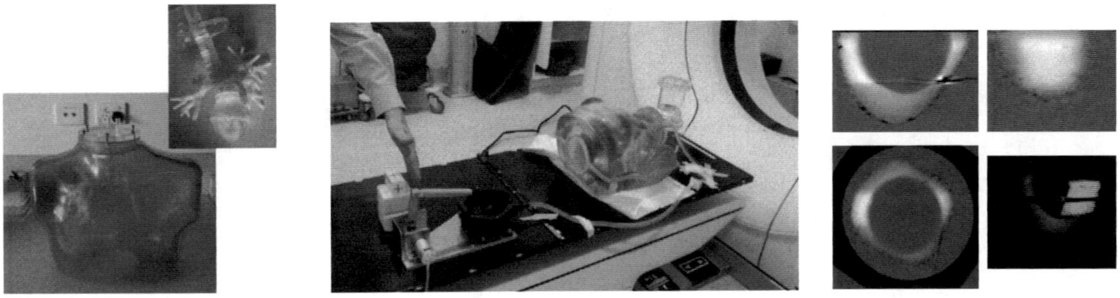

(a) **(b)** **(c)**

Figure 5. Thorax phantom in which the cardiac dosimeter (figure 1) is suspended (a) and phantom on a 4D CT scanner connected to a hydraulic actuator to simulate pumping of the heart (b) and dose maps after irradiation with a square photon beam generated with a clinical linear accelerator (c).

4. Conclusion
A new flexible 3D dosimeter is proposed that can be cast in an anthropomorphic shape and which is easy to fabricate and non-toxic. The new dosimeter can be readout optically using a dual wavelength optical CT scanner. The dosimeter has great potential as a dosimetric tool for dose verification in image guided tracking of radiosurgery ablation treatment.

5. References
[1] Baldock C *et al* 2010 *Phys. Med. Biol.* **55** R1-63
[2] Vandecasteele J *et al* 2013 *Phys. Med. Biol.* **58** 19-42
[3] Vandecasteele J *et al* 2013 *Phys. Med. Biol.* **58** 43-61
[4] Vandecasteele J *et al* 2013 *Phys. Med. Biol.* **58** 63-85
[5] Gore J C *et al* 1996 *Phys. Med. Biol.* **41** 2695-704
[6] Adamovics J A *et al* 2004 *Med. Phys.* **30** 1349
[7] Jordan K *et al* 2009 *Phys. Med. Biol.* **54** 6773-89
[8] Vandecasteele J *et al* 2013 *Phys. Med. Biol.* **58** 6241-62
[9] Berger M J *et al* 1998 XCom: Photon Cross Sections Database http://www.nist.gov/pml/data/xcom/
[10] Sharma A *et al* 2010 *Heart Rhytm* **7** 802-10

PRESAGE® as a new calibration method for high intensity focused ultrasound therapy

M Costa[1,2], C McErlean[3], I Rivens[1], J Adamovics[4], M O Leach[3], G ter Haar[1] and S J Doran[3,5]
Department of Physics, The Institute of Cancer Research, London, UK
[2]Fundacao para a Ciencia e a Tecnologia, Lisbon, Portugal
[3]CRUK Cancer Imaging Centre, Institute of Cancer Research, London, UK
[4]Department of Chemistry and Biology, Rider University, Lawrenceville, NJ, USA
[5]Department of Physics, University of Surrey, Guildford, UK

E-mail: Marcia.Costa@icr.ac.uk

Abstract. High Intensity Focused ultrasound (HIFU) is a non-invasive cancer therapy that makes use of the mainly thermal effects of ultrasound to destroy tissue. In order to achieve reliable treatment planning, it is necessary to characterise the ultrasound source (transducer) and to understand how the wave propagates in tissue and the energy deposition in the focal region. This novel exploratory study investigated how HIFU affects PRESAGE®, an optical phantom used for radiotherapy dosimetry, which is potentially a rapid method of calibrating the transducer. Samples, of two different formulations, were exposed to focused ultrasound and imaged using Optical Computed Tomography. First results showed that, PRESAGE® changes colour on ultrasound exposure (darker green regions were observed) with the alterations being related to the acoustic power and sample composition. Future work will involve quantification of these alterations and understanding how to relate them to the mechanisms of action of HIFU.

1. Introduction

High Intensity Focused Ultrasound (HIFU) is a non-invasive thermal therapy for treating cancer, which involves the focusing of 0.5-3 MHz ultrasound waves in tumour tissue, from outside the body, to activate cell death pathways. Two mechanisms of action are involved in HIFU treatments: thermal and mechanical. Thermal effects, which are the dominant cause of cell death, arise from the absorption of acoustic energy in tissue, resulting in an increase of temperature, which creates cigar-shaped thermally ablated lesions [1].

The most notable mechanical effect of HIFU is cavitation. This involves the formation and/or activity of gas bubbles in the tissue. If temperatures exceed 100°C then boiling bubbles may be created – this is thermal cavitation. This can cause an undesirable temperature distribution with an irregular shape [2]. A second effect is acoustic cavitation. Here the ultrasonic pressure cycles either nucleate micron-sized bubbles or cause pre-existing bubbles to grow and oscillate. If bubbles become too large, they can implode in a violent, highly localised, collapse causing mechanical disruption of tissue on a subcellular level [3]. However, the effects of HIFU are material (tissue) dependant, making it a challenge to predict the bio-effects and, consequently, to perform treatment planning. It is, therefore, necessary to characterize both the acoustic field at different transducer drive power levels and the tissue attenuation and the absorption within the focal region [4].

PRESAGE® is a volumetric, chemical dosimeter, composed of a solid polyurethane matrix into which is incorporated a leuco-dye reporting agent, which changes colour when exposed to ionising radiation [5-7]. This change can be measured quantitatively in 3D using Optical Computed Tomography (CT). Whilst it has been previously observed that variations in temperature modulate the response of PRESAGE® to radiation [8], the direct production of a spatially resolved colour change by local temperature variations is a novel development. In this work we have tested the viability of using PRESAGE® as a phantom for HIFU beam calibration. Initial results show that colour changes on HIFU exposure and that this is dependent on the ultrasound parameters chosen and the sample composition. PRESAGE® could, therefore, provide a rapid method for performing HIFU calibration.

2. Methods

Four cylindrical samples of PRESAGE® (2.2 cm diameter and a minimum of 3 cm height), provided by Heuris Pharma (Skillman, NJ), were exposed to HIFU for 5 seconds, using a single-element transducer (Sonic Concepts, 64 mm outer diameter and 63 mm focal length), driven at 1.7 MHz. For these exploratory experiments, two PRESAGE® formulations with different elastic moduli were selected ("hard" polyurethane SO206 standard PRESAGE formulation with density 1.07 $g.cm^{-3}$ (speed of sound of 2298 $m.s^{-1}$ and attenuation coefficient of 9.2 $dB.cm^{-1}$ at 1.7 MHZ) and Shore hardness 80D; "soft" polyurethane CF 95 with density 1.04 $g.cm^{-3}$ and Shore hardness 95A (speed of sound of 2130 $m.s^{-1}$ and attenuation coefficient of 23.3 $dB.cm^{-1}$ at 1.7 MHZ)). Two samples of each composition were exposed to free field peak negative pressures of 9.95 MPa and 7.87 MPa (measured in water). Each sample was placed in a tank filled with degassed water. The Presage long axis was aligned with the acoustic beam axis, with the focal plane set 5 mm behind the front face, as represented in figure 1. The samples were imaged using Optical CT before and after exposure using the scanner described in [9] and 3D reconstruction was performed using in-house software.

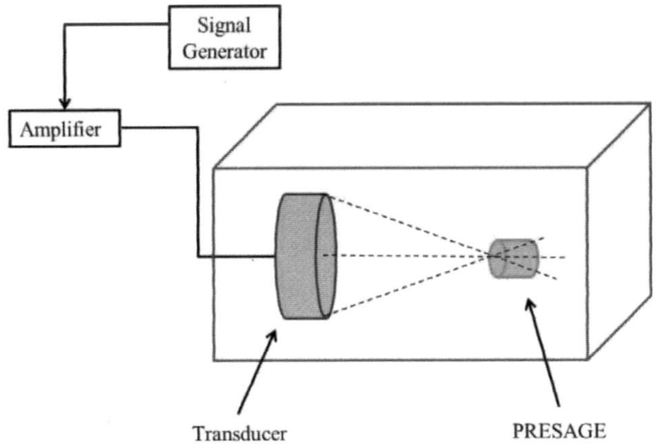

Figure 1. Experimental setup for the HIFU exposure of PRESAGE®. The HIFU transducer was driven using an amplified voltage produced by a signal generator. PRESAGE® long axis was aligned with the acoustic beam axis, with the focal plane set 5 mm behind the front face. For the peak negative pressure values used, a 10 mm long and 2 mm wide focus was expected. The transducer and PRESAGE® were immersed in degassed water to couple the ultrasound between the probe and the sample while supressing the formation of acoustic cavitation bubbles.

3. Results and discussion

All HIFU exposures resulted in an observable colour change in the PRESAGE® (as exemplified in figure 2). These changes were assessed using the optical CT images, represented in figure 3.

Figure 2. Colour changes observed when PRESAGE® was exposed to 7.87 MPa peak negative pressure HIFU. The left sample is a 'soft' one made of CF95 polyurethane and the right is the "hard" SO206 polyurethane.

Journal of Physics: Conference Series **573** (2015) 012026 doi:10.1088/1742-6596/573/1/012026

Figures 3 and 4 show a significant difference in size and shape of the HIFU "lesions" obtained with the different acoustic pressures used and the different sample composition. As expected, higher acoustic pressures resulted in larger "lesions", as more energy is deposited. However, the shapes are also different. Comparing the optical CT images (figures 3A-C) with the beam profile for the Sonic Concepts transducer, (figure 3E), suggests that in figure 3A the focal region can be seen (corresponding to the red region in figure 3E), while in figure 3C the "lesion" seems to be shaped more like the wider (blue) beam profile. Furthermore, in figure 3C, some bubble activity may also be taking place (red arrows pointing for areas of higher greyscale intensity in the image). Further studies will be required to understand the underlying effects causing the changes observed in order to use PRESAGE® as a quantitative HIFU calibration phantom, since one of the key aspects in treatment planning is to distinguish between thermal and mechanical mechanisms. Both "lesions" in figure 3 appear angled, suggesting that the probe might have been slightly tilted in the tank.

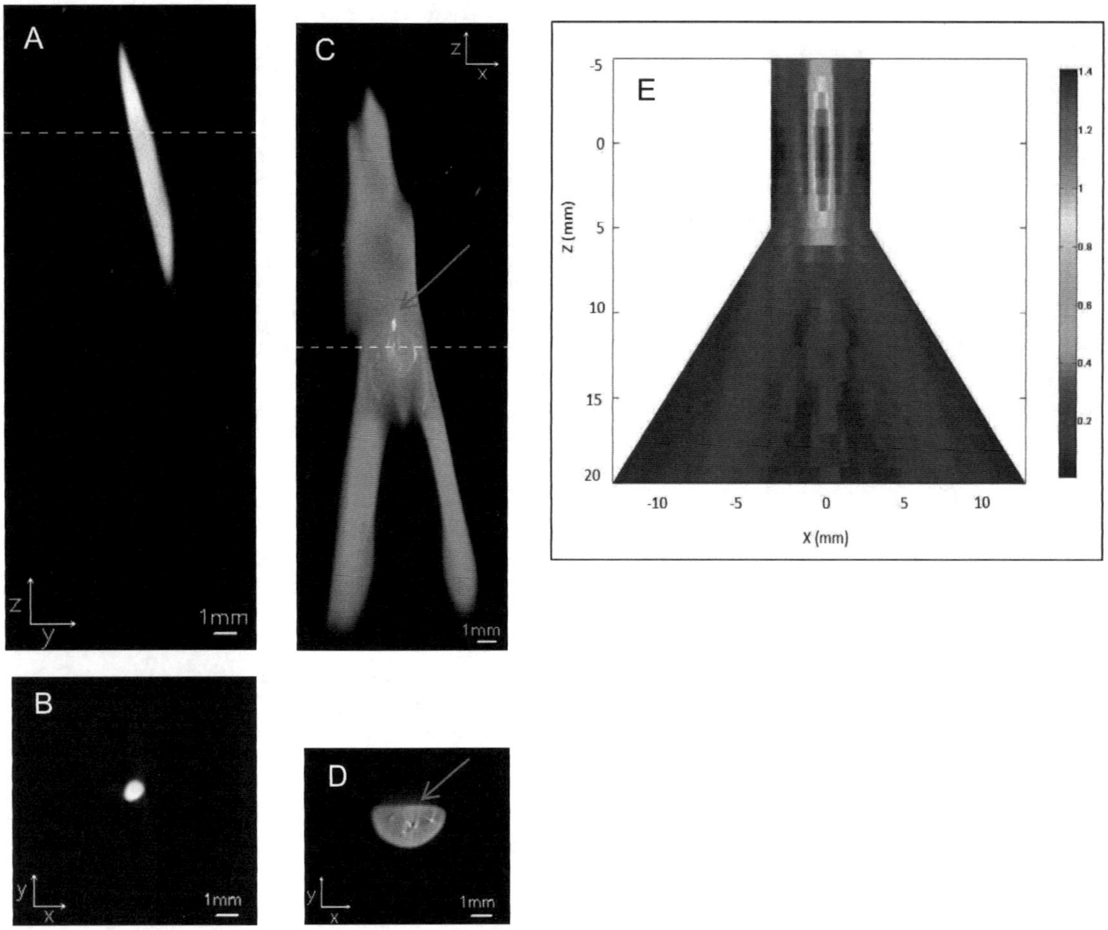

Figure 3. Optical CT images of PRESAGE® samples exposed to HIFU. z axis is along the direction of ultrasound wave propagation. A: Maximum Intensity Projection (MIP) through a "soft" sample (7.87 MPa ultrasound exposure) in the yz plane (long axis of the sample) showing a roughly cigar shaped lesion. The corresponding reconstruction in the xy plane for the marked slice (dashed line) is shown in B. C: MIP through a "soft" sample (9.95 MPa ultrasound exposure) in the xz plane. The higher power HIFU beam shows a different shape to that in A. Bubbles might have formed inside the PRESAGE® (red arrows) possibly due to acoustic or thermal cavitation in the xy plane reconstruction shown in D. E: beam profile of the transducer obtained in water, as described in [4]. Note: The scale for figure 3E is different from that used in the optical CT images.

Significant differences in beam shape and optical intensity were also observed between "hard" and "soft" samples. Figure 4 shows a clearly different "lesion" from the one in figure 3A, although both samples were exposed to the same acoustic pressure. As in figure 3C, the beam profile seems to have further spread around the focal region. As these were the first experiments of their kind, spare samples from an ionising radiation study were used, but for future studies, it will be important to have a customised and standardised recipe for HIFU experiments, since the results are highly dependent on the composition of the sample. Ideally the PRESAGE® would have a density and sound speed similar to that of water (to optimise acoustic coupling); and acoustic attenuation and absorption and thermal properties (specific heat capacity and thermal conductivity) similar to soft tissues, to mimic treatment exposure conditions. The final "lesion" observed in a "hard" sample exposed to a 9.95 MPa (not shown) was similar to that shown in figure 4.

a. *b.*

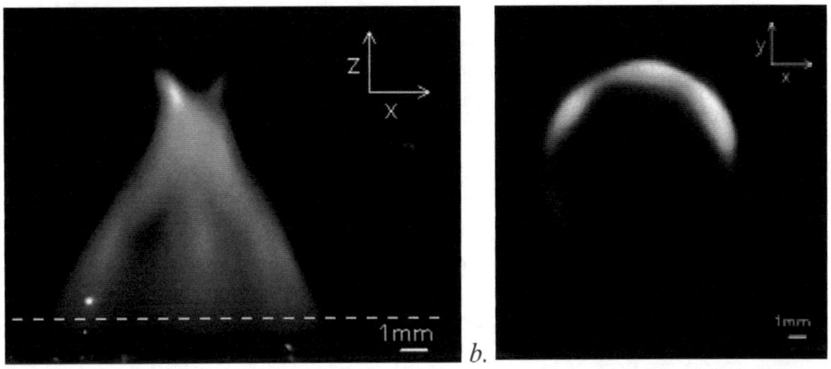

Figure 4. (a) MIP reconstruction in the xz plane for a "hard" sample (7.87 MPa exposure). The optical CT shows a different shape to that seen in figure 3A. A corresponding xy slice is shown in (b). This shows a 'half-moon' profile that was also observed in other samples. We would expect a more circular beam profile in water for the transducer that was used, but these shapes might be observed if the acoustic field and/or PRESAGE® sample have some inhomogeneities.

4. Conclusions

Encouraging initial results show a readily measurable HIFU effect on PRESAGE®, which varies with both input power and PRESAGE® composition. Further characterisation of the thermal (specific heat capacity and thermal conductivity) and acoustic (density, speed of sound and attenuation) properties of PRESAGE® is thus the next step. This will then allow us to perform simulations of wave propagation and choose the best composition for future imaging experiments. Furthermore, an existing passive cavitation detection system will be used to determine whether acoustic cavitation is occurring and hence whether the mechanism of colour change is primarily thermal or mechanical. An ultimate goal will be the determination of a quantitative relationship between optical CT image intensity and acoustic HIFU exposure parameters.

5. References

[1] Zhou Y F 2011 *World J. Clin. Oncol.* **2** 8-27
[2] Damianou C *et al* 2010 *J. Biomedical Science and Engineering* **3** 763-75
[3] Coussios C C *et al* 2007 *Int. J. Hyperthermia* **2** 8-27
[4] Zhou Y *et al* 2006 *J. Acoust. Soc. Am.* **120** 676-85
[5] Guo P Y *et al* 2006 *Med. Phys.* **33** 1338-45
[6] Gorjiara T *et al* 2011 *Med. Phys.* **38** 2265-74
[7] Gorjiara T *et al* 2012 *Med. Phys.* **39** 7071-9
[8] Skyt P S *et al* 2012 *Med. Phys.* **39** 7232-36
[9] Baldock C *et al* 2010 *Phys. Med. Biol.* **55** R1-63

Availability of a containerless polymer gel detector and a gelatin container

Takahiro Tominaga, Munenori Yoshioka, Shin-ichiro Hayashi, Shuji Usui and Mitsutoshi Tada
Department of Clinical Radiology, Hiroshima International University
555-36 Kurose-Gakuendai, Higashi-Hirosima, Hiroshima 739-2695 Japan

E-mail: t-tomi@hs.hirokoku-u.ac.jp

Abstract. We considered an availability of the polymer gel detector without container but with a plastic wrap under assumption of the low oxygen transmissivity of a sheet of plastic wrap. And a gelatin container was also examined for a gel detector. These samples can be made easily and this containerless polymer gel detector well works without any artifacts by means of wrapping with a thin plastic sheet. Nevertheless, there is still room for improvement on preventing oxygen contamination. Combination with a gelatin container and a polymer gel detector and/or Gafchromic films has a various potential for extension of 3D dosimetry.

1. Introduction

Polymer gel detectors have been applied not only for measurements of dose distributions of X-ray, proton and heavy ion beams in radiation therapy [1], but also in boron neutron capture therapy to measure the dose sensitivity in a gel doped with boron [2], and in the dosimetry of radionuclide therapy [3]. This extensive range of application is attributed to the unique dosimetric properties of polymer gel detectors; their tissue equivalency, radiation sensitivity, spatial resolution and their potential to be a 3D dosimeter [4-6]. Nonetheless, polymer gel detectors have a number of limitations to be considered in executing the measurements. One of them is the need for supporting container to protect monomers in gel from oxygen contamination [7] and to persist the shape of gel. Well-sealed glass [8] and Barex vials are most commonly used as a detector's container, but these compositions are not tissue equivalency, and sometime lead to artifacts and deficiency of sensitivity at container proximity [9]. In this study, we examined applying a thin plastic sheet to pack gel instead of a solid container. Recent food preservative plastic wrap is thin and has a high sealing ability. Because of that, the polymer gel detector, which has no container but is wrapped by only a thin plastic sheet, is investigated its availability as a dosimeter. Further the potential of a gelatin container for 3D dosimetry is studied.

2. Material and methods

The procedure to fabricate a containerless polymer gel detector and a gelatin container are described. In this work standard polyacrylamide base gel (PAGAT) was used.

2.1. *Containerless polymer gel detector*

First a container is prepared. Cooking oil is applied all over the inner side of a container, and a sheet of plastic wrap is stuck on it (figure 1a). This process makes it easy to remove gel from a container after its solidification and prevent from oxygen contamination. Oxygen permeability of this wrap (Saran Wrap™, Asahi KASEI Co.) is 60 cc/m^2·day·atom, and thickness is 11μm. Then pour polymer-gel solution (PAGAT) filled the container. The surface is sealed in wrap. (figure 1b) The lump of polymer-gel is removed with wrap from the container just before irradiation (figure1c).

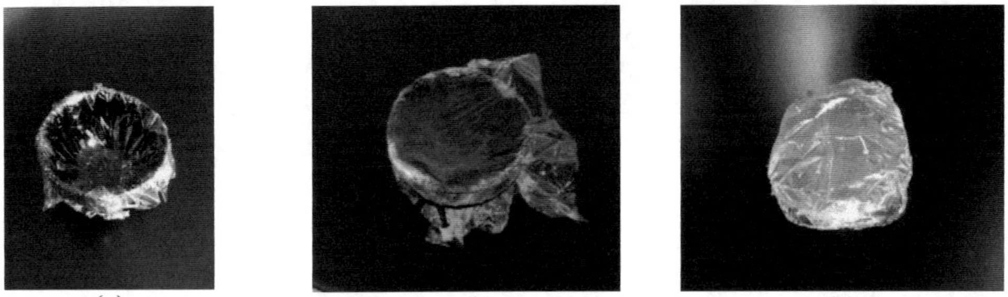

(a) (b) (c)

Figure. 1 The procedure to fabricate a containerless polymer gel detector.

2.2. *Gelatin container for polymer gel detector.*

Figure 2 shows the process of making a gelatin container. The lump of gelatin is made before making a container in the same procedure as having described in *2.1.* A bigger container is prepared and cooking oil is applied all over the inner side of a container and a sheet of plastic wrap is stuck on it. Then pour gelatin into the container, holding a lump of gelatin tenaciously at the centre of the container (figure 2a). On the next day a solidified gelatin container is removed from the outer container, and a lump of gelatin is removed from at the centre of gelatin container (figure 2b). A sheet of plastic wrap is stuck on the space where a lump of gelatin located. Then pour polymer gel solution filled that space. The surface is sealed in wrap.

(a) (b)

Figure. 2 The procedure to fabricate a gelatin container.

2.3. *Irradiation and Magnetic resonance imaging*

Fabricated gel samples were irradiated with 5 Gy of 6 MV photons using a linear accelerator to examine uniformity of the sensitivity. The sample made from one batch were irradiated on the next day, and the sample from another batch were irradiated one week after fabrication to examine the time degradation of sealing. MRI was undertaken approximately 24 h after the irradiation by 0.3T HITACHI AIRIS II scanner using head coil with a double spin-echo sequence at echo times TE_1=20 ms, TE_2=200 ms with a repetition time TR=5 sec.

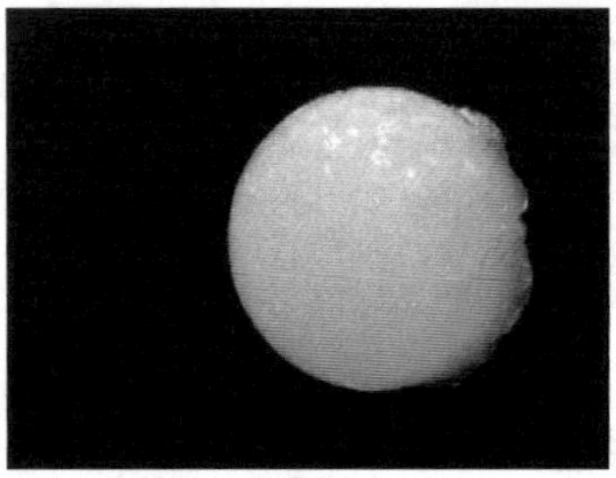

(a)

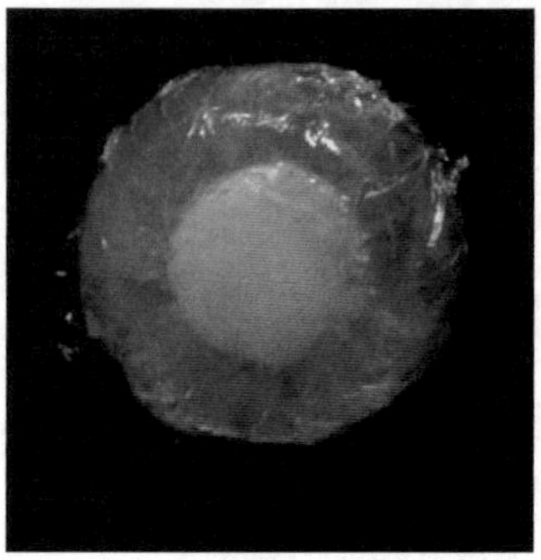

(b)

Figure 3. Containerless polymer gel detector (a) and a gelatin container (b) after irradiation

3. Results and discussions

Two consecutive magnetic resonance images (R2: the transverse nuclear magnetic relaxation rates) of a containerless polymer gel detector and a gelatin container are shown in figure 4.

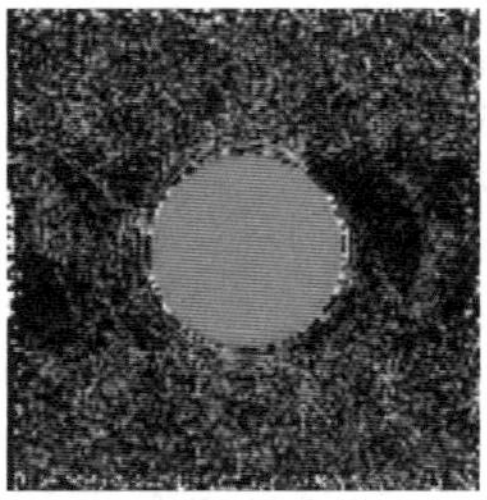

(a)

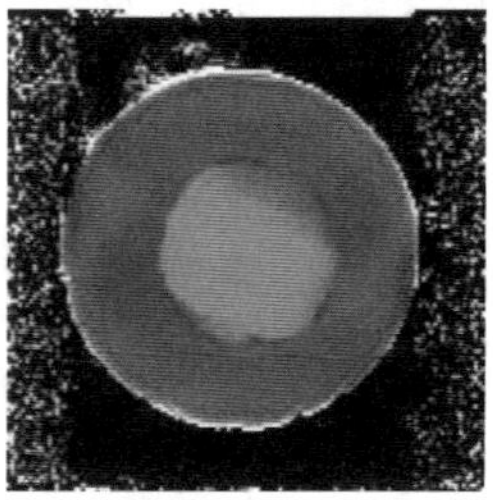

(b)

Figure 4. R2 images of a containerless polymer gel detector (a) and a gelatin container (b).

Line profiles of each detector are shown in figure 5 and figure 6 which correspond to the profile of the detector irradiated on the next day and one week after fabrications respectively.

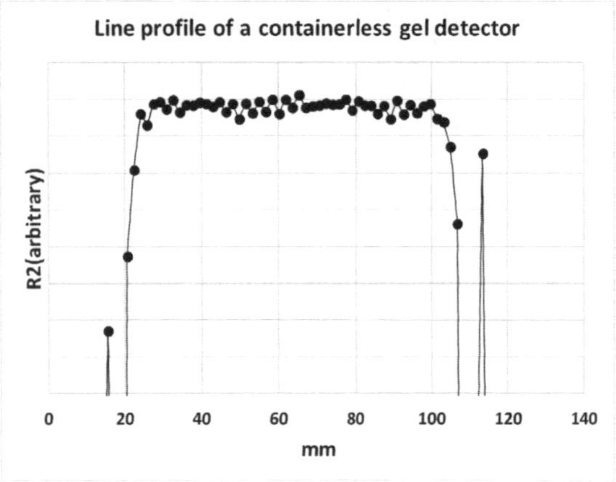

(a)

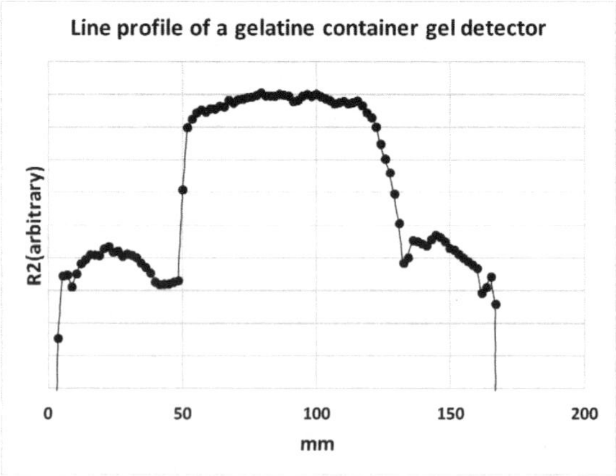

(b)

Figure 5. Line profile of a containerless polymer gel detector (a) and a gelatin container (b) irradiated on the next day.

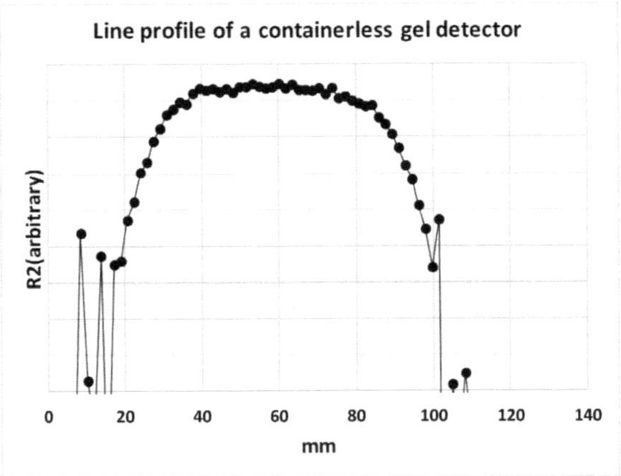

(a)

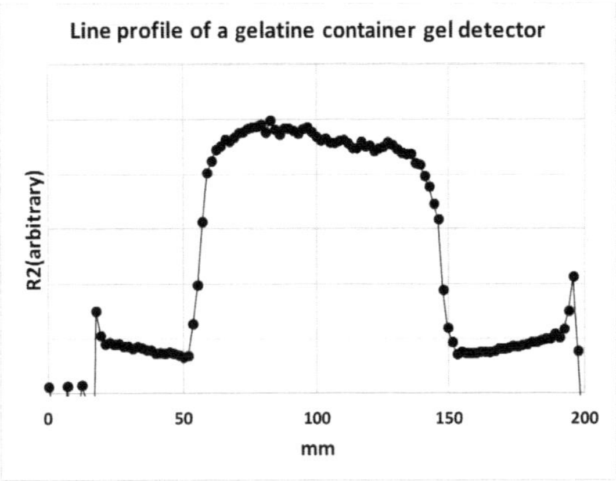

(b)

Figure 6. Line profiles of a containerless polymer gel detector (a) and a gelatin container (b) irradiated after one week.

As these profiles show, plastic wrap is useful to protect the gel detector from oxygen contamination at least one day, and the effect of the time degradation of sealing is seen as the deficiency of sensitivity at the surface in figure 6.

4. Summary

We examined an availability of the containerless polymer gel detector, and a gelatin container was applied to the polymer gel detector. This containerless polymer gel detector, which is only wrapped by a sheet of plastic wrap, well works without any artifacts and deficiency of sensitivity. The advantage of this containerless detector is; reducing a modality related error in the analysis and less ambiguity in constructing the geometry of computer simulations to evaluate dose measurements. And the combination of this detector with a gelatin container has also same advantages. Nevertheless, there is still room for improvement on preventing oxygen contamination as for the combination with gelatin container. The idea of gelatin container can be applied to another radiation detector, for instance as a holder of Gafchromic films. And the variable density of gelatin with additive may make the application of gelatin container wider extensive range, for instance as a phantom with the density of lung equivalency. As a continuation to this initial work, the following investigation are being undertaken; the examination of dose-response characteristics of a containerless polymer gel detector with much circumstance (esp. detector's size dependence in sensitivity), and comparison with Monte Carlo simulations and a treatment planning system in radiotherapy.

5. References

[1] Baldock C *et al* 2010 *Phys. Med. Biol.* **55** R1-63
[2] Farajollahi A R *et al* 2000 *Phys. Med. Biol.* **45** N9-14
[3] Braun K *et al J. Phys: Conf Ser.* **164** 01205
[4] Venning A J *et al* 2004 *J. Phys.: Conf. Ser.* **3** 155-8
[5] Baldock C 2006 *J. Phys.: Conf. Ser.* **56** 14-22
[6] Mather M L *et al* 2003 *Ultrasonics* **41** 551-9
[7] Hepworth S J *et al* 1999 *Phys. Med. Biol.* **44** 1875-84
[8] Michael G J *et al* 2000 *Phys. Med. Biol.* **45** N133-8
[9] Vandecasteele J *et al 2013 Phys. Med. Biol.* **58** 63-85

Enhancement in dose sensitivity of polymer gel dosimeters composed of radiation-crosslinked gel matrix and less toxic monomers

A Hiroki[1], S Yamashita[2] and M Taguchi[1]
[1]Environmental Radiation Processing Group, Quantum Beam Science Center, Japan Atomic Energy Agency, Takasaki, Gunma, 370-1292, Japan
[2]Nuclear Professional School, School of Engineering, The University of Tokyo, Tokai-mura, Naka-gun, Ibaraki, 319-1188, Japan

E-mail: hiroki.akihiro@jaea.go.jp

Abstract. Polymer gel dosimeters based on radiation-crosslinked hydroxypropyl cellulose gel were prepared, which comprised 2-hydroxyethyl methacrylate (HEMA) and polyethylene glycol #400 dimethacrylate (9G) as less toxic monomers and tetrakis (hydroxymethyl) phosphonium chloride (THPC) as an antioxidant. The dosimeters exposed to ^{60}Co γ-rays became cloudy at only 1 Gy. The irradiated dosimeters were optically analyzed by using a UV-vis spectrophotometer to evaluate dose response. Absorbance of the dosimeters linearly increased in the dose range from 0 to 10 Gy, in which dose sensitivity increased with increasing 9G concentration. The dose sensitivity of the dosimeters with 2 wt% HEMA and 3 wt% 9G was also enhanced by increment in THPC.

1. Introduction

Most of the polymer gel dosimeters [1] utilize gelatin and either methacrylic acid or acrylamide [2, 3]. Since both monomers are considered harmful, new recipes for polymer gel dosimeters using less toxic monomer, *N*-isopropyl acrylamide (NIPAM), have been reported [4]. Furthermore, NIPAM polymer gel dosimeters with improved dose sensitivity have been reported by Chain *et al* [5]. Recently, Hayashi *et al* reported that enhancements of dose sensitivity were achieved by adding inorganic salts [6]. On the other hand, Hiroki *et al* reported that the dose sensitivity of dosimeters based on gellan gum and less toxic monomers increased with the initial monomer concentration [7]. In these gel dosimeters, the solubility and compatibility of the chemicals become a hindrance to the increasing dose sensitivity as well as the preparation of transparent polymer gel dosimeters.

Gels of polysaccharide derivatives such as carboxymethyl cellulose and hydroxypropyl cellulose (HPC) were prepared by γ-rays or electron beam irradiation to their highly concentrated aqueous solution without any crosslinker. The prepared gels have been studied to apply to a soil improvement material in agriculture and a radiation dosimeter in radiation therapy, using a high water-absorption [8-10]. It is possible to prepare the polymer gel dosimeter with a high monomer concentration, since the gel largely absorbs not only water, but also organic solvent.

This study aims to enhance in dose sensitivity of polymer gel dosimeters based on a radiation-crosslinked gel matrix. HPC gel was selected as gel matrix for the dosimeter because of transparent and thermostable material in comparison with gelatin gel. Polymer gel dosimeters comprising less toxic methacrylate monomers, THPC, and HPC gel matrix were prepared, and then exposed to γ-rays.

Journal of Physics: Conference Series **573** (2015) 012028 doi:10.1088/1742-6596/573/1/012028

Dose response was evaluated by measuring absorbance of the dosimeters. Effects of the composition ratio of monomers and THPC concentration on the dose sensitivity of the dosimeters were investigated.

2. Materials and methods

2.1. Preparation of polymer gel dosimeters

HPC was dissolved in ultrapure water at the concentration of 20 wt%. The HPC aqueous solution as a paste-state was formed to 1.0 mm thickness by pressing, and then was irradiated to a dose of 10 kGy with electron beam (2 MeV, 2 mA) at the Takasaki Advanced Radiation Research Institute, Japan Atomic Energy Agency. The obtained HPC gel membrane was immersed into an excess amount of distilled water to remove uncrosslinked HPC, and then vacuum-dried. The dried HPC gels were immersed into monomer aqueous solutions that consist of 0.8-4 wt% HEMA, 1-6 wt% 9G, and 0-0.24 wt% THPC. The swollen HPC gels were vacuum packed in PE/nylon package to obtain polymer gel dosimeters. The prepared samples were stored in a refrigerator until γ-irradiation.

2.2. Irradiation

The prepared samples were irradiated up to 10 Gy by using a ^{60}Co g-ray source at the Takasaki Advanced Radiation Research Institute, Japan Atomic Energy Agency. Dose rate was adjusted by varying the distance between the samples and the source, which was in the range of 3.0 to 30 Gy h^{-1}. The temperature inside the irradiation room was about 20°C and was controlled with an air conditioner. After irradiation, all samples were stored in the refrigerator.

2.3. Analysis of irradiated samples

The absorbance of the polymer gel dosimeters were measured after 24 hours post-irradiation. The optical analysis was carried out using an UV-vis spectrophotometer (Hitachi High-Technologies Corporation, U-3310) at 25°C. Dose sensitivity of the dosimeter was estimated from the absorbance at 660 nm as a function of the dose

3. Results and Discussion

Figure 1 shows photographs of non- and γ-irradiated polymer gel dosimeters. They became cloudy by g-irradiation at 1 Gy. It is clearly seen that the cloudiness of the polymer gel dosimeters prepared by using a monomer solution with 2 wt% HEMA, 3 wt% 9G, and 0.16 wt% THPC increased with the increasing dose.

Figure 1. Photographs of non- and γ-irradiated polymer gel dosimeters consisting of HEMA, 9G, and THPC with HPC gel. HEMA, 9G, and THPC concentrations of the monomer solution were 2 wt%, 3 wt%, and 0.16 wt% respectively.

Figure 2 shows the absorbance of the polymer gel dosimeters with various composition ratios between HEMA and 9G as a function of dose. The absorbance of the polymer gel dosimeters increased approximately linearly in the dose range from 0 to 10 Gy. Dose sensitivity defined as the initial increment in absorbance per unit dose increased with an increase in 9G concentration, reached about 0.04 Abs Gy^{-1} at the dosimeter with 2 wt% HEMA, 3 wt% 9G. Maryanski *et al* also reported that the dose sensitivity of BANG polymer dosimeter significantly depended on the composition ratio between acrylamide (AAm) and *N,N'*-methylene bis-acrylamide (MBAAm) as a crosslinker [11]. A crosslinker induces an enhancement of gel fraction because it promotes polymerization [12]. 9G as a crosslinker would promote the formation of a copolymer gel comprised of HEMA and 9G. Small particles are produced with irradiation in 9G-rich condition in comparison with HEMA-rich one [10]. It is considerable, therefore, that the polymer conversion and the aggregation size enhance the increase in the absorbance, resulting in the increase in the dose sensitivity.

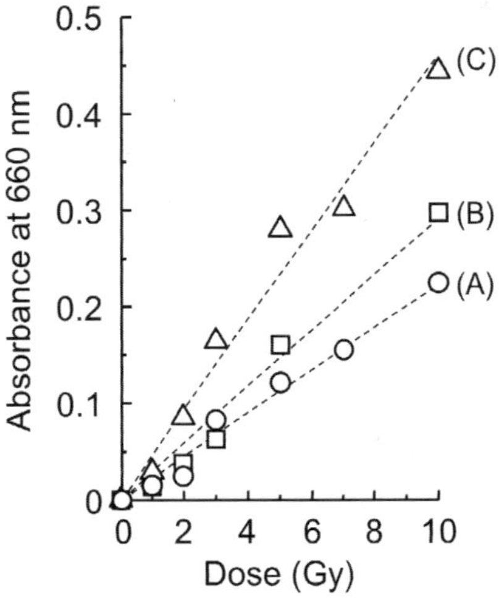

Figure 2. Increment of the absorbance of the polymer gel dosimeters prepared with aqueous solutions consisting of HEMA, 9G, and THPC. HEMA and 9G concentrations are 4 wt% and 1 wt% (A), 3 wt% and 2 wt% (B), and 2 wt% and 3 wt% (C), respectively. THPC concentration is the same in 0.16 wt%.

To induce higher absorbance in the polymer gel dosimeter, it is preferable to increase in polymerization product scattering light. The polymer gel dosimeters with various total monomer concentrations (2, 5, and 10 wt%) at composition ratio of 2/3 were prepared. As a result of the irradiation at 10 Gy, the absorbance of the dosimeters increased with total monomer concentration, reaching about 1.3 at 4 wt% HEMA and 6 wt% 9G.

The polymerization reaction is inhibited in oxygen-contaminated gel, since oxygen scavenges free radicals produced by water radiolysis. THPC as an antioxidant plays an important role for reduction of the oxygen effects. The absorbance of the dosimeters with various concentrations of THPC increased in the dose range from 0 to 10 Gy. Figure 3 shows relationship between THPC concentration and the dose sensitivity of the polymer gel dosimeters with 2 wt% HEMA and 3 wt% 9G. The dose sensitivity increased with the initial THPC concentration. Jirasek *et al* reported that optimum THPC concentration to maximize the dose response in the PAGAT dosimeter is about 4.5 mM because of the reactions with gelatin, AAm, and MBAAm as gel constituents at high concentration [13]. This is corresponded to about 0.09 wt% THPC. In the dosimeter consisting of HEMA and 9G with HPC gel, there is no maximum of the dose sensitivity until 0.24 wt% THPC. This would be due to the effects of the matrix and monomers.

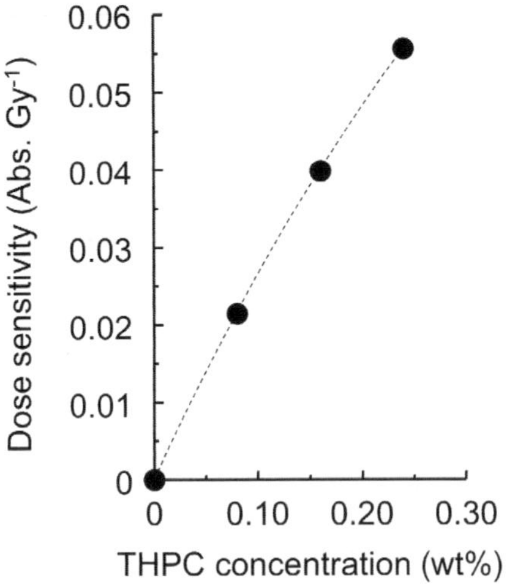

Figure 3. Relationship between THPC concentration and dose sensitivity of the prepared polymer gel dosimeters with 2 wt% HEMA and 3 wt% 9G.

4. Conclusion

Polymer gel dosimeters that consist of HEMA, 9G, and THPC with radiation-crosslinked HPC gel showed the cloudiness at 1 Gy to be observable easily by eyes. The absorbance of the polymer gel dosimeter increased linearly in the range of dose from 0 to 10 Gy. The dose sensitivity increased with 9G concentration. The dose sensitivity could be also enhanced by adjusting the THPC concentration, reached about 0.056 Abs. Gy^{-1} at the dosimeter with 2 wt% HEMA, 3 wt% 9G, and 0.24 wt% THPC.

5. Acknowledgement

This work was partially supported by JSPS KAKENHI Grant Number 23710073, 26460737.

6. References

[1] Baldock C *et al* 2010 *Phys. Med. Biol.* **55** R1-63
[2] Maryanski M J *et al* 1994 *Phys. Med. Biol.* **39** 1437-55
[3] Baldock C *et al* 1998 *Phys. Med. Biol.* **43** 695-702
[4] Senden R J *et al* 2006 *Phys. Med. Biol.* **51** 3301-14
[5] Chain J N M *et al* 2011 *Phys. Med. Biol.* **56** 2091-102
[6] Hayashi S *et al* 2012 **81** 884-8
[7] Hiroki A *et al* 2013 *Phys. Med. Biol.* **58** 7131-41
[8] Hiroki A *et al* 2011 *Trans. Mater. Res. Soc. Jpn.* **36** 397-400
[9] Hiroki A *et al* 2013 *J. Phys.: Conf. Ser.* **444** 012028
[10] Yamashita S *et al* 2014 *Radiat. Phys. Chem.* **101** 53-8
[11] Maryanski M J *et al* 1997 *Phys. Med. Biol.* **42** 303-11
[12] Sajjadi S *et al* 1996 *Polymer* **37** 4141-8
[13] Jirasek A *et al* 2006 *Phys. Med. Biol.* **51** 1891-906

8th International Conference on 3D Radiation Dosimetry (IC3DDose) IOP Publishing
Journal of Physics: Conference Series **573** (2015) 012029 doi:10.1088/1742-6596/573/1/012029

Do saccharide doped PAGAT dosimeters increase accuracy?

B Berndt[1], P S Skyt[1,2], L Holloway[3,5], R Hill[4], A Sankar[3] and Y De Deene[1,6]
[1]Institute of Medical Physics, University of Sydney, Sydney, Australia
[2]Department of Medical Physics, Aarhus University Hospital, Aarhus, Denmark
[3]Liverpool Cancer Therapy Centre, Sydney, Australia
[4]Chris O'Brien Lifehouse, Sydney, Australia
[5]Macarthur Cancer Therapy Centre, Sydney, Australia
[6]Department of Engineering, Macquarie University, Sydney, Australia

E-mail: yves.dedeene@mq.edu.au

Abstract. To improve the dosimetric accuracy of normoxic polyacrylamide gelatin (PAGAT) gel dosimeters, the addition of saccharides (glucose and sucrose) has been suggested. An increase in R_2-response sensitivity upon irradiation will result in smaller uncertainties in the derived dose if all other uncertainties are conserved. However, temperature variations during the magnetic resonance scanning of polymer gels result in one of the highest contributions to dosimetric uncertainties. The purpose of this project was to study the dose sensitivity against the temperature sensitivity. The overall dose uncertainty of PAGAT gel dosimeters with different concentrations of saccharides (0, 10 and 20%) was investigated. For high concentrations of glucose or sucrose, a clear improvement of the dose sensitivity was observed. For doses up to 6 Gy, the overall dose uncertainty was reduced up to 0.3 Gy for all saccharide loaded gels compared to PAGAT gel. Higher concentrations of glucose and sucrose deteriorate the accuracy of PAGAT dosimeters for doses above 9 Gy.

1. Introduction

In the last decades, normoxic polymer gels have been investigated as a device for three dimensional dosimetry [1]. Radiation-induced polymerisation of contained monomers reduces the spin-spin relaxation time (T_2) in a nuclear magnetic resonance (NMR) read-out. This increase in R_2 is proportional to the dose delivered to the polymer gel during irradiation [2]. It has been stated that normoxic PolyAcrylamide Gelatin gels fabricated at ATmospheric conditions (PAGAT-gels) show a superior performance with respect to spatial integrity, temperature sensitivity, dose-rate dependence and radiological water equivalence compared to other polymer gels [3, 4]. Nevertheless, an extensive study of the sources of uncertainty of PAGAT polymer gel dosimeters has pinpointed the temperature of the PAGAT gel during scanning as a major source of inaccuracies [5]. This current study focuses on the improvement of the accuracy of PAGAT gels by adding saccharides. The uncertainty was determined taking into account both the change in dose sensitivity and the temperature sensitivity. According to Yoshioka *et al* [6], the dose sensitivity of PAGAT gels is improved by adding sucrose in different concentrations. Furthermore, studies conducted by Zhu *et al* [7] and Ding [8] confirmed an increase in dose sensitivity of methacrylic acid-based polymer gels (MAGIC) by adding glucose. However, the increase in dose sensitivity by adding saccharides does not necessarily improve the overall dosimetric accuracy, if the temperature sensitivity of the gel during read-out is also increased proportionally. The temperature sensitivity is therefore as important as the dose sensitivity. Therefore,

PAGAT gels with saccharides additives were irradiated to a range of doses, in order to investigate whether the expected increase in dose sensitivity is correlated with an overall reduction of the dose uncertainty caused by temperature dependence. A NMR-relaxometer was utilized in order to obtain overall relaxation times.

2. Materials and methods

2.1. Gel fabrication

Thirty PAGAT gel samples were fabricated and divided into five sets. The first set of samples consisted of pure PAGAT gel for comparison reasons. In a second and third set, 10% and 20% glucose were added, respectively. In a fourth and fifth set 10% and 20% sucrose were added. The remaining ingredients of the gel were: 6% (w/w) gelatin, 3% (w/w) N,N'-Methylene-bis-acrylamide (Bis) as a crosslinker monomer, 3% (w/w) acrylamide (Aam) and 10 mM tetrakis(hydroxymethyl) phosphonium chloride (THP) as an antioxidant. A stock solution was produced with these ingredients according to the procedure described in [9]. After splitting the stock solution into five equal parts, the corresponding amounts of water, sucrose and glucose were added. The THP was then added and the liquid gel was poured into 6 glass tubes for NMR-measurements. The upper part of the glass tubes were subsequently filled with nitrogen gas to prevent oxygen from penetrating the gel, inhibiting polymerisation. All the samples were stored in the refrigerator between 0.5 and 2 hours before irradiation.

2.2. Irradiation

A Varian 21 iXs linear accelerator (Varian Medical Systems) producing 6 MV photons was used to irradiate all gel samples. Each sample was irradiated with the tube axis perpendicular to the beam axis using a 12×12 cm^2 field at a depth of 6.8 cm in a special solid water equivalent phantom. The source to surface distance (SSD) was 100 cm and the sequentially delivered doses (D) were 2, 4, 6, 9 and 13 Gy for the different samples with a dose rate of 4.84 Gy/min. The sixth sample of each set was left unirradiated.

2.3. NMR T_2-measurements

The overall T_2-value of each test tube at different temperatures was measured using a NMR-minispec mq20 analyzer (Bruker Optik GmbH) with a magnetic field strength of 0.47 T. The echo time spacing was chosen to be 0.6 ms for all measurements. Between 1,000 and 10,000 data points were acquired depending on the corresponding spin-spin relaxation time (T_2). An average of 4 scans of the measured magnetisation was fitted to a mono-exponential function to obtain a T_2-value. The sample temperature was controlled before and during the read-out process using a thermostatic water bath. The temperature within the gel sample was measured with a fiber optic thermometry system (Neoptix, Inc.). The accuracy of the temperature recording was $\pm$ 0.2 °C. Each sample was measured twice for each water bath temperature. In a time period between 16 and 45 hours after irradiation, all samples were scanned at subsequent temperatures ($T \approx 6$, 14, 18, 23 and 28 °C). At the end, all test tubes were scanned again at the lowest temperatures in order to detect possible changes in T_2 caused by post-polymerisation of gels.

3. Results and discussion

After an interpolation of R_2-values for specific temperature values to facilitate comparison, the dose-R_2 response relations of five different gels at room temperature ($T = 20$ °C) were plotted and are shown in figure 1. The fitting of the R_2-values indicates a mono-exponential dose-R_2 relation: $R_2(D) = A_D \cdot \exp(B_D \cdot D) + C_D$, where A_D, B_D and C_D are exponential fitting parameters. A significant increase in dose-R_2 response for samples with glucose or sucrose compared to conventional PAGAT gel samples is observed. The slope of the dose-R_2 response around 0 Gy is especially increased for samples with 20% glucose and sucrose. At 20 °C, the slope is increased by a factor of approximately 9

8th International Conference on 3D Radiation Dosimetry (IC3DDose) IOP Publishing
Journal of Physics: Conference Series **573** (2015) 012029 doi:10.1088/1742-6596/573/1/012029

for 20% glucose, a factor of 7 for 20% sucrose and a factor of 3 for 10% glucose and 10% sucrose compared to PAGAT gel. The temperature dependence of the dose-R_2 response relation can be seen in figure 2 for PAGAT gel and PAGAT gel with 20% glucose.

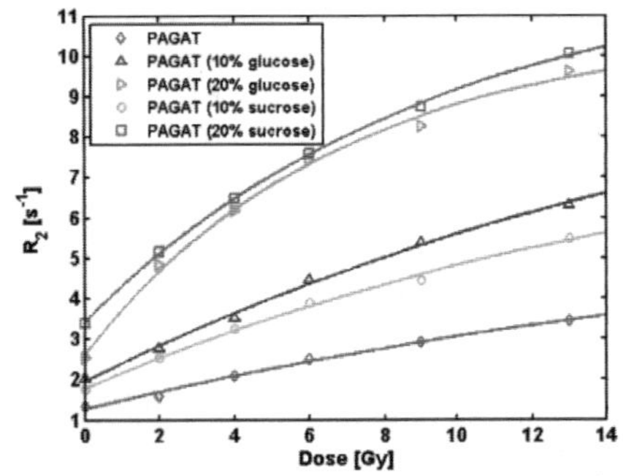

Figure 1. R_2-dose plots at room temperature (T = 20 °C) for different types of gels fitted to mono-exponential equations.

Besides other less dominant sources of uncertainty which were not considered in this experiment, the mono-exponential temperature dependence of R_2 ($R_2(T) = A_T \cdot \exp(B_T \cdot T) + C_T$, with fitting parameters A_T, B_T and C_T) leads to an uncertainty of the subsequently determined dose in an irradiated 3D phantom. This dose uncertainty due to temperature changes (U_D) during the read-out process at a specific temperature can be calculated according to equation (1) and is dependent on the delivered dose.

$$\square_{\square}(\square,\square) = \frac{\Delta\square}{\Delta\square}(\square = \square_0) = \left(\frac{\square\square_2(\square,\square)}{\square\square} \Big/ \frac{\square\square_2(\square,\square)}{\square\square}\right) = \frac{\square_{\square} \cdot \square_{\square} \cdot \exp(\square_{\square} \cdot \square_0)}{\square_{\square} \cdot \square_{\square} \cdot \exp(\square_{\square} \cdot \square)} \quad (1)$$

The absolute (U_D) and relative uncertainty ($U_{D,rel} = U_D/D$) attributed to a temperature change of $\Delta T = 1$ °C are shown in figure 3 for different gels. In the lower dose region (up to 9 Gy), the dose uncertainty is decreased by adding glucose and sucrose. The lowest absolute uncertainty can be assigned to samples with 20% glucose and is around 0.09 and 0.52 Gy between delivered doses of 0 to 6 Gy, whereas the absolute dose uncertainty of PAGAT gel in this low dose range is between 0.43 and 0.83 Gy. U_D of the 20% glucose as well as of the 20% sucrose samples exceeds the uncertainty of PAGAT gel for doses above 9 Gy. PAGAT gel with 10% sucrose seems to have the lowest and most stable absolute dose uncertainty in the whole evaluated dose range. By adding 10% sucrose the dose uncertainty is reduced by between 0.18 Gy for a delivered dose of 2 Gy and 0.32 Gy for a delivered dose of 13 Gy compared to PAGAT gel.

No significant change in R_2-values was observed between 16 and 43 hours after irradiation. Therefore, observed changes in R_2 over time as a result of temporal instability were not considered in the analysis.

Another temperature dependence experiment with samples containing different gelatine concentrations (results not shown here) showed that the relative change in R_2 is considerably increased by adding a certain amount of gelatin. It is expected that the gelling agent plays a significant role in temperature sensitivity.

4. Conclusion

PAGAT gels with sucrose and glucose as additives showed a significant increase in dose response compared to conventional PAGAT gel. However, the temperature sensitivity during the read-out process is increased significantly. At lower doses, a small decrease in temperature related relative dose uncertainty is found. For higher doses, samples with a high sucrose or glucose concentration (20%) demonstrate an even higher dose uncertainty than PAGAT gel. In summary, although a higher dose sensitivity of PAGAT gels with saccharides additives was observed, adding glucose and sucrose does not correspond to a considerable improvement as compared to conventional PAGAT gel dosimeters.

8th International Conference on 3D Radiation Dosimetry (IC3DDose) IOP Publishing
Journal of Physics: Conference Series **573** (2015) 012029 doi:10.1088/1742-6596/573/1/012029

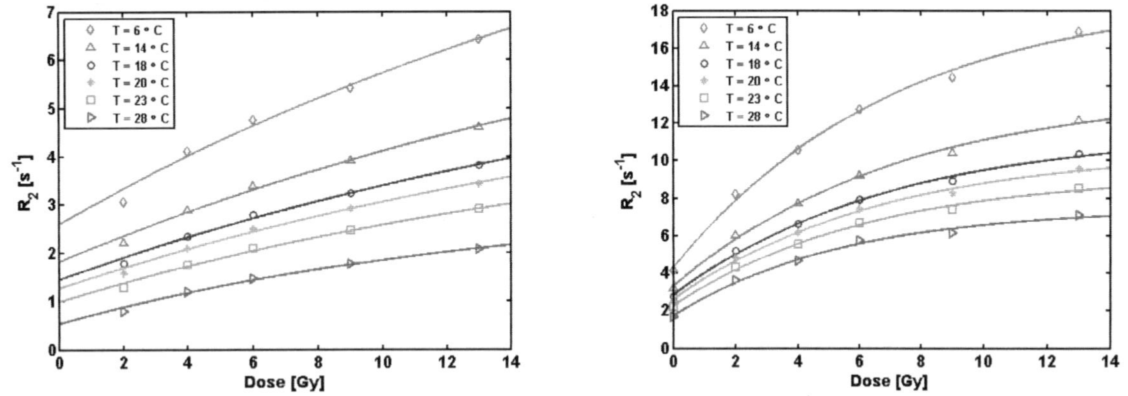

(a) PAGAT gel (b) PAGAT gel with 20% glucose

Figure 2. Temperature dependence of R_2-dose response curves for (a) PAGAT gel and (b) PAGAT gel with 20% glucose.

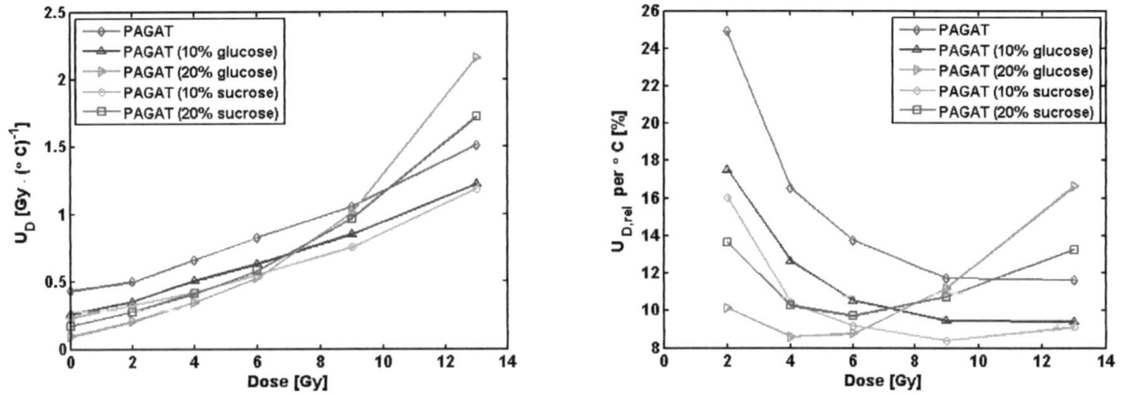

(a) Absolute dose uncertainty (U_D) (b) Relative dose uncertainty ($U_{D,rel}$)

Figure 3. (a) Absolute and (b) relative dose uncertainty in determined dose caused by temperature variation around room temperature as a function of delivered dose. (Lines are drawn to guide the eye).

5. References

[1] De Deene Y *et al* 2002 *Phys. Med. Biol.* **47** 3441-63
[2] Baldock C *et al* 2010 *Phys. Med. Biol.* **55** R1-63
[3] De Deene Y *et al* 2006 *Phys. Med. Biol.* **51** 653-73
[4] Hill R *et al* 2010 *Med. Phys.* **37** 4355-63
[5] Vandecasteele J *et al* 2013 *Phys. Med. Biol.* **58** 63-85
[6] Yoshioka M *et al* 2010 *J. Phys. Conf. Series* **250** 012016
[7] Zhu X *et al* 2010 *Med. Phys.* **37** 183-8
[8] Ding X 2012 *High field magnetic resonance imaging-based gel dosimetry for small radiation fields* Ph.D. thesis Wake Forest University Graduate School of Arts and Science
[9] De Deene Y *et al* 2000 *Phys. Med. Biol.* **45** 859-78

Methylthymol blue in Fricke gels

K I Penev[1] and K Mequanint[1,2]
[1]Department of Chemical and Biochemical Engineering, University of Western Ontario, London, Ontario, N6A 5B9, Canada
[2]Biomedical Engineering Program, University of Western Ontario, London, Ontario, N6A 5B9, Canada

E-mail: kmequani@uwo.ca

Abstract. The initial trial of methylthymol blue (MTB) as a chelator for ferric iron in Fricke gel dosimeters, used for three-dimensional (3D) dosimetry in cancer radiotherapy, is reported. MTB is a structural analogue of the conventionally used xylenol orange (XO); however, the absorbance spectrum of the ferric-MTB complex is shifted to higher wavelengths, which should allow for lower amount of light scattering during gel scanning. In this study, two gelatin substrates, two sources of XO and one source of MTB have been compared. The MTB-containing gels exhibited similar dose response and diffusion coefficient to the XO-containing gels at their wavelengths of maximum absorption (620 and 585 nm, respectively). In addition, the MTB gels gave an excellent dose response at 633 nm, which is an important wavelength that is already used with other 3D dosimeters.

1. Introduction

The principle of the ferrous sulfate – gelatin dosimeter (Fricke gel) is based on the ionizing radiation-induced oxidation of ferrous iron (Fe^{2+}) to ferric iron (Fe^{3+}) under acidic conditions in a gel material that retains the three-dimensional (3D) distribution of the delivered dose. Xylenol orange (XO) is a metal chelator that is typically added to Fricke gels to limit the diffusion and allow optical readout of the dose delivery [1, 2]. Other chelators have been tested to generally less favorable diffusion and dose response properties [2]. However, to our best knowledge, structural analogues of XO have never been tested in the Fricke gel dosimeter. We performed a structural search within the SciFinder® chemical database and found several potential compounds that could substitute XO in Fricke gels. Among them methylthymol blue (MTB) was found to be most promising: (1) the compound is very similar but larger than XO, which means that the diffusion of the Fe^{3+}-MTB complex may be slower than the Fe^{3+}-XO diffusion, (2) MTB is readily available as a commercial product, and (3) the maximum absorption Fe^{3+}-MTB is shifted to the red at 620-625 nm, compared to 585-587 nm for Fe^{3+}-XO [3]. Ultimately this means that the optical scattering in ferrous – MTB – gelatin (FMG) gels should be lower than that in ferrous – XO – gelatin (FXG) gels, and lower scattering means better image quality [4,5]. This report presents a comparison between XO and MTB as chelators in Fricke – gelatin gels.

2. Experimental

Two sources of gelatin were used in this study: 300 Bloom porcine skin gelatin (G_P) (Sigma-Aldrich, Canada, product no: G1890), and 300 Bloom bovine bone gelatin (G_B) (Rousselot International, Peabody, MA, USA), identical with the Eastman gelatin used in [5]. Xylenol orange as a disodium salt was supplied from Sigma-Aldrich (Fluka brand, product no: 52097), designated as X_S, and from TCI

America (Portland, OR, USA, product no: B0477), designated as X_T. Methylthymol blue, as a sodium salt was supplied for TCI America (product no: B0478), designated as M_T. Sulfuric acid (SA) and ferrous ammonium sulfate hexahydrate (FAS) were purchased from Caledon Labs (Georgetown, ON, Canada). All compounds were used without further purification in the preparation of Fricke gels in all possible combinations between gelatin substrate (G_P or G_B) and chelate (X_S, X_T or M_T) to the following common formulation (in mmol/L): SA:FAS:(XO or MTB) = 25:0.1:0.1 in 4% (w/v) gelatin. The gels were poured into polymethylmethacrylate cuvettes with 10 mm optical path length and left to solidify in at 10 °C overnight before irradiation on the next day.

Irradiations were performed on a Co-60 source (Eldorado 6, Atomic Energy of Canada Ltd.) at the London Regional Cancer Program (London, Canada). Cuvettes of each composition were exposed to 0, 5.0 and 20.0 Gy, and, semi-shielded, to 10.0 Gy for diffusion measurements. Optical scans were performed on a microplate reader (Infinite 200 Pro, Tecan Group Ltd., Switzerland). The semi-irradiated cuvettes were scanned at twelve points along the long axis of the cuvettes at three different times, starting one hour after irradiation, as previously described [6]. The diffusion profiles were fitted to the complementary error function to determine their 'curvature parameters' (*n*), as in [7]. The diffusion coefficients (*D*) were determined from the slope of *n* against time (*t*).

$$D = \frac{\sqrt[3]{4}-1}{4\ln(2)}\frac{dn}{dt} \approx 0.212\frac{dn}{dt} \qquad (1)$$

3. Results and discussion

Figure 1 depicts the typical behavior of the gels in porcine gelatin. As seen, the initial absorbance in the green part of the spectrum for FMG gels is much lower than that of FXG gels; however, after irradiation, the signal is stronger and red-shifted, forming the saturated green color seen on the right.

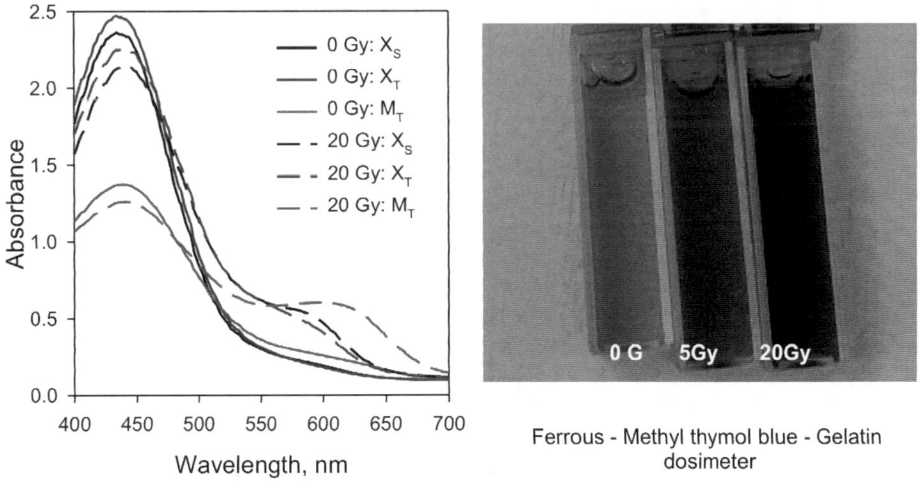

Figure 1. Dose responses to 20 Gy in 4% porcine skin gelatin and 25 mmol/L sulfuric acid.

Figure 2 presents the background absorbance, dose sensitivity, and the non-linearity of the dose response over a wide range of wavelengths (550 to 650 nm). In both porcine and bovine gelatin, the gels containing XO sourced from TCI America show the lowest sensitivity and linearity; nevertheless, even with these gels the error of linearity is less than 1% around the Fe^{3+}-XO absorbance maximum (585 nm). The Sigma-Aldrich – sourced XO could be used at a wide range of wavelengths but the sensitivity of the dosimeter drops fast at wavelengths over 585 nm. By contrast, the FMG gels show both high sensitivity and good linearity on both sides of the absorbance maximum at 620 nm, and could potentially be used at 633 nm – a wavelength that can be generated by He-Ne lasers and is already used for scanning of 3D dosimeters (*viz*. PRESAGE®).

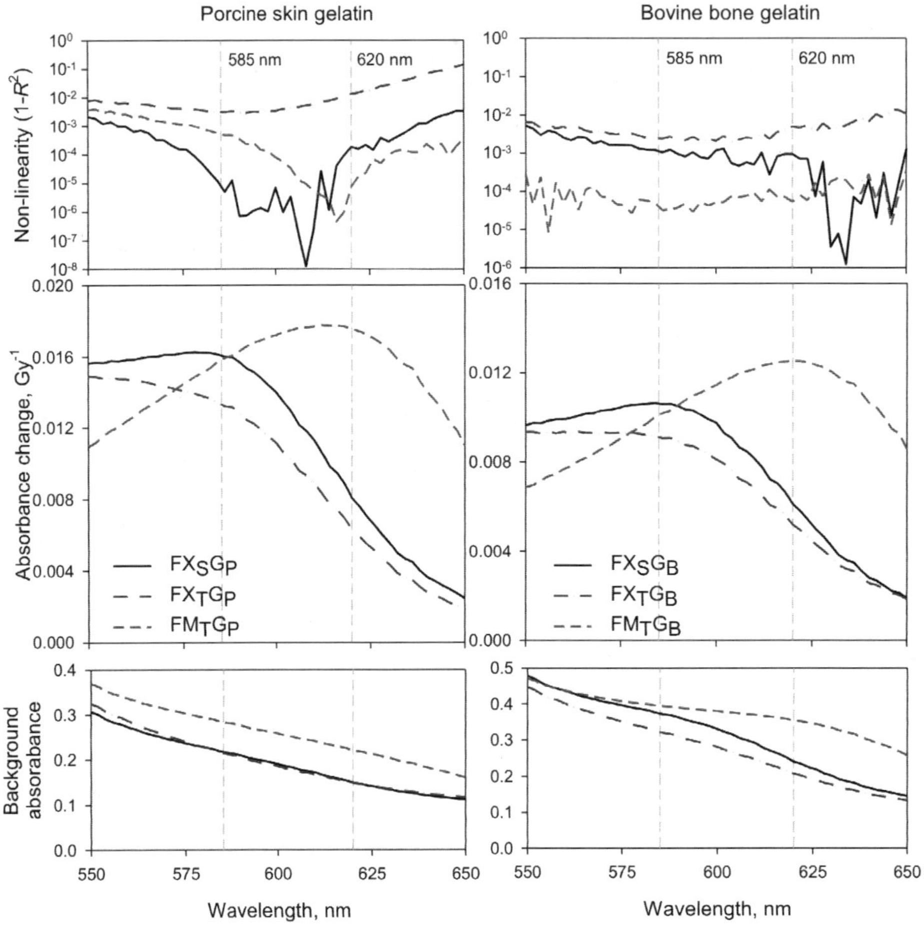

Figure 2. Background absorbance, dose sensitivity, and non-linearity of dose response, expressed as the complement of the correlation coefficient to unity ($1\text{-}R^2$).

Figure 3 (left) presents the diffusion profiles for the FMG gels, at three times with t = 0 at one hour after irradiation. As expected, the diffusion slopes become shallower with time; hence, the curvature parameters (n) increase. The linear dependences between n and time (figure 3, left) give the diffusion coefficients, according to equation (1).

Table 1 gives a snapshot of the data from figure 2 together with the estimated diffusion coefficients from figure 3. The diffusion properties of dosimeters do not show a consistent behavior between the porcine and the bovine gels, and since the experiments were not conducted in multiples definitive conclusions are hard to make. However, the average diffusion coefficients of the porcine and the bovine gels (0.573±0.069 and 0.341±0.011 mm^2h^{-1}, respectively) are significantly different (p = 0.005). In a previous study we found the diffusion coefficient of an FX_SG_P gel with 25 mM SA to be 0.476±0.011 mm^2h^{-1} [6]. This value is not statistically different from the average porcine gel diffusion from this report (p = 0.094). Based on these results, we can conclude that the diffusion in FMG gels is not lower than that in FXG gels (as hoped), but it is of equal magnitude or slightly higher.

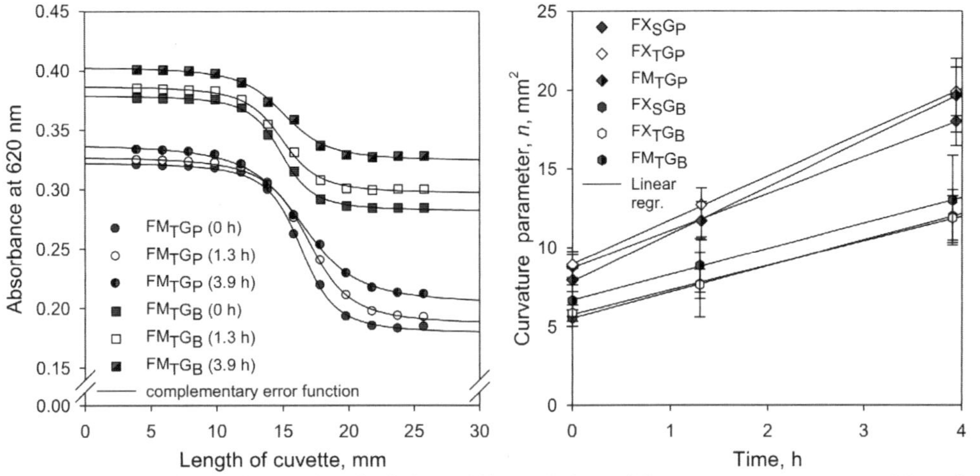

Figure 3. Diffusion profiles in the FMG gels (left) and linear fitting of the curvature parameters (n) against time for all tested compositions (right).

Table 1. Properties of the dosimeters are characteristic wavelengths.

Gel code	Gelatin type	Chelator	Wavelength (nm)	Bckgr. abs. $\times10^3$	Abs. change $\times10^3$ (Gy^{-1})	$1 - R^2$	Diffusion coeff. (mm^2h^{-1})
FX_SG_P	porcine	XO_{Sigma}	585	218	16.1	5×10^{-6}	0.497
FX_TG_P	porcine	XO_{TCI}	585	215	13.3	0.003	0.589
FM_TG_P	porcine	MTB_{TCI}	620	221	17.5	7×10^{-6}	0.632
			633	196	15.7	1×10^{-4}	n.d.
FX_SG_B	bovine	XO_{Sigma}	585	373	10.6	0.001	0.351
FX_TG_B	bovine	XO_{TCI}	585	322	9.1	0.002	0.329
FM_TG_B	bovine	MTB_{TCI}	620	354	12.5	5×10^{-5}	0.341
			633	323	11.7	2×10^{-4}	n.d.

4. Conclusions

The substitution of xylenol orange with methylthymol blue in Fricke-gelatin gels produced dosimeters with comparable properties, and an absorbance maximum shifted to the red. Due to that, the new dosimeters are expected to cause less optical scatter in optical computed tomography systems and may be used with existing light sources, set at 633 nm for the scanning of other 3D dosimeters. In addition, there was a noticeable difference in the behaviors of dosimeters prepared with porcine skin gelatin and the bovine bone gelatin. It is possible that the gelatin substrates are hydrolyzed to different extents by the presence of sulfuric acid, which may account for their different properties. We plan on examining those effects in the future. Further research on the ferrous – methylthymol blue – gelatin gel will focus on the stability and reusability of the dosimeter, the effects of the sulfuric acid concentration, and presence of ferrous iron stabilizing dopants.

5. Acknowledgments

We would like to acknowledge the financial support of the Mitacs Elevate program, the Ontario Centres for Excellence and the Natural Sciences and Engineering Research Council of Canada, as well as Modus Medical Devices, Inc. (London, Ontario, Canada).

6. References

[1] Appleby A and Leghrouz A 1991 *Med. Phys.* **18** 309
[2] Rae W I D *et al* 1996 *Med. Phys.* **23** 15
[3] Yoshino T *et al* 1979 *J. Inorg. Nucl. Chem.* **41** 1011

[4] Jordan K and Sekimoto M 2010 *J. Phys.: Conf. Ser.* **250** 012048
[5] Olding T *et al* 2010 *J. Phys.: Conf. Ser.* **250** 012028
[6] Penev K I and Mequanint K 2013 *Phys. Med. Biol.* **58** 1823
[7] Kron T *et al* 1997 *Magn. Reson. Imaging* **15** 211

NMR spectroscopy of saccharide-doped PAGAT dosimeters

P S Skyt[1,2], B Berndt[1], L Holloway[1,3,4,5], R Hill[6] and Y De Deene[1,7,8]
[1]Institute of Medical Physics, University of Sydney, Australia
[2]Department of Medical Physics, Aarhus University Hospital, Denmark
[3]Liverpool and Macarthur Cancer Therapy Centres and Ingham institute, Australia
[4]Centre for Medical Radiation Physics, University of Wollongong, Australia
[5]University of New South Wales, Australia.
[6]Chris O'Brien Lifehouse, Sydney, Australia
[7]Department of Engineering, Macquarie University, Australia
[8]Department of Radiotherapy, Ghent University Hospital, Belgium

E-mail: skyt@phys.au.dk

Abstract. The aim of this study was to investigate the chemistry of the PAGAT dosimeters when doped with saccharides and irradiated, using NMR spectroscopy. Three batches of PAGAT gel dosimeters were manufactured. Two of them were doped with 20 % glucose and sucrose, respectively. For each batch, one sample was left unirradiated while the remaining samples were irradiated to different doses. After irradiation, NMR spectra were obtained which clearly showed the composition of the dosimeter and the change in monomer concentration caused by irradiation. In addition, it revealed that the saccharides did not directly participate in the chemical process before and during irradiation but the addition of saccharides resulted in a higher consumption rate of the monomers.

1. Introduction

The radiation sensitivity of polymerizing dosimeters is based on polymerization of monomers. This radiation-induced polymerization reaction reduces the spin-spin relaxation time (T_2) and dose read out can therefore be performed using nuclear magnetic resonance (NMR) [1]. This results in a relation between the relaxation rate ($R_2=1/T_2$) and the delivered dose. One such dosimeter system is the normoxic PolyAcrylamide Gelatine gel fabricated at ATmospheric conditions (the PAGAT dosimeter). Studies have shown that the response to irradiation of such dosimeter can be increased by adding saccharides [2], however, the chemistry has, to our knowledge, not yet been explored. The aim of this study was to investigate the difference in chemistry between PAGAT and saccharide-doped PAGAT dosimeters before and after irradiation by use of NMR spectroscopy.

2. Materials and Methods

2.1. Gel fabrication

Three batches of gel dosimeters were fabricated: PAGAT, glucose-doped PAGAT, and sucrose-doped PAGAT. They all consisted of 6% (w/w) gelatine, the monomers 3 % (w/w) acrylamide (AA) and 3 % (w/w) N,N'-methylene-bis-acrylamide (BIS), and the antioxidant 10 mM tetrakis(hydroxymethyl) phosphonium chloride (THP). All compounds were dissolved in deuterium oxide containing 0.05 % (w/w) 3-(trimethylsilyl)propionic-2,2,3,3-d_4 acid sodium salt (TMS). In the batches containing saccha-

Journal of Physics: Conference Series **573** (2015) 012031 doi:10.1088/1742-6596/573/1/012031

rides a corresponding amount of deuterium oxide was replaced with 20 % (w/w) glucose or 20 % (w/w) sucrose. After mixing the compounds the solution for each batch was poured into three 5 mm diameter NMR spectrometry tubes and placed in a refrigerator.

2.2. Irradiation

For each gel batch two samples were irradiated to 9.3 Gy and 40 Gy, respectively, while one was left unirradiated. Irradiation was performed using an Elekta linear accelerator set at 6 MV with the samples placed in a water equivalent phantom which was placed at a SSD of 90 cm.

2.3. NMR spectrometry and data analysis

Each sample was measured using a 400 MHz NMR spectrometer (Bruker Optik GmbH) at a temperature of 298 K and without sample rotation. Deuterium oxide was used as locking signal and the spectra were recorded with 64 averages. Subsequently, the NMR data was Fourier transformed and phase corrected. The TMS peak was then used as origin of the frequency axis. A series of peaks were analysed by fitting Lorentz expressions with baseline corrections to obtain peak areas and widths. To compare peak areas across spectra they were normalized to that of the TMS peak in each spectrum assuming that it was not affected by the chemistry in the dosimeter. Subsequently, the change in peak area and width at specific dose levels were compared to that of the unirradiated samples.

3. Results

The obtained NMR spectra for all samples are shown in figure 1. The specific compounds have been identified by recording NMR spectra of pure compounds dissolved in deuterium oxide (data not shown). The monomers have been identified as nitrogen-hydrogen peaks in the first section of the figure and as multiple carbon-hydrogen peaks in the second section. A peak of residual water is observed in section three while multiple peaks from gelatine is observed in section four and five. In addition, the double peak near 7.3 ppm is believed to be of gelatine. In the spectra of the irradiated samples the monomers are considerably reduced and all peaks appear to be wider.

When adding glucose (middle panel of figure 1) or sucrose (lower panel) the corresponding saccharide peaks appear in section three and four while the resonances of the remaining compounds are similar to that of PAGAT (upper panel). In addition, all peaks have become considerably wider, which is especially evident at the water peak. The effect of dose is similar to that of the PAGAT samples with clear consumption of the monomers, but is more pronounced in the saccharide samples.

The change in peak area when applying dose is shown in figure 2. High monomer consumption is observed depending on dose even though the area of both saccharides is approximately unchanged. The gelatine area decreases up to 30% while the area of the water peak does not show a clear trend.

The changes in peak width upon irradiation relative to that of the unirradiated samples are shown in figure 3. The width of the BIS peak is highly dependent on dose. AA shows a lower increase in width for the saccharide samples and decreases for the PAGAT sample. The width of the water peak is unchanged while the width of gelatine and saccharide peaks increase.

4. Discussion

The individual chemical compounds of the PAGAT dosimeter were observed in the NMR spectra and the identified peaks were consistent with the literature [3]. A dose dependent decrease in monomer peak area was clearly observed in the spectra indicating consumption of both monomers. However, this did not result in the appearance of new peaks from the reaction products probably due to low mobility of polymers, hence, line broadening [3]. This might also explain the decrease of gelatine peak area as the gel matrix becomes entangled with the polymer clusters.

Upon addition of saccharides to the PAGAT dosimeter existing resonances were unchanged, i.e. no change in chemistry was observed. However, all peaks were wider with saccharide additives which might be caused by the increased viscosity in the saccharide samples. In addition, a decrease in molecular mobility may be related to an increased rigidity of the gel matrix by the saccharides [4].

Journal of Physics: Conference Series **573** (2015) 012031 doi:10.1088/1742-6596/573/1/012031

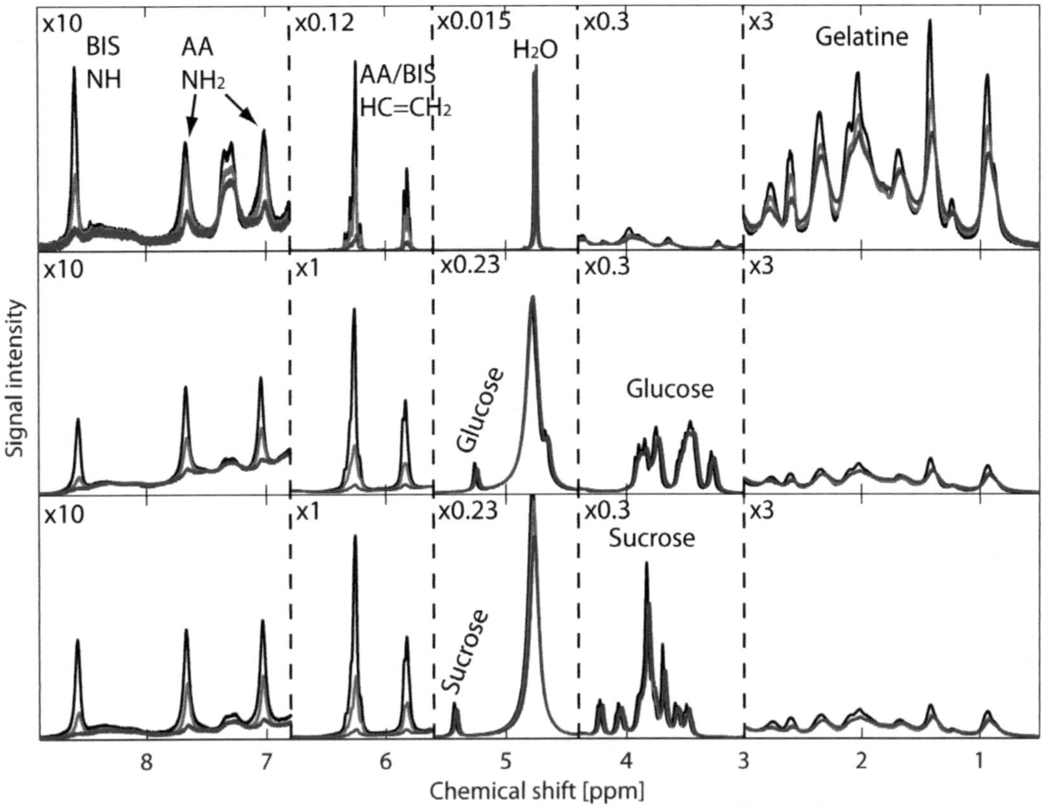

Figure 1. NMR spectra of PAGAT (upper graph), glucose-doped PAGAT (middle graph) and sucrose-doped PAGAT (bottom graph). Each graph is divided into five sections that each are scaled for clarity. The scaling factor is shown in the upper left corner of each section. The data are shown for unirradiated samples (black), samples irradiated to 9.3 Gy (red), and 40 Gy (blue).

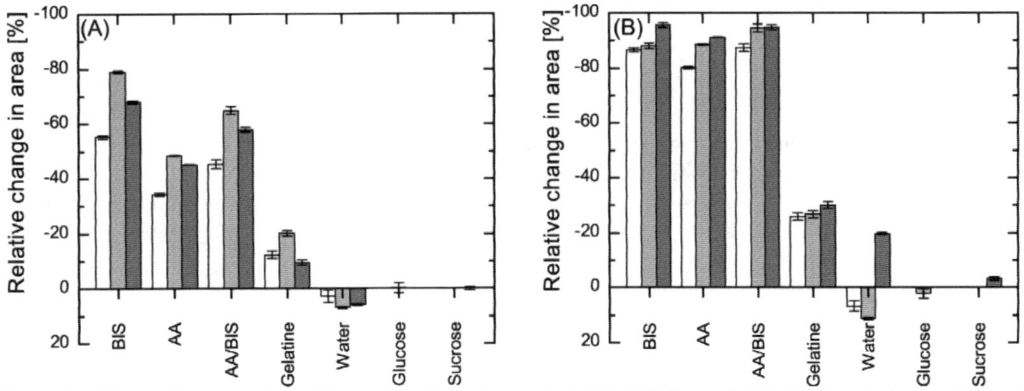

Figure 2. Change in area of specific peaks from figure 1 at (A) 9.3 Gy and (B) 40 Gy relative to the peak area of the unirradiated samples. The data are shown for PAGAT (white), glucose-doped PAGAT (light gray) and sucrose-doped PAGAT (dark gray). Note the inverted scale. The error bars originate from the fitting procedure. The areas correspond to the peaks in figure 1 as follows: BIS: 8.6 ppm. AA: 7.7 and 7 ppm. AA/BIS: 6.25 and 5.8 ppm. Gelatine: 7.3, 2.35, and 0.9 ppm. Water: 4.7 ppm. Glucose: 5.25 ppm and the glucose peaks in section 4. Sucrose: 5.4 ppm and the sucrose peaks in section 4.

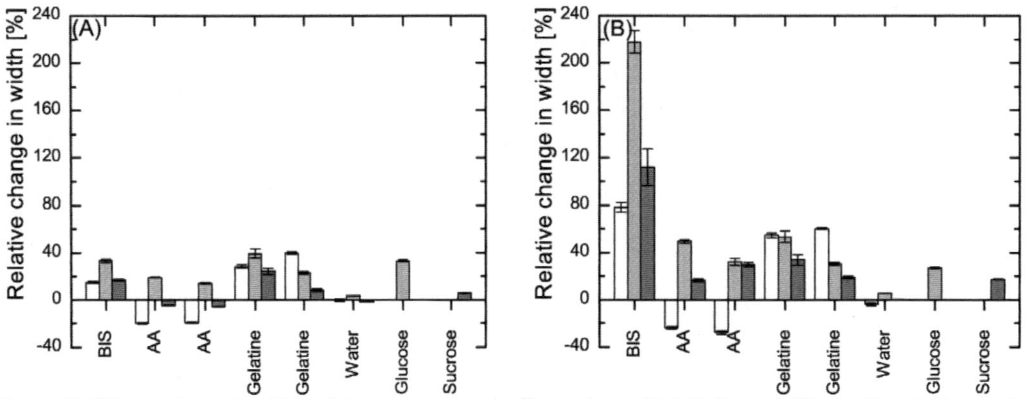

Figure 3. Change in peak width of the resonances in figure 1 at (A) 9.3 Gy and (B) 40 Gy relative to the peak width of the unirradiated samples. The data are shown for PAGAT (white), glucose-doped PAGAT (light gray) and sucrose-doped PAGAT (dark gray). The error bars originate from the fitting procedure. The peak widths correspond to the peaks in figure 1 as follows: BIS: 8.6 ppm. AA: 7.7 and 7 ppm, respectively. Gelatine: 7.3 and 0.9 ppm, respectively. Water: 4.7 ppm. Glucose: 5.26 ppm. Sucrose: 5.4 ppm.

Especially the width of the BIS peak increased when dose was applied. This might be due to the creation of polymer particles which decrease mobility. The width of the AA peaks also increased with dose for the saccharide samples but only slightly compared to BIS, probably because of its smaller molecular size. The width of the AA peaks for the PAGAT sample without saccharides decreased with dose but the mechanism behind this is unknown.

The consumption rate of BIS with dose was higher than that of AA in all samples. This was expected since BIS has two polymerization sites compared to one on AA. The area of the saccharide peaks seemed to be independent of dose indicating that they are not directly involved in reactions with the monomers upon irradiation. Glucose however, has been shown to lower the activation energy of a polymerization process for other monomers [5] and the mechanism might be similar here. However, an increase in width of the saccharide peaks was seen. The increase was comparable to both that of the AA and the gelatine peaks but since the saccharides were not consumed upon irradiation it might be attributed to the decreased mobility as a result of the polymerization.

5. Conclusion

The compounds of the PAGAT dosimeter were identified and the consumption of monomers with dose was clearly visible in the NMR measurements. The spectral data indicated that the addition of saccharides did not change the chemistry of the dosimeter. However, the radiation induced consumption as well as the change in peak widths of the monomers was higher for the saccharide-doped dosimeters. The mechanism behind this is not fully understood and further studies are needed to conclude which processes are responsible for the increase in dose response when adding saccharides.

6. Acknowledgements

Supported by The Danish Cancer Society and by CIRRO - The Lundbeck Foundation Center for Interventional Research in Radiation Oncology and The Danish Council for Strategic Research.

7. References

[1] Baldock C *et al* 2010 *Phys. Med. Biol.* **55** R1-R63
[2] Yoshioka M *et al* 2010 *J. Phys. Conf. Series* **250** 012016
[3] Lepage M *et al* 2000 *J Appl Polym Sci* **79** 1572
[4] Oakenfull D *et al* 1986 *Food Hydrocolloids* **1** 163
[5] Hu J *et al* 2004 *Polym Int* **53** 1003

High resolution 3D dosimetry for microbeam radiation therapy using optical CT

C McErlean[1], E Bräuer-Krisch[2], J Adamovics[3,6], M O Leach[1,4] and S J Doran[1,5]
[1]CRUK Cancer Imaging Centre, Institute of Cancer Research, London, UK
[2]European Synchrotron Radiation Facility, Grenoble, France
[3]Department of Chemistry and Biology, Rider University, Lawrenceville, NJ, USA
[4]Clinical Magnetic Resonance Research Group, The Royal Marsden NHS Trust, UK
[5]Department of Physics, University of Surrey, Guildford, UK
[6]Heuris Pharma, LLC, 412 Sunset Rd., Skillman, NJ, 08558 USA

E-mail: ciara.mcerlean@icr.ac.uk

Abstract. Optical Computed Tomography (CT) is a promising technique for dosimetry of Microbeam Radiation Therapy (MRT), providing high resolution 3D dose maps. Here different MRT irradiation geometries are visualised showing the potential of Optical CT as a tool for future MRT trials. The Peak-to-Valley dose ratio (PVDR) is calculated to be 7 at a depth of 3mm in the radiochromic dosimeter PRESAGE®. This is significantly lower than predicted values and possible reasons for this are discussed.

1. Introduction

Microbeam Radiation Therapy (MRT) is a new therapy that makes use of the fact that spatially fractionated dose is very well tolerated by normal tissue while being lethal to tumours [1]. Treatments involve delivering dose in narrow, well collimated x-ray beams the order of 50μm wide. Dosimetry of this therapy is challenging as there is no commercially available technique that can deliver the high resolution and high dynamic range measurements required [2]. Many developments have been made in the area of 'multiple single point' dosimeters, such as MOSFETs, and planar dosimetry including radiochromic film. However, with increasingly complex dose distributions there is a need for a truly 3D dosimetry technique for this therapy [3]. A 3D dataset provides advantages such as the ability to reconstruct in multiple planes, providing easier visualisation of any alignment problems. Furthermore, using a 3D dosimeter it is possible to provide 'end-to-end' quality assurance of the entire therapy process, including planning Computed Tomography (CT) scans and Treatment Planning calculations. Optical CT has been previously shown to be a potential candidate to fill this gap [4]. This type of dosimetry would be complementary to existing ultra-high resolution planar techniques.

The construction and characterisation of a version of our Optical CT microscope has been previously reported [4]. The Modulation Transfer Function (MTF) and dose response were measured through the use of specialised test objects [5]. It was observed that although quantitative measurements are possible at the millimetre scale, at high spatial frequencies the lower MTF results in an underestimation of the dose [6].

Here we present some initial results from a series of irradiations on the ID-17 Biomedical Beamline at the European Synchrotron Radiation Facility (ESRF) in Grenoble. The goals were to assess the visualisation of different MRT irradiation geometries with Optical CT and secondly, to test whether

Journal of Physics: Conference Series **573** (2015) 012032 doi:10.1088/1742-6596/573/1/012032

quantitative measurements could be made on the microbeams themselves. The Peak to Valley Dose Ratio (PVDR) is of significant interest in the MRT community as it is a metric which influences how safe and effective an MRT treatment will be. If the PVDR could be measured with our technique in 3D this would be a major advantage in bringing MRT closer to patient trials. However, given previous findings we expect to have to adapt our optical system in order to measure this value accurately.

2. Methods

Samples of PRESAGE® [7], a radiochromic dosimeter, were provided by Heuris Pharma (Skillman, NJ) in the form of 9.7mm and 22mm diameter cylinders and a series of cuvettes. The cuvettes were irradiated with a range of doses and their optical density was measured with a spectrophotometer. Several cylindrical samples were irradiated with a calibration test pattern of a range of doses allowing the Optical CT system dose response to be measured, as in [4].

The cylindrical samples were scanned using an improved version of the Optical CT system described in [4]. In previous work, the large datasets involved in CT images were a limiting factor during both acquisition and processing. In the new system the camera and computer used to acquire and process images have been upgraded. High quality images were acquired quickly using an Andor Zyla sCMOS camera, with a large pixel array (5.5 Megapixels) and a 100fps framerate. New CMOS technology means that high quality cameras are more affordable and the high pixel density and fast readout rate overcome previous limitations in projection matrix size. Images were saved and processed using a Dell computer with 256GB of RAM, allowing very large datasets to be held in memory at once. This means the time taken for acquisition and reconstruction is significantly improved over previous systems.

2.1. Visualisation

A number of samples were irradiated using the different geometries possible with MRT including crossfire, interlaced and pencil beams. Projection images were acquired in a fast, 'low-resolution' scan with 512 × 512 pixel images at 1000 angular positions. The goal of this exercise was to test whether these fast scans, under 3 minutes acquisition, could provide useful information which could be used to correct MRT alignment during treatment. The projections were reconstructed to 512^3 matrix with isotropic voxels, size 51μm for the 22mm samples and 25μm for the 9.7mm samples. The reconstructions were carried out using in-house software written in IDL (Exelis Visual Information Solutions, Boulder, CO).

2.2. PVDR Measurement

A 22mm sample was irradiated end-on with a 10x10mm array of 50μm microplanar beams, with a centre-to-centre distance of 400μm. This sample was scanned with 3200 projections, chosen to satisfy the Nyquist sampling limit, each of matrix size 1800 × 1800. Although the results presented here were reconstructed non-optimally to slices of 512 × 512 pixels, with non-isotropic $(45 \times 45 \times 12.8)(\mu m)^3$ voxels, we are currently working on improved acquisition and reconstruction techniques that will make full use of the large pixel density of the camera and lead to still higher spatial resolution.

3. Results and Discussion

The cuvette optical density measurements verified the linear response of PRESAGE® to dose and the Optical CT system's dose response was found to be linear up to a dose of 100Gy. We are working on methods of extending the dynamic range of the system to avoid saturation at high doses.

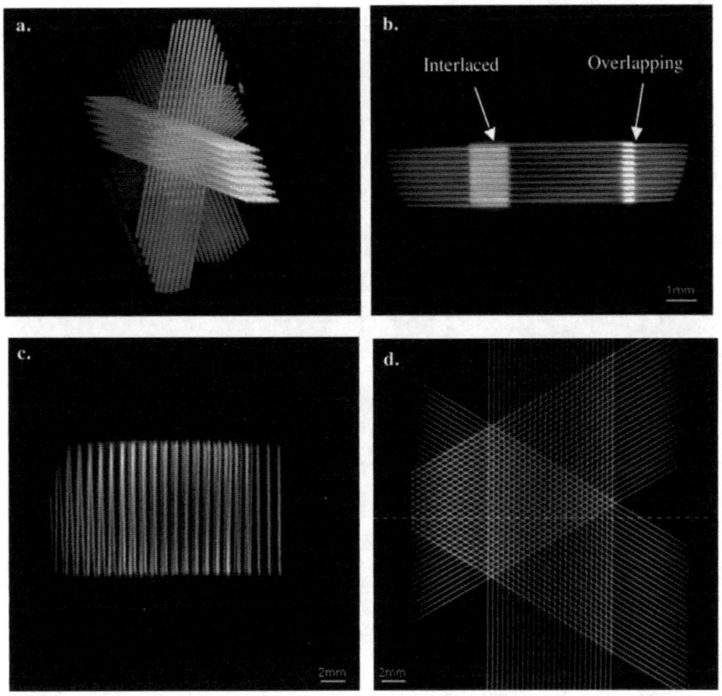

Figure 1. (a) Maximum intensity projection image of 4 microplanar arrays of 7 × 7 pencil beams, peak dose 300Gy, reconstructed to a 512^3 array. This treatment was well aligned in the centre of the sample and the individual beams are easily visualised. (b) A reconstructed slice through an interlaced pattern, 200Gy peak dose. This slice clearly demonstrates that the MRT alignment was incorrect, producing overlapping beams on the right instead of the interlaced seen on the left. (c) An xy slice through a 3-port crossfire irradiation, 200Gy peak dose, and (d) the corresponding xz reconstruction that demonstrates the misalignment problem more clearly than the xy slice (marked by the dashed line). This emphasises the usefulness of having a 3D dataset to perform MRT QA.

3.1. Visualisation

Different MRT irradiation geometries were visualised with some results shown in figure 1. From the reconstructed 3D maps the dose was well visualised and it was relatively easy, compared with other dosimetry methods, to see where misalignments occurred. Overall, these quick 'low resolution' scans (512 × 512 pixels, 1000 projections) would be useful for MRT technicians.

3.2. PVDR Measurement

Profiles were measured from the median value of 5 reconstructed slices, giving an effective slice thickness of 65µm (see figure 2). The median value along each beam was calculated to give a PVDR of 7 at a depth of 3mm. This is an improvement over previous Optical CT measurements, however, it is significantly lower than the values predicted by Monte Carlo simulation, which are of order 70 [8]. Although our pixel size is smaller than the microbeams, it has been shown that in order to avoid under-sampling the peak dose, a resolution of better than 5 times the smallest feature is required meaning we need a resolution of 10µm. For Optical CT, the entire sample must reside within the

Depth of Field (DoF) for all points to be in focus during rotation. However, there is a trade-off between magnification and DoF, limiting the resolution achievable for a given sample size. This DoF problem will be crucial in any efforts to extend the resolution of our system.

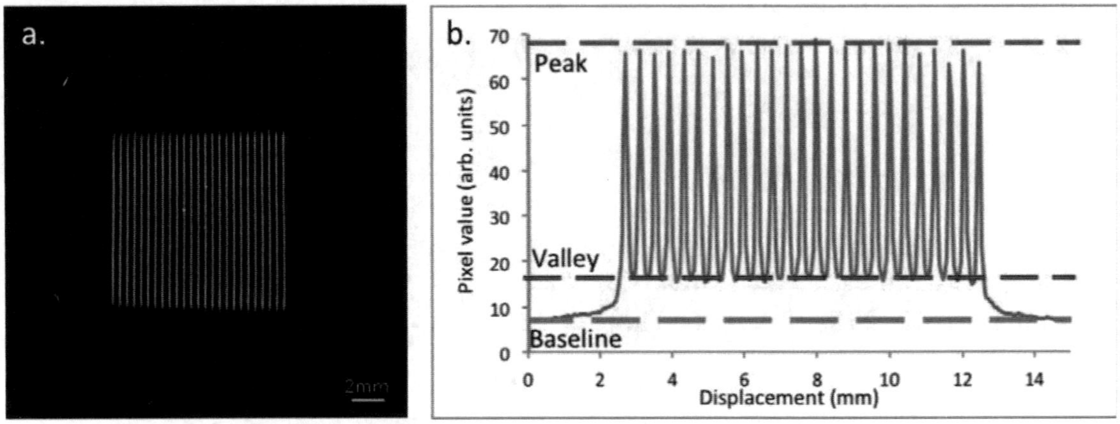

Figure 2. (a) Slice through 10x10mm microplanar array of 50µm beams, 400µm CTC distance and 100Gy peak dose. (b) The corresponding profile where the baseline value is calculated by taking the average of an unirradiated area of the sample.

4. Conclusions

We have demonstrated here that Optical CT is a viable technique for QA of the geometrical aspects of MRT treatments. It is, arguably, the only method currently available to perform this task adequately in 3D, something that will increase in importance as MRT develops in a more conformal direction. With new technology, scans and reconstructions can be accomplished very quickly with a small, comparatively inexpensive scanner. Using this upgraded scanner, together with GPU-based reconstruction, it will be possible to acquire and reconstruct ultra-high resolution 3D datasets within 10 minutes, which is an acceptable time frame for use as a QA tool during treatment at the beamline.

It is clear that we need to extend the dynamic range and resolution of the system if quantitative measurements are to be made at microbeam level. Several approaches have been identified. These include: (i) accurate measurement of the point spread function and deconvolution of this known blur; (ii) sub-voxel reconstructions of linear features based on prior knowledge; (iii) CT zoom imaging; (iv) overcome the depth-of-field problem by extending a method introduced for Optical CT for tissue samples, in which the sample is scanned through the DoF such that each point in the sample is in focus at some time during the scan [9].

5. Acknowledgments

We acknowledge the support received from the CRUK and EPSRC Cancer Imaging Centre in association with the MRC and Department of Health (England) grant C1060/A10334, and NHS funding to the NIHR Biomedical Research Centre.

6. References

[1] Anschel D J *et al* 2011 *Neurosurg. Rev.* **34** 133-42
[2] Bräuer Krisch E *et al. AIP Conf. Proc.* **89** 1266
[3] Baldock C *et al* 2010 *Phys. Med. Biol.* **55** R1-63
[4] Doran S J *et al* 2013 *Phys. Med. Biol.* **58** 6279-7
[5] Abdul Rahman A T *et al* 2011 *Phys. Med. Biol.* **56** 4177-99
[6] Doran S J *et al* 2013 *J. Phys.: Conf. Ser.* **444** 012074
[7] Gorjiara T et al 2011 *Med. Phys.* 38 2265-74
[8] Martínez-Rovira I *et al* 2012 *Med. Phys.* **39** 119-31
[9] Fauver M *et al* 2005 *Opt. Express* **13**, 4210–23

Stray light reduction in optical computed tomography using a convergent cone-beam source

Kevin Jordan[1,2], Kurtis Dekker[1] and Jerry Battista[1,2]
[1]London Regional Cancer Program, London Health Sciences Centre,
[2]Department Medical Biophysics, Western University, London, Ontario, Canada

E-mail: kevin.jordan@lhsc.on.ca

Abstract. The planar diffuser light source for a cone-beam optical CT scanner was replaced with a filtered LED and large Fresnel lens. The source was focused on a camera and convergent cone-beam images were acquired. Images are sensitive to mismatches in the refractive index of vessels, samples and matching liquid. For PETE jars and water solutions, a loss of approximately 30% of the projections was demonstrated. Teflon PFA cylinders provided better refractive index matching and more accurate reconstructions. The convergent cone-beam source dramatically increased imaging efficiency by a factor of 1000 and reduced stray light levels by confining illumination to image forming rays.

1. Introduction

At the first Dosgel meeting, Wolodzko et. al. introduced an optical cone-beam CT geometry that involved recording transmission images of hydrogels in a refractive index matching liquid with a digital camera [1]. The light source consisted of a planar diffuser back-illuminated with white light. The projection images were reconstructed using iterative algorithms developed previously for nuclear medicine imaging applications. The diffuse planar source and cone-beam geometry could be easily and inexpensively scaled to image vessels of any size [2]. This simple design has since been refined by using arrays of red or yellow LED's to provide a stable narrow band light source, and higher performance CCD cameras for recording the projection images. A commercial scanner was designed by Modus Medical Devices Inc. and is marketed as the Vista scanner. With this system, sources of stray light were isolated and reported by several researchers. External reflections from the cylindrical vessels, reflections from the bandpass filter at the camera lens, and stray light near the optical axis were reported [3]. The external reflection was minimized by reducing the camera aperture to collect light from slightly larger than the cylinder diameter. Tilting the bandpass filter reduced reflections in projection data. However, additional stray light remained near the optic axis. Methods to characterize stray light were reported using grids and beam-stops located at the diffuser face, [4]. For conditions representative of 4% gelatin ferrous xylenol orange hydrogels, stray light accounted for 10-30% of the detected light. Stray light depends on projection angle and would need to be measured and subtracted from each transmission image. Arrays of slot apertures between the source and object provided collimation reducing incident stray light and provided a technique for recording high resolution stray light maps [5]. The performance of the Vista scanner reading *polymer* gels [6] was found to be limiting at the doses typically used for MRI readout [7]. However, at lower doses, adequate optical contrast was observed for qualitative 3D dosimetry. Further evaluations by Olding reported that stray light measurements with a pinhole grid and interpolation provided more accurate reconstructions [8].

In principle, *single ray - single small detector geometry* (i.e. first generation CT) is best for stray light rejection. However, early scanners were slow due to speed at which the mirrors and detectors could be moved. Laser beam steering with mirrors on galvanometers coupled with mirrors or lenses provided much faster instruments [9]. Telecentric geometries have been successfully implemented [10, 11]. Fan-beam geometry with a detector array and multiple-aperture collimator has also been investigated [12].

In this report, an alternative cone-beam geometry is introduced. The diffuse planar light is replaced with a source consisting of a single small LED and a large Fresnel lens. This system (figure 1) provides a high degree of collimation and consequent dramatic increase in imaging efficiency. The primary to stray light ratio in the projection images will also be much greater than with a diffusive planar source and hence improves the accuracy of dose reconstructions.

2. Instrumentation

Two light sources were used. The first was shown previously and consists of a plastic sheet diffuser located 80 cm from the object back-illuminated by three Luxeon III, 3W, yellow LEDs operating at 600 mA and bandpass filtered [5]. The second source is shown in figure 1. It consists of the same type of LED but operating at 0.8 mA. The LED light is collimated with a composite optic to a beam divergence of 20 degrees, filtered through a 10 nm bandpass filter (590 nm) and then focused to a spot, approximately 2 mm in diameter, on a white diffuser. A 25 cm diameter, Fresnel lens is located ~ 30 cm from the rotation axis of the scanner and the source position is adjusted to provide a cone of light focused at the camera aperture.

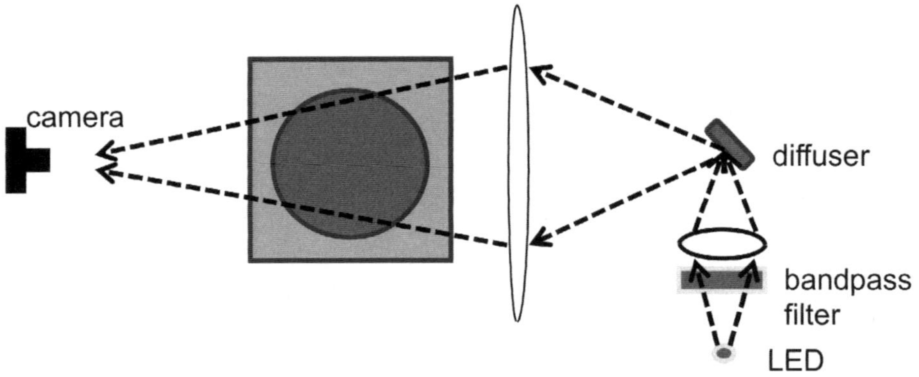

Figure1. Top view schematic of modified convergent cone-beam optical CT scanner.

Test solutions consisted of carbon black nanoparticles dispersed in Triton X100 micelles in water. Visually the solution scatter is comparable to 4% gelatin hydrogel. The reference solution was water. These solutions were placed in either PETE jars or PFA Teflon cylinders. Water was used at the refractive index matching liquid outside the sample vessels. Each scan consisted of 512 projections over 360 degrees of sample rotation. Filtered back projection with Ram-Lak filter and isotropic 0.25 mm voxels were used for reconstructions.

3. Results

Camera shutter times were 20 times shorter with the Fresnel cone-beam source. The central axial slice from the PETE jar scan with the planar diffuse source is shown in the left panel of figure 2.

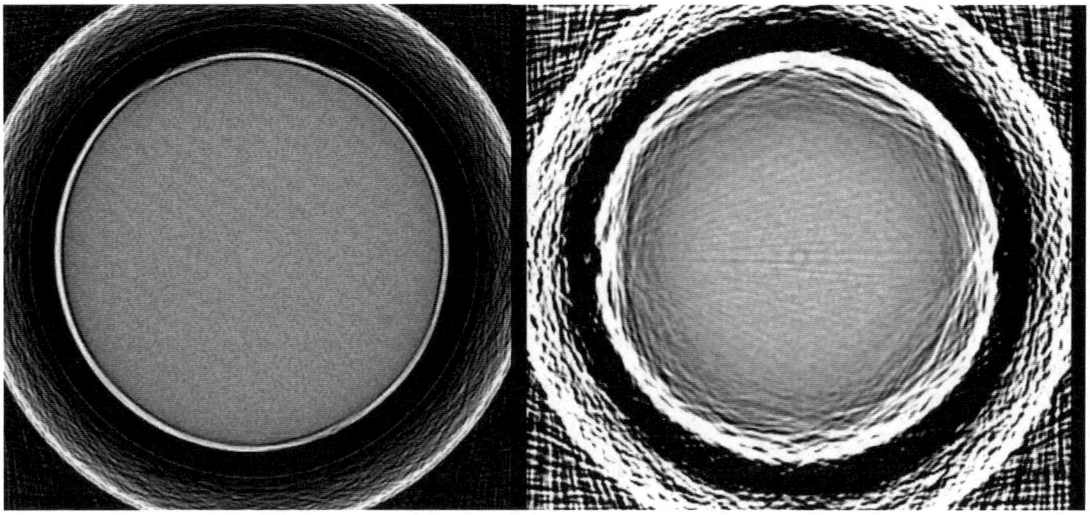

Figure 2. Transverse central slices of carbon black solution in PETE jar: left, diffuse planar source, right, convergent-cone source. Similar intensity window and level limits.

The reconstructed attenuation coefficient for the solution is uniform to within 10% of the central mean value over the central 85% of the jar with lower values closer to the wall. The central attenuation coefficient is typically within 3% of independent spectrometer measurements. There is a bright circular band just inside the jar wall and the wall itself is not clearly reconstructed. Figure 2 right panel is the reconstruction with the convergent-cone source. In contrast, it is noisy and highly distorted. The central attenuation coefficient is approximately 35% low and the outer 30% is only noise.

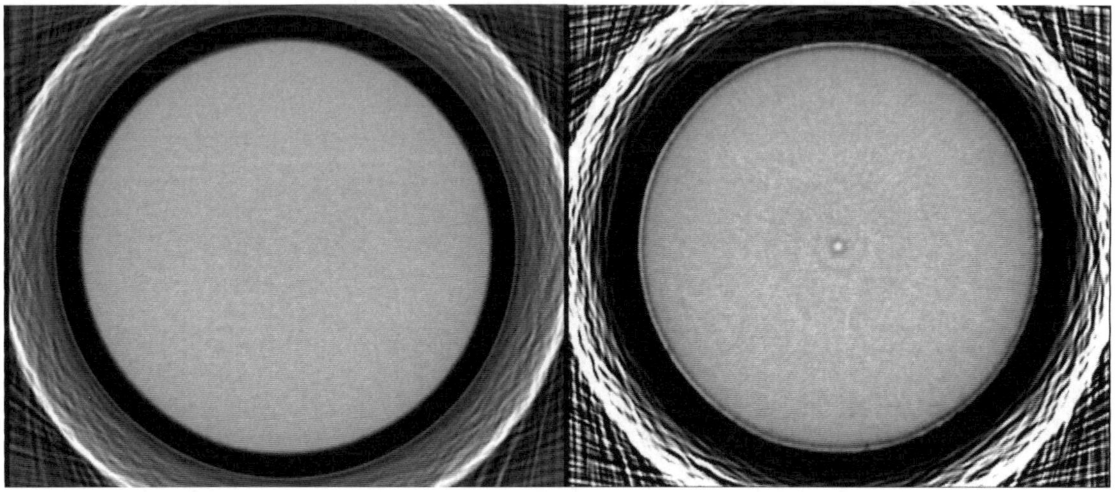

Figure 3. Transverse central slices of carbon black solution in PFA Teflon cylinder: left, diffuse planar source, right convergent-cone source. Same intensity window and level limits.

However, the PFA cylinder reconstructions are similar for both sources, as seen in figure 3. The reconstructions are uniform to greater than 90% of the diameter and attenuation coefficients are within 3% of independent spectrometer readings. In the convergent-cone source reconstruction the cylinder walls is clearly observed as well as an artifact near the optic axis. This demonstrates higher spatial resolution is observed with the convergent-cone source. Also, noise is lower in the diffuse source reconstruction, consistent with a blurring effect due to more stray light.

4. Discussion
The shorter shutter times and the lower drive current for the LED, provide an estimated improvement in imaging efficiency of ~1000 and an increase in the primary to stray light ratio for the projection images with the convergent-cone source. The very low LED drive current minimizes thermal stabilization time and provides the opportunity for precise intensity modulation of the source. Figure 2 shows dramatic differences in the PETE jar reconstructions for the two sources. The planar diffuse source can be considered as a 2D array of point sources illuminating the object from many angles. But only a few of the rays form the image. In contrast, the convergent-cone source is effectively a small source and most of the incident rays contribute to image formation. Because of the large refractive index mismatch between the PETE jar and water, projection images are distorted. In certain regions of the images there are no primary rays leading to missing data. Filtered back projection reconstruction is particularly sensitive to missing data and this is seen with the convergent-cone source scan. In contrast, the diffuse source provides rays at many incident angles besides the primary rays. For this reason, the diffuse source images appear “better” because of the incident stray light which fills in areas with no primary rays. For the PETE jar and hydrogel dosimeter combination, iterative reconstruction algorithms and innovative data collection methods will be required to get useful data with the convergent-cone source. However, Teflon cylinders have a similar refractive index to most hydrogel dosimeters and images are much less distorted. In this case, the lower stray light with the convergent-cone source results in a reconstruction with higher spatial resolution and more noise. This source shows promise for extending the performance of optical CT dosimetry.

5. Acknowledgements
K Dekker is supported by a Queen Elizabeth II Scholarship and the Ontario Research Fund (OCAIRO Project). Disclosure: The authors have a research agreement with Modus Medical Devices Inc. concerning optical CT and dosimetry applications.

6. References
[1] Wolodzko J G *et al* 1999 *Med. Phys.* **26** 2508-2513
[2] Jordan K *et al* 2006 *J. Phys.: Conf. Ser.* **56** 214
[3] Jordan K *et al* 2006 *J. Phys.: Conf. Ser.* **56** 217
[4] Jordan K *et al* 2009 *J. Phys.: Conf. Ser.* **164** 012028
[5] Babic S *et al* 2012 *Phys. Med. Biol.* **57** N377-89
[6] Baldock C *et al* 2010 *Phys. Med. Biol.* **55** R1-63
[7] Bosi S *et al* 2007 *Phys. Med. Biol.* **52** 2893-903
[8] Olding T *et al* 2010 *Phys. Med. Biol.* **55** 2819-40
[9] Krstajic N *et al* 2007 *Phys. Med. Biol.* **52** N257-63
[10] Krstajic N *et al* 2007 *Phys. Med. Biol.* **51** 2055-75
[11] Thomas A *et al* 2011 *Med. Phys.* **38** 4846-57
[12] Campbell W *et al* 2013 *Med. Phys.* **40** 061712-1

Pushing the boundaries of spatial resolution in dosimetry using polymer gels and radiochromic films

G Heilemann[1], D Georg[1,2] and A Berg[3]
[1]Department of Radiation Oncology & Comprehensive Cancer Center, Vienna, Medical University of Vienna/AKH Vienna, Austria
[2]Christian Doppler Laboratory for Medical Radiation Research for Radiation Oncology, Medical University of Vienna, Austria
[3]MR Centre of Excellence, Center for Biomedical Engineering and Physics, Medical University of Vienna, Vienna, Austria

Email: gerd.heilemann@meduniwien.ac.at

Abstract. Advanced radiotherapy and brachytherapy techniques are raising the bar for detectors with respect to high spatial resolution. Dosimetry based on most point-like dosimeters, e.g. diamond detectors or small volume ionization chambers cannot be used efficiently and accurately for detecting 2 or 3D-dose variations at millimeter scale. Hence radiochromic films and polymer gels with high two/three-dimensional resolution provide a good verification tool for measuring dose distributions of very small collimated beams. In this study the performance of film and gel detectors in detecting the very fine dose distributions generated from collimation holes of four different sizes is investigated. Pencil beams with diameters down to 0.455 mm could be resolved by both detector types comparably.

1. Introduction

Current advances in stereotactic radiosurgery, intensity-modulated radiotherapy and image-guided radiation therapy as well as steep dose gradients of brachytherapy applicators call for an increasingly high spatial resolution in dosimetry [1-8]. While conventional radiation detectors (e.g. small volume ionization chambers, diodes or diamond detectors) show significant uncertainties due to their relatively large sensitive volumes [9] or non-equilibrium conditions as a consequence of non-water equivalence of the detection material or electrical equipment, polymer gels [10] and radiochromic films are well suited for this application because of their intrinsically high resolution and their near water equivalence. This study will compare the dose imaging results of these detectors with respect to collimated pencil beams with field sizes of 0.455 mm – 1.925 mm.

2. Materials and Methods

2.1. Polymer gel preparation and calibration

The normoxic polymer gel was manufactured in-house at normal atmosphere oxygen concentration levels. It consists of gelatin, methacrylic acid, ascorbic acid, copper-sulfate and distilled water [10. 11].

The calibration was performed in a single-scan approach as proposed by Oldham *et al* [12]. The BAREX container was placed in a water phantom with a source to water surface distance (SSD) of

100 cm in a 6 MeV electron beam and irradiated to a known dose, which was validated with a Roos chamber (PTW-Freiburg, Germany). From the well known depth dose curves of the 6 MeV electron beam provided by an Elekta Synergy linear accelerator (LINAC) a number of calibration data points were obtained in a range of 0 – 30 Gy.

The MR imaging was performed on a 7 T scanner (Siemens, Germany) equipped with a MR-microscopy imaging gradient system [13, 14]. The achieved in-plane pixel size was 156 × 156 μm^2 with a slice thickness of 1000 μm. All gel containers were scanned following identical protocols using a multiple spin-echo sequence with 20 equidistant echoes (TE = 10 ms). For the evaluation T2 images were extracted. The pixel values from the T2 weighted images are transformed to relaxation rates R2. The data points of the calibration are plotted on a dose versus R2 plot and a polynomial fit was applied to convert the R2 images into dose maps.

2.2. *Radiochromic film and calibration*

For this study Gafchromic EBT3 (ISP Technologies Inc., Wayne, NJ, USA) are used from the same lot (# 11041302) following published guidelines [15]. To account for post-exposure darkening [16] the films were scanned 24 h after irradiation. All films were scanned on an EPSON Perfection V700 scanner using Epson Scan software with a color depth of 48 bit and a resolution of 150 dpi resulting in a pixel size: 0.169 mm. The pixel scanner values (SV) of the more sensitive red color channel [17] were transformed to net optical density (netOD) values using the following formula [18]:

$$netOD = log\frac{SV_{BG}}{SV}$$

where SV_{BG} is the average pixel grey value from red color channel of the un-irradiated film and SV being the averaged scan value in a 2 × 2 cm^2 region of interest of the irradiated calibration films.

For calibration 4 × 4 cm^2 film pieces were placed in a solid water slab phantom and exposed to eight dose steps from 0 to 25 Gy in a Co-60 photon beam. The mean netOD of each calibration film was related to dose and fitted using a fourth order polynomial.

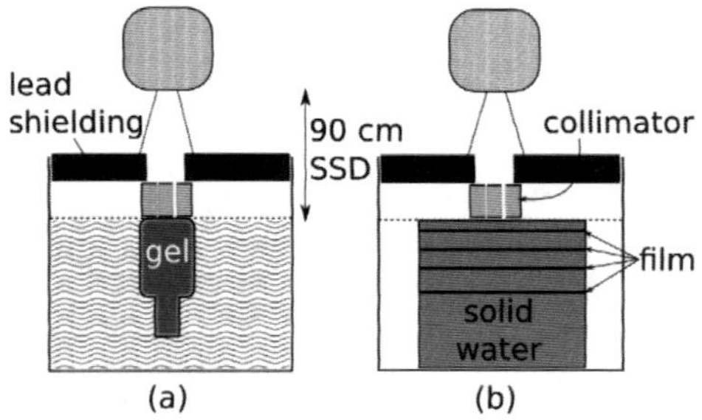

Figure 1. Setup for the collimated pencil beam for the polymer gel (a) and radiochromic film (b).

2.3. *Measurements with pencil beam collimator*

The polymer gel filled BAREX containers were positioned upside down in a water phantom at a SSD (source surface distance) of 100 cm with the bottom of the container at water level (see figure 1). A block of 10 mm thick lead (see figure 2) with four holes of different diameters (1.925 mm, 1.540 mm, 0.910 mm and 0.455 mm) was used to collimate a 6 MV and a 10 MV photon beam from a LINAC with a 5 × 5 cm^2 field size. To further avoid dose contributions from aside, additional lead absorbers were set next to the central leaden hole-absorber as shown in figure 1.

A similar setup was used for the film irradiation with identical SSD (= 90 cm). Six film strips of 4 × 4 cm^2 were positioned between slabs of solid water of 2 mm, 5 mm and 10 mm thickness. The first film was located at a depth of 2 mm. The central axis of the pencil-beam collimator block was placed above the center of the films on top of the first solid water slab. Similar measures as for the polymer gel were taken to avoid scatter irradiation.

The LINAC was set to deliver 4000 MU and 3000 MU for the polymer gels and radiochromic films, respectively. These settings were chosen in order to achieve doses in the high sensitivity range of the polymer gel and film dosimetry.

3. Results and Discussion

The netOD, originating from the scanned extracted red-channel of an irradiated film and a T2 weighted image of the polymer gel, both in a depth of 12 mm, are shown in figure 2. The dark spots in both images indicate that the pencil beams at all sizes down to 0.4 mm can be identified in both detectors. Because of the limited thickness of the absorber both detectors receive a level of dose outside the collimation holes. For the film the background dose is 20.50 ± 0.14 Gy and for the polymer gel it is 26.93 ± 0.37 Gy. In the peak of the largest pencil beam (1.925 mm) doses reach up to 24.69 Gy and 31.44 Gy, respectively.

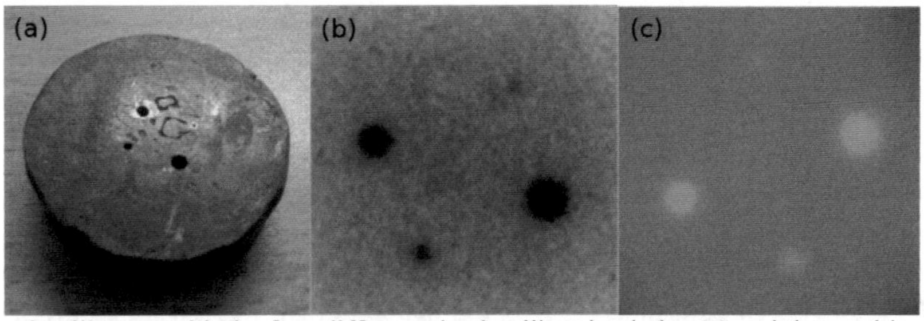

Figure 2. The lead collimator with the four different sized collimation holes **(a)** and the resulting scans of EBT3 film (b) and R2 map of the MR scan (c) of a slice in 12 mm depth for the polymer gel.

Profiles perpendicular to the incident beam in a depth of 12 mm from the surface for all four pencil beams are shown in figure 3. The calculated dose data from the measurements obtained with the polymer gel and radiochromic film are fitted to a Gaussian equation using MATLAB. The baseline was considered to be the average dose value in the vicinity of the normal distribution outside the primary pencil beam. The fitted curves were used to center the data points around the presumable center of the collimation holes. The normalized relative dose profile of the smallest pencil beam shows a very noisy signal in which the dimension of noise in the background dose level is up to 40 % (film) and 20 % (gel) of the dose signal in the peak of the pencil beam.

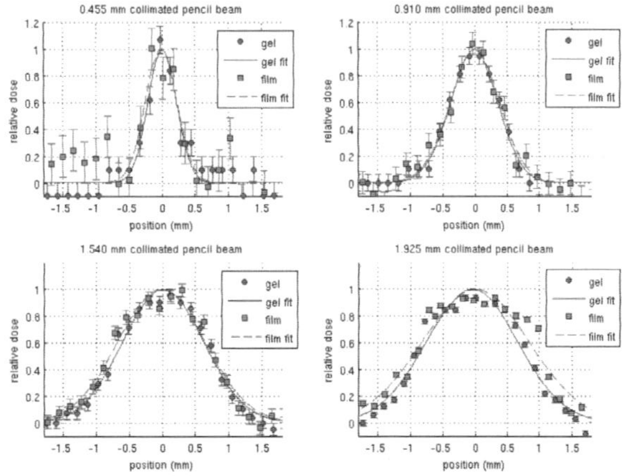

Figure 3. Comparison of dose profiles obtained using polymer gels and EBT3 films across collimated pencil beams of various widths. Error bars indicate the average standard deviation in the background signal outside the

peak.

Table 1 shows the full width at half maximum (FWHM) values of all four pencil beam sizes and compares the FWHM obtained with gel and film to the nominal sizes of the holes in the collimator block. While for the three smallest pencil beams the FWHM values with both detectors are larger than the actual size of the holes, in case of the largest pencil beam (1.925 mm) the Gaussian fitting algorithm underestimated the width of the hole. However, as can be seen in figure 3, the single Gaussian was not the best fitting algorithm. The gel measurements yielded generally narrower pencil beams (figure 3). This tendency of smaller FWHM values might, however, be in the range of the level of uncertainty and hence be caused by the noise in the data. Further studies need to be performed to confirm this trend in the future. This is also represented by lower FWHM values in table 1.

Table 1: FWHM values for the different sized pencil beams across the x-axis perpendicular to the incident beam with the first standard deviation according to the error bars shown in figure 3.

Nominal size of collimator (mm)	FWHM (mm) Polymer gel	Radiochromic film
0.455	0.513 ± 0.045	0.566 ± 0.047
0.910	0.875 ± 0.010	0.920 ± 0.031
1.540	1.510 ± 0.007	1.551 ± 0.024
1.925	1.755 ± 0.051	1.953 ± 0.010

The maximum dose modulation in terms of difference in signal between dose in the peak and background was typically slightly higher with polymer gels than in films. The difference might be related to the edge enhancement effect [19] seen for these types of gels [7]. The largest pencil beam (1.925 mm) yielded dose modulations of $\Delta D_{gel,1.925mm} = 4.36$ Gy and $\Delta D_{film,1.925mm} = 4.04$ Gy. For the smallest pencil beam (0.455 mm) the corresponding dose modulations were: $\Delta D_{gel,0.455mm} = 2.50$ Gy and $\Delta D_{film,0.455mm} = 1.30$ Gy almost twice as high for the gel compared to the film.

4. Conclusion

Very small sized photon pencil beams down to about 0.455 mm lateral size can be visualized with MR-micro-imaging based polymer gel dosimetry not only in two dimensions at a certain depth with radiochromic films but continuously with 1 mm depth distance in 3-dimensions using a MR-μ-imaging insert to a high-field human scanner. The dosimetric resolution (voxel volume of $156 \times 156 \times 1000$ μm^3) is comparable to that of a commonly used radiochromic film and flatbed scanner system (pixel size of 169×169 μm^2). Quantitative data on absolute dose levels in the center of the small sized pencil beams might be influenced by edge enhancement and noise in the film scanner system.

5. References

[1] Pappas E *et al* 2001 *Phys. Med. Biol.* **46** 783-97
[2] Berg A *et al* 2001 *Med. Phys.* **28** 833
[3] Pappas E *et al* 2006 *Med. Phys.* **33** 3700
[4] Pappas E *et al* 2008 *Med. Phys.* **35** 4640
[5] Berg A *et al* 2004 *Phys. Med. Biol.* **49** 4087-108
[6] Tseng Y J *et al* 2002 *Magn. Reson. Imaging* **20** 495-502
[7] Bayreder C *et al* 2008 *Med. Phys.* **35** 1756
[8] Brown S *et al* 2008 *Appl. Radiat. Isotopes* **66** 1206-12
[9] Low D A *et al* 2003 *Med. Phys.* **30** 1706
[10] Baldock C *et al* 2010 *Phys. Med. Biol.* **55** R1-63
[11] Fong P M *et al* 2001 *Phys. Med. Biol.* **46** 3105-13

[12] Oldham M *et al* 1998 *Phys. Med. Biol.* **2709** 2709-20
[13] Berg A *et al* 2013 *J. Phys. Conf. Ser.* **444** 012054
[14] Berg A *et al Proc. ISMRM* **2010** 1048
[15] Niroomand-Rad A *et al* 1998 *Med. Phys.* **25** 2093-115
[16] Andrés C *et al* 2010 *Med. Phys.* **37** 6271
[17] Fiandra C *et al Med. Phys.* **33** 4314
[18] Ferreira B C *et al* 2009 *Phys. Med. Biol.* **54** 1073-85
[19] Vergote K *et al* 2004 *Phys. Med. Biol.* **49** 4507-22

Inclusion type radiochromic gel dosimeter for three-dimensional dose verification

Shuji Usui, Munenori Yoshioka, Shin-ichiro Hayashi and Takahiro Tominaga
Department of Clinical Radiology, Hiroshima International University, Higashi-Hiroshima, Hiroshima 739-2695, Japan

E-mail: s-usui@hs.hirokoku-u.ac.jp

Abstract. For the verification of 3D dose distributions in modern radiation therapy, a new inclusion type radiochromic gel detector has been developed. In this gel, a hydrophobic leuco dye (leucomalachite green: LMG) was dissolved in water as an inclusion complex with highly branched cyclic dextrin. The radiation induced radical oxidation property of the LMG gel with various sensitizers was investigated. As a result, the optical dose responses were enhanced by the addition of bromoacetic acid and manganese (II) chloride. Unfavorable auto-oxidation of the gel was reduced when it was stored at 4°C.

1. Introduction

Radiochromic gel dosimeters are one of the promising tools for the verification of 3D dose distributions [1] in modern radiation therapy [2]. Among them, a micelle gel dosimeter has several advantages: water equivalency, no or low diffusion property, insensitivity to oxygen, lower toxicity, and suitability for optical CT read-out [3-7]. The micelle gel consists of leuco dye, surfactant, radical initiator, and gelling agent. In the gel, hydrophobic leuco dye is dissolved in water with a surfactant. However, formation of micelles is affected by surfactant concentration, ambient temperature, pH, and other chemicals [8].

To dissolve hydrophobic compound in water, it is also known to use inclusion compounds such as cyclic dextrins. Since it affords a prepared stable hydrophobic cavity, we focused on an inclusion type gel dosimeter. In this paper, we will describe the dose response of a newly developed radiochromic gel consisting of leucomalachite green (LMG) and highly branched cyclic dextrin (HBCD) [9, 10] as an inclusion host. The HBCD produced by the cyclization reaction of branching enzyme from amylopectin is highly soluble in water and has hydrophobic cavities.

2. Materials and Methods

2.1. Gel manufacturing

A typical LMG gel contains water, 4 wt% HBCD (Ezaki Glico Co), 2 wt% k-carrageenan (Marine Science, KK-9), 4 mM bromoacetic acid (Tokyo Kasei Co), 0.3 mM LMG (Sigma Aldrich), 20 mM chloroform (Nacalai Tesque Inc), and 6 mM $MnCl_2$ (Katayama Chemical). After k-carrageenan was dissolved in a two-third portion of water at 80°C, the solution was cooled to 55°C (solution A). In the remaining one-third portion of water HBCD was dissolved. Then 0.1 M-bromoacetic acid (BAA) and LMG-$CHCl_3$ were successively added and stirred for 30 min to create a clear solution. The solution was combined with 0.1 M-$MnCl_2$ and warmed to 55°C. Finally it was mixed with the above solution

A. The combined solution was poured into 1 × 1 × 4.5 cm^3 of PMMA cuvettes for optical measurements and stored at 10°C before irradiations. A variety of LMG gels in this study were summarized in Table 1. Trichloroacetic acid (TCAA) or bromoacetic acid was chosen as a radical initiator. Several transition metal salts listed in Table 1 were chosen as a sensitizer.

Table 1. Compositions of LMG gels in this study.

composition	LMG-TCAA	LMG-BAA		
	$MnCl_2$	$MnCl_2$	MnX	Metal Cl_2
water				
k-Carrageenan [wt%]	2	2	2	2
HBCD [wt%]	4	4	4	4
TCAA [mM]	4	-	-	-
BAA	-	4	4	4
LMG [mM]	0.3	0.3	0.3	0.3
$CHCl_3$ [mM]	5	5	5	5
$MnCl_2$ [mM]	0, 6	0, 2, 4, 6, 8, 10	-	-
$MnSO_4$ or $Mn(NO_3)_2$ [mM]	-	-	6	-
$CoCl_2$, $NiCl_2$ or $CuCl_2$ [mM]	-	-	-	6

HBCD: highly branched cyclic dextrin
TCAA: trichloroacetic acid, BAA: bromoacetic acid

2.2. Irradiation
The cuvette samples were irradiated with a medical linear accelerator (Clinac iX, Varian) using 6 MV photon beams. A source sample distance of 100 cm and a field size of 20 × 20 cm^2 at the isocenter was used to guarantee the beam flatness. The doses of 1, 2, 5, 10 and 20 Gy were delivered to each sample with a constant dose rate of 300 cGy/min at 20°C. The dose rate was changed from 100 to 600 cGy/min to examine the dose rate dependency.

2.3. Optical density measurement
The optical densities of the gel samples were measured with a spectrophotometer (Shimadzu UV-1200) at 630 nm before and after irradiation. The optical dose responses of the LMG gels were evaluated by subtraction of the optical density of before and after irradiated samples.

3. Results and Discussion
By using HBCD as the host compound, hydrophobic leucomalachite green was easily dissolved in water with chloroform. Auto-oxidation of leuco dye is an unfavorable property of radiochromic gels because it lowers the dose sensitivity [7]. While dissolving LMG in water, a significant coloring was observed above 60°C. Therefore, the solution was kept at room temperature (20°C) before it was combined with a gelling agent solution. We used k-carrageenan as a gelling agent, because both gelatin and agarose yielded a somewhat cloudy gel in the presence of 4 wt% of HBCD.

The initial optical densities of the gels were different between batches because the temperature histories during the preparations were not exactly the same for each of the samples, however, their optical dose responses were almost the same when the initial value was subtracted from the irradiated one.

In contrast to LMG micelle gel dosimeters, our inclusion type LMG gel showed no optical dose responses up to 20 Gy without transition metal salts. When adding manganese (II) chloride in the gel, dose responses were observed. The dose sensitivity increased with increasing $MnCl_2$ concentration (2-8 mM). The gels having other manganese (II) salts, sulfate or nitrate, also showed the same dose responses as was found in the gel with manganese (II) chloride. On the other hand, addition of cobalt (II), nickel (II), or cupper (II) chloride was ineffective in the dose response. Although the exact

Journal of Physics: Conference Series **573** (2015) 012035 doi:10.1088/1742-6596/573/1/012035

mechanism is not known at this stage, we speculate that manganese (II) ion is quite stable and once it is oxidized to a higher oxidation state, it acts as a strong oxidizing agent.

The effect of halocarbon radical initiator was then examined and it was found that the gel with BAA showed a nearly 2 times higher sensitivity (8.5×10^{-3} cm^{-1} Gy^{-1}) than that of TCAA (4.0×10^{-3} cm^{-1} Gy^{-1}) (figure 1) or those of the reported radiochromic micelle gels [3, 6]. Since bond dissociation energy of C-Br is lower than that of C-Cl, BAA effectively generates a bromine radical which abstracts hydrogen atom from leuco dye.

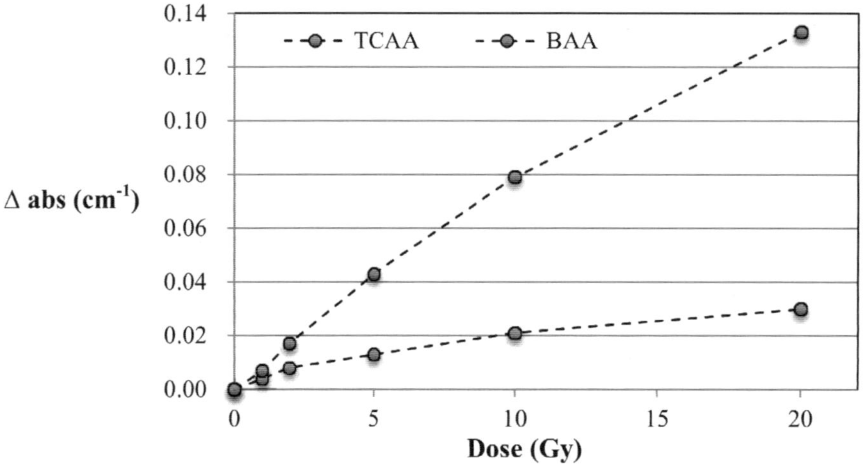

Figure 1. Effect of initiators on the dose sensitivity of LMG-$MnCl_2$ (6 mM) gels. The dose sensitivities (TCAA: 4.0×10^{-3} cm^{-1} Gy^{-1}, BAA: 8.5×10^{-3} cm^{-1} Gy^{-1}) were calculated at the linear part (<5Gy).

The optical densities were increased by increasing the post-irradiation time due to the auto oxidation of leuco dye. When samples were stored at room temperature (ca. 20°C), the dose sensitivities also increased (figure 2). This auto oxidation can be reduced by storage of the samples in a refrigerator (4°C). The time-dependences of the optical density were 3.0×10^{-3} $cm^{-1}h^{-1}$ at 20°C and 0.49×10^{-3} $cm^{-1}h^{-1}$ at 4°C, respectively.

Dose rate dependency was observed in our inclusion type gel (figure 3), as was previously reported on radiochromic micelle gels [6, 7]. Using 100 cGy/min dose rate resulted ca. 17% (at 4 Gy) and 21% (at 8 Gy) higher optical dose responses than those of 600 cGy/min.

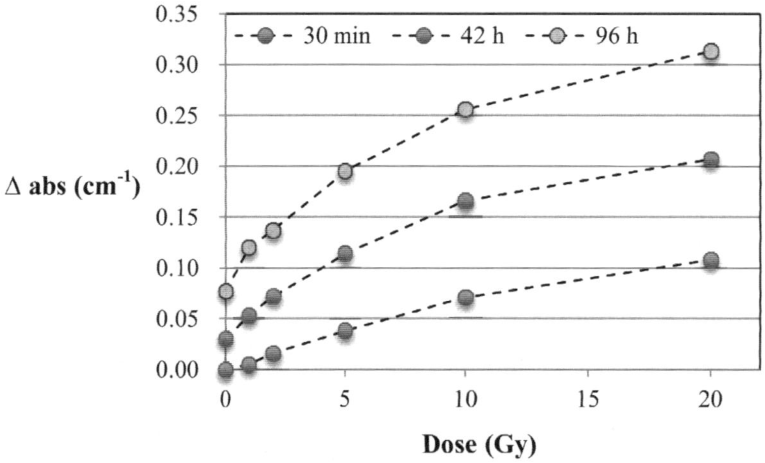

Figure 2. Dose response change of LMG-BAA-$MnCl_2$ (4 mM) gel by post-irradiation times.

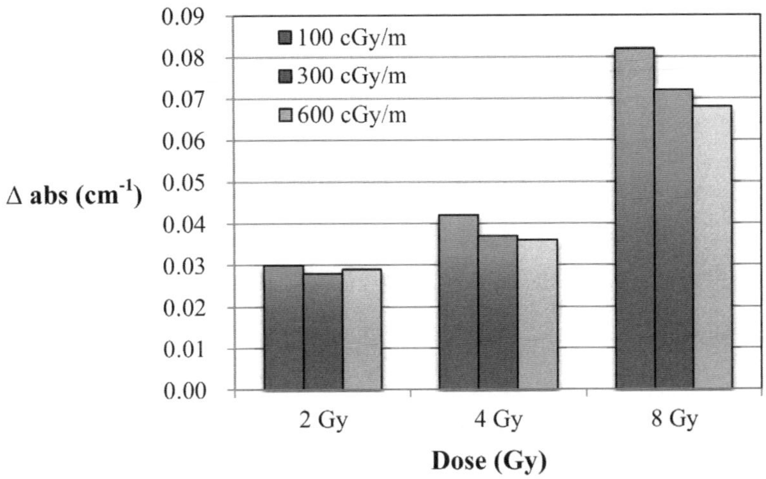

Figure 3. Dose rate dependence of LMG-BAA-$MnCl_2$ (6 mM) gel.

4. Conclusion

We have newly developed an inclusion type radiochromic gel detector consisting of LMG-HMBC, BAA, and $MnCl_2$. It can be easily prepared and has relatively high dose sensitivity. The auto oxidation property was reduced by storage in a refrigerator (4°C). However, the unfavorable dose rate dependency remains to be overcome for clinical applications.

5. References

[1] Baldock C *et al* 2010 *Phys. Med. Biol.* **55** R1-63
[2] Jordan K 2010 *J. Phys. Conf. Ser.* **250** 012043
[3] Jordan K *et al* 2009 *Phys. Med. Biol.* **54** 6773-89
[4] Babic S *et al* 2009 *Phys. Med. Biol.* **54** 6791-808
[5] Babic S *et al* 2009 *Phys. Med. Biol.* **54** 2463-81
[6] Vandecasteele J *et al* 2011 *Phys. Med. Biol.* **56** 627-51
[7] Vandecasteele J *et al* 2013 *Phys. Med. Biol.* **58** 6241-62
[8] Skyt P S *et al* 2013 *J. Phys. Conf. Ser.* **444** 012036
[9] Takata H *et al* 2006 *Seibutsu-kogaku* **84** 61-6
[10] Fukami T *et al* 2006 *J. Incl. Phenom. Macrocycl. Chem.* **56** 61-4

Radiation-chemical and optical properties of a radio-fluorogenic gel

Tiantian Yao, Alessia Gasparini, Antonia G Denkova and John M Warman
Delft University of Technology, Department of Radiation Science and Technology, Section of Radiation and Isotopes for Health, Mekelweg 15, 2629 JB Delft, The Netherlands

E-mail: a.gasparini@tudelft.nl

Abstract. The radiation-induced polymerization and fluorescence intensity of a radio-fluorogenic medium consisting of tertiary-butyl acrylate (TBA) with *ca* 100 ppm maleimido-pyrene (MPy) display a super-linear dependence on dose and a close to inverse square root dependence on dose rate over the range from 2 to 30 Gy/min. In contrast with the fluorescence, the clarity and optical absorption remain unchanged on irradiation up to at least 17% monomer conversion for which the medium is a rigid gel.

1. Introduction

A radio-fluorogenic (RFG) gel becomes fluorescent in UV light after exposure to high-energy radiation. This radiation-induced fluorescent property is *permanent*, which distinguishes it from *transitory* optical emission or scintillation effects. We have shown that a quasi-rigid RFG gel, consisting of tertiary-butyl acrylate (TBA) with *ca* 100 ppm maleimido-pyrene (MPy), can be used to make fixed fluorescent images of complex radiation fields, including those produced by overlapping X-ray beams [1], 3MeV electrons [2], 80 MeV protons [3] and an Ir-192 brachytherapy seed [4]. In the last case we demonstrated that RFG images could be video-recorded *in situ* during seed insertion and retraction. Having shown *qualitatively* the potential of a TBA/MPy gel to produce spatially resolved images of dose deposition, we have initiated a programme of research into the fundamental radiation chemistry and photophysics of the medium with the ultimate goal of providing a *quantitative* fluorescence-based method of *in-situ* 3D dosimetry. The work presented here is a follow-up to a preliminary investigation of the dose and dose-rate dependence of polymer-gel formation on irradiation of pure TBA [5].

2. Experimental

The radio-fluorogenic medium studied here consists of >99% TBA, a monomeric acrylate that rapidly polymerizes when exposed to high-energy radiation, and *ca* 100 ppm of the fluorogenic compound MPy. MPy is initially non-fluorescent but becomes fluorescent when incorporated into a growing chain of TBA. Both compounds are commercially available and inexpensive. Details on material suppliers and physical properties are to be found in [1]. All samples were de-aerated prior to irradiation and prepared in a nitrogen-flushed glove box. Two cobalt-60 gamma-ray sources were used to irradiate samples: a GC200, dose rate *ca* 2 Gy/min and a GC220, dose rate *ca* 30 Gy/min. The actual dose rate on the measurement days was corrected for the natural decay of cobalt-60.

The fractional conversion of monomer to polymer was determined gravimetrically by weighing irradiated samples before and after vacuum evacuation of remaining monomer, as described in [5]. For monitoring optical properties, the samples were contained in 1 cm square screw-capped, borosilicate glass cuvettes, and irradiated as described in the next section. The optical absorptive and emissive properties of the samples were monitored using a Kontron Instruments UV/vis spectro-photometer and an Ocean Optics fibre-optic spectro-fluorimeter with LED excitation at 365 nm.

3. Results and Discussion

3.1. Monomer to polymer conversion, C_M

In figure 1 the fractional conversion of monomer to polymer is plotted as a function of dose for pure TBA and for a $0.6x10^{-3}$ mol/L (*ca* 100 ppm) solution of MPy in TBA. The data were obtained using the GC200 and GC220 sources with dose rates differing by a factor of approximately 15. For free-radical chain-polymerization with chain termination controlled by second order radical-radical combination and disproportionation, C_M is expected to be a linear function of dose, D, divided by the square root of the dose rate, D'. This behaviour is approached for low values of C_M as shown by the dashed lines in figure 1. An increasing positive deviation from linearity is however apparent. This super-linear dependence can be described by the relationship $C_M = K_C.a/(1 - Ba)$ with $a = D/\sqrt{D'}$ [5]. The full-line fit to the pure TBA data in figure 1**A** was calculated using values of $K_C = 0.143\ Gy^{-0.5}s^{-0.5}$ and $B = 1.00x10^{-3}\ Gy^{-0.5}s^{-0.5}$ (for C_M in %) [5]. The same values were used to calculate the full line in figure 1**B**. As can be seen the fit to the high dose rate data for the MPy solution is good indicating that the presence of 100 ppm MPy does not significantly influence the polymerization kinetics. At the lower dose rate the C_M values are slightly lower than the fit line. This may be due to a small contribution from a first-order chain termination reaction that becomes more important at lower doses for which the free-radicals live longer.

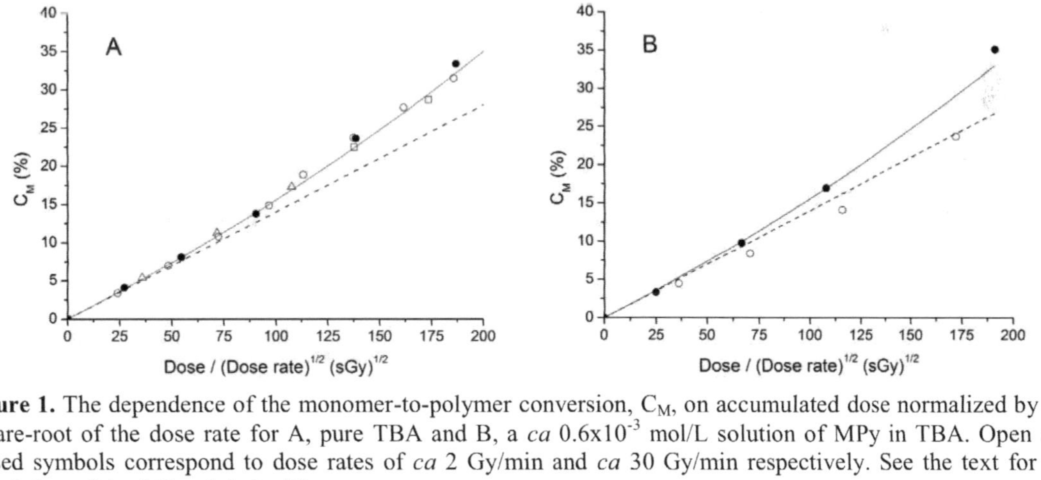

Figure 1. The dependence of the monomer-to-polymer conversion, C_M, on accumulated dose normalized by the square-root of the dose rate for A, pure TBA and B, a *ca* $0.6x10^{-3}$ mol/L solution of MPy in TBA. Open and closed symbols correspond to dose rates of *ca* 2 Gy/min and *ca* 30 Gy/min respectively. See the text for the calculation of the full and dashed lines.

3.2. Optical absorption and emission spectra

In figure 2 are shown the optical absorption spectra of a saturated solution of MPy in TBA (*ca* $1.8x10^{-3}$ mol/L) before irradiation and after a GC200 dose of 20 Gy. For this dose the medium has changed from a free-flowing liquid to a rigid gel that shows little tendency to flow on a timescale of an hour or more [5]. Also, 17% of the monomer has been converted to polymer with an average chain length on the order of 10^5 monomer units. After irradiation the gel remains as clear as the original solution with no indication of the creation of additional absorptions or turbidity. This independence of the absorptive and light-scattering properties of an RFG gel on dose is particularly relevant to the interpretation of 3D images since it means that one can take the attenuation with depth of the UV excitation light and the self-absorption of the emitted fluorescence within the medium to be constant and uniform.

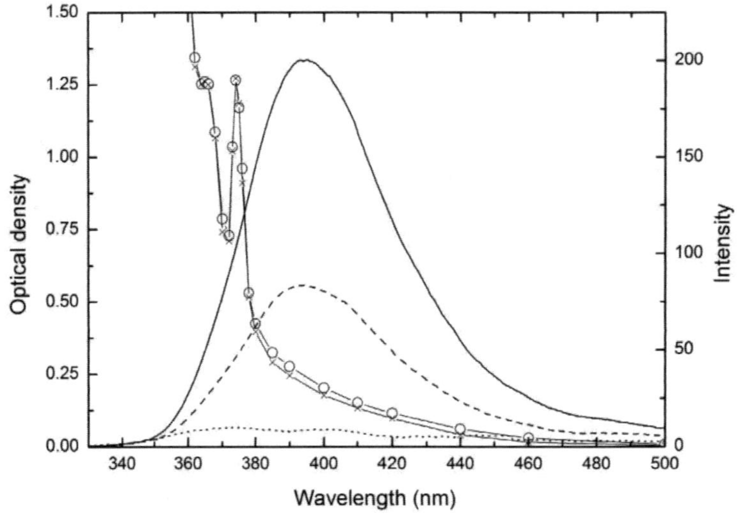

Figure 2. Circle and cross data points: the optical density of a $1.8x10^{-3}$ mol/L solution of MPy in TBA before and after irradiation with a dose of *ca* 20 Gy respectively. Line spectra: the fluorescent emission of a $0.6x10^{-3}$ mol/L solution of MPy in TBA before irradiation (dotted line) and after doses of 8 Gy (dashed line) and 16 Gy (full line).

In contrast to the optical absorption, the fluorescent emission changes dramatically on irradiation, from close to zero initially, to a readily measureable, blue emission after only a few Gy. This is due to the fact that the S_0 to S_1 absorptive transition of MPy is controlled mainly by the aromatic pyrene moiety, which remains unchanged on co-polymerization, while the S_1 to S_0 emissive transition is controlled by the maleimido moiety, which changes dramatically on co-polymerization to a succinimido moiety, which no longer quenches the normal pyrene fluorescence.

3.3. Dose and dose rate dependence of the fluorescence

In figure 3 the intensity of the fluorescence at the spectral maximum (395 nm) is plotted against $D/\sqrt{D'}$ for a $0.6x10^{-3}$ mol/L MPy solution. The data display a similar dependence to that shown in figure 1**B** for the monomer-to-polymer conversion. The full line drawn in figure 3 is a best fit to the high dose rate data using the relationship $I = K_I.a/(1 - Ba)$. The value of B, which describes the upward curvature of the data, was fixed to that used in figure 1, *i.e.* $1.00x10^{-3}$ $Gy^{-0.5}s^{-0.5}$. The fluorescence is seen to follow the same super-linear behaviour as found for the monomer conversion. As for the C_M measurements for the solution, the values for the lower dose rate tend to be somewhat lower than for the high dose rate. An investigation of the dependence of the fluorescence on MPy concentration is presently underway in order to clarify a possible secondary influence of MPy on the polymerization kinetics.

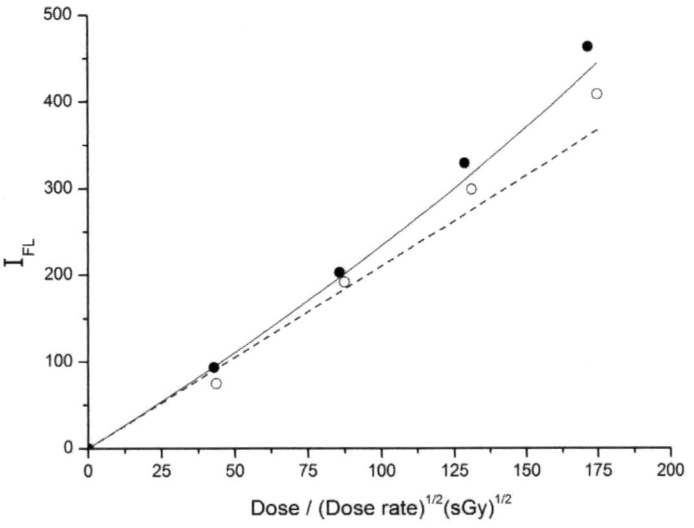

Figure 3. The intensity at 395 nm of the fluorescent emission of a $0.6x10^{-3}$ mol/L solution of MPy in TBA as a function of the accumulated dose normalized by the square-root of the dose rate. Open and closed symbols for dose rates of *ca* 2 Gy/min and *ca* 30 Gy/min respectively. For the calculation of the full and dashed lines see the text.

4. Conclusions

We have shown that the radiation-induced polymerization of tertiary-butyl acrylate results in a dependence of the monomer-to-polymer conversion, C_M (%), on dose, D (Gy), and dose rate, D' (Gy/s), of the form $C_M = 0.143D/\sqrt{D'}/[1 - 0.001D/\sqrt{D'}]$ for C_M 30% or less and dose rates from *ca* 0.03 to *ca* 0.5 Gy/s. From the value of C_M after irradiation of TBA for a time t at a dose rate D', the actual value of the dose rate can be determined from $D' = [C_M/(0.143 + 0.001C_M)t]^2$.

We have also shown that the intensity of the radiation-induced fluorescence of TBA containing a low concentration of the fluorogenic compound MPy parallels closely the dose dependence of C_M. Making fixed fluorescent images of complex radiation fields in the quasi-rigid gel state of TBA ($C_M >$ 10% [5]) under exactly the same conditions of UV excitation as for homogeneously irradiated, gamma-ray calibration samples should make it possible to determine the local dose rate. Parallel research is also being carried out on radio-chromic films, which are close to dose rate independent, with a view to combining the two methods, both of which have sub-millimetre spatial resolution.

For quantitative measurements corrections for absorption of the excitation light and the emitted fluorescence in larger volumes have to be made. Taking the spectral data shown in figure 2 into account and the UV wavelength of the collimated LED sources presently used (wavelength maximum 384 nm), we consider phantom volumes on the order of a litre to be do-able. This could be extended by using a long-wavelength, double-photon method of photo-excitation.

5. References

[1] Warman J M *et al* 2011 *Phys. Med. Biol* **56** 1487-508
[2] Warman J M *et al* 2013 *Rad. Phys. Chem.* **84** 129-35
[3] Warman J M *et al* 2013 *Rad. Phys. Chem.* **85** 179-81
[4] Warman J M *et al* 2011 *Advanced Materials* **23** 4953-5
[5] Yao T *et al* 2014 *Rad. Phys. Chem.* **97** 147-52

Opportunities for improving the performance of LCV micelle gel dosimeters: I. Preliminary investigation

A T Nasr[1], KM Alexander[2], L J Schreiner[2,3] and K B McAuley[1]
[1]Dept. Chemical Engineering, Queen's University, Kingston, Canada, K7L 3N6
[2]Dept. of Physics, Queen's University, Kingston, Ontario, Canada, K7L3N6
[3]Cancer Centre of Southeastern Ontario, Kingston, Canada, K7L 5P9

E-mail: kim.mcauley@chee.queensu.ca

Abstract. The effects of the various components of leuco crystal violet (LCV) micelle gels on dose sensitivity and initial colour are tested. Dose sensitivity and gel turbidity are influenced by tri-chloro acetic acid (TCAA) concentration, with the highest dose sensitivity obtained at ~21.5 mM. Increasing Triton x-100 (Tx100) concentration improved dose sensitivity but, unfortunately, increased initial colour of the gel. Using Cetyl Trimethyl Ammonium Bromide (CTAB) in place of Tx100 produces gels that are nearly colourless prior to irradiation, but reduces the dose sensitivity. The effects of chlorinated species on dose sensitivity are tested using 2,2,2-trichloroethanol (TCE), chloroform, and 1,1,1-trichloro-2-methyl-2-propanol hemihydrate (TCMPH). TCE gives the largest improvement in dose sensitivity and is recommended for use in micelle gel dosimeters because it is less volatile and safer to use than chloroform. A new gel containing CTAB as the surfactant and TCE gives improved dose sensitivity compared to previous LCV micelle gels.

1. Introduction

Radiochromic micelle gel dosimeters, developed by Jordan and co-workers for optical readout, are promising because they have the potential to produce accurate 3D dosimetry results [1] with excellent spatial stability [2-4]. However, current micelle gel dosimeters have relatively lower dose sensitivity compared to Fricke gel dosimeters [3]. Leuco dye micelle gels are also sensitive to temperature during irradiation and are sensitive to light exposure during storage [2, 3]. Among the two radiochromic micelle gels developed by Jordan and co-workers, LCV micelle gels are recommended over leuco malachite green (LMG) micelle gels because the LCV gel dosimeters exhibited linear dose response, lower effective diffusion, and higher dose sensitivity [3]. The recommended LCV micelle gel recipe reported by Babic *et al* [3] contains five components: 4.0 wt% gelatin, 4.0 mM Tx100 surfactant, 25.0 mM TCAA, 1.0 mM LCV, and ~ 96.0 wt% water. This recipe is referred to as Jordan's standard LCV gel throughout this article. To develop this recipe, Babic *et al* [3] studied the effects of varying the concentrations of Tx100, LCV, and TCAA on dose sensitivity. They found that dose sensitivity could be improved by increasing Tx100 concentration. They also showed that a concentration of ~25.0 mM TCAA maximizes the dose sensitivity. The influence of TCAA on dose sensitivity may arise from two factors, which were discussed by Vandecasteele *et al* [4]; TCAA contains chlorine atoms that may have a sensitizing effect and is an acid that influences pH.

Vandecasteele *et al* [4] reported difficulties in making LCV micelle gels using Tx100 and replaced it with sodium dodecyl sulfate (SDS). When attempting to make radiochromic micelle gels using diacetylenes as reporter molecules, Nasr *et al* [5] tested Tx100, SDS and CTAB as surfactants. Precipitation problems were observed for several diacetylene gels produced using Tx100 and SDS, while transparent stable gels were obtained using CTAB. Vandecasteele *et al* [4] also studied the effect of adding chloroform to LMG micelle gels (made with SDS) and reported an improvement in dose sensitivity. This result is not surprising because chloroform is an important component used in some PRESAGE gels [6, 7] and chloroform and other chlorinated molecules are effective in increasing dose sensitivity in other dosimetry applications, due to a variety of chemical reactions [8].

The purpose of the current study is to gain a deeper understanding of the effects of various gel components on dose sensitivity, initial colour prior to irradiation, and gel transparency of LCV micelle gels. First, the effect of TCAA concentration on dose sensitivity and gel transparency is studied. Next, the effect of surfactant type and concentration on dose sensitivity and initial colour is studied using Tx100 and CTAB. Finally, the effects of three halogenated hydrocarbons (chloroform, TCE and TCMPH) on dose sensitivity are investigated.

2. Materials and Methods

2.1. Gel preparation, irradiation, and scanning

The following manufacturing procedure was followed to manufacture gels:

1- Add gelatin (Type A, 300 Bloom porcine skin) into 75 wt% of the total deionized water in the recipe and let swell for 20 minutes. Then stir while heating to 45 °C until gelatin dissolves.
2- Add surfactant (Tx100 or CTAB) and TCAA to the remaining 25 wt% of the deionized water and stir at room temperature for 5 minutes until dissolved.
3- Add LCV to the mixture from step 2 and stir at room temperature for 5 minutes.
4- Pour LCV mixture from step 3 (at room temperature) into gelatin mixture from step 1 (at 45 °C) and stir for 2 minutes.

Samples of the final mixture, produced using a variety of recipes, were poured into 4.5 ml polystyrene cuvettes with 10 mm light path and sealed using a cap and Parafilm. pH was measured using a pH electrode. All gels were refrigerated at 4 °C for ~ 24 h and then placed in a water bath at 22 °C for 20 minutes before irradiation. A Varian Clinac 6EX linear accelerator was used to irradiate the cuvettes to different doses of 0, 5, 10, 15, 20, and 40 Gy at a dose rate of 400 cGy min^{-1}. Absorption spectra were measured ~20 minutes after irradiation using a SpectroVis Plus spectrophotometer over the wavelength region of 380-900 nm. Absorbance measurements were calibrated using a reference cuvette filled with deionized water. Further details are provided in [9].

2.2. Effect of TCAA concentration on dose sensitivity and gel transparency

Gels were made with different TCAA concentrations of 14.5, 21.5, 25.0, 27.5, and 33.5 mM. Gelatin and Tx100 concentrations were fixed at 4.0 wt% and 4.0 mM, respectively (as in Jordan's standard LCV gel) and the LCV concentration was fixed at 0.25 mM.

2.3. Effect of surfactant type and concentration on dose sensitivity of LCV micelle gels

Gels were manufactured using Tx100 (4.0, 19.0, and 39.0 mM) and CTAB (at 2.5, 9.0, and 17.0 mM). TCAA, LCV, and gelatin concentrations were fixed at 25.0 mM, 1.0 mM, and 4.0 wt%, respectively.

2.4. Effect of halogenated species on dose sensitivity of LCV micelle gels

Chloroform, TCE, and TCMPH were tested as chlorinated additives to improve dose sensitivity of LCV micelle gels. TCE and TCMPH were selected because of their lower volatility and lower toxicity compared to chloroform. LCV micelle gels were made at two different concentrations (10.0 and 90.0 mM) of each of the three chlorinated species. All gels were manufactured as described in section 2.1 except that, in step 2, the halogenated compound was added to the surfactant and TCAA solution and mixed for ~10 minutes before LCV was added. Concentrations of gelatin, Tx100, LCV and TCAA were kept at 4.0 wt%, 4.0 mM, 1.0 mM, and 25.0 mM, respectively as in Jordan's standard LCV gel.

Journal of Physics: Conference Series **573** (2015) 012037 doi:10.1088/1742-6596/573/1/012037

An additional gel was produced using 60 mM TCE in a gel where Tx100 was replaced by17.0 mM CTAB.

3. Results and Discussion

3.1. Effect of acid concentrations (pH) on dose sensitivity and gel transparency

Results in table 1 show that TCAA strongly influences dose sensitivity in a nonlinear fashion, with the highest dose sensitivity obtained at a TCAA concentration of 21.5 mM (gel 1.3). A similar maximum in dose sensitivity was observed by Babic *et al* [3]. LCV micelle gels manufactured at low acid concentration (high pH) were turbid (gels 1.1, 1.2). This result reveals that the transparency of LCV micelle gels is influenced by acid concentration (or pH).

Table 1. Dose sensitivity of LCV micelle gels at different concentrations of TCAA. Concentrations of gelatin, LCV, and Tx100 were fixed at 4 wt%, 0.25 mM, and 4 mM, respectively. Absorbance values were measured at 590 nm.

#	TCAA Conc. [mM]	LCV [mM]	Transparency	Initial Colour	Initial Absorbance	Dose Sensitivity [$Gy^{-1}.cm^{-1}$]	pH
1.1	14.5	0.25	Turbid	----	0.185	$3.87 \cdot 10^{-3}$	4.04
1.2	18.0	0.25	Turbid	----	0.122	$5.37 \cdot 10^{-3}$	3.75
1.3	21.5	0.25	Clear	No	0.030	$6.38 \cdot 10^{-3}$	3.63
1.4	23.0	0.25	Clear	Light Blue	0.036	$5.64 \cdot 10^{-3}$	3.43
1.5	25.0	0.25	Clear	No	0.030	$4.79 \cdot 10^{-3}$	3.36
1.6	27.5	0.25	Clear	No	0.019	$4.33 \cdot 10^{-3}$	3.15
1.7	33.5	0.25	Clear	No	0.018	$1.93 \cdot 10^{-3}$	2.63

3.2. Effect of surfactant type and concentration on dose sensitivity and initial colour

The dose sensitivity of LCV gels increased with increasing Tx100 concentration, as shown in table 2. Unfortunately, the initial colour of the un-irradiated gels also increased with increasing Tx100. Using CTAB at different concentrations resulted in gels with no noticeable initial colour, but did not improve the dose sensitivity. Similar gels prepared using SDS at different concentrations (not shown) resulted in viscous, turbid and sticky mixtures that are not suitable for optical scanning [9].

Table 2. Dose sensitivity of LCV micelle gels made using Tx100 and CTAB. Surfactant concentrations were selected so that the number of micelles/L in the gels would be roughly 1, 5.5, and 10 times the number of micelles in Jordan's standard LCV gel. Gelatin, TCAA, and LCV concentrations were fixed at 4.0 wt%, 25 mM, and 1.0 mM, respectively. Gel 2.2 is a replicate of Jordan's standard LCV gel. Absorbance values were measured at 590 nm.

#	Surfactant	Conc. [mM]	Transparency	Initial Colour	Initial Absorbance	Dose Sensitivity [$Gy^{-1}.cm^{-1}$]	pH
2.1	--		clear	No	0.020	$5.50 \cdot 10^{-3}$	3.40
2.2	Tx100	4.0	clear	Light Blue	0.049	$7.65 \cdot 10^{-3}$	3.36
2.3	Tx100	19.0	clear	Dark Blue	0.116	$1.09 \cdot 10^{-2}$	3.44
2.4	Tx100	38.0	clear	Dark Blue	0.207	$1.37 \cdot 10^{-2}$	3.43
2.5	CTAB	2.5	clear	No	0.020	$5.62 \cdot 10^{-3}$	3.37
2.6	CTAB	9.0	clear	No	0.025	$5.57 \cdot 10^{-3}$	3.39
2.7	CTAB	17.0	clear	No	0.026	$5.80 \cdot 10^{-3}$	3.39

3.3. Effect of halogenated hydrocarbons on dose sensitivity of LCV micelle gels

Results in table 3 show that dose sensitivities of LCV micelle gels were highest for gels containing TCE. Moreover, the dose sensitivity of LCV micelle gels increased with increasing TCE concentration. Using chloroform did not affect the dose sensitivity of LCV micelle gels significantly compared to gel (3.7) which is a standard Jordan LCV gel. When TCMPH was used at the low concentration of 10.0 mM, dose sensitivity was higher than that of Jordan's standard LCV gel.

However, when the TCMPH concentration increased to 90.0 mM, dose sensitivity decreased below that of Jordan's standard LCV gel.

Gel (3.8), which contained 60.0 mM TCE and 17.0 mM CTAB (in place of Tx100) was colourless and had the highest dose sensitivity of any gel in this study, except for gels (2.3) and (2.4), which were dark blue prior to irradiation. TCE will be used as a gel additive in our on-going research because it gives the best dose sensitivity results of the three additives tested.

Table 3. Effects of adding three different halogenated compounds on dose sensitivity of LCV micelle gels. Concentrations of LCV, TCAA, Tx100 and gelatin are the same as in Jordan's standard LCV gel, except for Gel 3.8, which contains 17.0 mM CTAB in place of Tx100. Gel 3.7 is a replicate of Jordan's standard LCV gel. Absorbance values were measured at 590 nm.

#	Halogenated hydrocarbon	Conc. [mM]	Transparency	Initial Colour	Initial Absorbance	Dose Sensitivity [$Gy^{-1}.cm^{-1}$]	pH
3.1	Chloroform	10.0	Clear	Light Blue	0.048	$8.08 \cdot 10^{-3}$	3.40
3.2	Chloroform	90.0	Clear	No	0.017	$7.09 \cdot 10^{-3}$	3.34
3.3	TCE	10.0	Clear	Light Blue	0.054	$9.07 \cdot 10^{-3}$	3.34
3.4	TCE	90 .0	Clear	Light Blue	0.088	$1.25 \cdot 10^{-2}$	3.39
3.5	TCMPH	10.0	Clear	Light Blue	0.055	$9.26 \cdot 10^{-3}$	3.42
3.6	TCMPH	90.0	Clear	Light Blue	0.041	$7.32 \cdot 10^{-3}$	3.43
3.7	----	----	Clear	Light Blue	0.055	$7.74 \cdot 10^{-3}$	3.48
3.8	TCE	60	Clear	No	0.030	$9.37 \cdot 10^{-3}$	3.29

4. Conclusions

This study was conducted to gain a deeper understanding of the effects of various components in LCV micelle gels on dose sensitivity, initial colour, and gel transparency. A maximum in the dose sensitivity vs. TCAA response curve is obtained using ~21.5 mM TCAA. Turbid gels are produced when a low concentration of TCAA (≤18.0 mM) is used. Dose sensitivity increases with increasing Tx100 concentration but, unfortunately, the initial colour of the un-irradiated gels also increases, which may cause complications during optical scanning in larger containers. Using CTAB to replace Tx100 surfactant produced colourless gels prior to irradiation, but reduced dose sensitivity (by~ 26%) when no chlorinated species (except TCAA) are included in the recipe. The effects of three halogenated compounds (chloroform, TCE and TCMPH) on dose sensitivity were tested. TCE is the most effective at improving the dose sensitivity and is safer to use than chloroform because it is less volatile and less toxic. Using this information, a new gel is proposed using CTAB as the surfactant and TCE as a sensitizer. This new gel has improved dose sensitivity compared to that in Jordan's standard LCV gel, and has no noticeable colour prior to irradiation. Additional information is provided elsewhere [9].

5. References

[1] Baldock C *et al* 2010 *Phys. Med. Biol.* **55** R1-63
[2] Jordan K and Avvakumov N 2009 *Phys. Med. Biol.* **54** 6773-89
[3] Babic S *et al* 2009 *Phys. Med. Biol.* **54** 6791-808
[4] Vandecasteele J *et al* 2011 *Phys. Med. Biol.* **56** 627-51
[5] Nasr A T *et al* 2013 *Phys. Med. Biol.* **58** 787-805
[6] Adamovics JA 2006 *United States Patent # US 7098463B2.*
[7] Adamovics J and Maryanski M 2006 *Radiation Protection Dosimetry.* **120** 107-12
[8] McLaughlin W 1970 *Manual on Radiation Dosimetry* ed. Holm NW and Berry RJ (Marcel Dekker INC. New York) p.129
[9] Nasr AT 2014 Ph.D. Thesis in Chemical Engineering *Queen's University*, Kingston, Ontario, Canada.

Journal of Physics: Conference Series **573** (2015) 012038 doi:10.1088/1742-6596/573/1/012038

Opportunities for improving the performance of LCV micelle gel dosimeters: II. Recipe optimization

A T Nasr[1], K M Alexander[2], T Olding[2,3], L J Schreiner[2,3] and K B McAuley[1]
[1] Dept. Chemical Engineering, Queen's University, Kingston, Canada, K7L 3N6
[2] Dept. of Physics, Queen's University, Kingston, Ontario, Canada, K7L3N6
[3] Cancer Centre of Southeastern Ontario, Kingston, Canada, K7L 5P9

E-mail: kim.mcauley@chee.queensu.ca

Abstract. Designed experiments and empirical models are used to optimize a Leuco Crystal Violet (LCV) micelle gel recipe to improve dose sensitivity and initial colour. The optimized recipe contains 0.75 mM LCV, 17.0 mM Cetyl Trimethyl Ammonium Bromide (CTAB), 120 mM 2,2,2-trichloroethanol (TCE), 25.0 mM tri-chloro acetic acid (TCAA), 4 wt% gelatin and ~96 wt% water. Dose sensitivity of the optimized gel is 1.5 times higher than Jordan's standard LCV gel. Spatial integrity of the 3D dose distribution information in 1L jar phantoms made using this recipe is maintained for more than two weeks.

1. Introduction

LCV micelle gel dosimeters can be read out using optical scanners and have the potential to produce accurate 3D dosimetry [1] results with excellent spatial stability [2-4]. Babic *et al.* [3] recommended a gel recipe containing 4.0 wt% gelatin, 4.0 mM triton x-100 (Tx100) surfactant, 25.0 mM TCAA, 1.0 mM LCV, and ~ 96.0 wt% water. This recipe is referred to as Jordan's standard LCV gel throughout this article. In our previous work, effects of components in LCV micelle gel were investigated [5]. We found that dose sensitivity increases with increasing Tx100 but, unfortunately, initial colour of the un-irradiated gels also increases. When Tx100 was replaced with CTAB, colourless gel was produced, but dose sensitivity decreased. TCE and other chlorinated compounds were tested as sensitizers for LCV micelle gels and TCE was chosen for further study because it was the most effective for improving sensitivity. Additional testing and optimization of the recipe are conducted in the current work.

Design of Experiments (DOE) techniques [6-7] involve applying formal statistical methods for planning and conducting experiments and for extracting, analyzing and interpreting data to draw conclusions and build models that can be used for optimization. DOE uses an efficient set of experimental runs to investigate the effects of the critical factors (inputs) on the responses (outputs) of interest. Rather than studying the influence of one input at a time, experiments are designed so that multiple factors are varied together, providing information about interactions between variables. In two-level full-factorial designs (a widely used DOE technique), each input variable is set at two different levels: low (coded as -1) and high (coded as 1). A two-level full factorial design with k factors is called a 2^k design, because 2^k experimental runs are required to examine all possible combinations of low and high values for the k factors (see Fig.1). Replicate points are often added at the centre of a 2^k design to estimate typical experimental error encountered in the experiment [6-7]. In the current paper, a 2^3 factorial experiment corresponding to figure 1 is conducted and linear

Journal of Physics: Conference Series **573** (2015) 012038 doi:10.1088/1742-6596/573/1/012038

regression models are developed to predict initial colour (absorbance at 590 nm) and dose sensitivity. This information is used to guide additional experimentation and to develop an optimized gel recipe

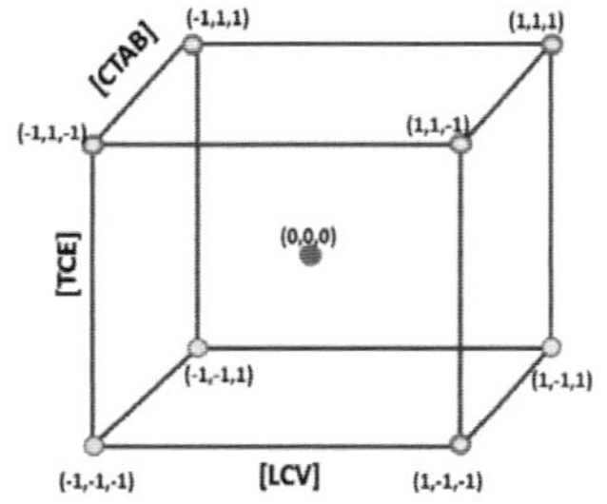

Figure 1. Two-level three-factor (2^3) full factorial design with central point. The three factors here are concentrations of LCV, TCE, and CTAB. Note that the low and high levels of the factors are shown in coded form.

2. Materials and Methods

2.1. Gel preparation, irradiation, and scanning

The following manufacturing procedure was followed to manufacture gels:

1- Add gelatin (Type A, 300 Bloom porcine skin) to 75 wt% of the total deionized water in the recipe and let swell for 20 minutes. Stir while heating to 45 °C until gelatin dissolves.
2- Add CTAB, TCE and TCAA to the remaining 25 wt% of the deionized water and stir at room temperature for 10 minutes.
3- Add LCV to the mixture from step 2 and stir at room temperature for 5 minutes.
4- Pour LCV mixture from step 3 (at room temperature) into gelatin mixture from step 1 (at 45 °C) and stir for 2 minutes.

Samples of the final mixture, produced using a variety of recipes, were poured into 4.5 ml polystyrene cuvettes with 10 mm light path and sealed using a cap and Parafilm. pH was measured using a pH electrode. All gels were refrigerated at 4 °C for ~24 h and placed in a water bath at 22 °C for 20 minutes before irradiation. A few samples were poured into 23×85 mm clear glass cylindrical vials and photographed. A Varian Clinac 6EX linear accelerator was used to irradiate cuvettes to doses of 0, 5, 10, 15, 20, and 40 Gy at a dose rate of 400 cGy min^{-1}. Absorption spectra were measured ~20 minutes after irradiation using a SpectroVis Plus spectrophotometer between 380 and 900 nm. Absorbance measurements were calibrated using a reference cuvette containing deionized water [8].

2.2. Designed experiments to optimize LCV micelle gels made with CTAB and TCE

Results from the full factorial design with five replicates at the central point (runs 1-13 in table 1, conducted in random order) were used to fit linear regression models with dose sensitivity and initial colour as response variables [8]. Runs (14-17) were conducted to extend the range and complexity of the models. Optimization calculations to find the best recipe were performed using Excel.

2.3. Preliminary experiments in large-jar phantoms

Two 1 L polyethylene terephthalate jars were filled with gel using the optimized gel recipe (gel 19). These replicate phantoms were irradiated using a Varian Trilogy linear accelerator with RapidArc to test their response to spatially non-uniform radiation. One gel was irradiated with a single arc VMAT test treatment plan up to a maximum dose of 1795.3 cGy in the gel. The second gel was used for calibration and was irradiated with a 1850 MU, 12 MeV 6 x 6 cm^2 field electron beam (SSD=100 cm) delivered to the top surface of the gel (jar lid removed). The phantoms were imaged using a Vista Optical CT scanner under 590 nm LED illumination. Reference scan images were taken of each gel before irradiation. After irradiation, data scan images were also taken. Each gel was carefully timed to ensure that post-irradiation imaging times were consistent. To convert optical density values to dose, central-axis depth-dose data were obtained from the gel and compared to Wellhofer ionization-chamber water-tank measurements. This calibration was applied to the VMAT plan gel irradiation.

Table 1. Influence of CTAB, LCV and TCE on initial colour and dose sensitivity of LCV micelle gels. Concentrations of gelatin and TCAA were fixed at 4.0 wt% and 25.0 mM, respectively. Absorbance values were measured at 590 nm.

#	CTAB [mM]	LCV [mM]	TCE [mM]	Turbidity	Initial Colour	Initial Absorbance	Dose Sensitivity [$Gy^{-1}.cm^{-1}$]	pH
1	9.0	0.75	40.0	Clear	No	0.026	$7.46 \cdot 10^{-3}$	3.32
2	9.0	0.75	80.0	Clear	No	0.026	$9.10 \cdot 10^{-3}$	3.31
3	9.0	1.25	40.0	Clear	No	0.025	$6.59 \cdot 10^{-3}$	3.29
4	9.0	1.25	80.0	Clear	No	0.025	$9.44 \cdot 10^{-3}$	3.32
5	25.0	0.75	40.0	Clear	No	0.028	$8.59 \cdot 10^{-3}$	3.30
6	25.0	0.75	80.0	Clear	No	0.028	$9.86 \cdot 10^{-3}$	3.29
7	25.0	1.25	40.0	Clear	No	0.029	$9.37 \cdot 10^{-3}$	3.29
8	25.0	1.25	80.0	Clear	Light Blue	0.042	$1.15 \cdot 10^{-2}$	3.33
9	17.0	1.0	60.0	Clear	No	0.030	$9.17 \cdot 10^{-3}$	3.27
10	17.0	1.0	60.0	Clear	No	0.029	$1.00 \cdot 10^{-2}$	3.22
11	17.0	1.0	60.0	Clear	No	0.030	$9.60 \cdot 10^{-3}$	3.30
12	17.0	1.0	60.0	Clear	No	0.030	$9.37 \cdot 10^{-3}$	3.29
13	17.0	1.0	60.0	Clear	No	0.029	$1.00 \cdot 10^{-2}$	3.30
14	25.0	1.25	120.0	Clear	Light Blue	0.045	$1.28 \cdot 10^{-2}$	3.30
15	33.0	1.25	100.0	Clear	Light Blue	0.04	$1.19 \cdot 10^{-2}$	3.30
16	33.0	1.5	100.0	Turbid	---	0.207	----	3.31
17	33.0	1.25	120.0	Clear	Light Blue	0.05	$1.30 \cdot 10^{-2}$	3.31
18	15.5	0.75	80.0	Clear	No	0.028	$1.03 \cdot 10^{-2}$	3.34
19	17.0	0.75	120.0	Clear	No	0.031	$1.13 \cdot 10^{-2}$	3.31
20	33.0	1.25	120.0	Clear	Light Blue	0.048	$1.38 \cdot 10^{-2}$	3.31

3. Results and Discussion

Initial linear regression models (not shown) were fitted using runs 1 to 13 [8]. The model was used to select additional runs (14 to 17) that permit quadratic terms to be estimated. Gel 16 was cloudy (see figure 2) and was not used in the parameter estimation. The following model equations were obtained:

$$DS = \mathbf{a} + \mathbf{b}[\text{CTAB}] + c[\text{LCV}] + \mathbf{d}[\text{TCE}] + \mathbf{e}[\text{CTAB}][\text{LCV}] + f[\text{CTAB}][\text{TCE}] + g[\text{LCV}][\text{TCE}] + h[\text{CTAB}][\text{LCV}][\text{TCE}] + i[\text{CTAB}]^2 + j[\text{LCV}]^2 + k[\text{TCE}]^2 \quad (1)$$

$$A = \mathbf{l} + \mathbf{m}[\text{CTAB}] + \mathbf{n}[\text{LCV}] + o[\text{TCE}] + \mathbf{p}[\text{CTAB}][\text{LCV}] + q[\text{CTAB}][\text{TCE}] + r[\text{LCV}][\text{TCE}] + s[\text{CTAB}][\text{LCV}][\text{TCE}] + t[\text{CTAB}]^2 + u[\text{LCV}]^2 + v[\text{TCE}]^2 \quad (2)$$

where DS is dose sensitivity and A is initial absorbance at 590 nm. Statistically significant parameters are bolded and units are consistent with those in table 1 [8], where $\mathbf{a} = 5.39 \times 10^{-3}$, $\mathbf{b} = 2.13 \times 10^{-4}$, $\mathbf{d} = 3.16 \times 10^{-5}$, $\mathbf{e} = 1.16 \times 10^{-4}$, $\mathbf{l} = -6.04 \times 10^{-3}$, $\mathbf{m} = 1.23 \times 10^{-3}$, $\mathbf{n} = 2.32 \times 10^{-2}$, and $\mathbf{p} = -5.41 \times 10^{-3}$. Equations 1 and 2 were used to search for an optimized recipe using the objective function:

$$J = \omega DS - A \quad (3)$$

where ω is a weighting factor that controls the relative importance of achieving high dose sensitivity and low initial absorbance. Selecting a large value of ω causes the optimizer to select gel recipes with high dose sensitivity at the expense of darker initial colour; low values of ω result in selection of less-coloured gels at the expense of dose sensitivity. Bounds were enforced during optimization: 9 mM $\leq$ $[CTAB] \leq$ 33 mM, 0.75mM $\leq$ $[LCV] \leq$1.25 mM, 40 mM $\leq$ $[TCE] \leq$ 120 mM to ensure that reasonable gel recipes would be selected. When ω= 3.362 Gy cm was selected so that the two terms in the objective function would have similar magnitudes, gel recipe 18 in table 1 was chosen when all parameters (significant and non-significant) in equations (1) and (2) were used in the models. The gel 19 recipe was selected when only significant parameters were used. When a higher value of ω= 33.62 Gy cm was selected so that dose sensitivity is more important to the optimizer than initial colour, gel recipe 20 was chosen using all of the parameters and also using only the significant parameters. Gels 18 to 20 were prepared and irradiated. As expected, gel 20 had the highest dose sensitivity in the current study, but this gel was noticeably light blue in colour as shown in figure 2. Gel 19 had higher dose sensitivity (~1.5 times the sensitivity of Jordan's standard LCV gel) and was nearly colourless, making it promising for 3D gel dosimetry. As a result, gel 19 was selected for further testing in large jars. Dose measurements for the VMAT plan gel showed that dose distribution information is maintained over more than two weeks (figures 3b-d).

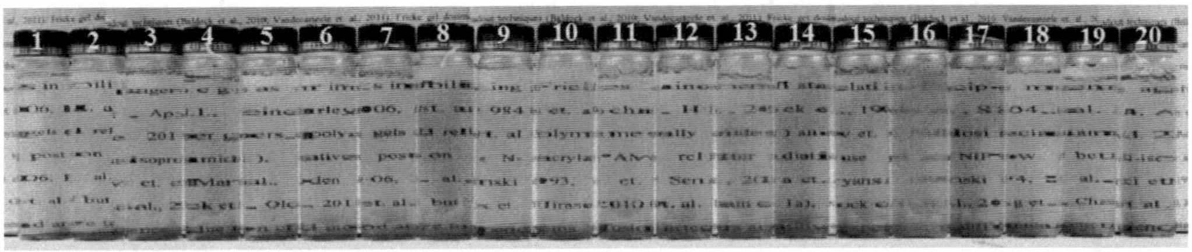

Figure 2. Photograph of un-irradiated vials containing gels manufactured at varying concentrations of CTAB, LCV, and TCE as described in table 1. Vial numbers 1 to 20 correspond to the sample numbers (1.1) to (1.20) in table 1. The vials were photographed in front of a page of text.

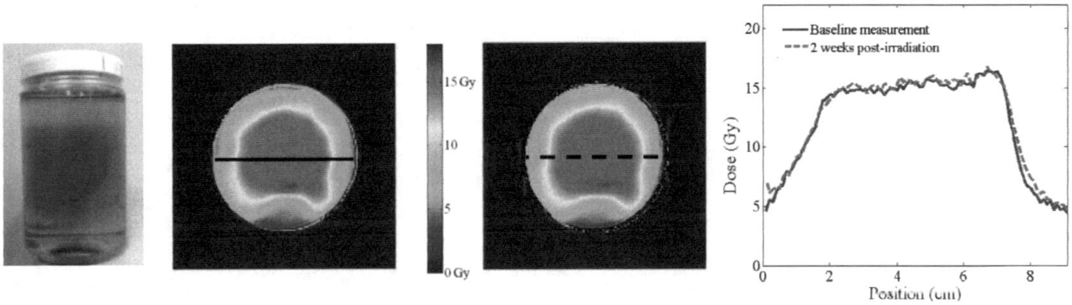

(a) (b) (c) (d)

Figure 3. (a) Photograph of VMAT test treatment plan irradiation in a jar of LCV micelle gel. (b) Dose map of one slice of the gel dosimeter imaged 30 min. after irradiation. (c) Dose map of the same slice imaged 2 weeks after irradiation. (d) Cross-plane profiles of slices in (b) and (c).

4. Conclusions

Optimization of the recipe for LCV micelle gels made with CTAB and TCE was aided by a designed experiment. The resulting optimal gel containing 0.75 mM LCV, 17 mM CTAB, 120 mM TCE, 25.0 mM TCAA and 4 wt% gelatin, showed improved dose sensitivity (~1.5 times higher than Jordan's standard LCV gel) and low initial colour. Experiments in 1 L jars showed that spatial dose distribution information is maintained for > two weeks. Further testing is required to quantify the spatial and temporal stability and to test dose resolution and dose-rate behaviour.

5. References

[1] Baldock C *et al* 2010 *Phys. Med. Biol.* **55** R1-63
[2] Jordan K and Avvakumov N 2009 *Phys. Med. Biol.* **54** 6773–6789.
[3] Babic S *et al* 2009 *Phys. Med. Biol.* **54** 6791–6808.
[4] Vandecasteele J *et al* 2011 *Phys. Med. Biol.* **56** 627–651.
[5] Nasr A T *et al* 2014. Proceedings of 8th International Conference on 3D Radiation Dosimetry (IC3DDose) 4-7 September 2014, Ystad, Sweden
[6] Montgomery DC 2005 *John Wiley & Sons, Inc USA.*
[7] Montgomery DC and Runger GC 2007 *John Wiley & Sons, Inc USA.* 4th edition, 500-617
[8] Nasr AT 2014 Ph.D. Thesis in Chemical Engineering *Queen's University*, Kingston, Ontario, Canada.

Journal of Physics: Conference Series **573** (2015) 012039 doi:10.1088/1742-6596/573/1/012039

Characterization of a reusable PRESAGE® 3D dosimeter

T Juang[1,2], J Adamovics[3,4] and M Oldham[2]
[1]Medical Physics Graduate Program, Duke University, Durham, NC, USA
[2]Duke University Medical Center, Durham, NC, USA
[3]Rider University, Lawerenceville, NJ, USA
[4]Heuris Pharma, LLC, 412 Sunset Rd., Skillman, NJ, 08558 USA

Email: mark.oldham@duke.edu

Abstract. This study investigates a reusable PRESAGE® 3D dosimeter (Presage-RU), which would improve cost-effectiveness and facilitate wider implementation of comprehensive, high resolution 3D dosimetry. Small (1x1x4.5 cm) and large (8 cm diameter, 4.5 cm length) sample dosimeters were irradiated multiple times to characterize dose response (i.e. radiation-induced change in optical density (ΔOD)), optical clearing rate, and dose distribution stability. Presage-RU exhibited an initial dose response sensitivity of 0.0119 ΔOD/(cm·Gy), a reduction in response with subsequent irradiations, and a small, permanent ΔOD (~1-6% of initial signal) following each irradiation. Dosimeters optically cleared at an exponential rate (average $T_{½}$ = 24.8±3.6 h), and were effectively cleared after ~5-8 days. 3D gamma analysis (3%/3mm, 10% dose threshold) of a 4-field box plan applied to the large dosimeter showed good agreement following initial irradiation (96.6% passing), but a reduction in passing rate (89.1% passing) with subsequent irradiation. Further study is warranted to fully assess and quantify the performance of Presage-RU for repeat irradiations.

1. Introduction
PRESAGE® is a radiochromic plastic which has demonstrated strong potential in high resolution 3D dosimetry [1-5]. Permanent radiochromic response has limited current PRESAGE® dosimeters to single use, and while optical clearing has been noted in several earlier PRESAGE® formulations [6-8], there has not yet been an extensive study to determine the feasibility of putting potentially reusable 3D PRESAGE® dosimeters into clinical use. A new PRESAGE® formulation (Presage-RU, Table 1) exhibits a reversible response that optically clears post-irradiation and allows for reirradiation, which opens up the potential for reusability. This would have the practical benefit of improving cost-effectiveness and facilitating the wider implementation of comprehensive, high resolution 3D dosimetry.

This study investigates the feasibility of Presage-RU for reusability by studying dose response, optical clearing rate, and the stability of relative dose distributions for repeat irradiations.

2. Methods
Multiple irradiation experiments were performed on Presage-RU samples of both small and large volumes to allow exploration of any volume effect. All irradiations (doses and details below) were performed at 6 MV and 600 MU/min on a clinical linac. All measurements were tracked for 14 days following each irradiation, and dosimeters were allowed to completely optically clear at room temperature in a dark room between irradiations.

Table 1. Comparison of the components in a standard, single-use PRESAGE® formulation and the new Presage-RU formulation.

Formulation	Polyurethane	Leuco Dye	Initiator	Other Components	Shore Hardness
Single Use PRESAGE®	91.10% Smooth-On Crystal Clear® 206	1.5% o-MeO-DEA LMG	0.40% CBr_4	2.0% DMSO 5.0% $C_6H_{12}O_2$	80D
Presage-RU	46.25% Heuris Inc. FC30 46.25% Heuris Inc. FC50	1.7% o-MeO-DEA LMG	0.40% CBr_4	2.0% DMSO 3.4% solvents	30-50A

2.1. Small volume measurements
Small volume dosimeter samples were made in optical cuvettes (1×1×4.5 cm). Characterization of dose response and optical clearing rates was performed on the small volume dosimeter samples, which were irradiated to known doses from 0-8 Gy. Subsequent change in optical density (ΔOD) was determined by measuring absorption before and after irradiation using a Spectronic Genesys 20 spectrophotometer, then subtracting the preirradiation value from the postirradiation value. Measurements of an unirradiated control sample were used to correct for systematic error at each time point. All samples were subject to 5 irradiations total, with the same dose delivered to each sample with each subsequent irradiation.

2.2. Large volume measurements
A large volume sample was made as a cylindrical dosimeter (8 cm diameter, 4.5 cm length). The larger cylindrical dosimeter was irradiated with a 4-field box plan (parallel opposed pairs of 4x4 cm^2 AP-PA beams and 4x2 cm^2 lateral beams) to a central dose of 20 Gy. High resolution (1 mm isotropic) 3D dose distributions were obtained through optical-CT imaging with the Duke Mid sized Optical-CT Scanner (DMOS) [2, 4, 9, 10] at multiple time-points. Measured relative dose distributions were compared to the corresponding calculated dose distribution from a commissioned treatment planning system (Eclipse, Varian Medical Systems). The Presage-RU dose was normalized to the central high dose region of the calculated distribution. Presage-RU's suitability for reusable dosimetry was evaluated with 3D gamma passing rates (3%/3mm criteria, 10% dose threshold) calculated for both the initial irradiation and one subsequent reirradiation. 3D gamma calculations and dose registrations were performed in CERR (Washington University).

Tracking measurements in the dosimeter (average ΔOD within the 98% isodose region) were also compared against the dose response and optical clearing rate determined from the small volume dosimeters.

3. Results

3.1. Dose Response
The initial irradiation yielded a change in optical density in small volumes of Presage-RU of 0.0119 ΔOD/(cm·Gy), followed by reduced responses in subsequent irradiations (figure 1), Table 2). A strong linearity of response ($R^2 \geq 0.993$) was observed for all irradiations.

3.2. Optical Clearing
Optical clearing in small volume Presage-RU samples was observed to be an exponential decay over time (figure 2). A summary of decay constants and half-life values (determined over 96 h and averaged across doses for each irradiation) can be found in Table 3. Presage-RU samples were effectively cleared after ~5-7 days. A small, permanent post-irradiation change in optical density (0.0001-0.0039 ΔOD/cm, ~1-6% of initial signal) was observed to persist after clearing. This does not

affect subsequent measurements as pre-irradiation measurements are used to determine ΔOD after irradiation.

Measurement tracking in the larger dosimeter similarly demonstrated a drop in response with re-irradiation and exponential decay in optical density (figure 3). Half-lives were 24.8 h (initial irradiation) and 27.7 h (re-irradiation), with effective dosimeter clearing after ~5-8 days.

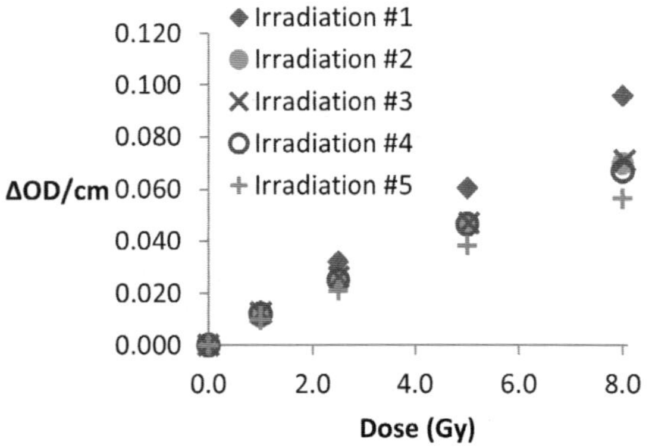

Figure 1. Change in optical density in small volume cuvettes following initial and subsequent irradiations. Fit values are summarized in table 1.

Table 2. Dose response sensitivity and linearity determined from multiple irradiations of PRESAGE-RU.

Irradiation (Iteration #)	**Dose Response** $\left(\frac{\Delta\square\square}{\square\square\cdot\square\square}\right)$	**Fit Linearity** (R^2)
1	0.0119	0.9996
2	0.0087	0.9966
3	0.0087	0.9960
4	0.0083	0.9930
5	0.0070	0.9943

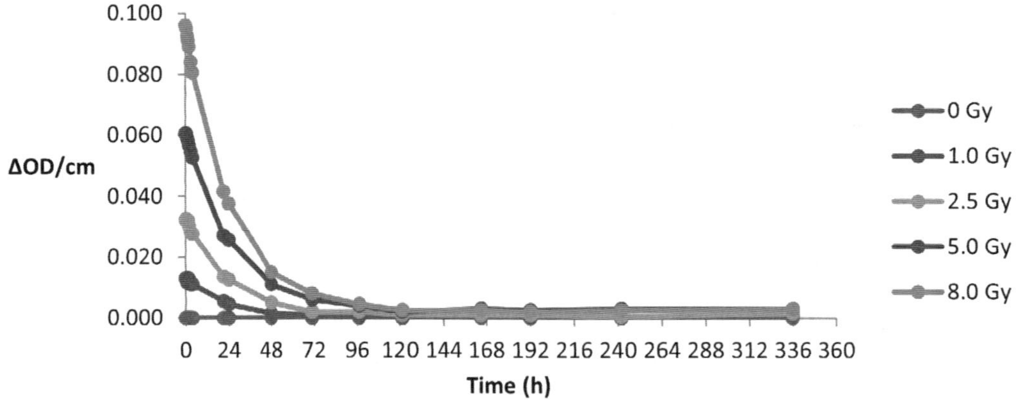

Figure 2. Optical clearing of small volume Presage-RU samples following initial irradiation.

Table 3. Optical clearing decay constants and half-lives.

Irradiation Iteration #	**Average λ (h^{-1})**	**St Dev λ (h^{-1})**	**Average $T_{1/2}$ (h)**	**St Dev $T_{1/2}$ (h)**
1	0.033	0.002	21.2	1.4
2	0.035	0.013	21.7	6.5
3	0.025	0.002	27.5	1.9
4	0.024	0.002	29.6	2.3
5	0.029	0.002	24.2	1.9
Average	0.029		24.8	
St Dev	0.004		3.6	

3.3. Relative Dosimetry

The 3D gamma passing rates were 96.6% for the initial irradiation and 89.1% for the subsequent irradiation. Gamma maps (figure 4) and line profiles showed that the passing rate decrease in the reirradiated distribution was primarily due to dose underestimation in the anterior portion of the dosimeter and dose overestimation in the lateral regions outside of the high-dose box.

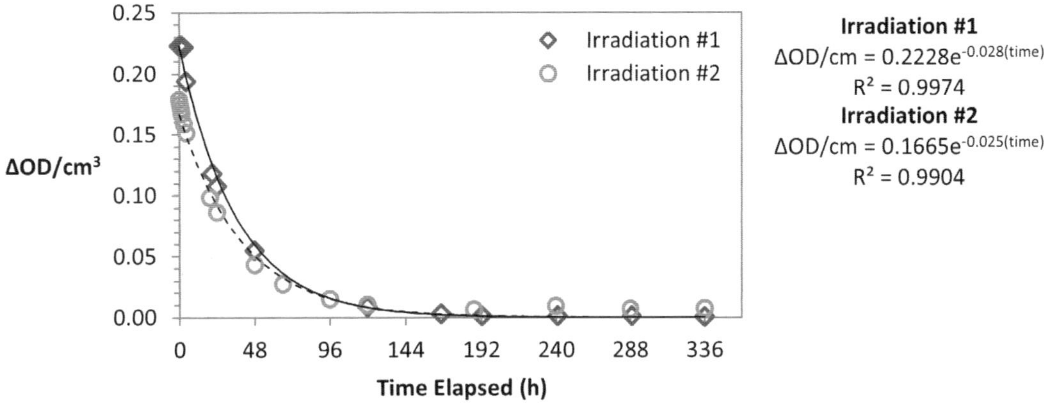

Figure 3. Optical clearing in a dosimeter as represented by average dose response over time in the 98% isodose region of the 4-field box distribution.

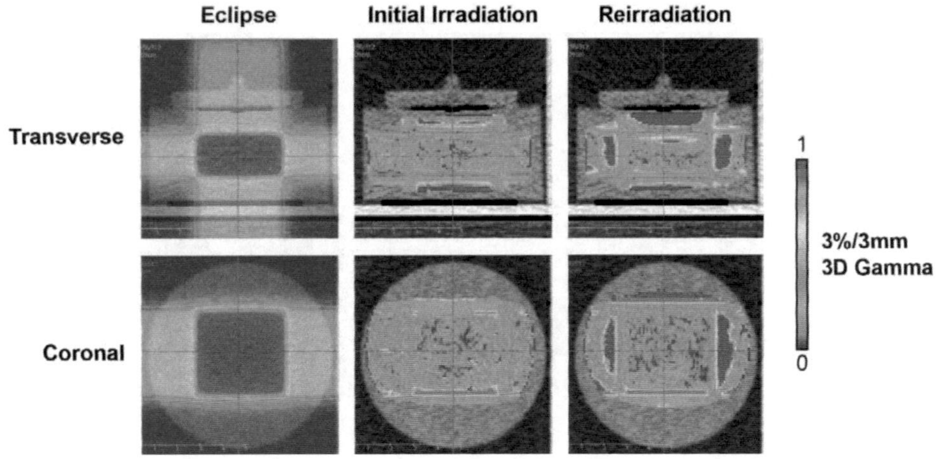

Figure 4. 3D gamma passing maps of initial dosimeter irradiation and subsequent reirradiation.

4. Conclusions

This preliminary study introduced Presage-RU, a potentially reusable radiochromic 3D dosimeter that can be imaged with high resolution optical-CT. Presage-RU demonstrated strong dose linearity and an average optical clearing half-life of 24.8±3.6 h over 5 irradiations. The dose response between the small and large volume Presage-RU dosimeters differed by 5.5% for the initial irradiation and 3.6% for a subsequent irradiation, but yielded similar optical clearing characteristics. A small permanent change in optical density (0.0001-0.0039 ΔOD/cm, ~1-6% of initial response) persists following each irradiation. Good agreement was seen between measured relative dose and calculated dose distributions upon initial irradiation, but further study is warranted to quantify the accuracy of Presage-RU for relative dosimetry following subsequent re-irradiations.

5. Acknowledgements
This work was supported by NIH R01CA100835.

6. References
[1] Sakhalkar H S *et al* 2009 *Med. Phys.* **36** 71
[2] Thomas A *et al* 2011 *Med. Phys.* **38** 4846
[3] Guo P Y *et al* 2006 *Med. Phys.* **33** 1338
[4] Oldham M 2014 *Advances in Medical Physics - 2014* 69-86
[5] Gorjiara T *et al* 2011 *Med. Phys.* **38** 2265-74
[6] Adamovics J and Maryanski M J 2006 *Radiat. Prot. Dosimetry* **120** 107-12
[7] Pierquet M *et al* 2010 *J Phys Conf Ser* **250** 1-4
[8] Juang T *et al* 2013 *J. Phys. Conf. Ser.* **444** 012029
[9] Juang T *et al* 2014 *Med. Phys.* **41** 071706
[10] Iqbal K *et al* 2014 *J. Appl. Clin. Med. Phys.* **15** 4531

Benchmarking the gamma pass score using ArcCHECK for routine dosimetric QA of VMAT plans

Sankar Arumugam[1], Tony Young[1], Aitang Xing[1], David Thwaites[2] and Lois Holloway[1,2,3,4]

[1]Liverpool and Macarthur Cancer Therapy Centres and Ingham Institute, New South Wales, Australia
[2]Institute of Medical Physics, School of Physics, University of Sydney, New South Wales, Australia
[3]Centre for Medical Radiation Physics, University of Wollongong, New South Wales, Australia
[4]University of New South Wales, New South Wales, Australia

E-mail: sankar.arumugam@sswahs.nsw.gov.au

Abstract. A minimum expected gamma (γ) pass rate for VMAT plan verification using ArcCHECK was established based on the RTTQA, TG119 test cases and 10 clinical plans with varying levels of complexity. The impact of the 'Measurement Uncertainty' parameter as available in the ArcCHECK software on γ pass rate was studied for both global and local γ analysis. Our results show that excluding measurement uncertainty adds tighter tolerance in local γ comparison. From the verification of our benchmark cases we established minimum expected γ pass rates of 85% and 88% for 2%/2mm global and 3%/3mm local tolerance criteria.

1. Introduction

Volumetric Modulated Arc Therapy (VMAT) has been shown to enable more efficient treatment delivery compared to conventional static gantry based Intensity Modulated Radiation Therapy (IMRT) delivery methods [1, 2]. The efficiency of VMAT comes at the cost of added complexity in the treatment planning and delivery stages. The complex nature of VMAT necessitates comprehensive three dimensional (3D) dosimetric validation of the plans to ensure dosimetric accuracy of planning and delivery [3]. Recently many electronic dosimetric systems have been introduced commercially that provide semi-3D or 3D dose information of the delivered treatment plans [4, 5].

ArcCHECK (Sun Nuclear Corporation (SNC), USA) is one such system which uses diode detectors arranged in a helical array in a perspex cylindrical phantom. In routine clinical practice the cumulative dose matrix measured by ArcCHECK is compared with the TPS calculated dose in similar geometry [4] using gamma analysis [6]. In SNC software in addition to the user defined tolerance criteria an additional optional parameter called 'Measurement Uncertainty' (e) is provided. The manufacturer defines this parameter as follows: "Measurement uncertainties include differences between the absolute calibration of the device and the standard dose value due to setup error, temperature change, accelerator output fluctuation, array calibration accuracy, and electronic measurement precision. The measurement uncertainty (e) is added to the percentage acceptance criterion defined by the user and is typically smaller than 1% for the relative comparison and close to 1% for the absolute comparison"

[7]. This parameter increases the dose tolerance criteria but does not impact on the distance tolerance criteria. The published γ pass rates for VMAT plans using ArcCHECK don't clearly specify whether this parameter is included in the analysis or not [2].

In this work we study the impact of this parameter on the local (L) and global (G) γ analysis of the ArcCHECK measured and TPS calculated dose matrices. Also we establish a benchmark pass rate with different types and level of tolerance criteria to ensure the accuracy of VMAT plans appropriate for our centre using the ArcCHECK dosimetric system.

2. Materials and Methods

The Pinnacle, v 9.2, (Philips Ltd, USA)Treatment Planning System (TPS) was used to generate VMAT plans using a 6 MV photon beam model for an Elekta Synergy (Elekta Ltd, Crawley, UK) Linear Accelerator (linac). Dose calculations for all plans in the study were performed using the adaptive convolution dose calculation algorithm. The Synergy accelerator used in this study was equipped with an Multi Leaf Collimator (MLCi) head and Integrity v1.1 linac control software.

2.1. Benchmark VMAT plans

Two sets of plans were considered to benchmark the dosimetric accuracy of VMAT plans. The first set included recommended test cases from the UK Radiotherapy Trial Quality Assurance (RTTQA) group and AAPM Task Group 119 (TG119). This includes the 3D treatment planning system (3DTPS) test cases from RTTQA and Mock Prostate, Mock Head and Neck, Multi Target, C shape-easy and C shape-hard from TG119 [8, 9]. In the second group 10 clinical cases with varying levels of complexity in target volume and dose objectives were selected retrospectively from clinical cases. This included 5 head and neck, 3 prostate and 2 pelvic nodes cases. The VMAT plans were generated with single full arc for simple cases and double full arcs for complex cases. RTTQA and TG119 recommended dose objectives were used for the test cases and for clinical cases the dose objectives prescribed as per RTOG guidelines and ICRU83 prescription methodology was used as the achievable aim [10].

2.2. Plan verification and analysis

The dosimetric accuracy of the generated plans was verified by the following three measurements: 1. point dose measurements in high, medium and low dose regions using a CC13 ion chamber in a CIRS phantom, 2. Coronal plane dose distribution verification using EBT3 gafchromic film placed in a solid water phantom, and 3. Cumulative dose measurement using the ArcCHECK dosimetric system.

The coronal film measurements were compared with the TPS calculated dose matrices using global γ analysis with 3%/3mm (3%G/3mm) tolerance criteria in RIT film dosimetry software. The ArcCHECK measured and TPS calculated dose matrices were compared using global and local γ analysis with 2%/2mm and 3%/3mm (2%G/2mm, 3%G/3mm and 2%L/2mm, 3%L/3mm) with and without the measurement uncertainty parameter included. In order to assess the impact of the measurement uncertainty (e) parameter on the tolerance criteria analysis was also performed with 1%G/2mm, 2%G3mm, 1%L/2mm and 2%L/3mm with the uncertainty parameter. All γ analysis was performed with a high dose threshold of 10% and gamma tolerance of 1.

3. Results and Discussion

The generated VMAT plans successfully achieved the dose objectives recommended by the RTTQA group and TG119 for all test cases except the C shape hard case. As the aim of the C-shape hard case is to test a system that is being pushed very hard this was not unexpected.

The point dose agreement in the CIRS phantom at high, medium and low dose regions of all studied plans resulted in a mean (1σ) percentage difference of -0.5 (1.3)%, 0.5 (1.6)% and -0.6 (3.4) % respectively. The coronal plane dose verification using film resulted in a mean (1σ) γ pass rate of 96.6 (2.7)%.

Figure 1 shows the local and global γ pass rates with 2%/2mm and 3%/3mm pass rates for studied VMAT plans. Figure 2 a, b and c show the gamma analysis results with 3%G/3mm with 'e',

3%G/3mm without 'e' and 2%G/3mm with 'e' for 3DTPS test case. Figures 3 a, b and c show similar results with local gamma analysis. In the figures the red and blue points show the failed points with high and low dose respectively. Table 1 shows the mean (1σ) γ pass rates with local and global γ analysis with and without measurement uncertainty in the analysis.

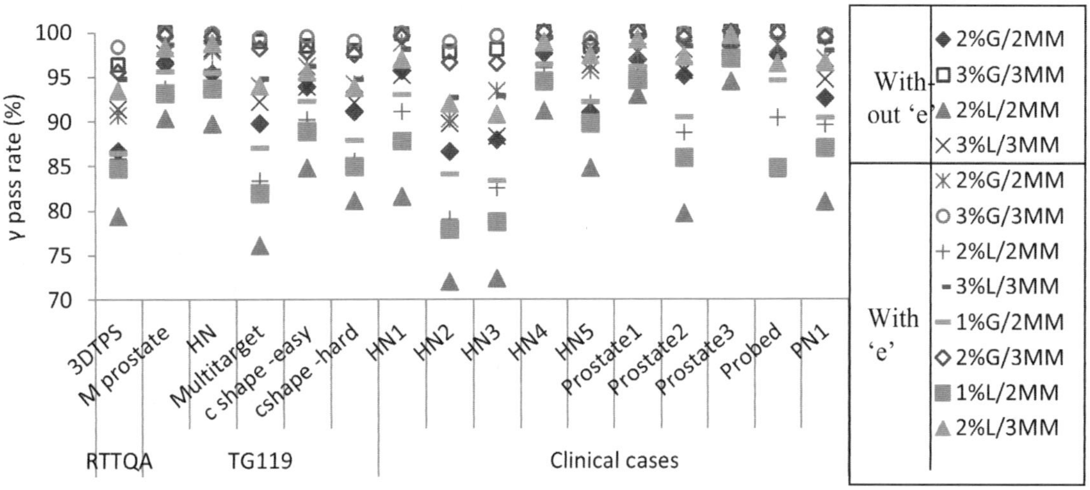

Figure1. Gamma pass rate with different tolerance criteria for studied VMAT plans using ArcCHECK.

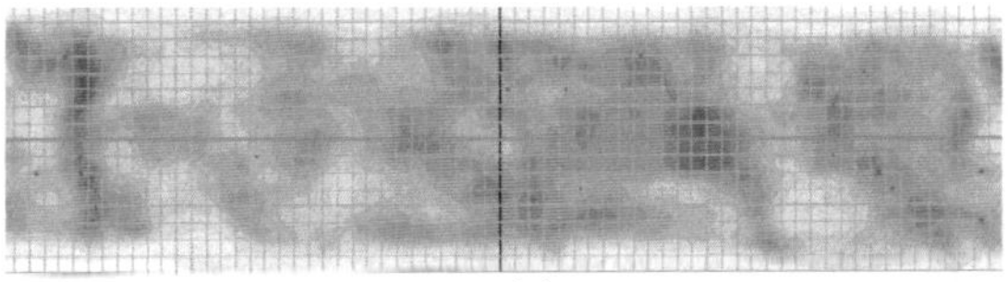

(a)

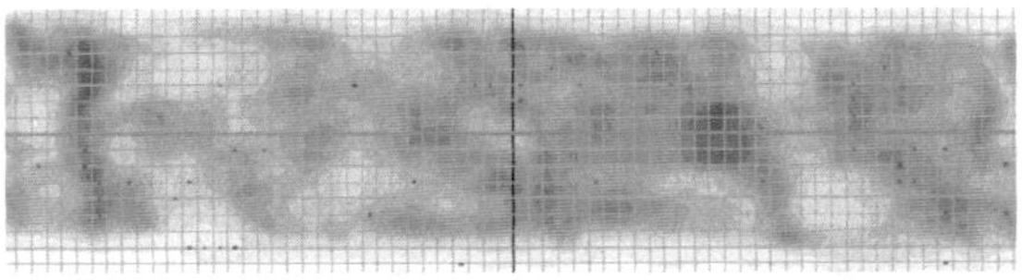

(b)

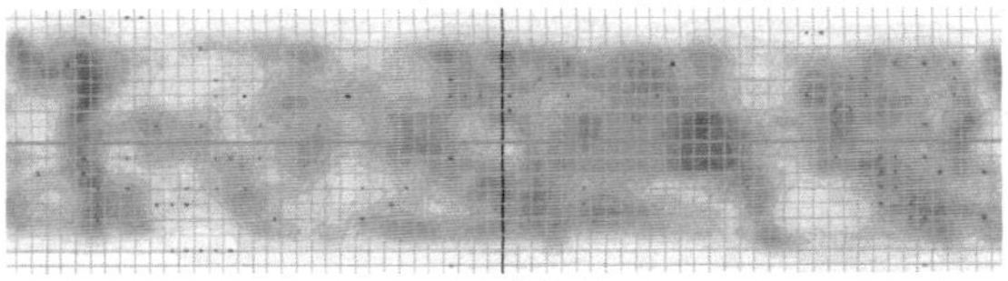

(c)

Figure 2. γ pass results for 3DTPS test plan with (a) 3%G/3mm with 'e', (b) 3%G/3mm without 'e' and (c) 2%G/3mm without 'e' tolerance criteria. Red and blue points show failed detector points with high (red) and low (blue) dose measurements.

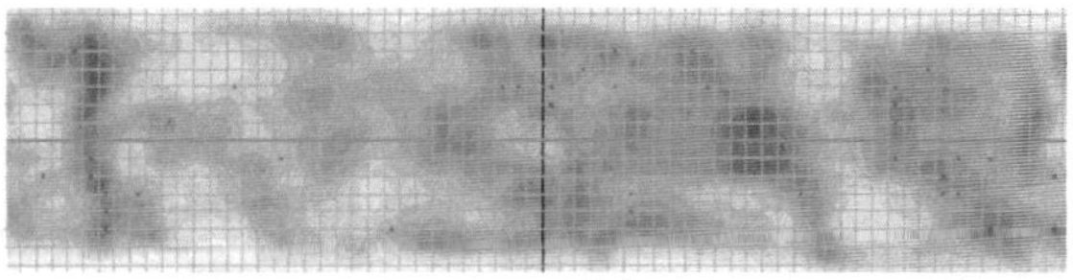

(a)

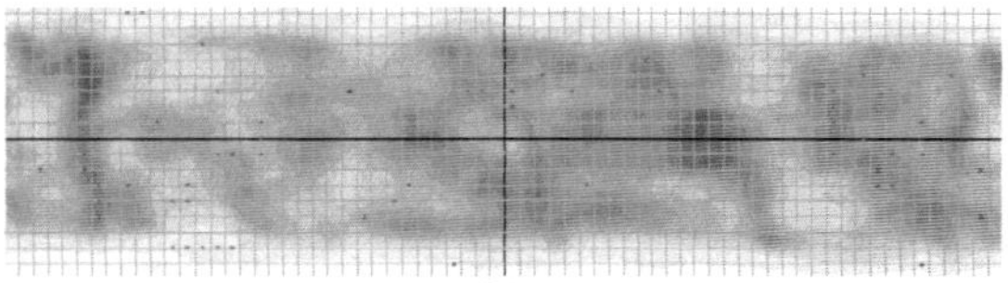

(b)

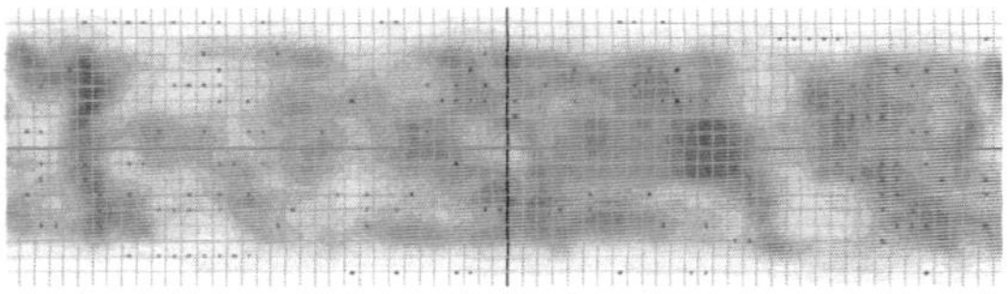

(c)

Figure3. γ pass results for 3DTPS test plan with (a) 3%L/3mm with 'e', (b) 3%L/3mm without 'e' and (c) 2%L/3mm without 'e' tolerance criteria. Red and blue points show failed detector points with high and low dose measurements.

Measurement uncertainty	Tolerance criteria			
	2%G/2mm	3%G/3mm	2%L/2mm	3%L/3mm
Included	96.1(3.0)	99.6(0.5)	89.8(5.6)	97.0(2.4)
Not included	93.1(4.6)	99.0(1.1)	83.7(7.1)	94.9(3.5)
	1%G/2mm	2%G/3mm	1%L/2mm	2%L/3mm
Included	91.5(4.7)	98.6(1.5)	87.9(5.9)	96.2(2.8)

Table 1. Mean (1σ) gamma pass rate of studied VMAT plans with different tolerance criteria.

In general including measurement uncertainty results in higher pass rates both in global and local γ comparisons due to the additional dose tolerance added to the user defined dose tolerance. The manufacturer states that the measurement uncertainty value is approximately 1% for the γ comparison in absolute dose mode. From our results the gamma pass rate for 2%/2mm and 3%/3mm tolerance criteria without measurement uncertainty included was not equal to the 1%/2mm and 2%/3mm criteria with measurement uncertainty (table 1, figures 1, 2 and 3). Also not including the measurement uncertainty in the global tolerance resulted in tighter criteria (reduced pass rate) whereas in local tolerance it resulted in more relaxed tolerance criteria (increased pass rate) (table 1).

While there is a lack of evidence between the correlation of acceptable gamma pass rate and clinical significance, 95% pixels passing the global γ analysis with 3%/3mm tolerance criteria (3%G/3mm) is considered to be widely acceptable in clinical practice. Many of the IMRT and VMAT credentialing bodies also follow this. Nelms *et al* show that 2%L/2mm has been shown to more closely reflect the clinically significant errors in treatment plans compared to the more generous 3%G/3mm criteria [11]. Similarly to the ArcCHECK software, Nelms *et al* also included measurement uncertainty in the analysis. Our results show that excluding measurement uncertainty adds a tighter tolerance in local γ comparison. From the verification of our benchmark cases we set a γ pass rate of mean-2σ as our minimum expected pass rate for clinical VMAT plans. Based on this we set minimum pass rate of 85% and 88% for 2%G2mm and 3%L/3mm tolerance criteria respectively noting that higher pass rates are expected for simple plans.

4. Conclusion

Exclusion of the measurement uncertainty parameter results in a tighter tolerance in local γ comparison of the ArcCHECK dose verification. Based on the verification of our benchmark plans we set γ pass rates of 85% and 88% as minimum expected pass rates for our clinical VMAT plans for 2%G/2mm and 3%L/3mm tolerance criteria.

5. References

[1] Otto K 2008 *Med. Phys.* **35** 310
[2] Rao M *et al* 2010 *Med. Phys.* **37** 1350
[3] Feygelman V *et al* 2011 *J. Appl. Clin. Med. Phys.* **12** 146
[4] McGarry C K *et al* 2013 *Med. Phys.* **40** 091707
[5] Low D A *et al* 2003 *Med. Phys.* **30** 2455
[6] ArcCHECK™ Reference Guide, Sun Nuclear Corporation, Florida, USA
[7] Tsang Y *et al* 2013 *Brit. J. Radiol.* **86** 1022
[8] Mynampati D K *et al* 2012 *J. Appl. Clin. Med. Phys.* **13** 5
[9] ICRU 2010 *Journal of the ICRU* **10**
[10] Nelms B E *et al* 2013 *Med. Phys.* **40** 111722

A self-sufficient method for calibration of Varian electronic portal imaging device

Baozhou Sun, Sridhar Yaddanapudi, Sreekrishna M Goddu and Sasa Mutic
Department of Radiation Oncology, Washington University in St. Louis, St. Louis, MO, USA

E-mail: bsun@radonc.wustl.edu

Abstract. Electronic portal imaging device (EPID) is currently used for dosimetric verification of IMRT fields and linac quality assurance (QA). It is critical to understand the dosimetric response and perform an accurate and robust calibration of EPID. We present the implementation of an efficient method for the calibration and the validation of a Varian EPID, which relies only on data collected with that specific device. The calibration method is based on images obtained with five shifts of EPID panel. With this method, the relative gain (sensitivity) of each element of a detector matrix is calculated and applied on top of the calibration determined with the flood-field procedure. The calibration procedure was verified using a physical wedge inserted in the beam line and the corrected profile shows consistent results with the measurements using a calibrated 2D array. This method does not rely on the beam profile used in the flood-field calibration process, which allows EPID calibration in 10 minutes with no additional equipment compared to at least 2 hours to obtain beam profile and scanning beam equipment requirement with the conventional method.

1. Introduction

Electronic portal imaging device (EPID) was initially developed for patient positioning, verification, and it is recently being utilized for dosimetric verification of IMRT treatments and linac quality assurance (QA), including dose output, beam profile, field size, MLC, wedge factor, etc. The EPID-based QA offers many advantages over the conventional methods: high spatial resolution, fast image acquisition, and digital output allowing a fast quantitative analysis. It provides an efficient and accurate tool for both patient and machine QA in radiation therapy [1].

Conventional methods for calibration of EPID includes dark-field and flood-field irradiations at a fixed panel position and requires a beam profile correction for off-axis ratio differences, which is typically collected with a scanning water tank. It is expected that some variations in detector's sensitivities are inevitable. Many different calibration techniques have been developed to compensate the variations. One of the methods to determine the relative gain of each detector element relies on comparison of the detector response when irradiated by the same fluence. By shifting or rotating the detectors, the gain of each detector relative to the central axis can be obtained. The general idea has been applied to calibrate various 2D array detectors including EPID [2-4]. Recently, Boriano *et al* [5] applied this approach to EPID calibration of Elekta iViewGT. In this paper we describe a self-sufficient procedure, related algorithm, results and verification for the calibration of Varian EPID devices.

2. Materials and Methods

A Varian Truebeam linear accelerator (Varian Medical Systems, Palo Alto, CA) equipped with amorphous silicon panel (*a*Si1000) was used in this study. The EPID has a resolution of 1024 x 768 pixels with a 40 x 30 cm^2 active detector area. The pixel size projected back to isocenter is 0.392 mm. All the irradiations were performed with a 6 MV beam of 100 Monitor Units (MUs). The standard calibration following manufacture's guideline was performed in service mode: i) An image of a dark-field was acquired while the beam was off; ii) A flood-field calibration was performed with an open beam covering the whole active detector area. This eliminates variation in sensitivity between different detector elements and areas of the panel; iii) A generic beam profile for off-axis correction provided by Varian was applied for the flood-field calibration.

The EPID was positioned at 108 cm source to detector distance (SDD) and the gantry was set at 0 (IEC scale). EPID images were acquired in integrated mode through Mosaiq (Elekta Medical Systems, Sunnyvale, CA) and saved in DICOM format. Data analysis and algorithm development were carried out using Matlab programming language (The Mathworks INC., Natick, MA). The calibration procedure consists of five consecutive deliveries with the same fluence. Each delivery was performed at the same SDD with an open field size of 22 x 22 cm^2. The first image was acquired when the EPID was in the central position with respect to the beam. The other four images were acquired with the EPID shifted along left, right, towards and away in the gantry table directions. Each shift was approximately 4 mm (10 pixels). Two different coordinate systems were defined in the algorithm: One, (x,y) is associated with the accelerator and the other, (i,j), is linked with the EPID system. During the five acquisitions of the calibration procedure, the center of the EPID has been placed respectively at the following coordinates (x, y): (0, 0), (0, 10), (0,-10), (10, 0), and (-10, 0). The fluence F (i, j), EPID readout I (i, j) and gain G (i, j) are related by equation where M is the maximum pixel number determined by the field size.

$$I(i,j) = F(i,j) \times G(i,j) \qquad (1)$$

When the EPID was at the central position, The EPID readout was given by

$$I_C(i,j) = F_C(i,j) \times G(i,j) \qquad (2)$$

When the EPID was made to right shift by 10 pixels, the EPID readout was expressed by

$$I_R(i+10,j) = F_R(i+10,j) \times G(i+10,j) \qquad (3)$$

It is assumed that the fluence does not change during the subsequent deliveries:

$$F_C(i,j) = F_R(i+10,j) \qquad (4)$$

By applying Eq.4 to Eq 2 and 3, one obtains

$$G(i+10,j) = \frac{I_R(i+10,j)}{I_C(i,j)} \times G(i,j) \quad \text{for } i = (1,\ 2,\ 3,\ldots,\ M) \qquad (5)$$

By repeating the same analysis, the gain for each detector element, as a 2D sensitivity map, can be related to the gain of the central region as a reference. In our study the reference region is G (i, j) = 1 for i = -10, -9, ..., 10 and j = -10, -9, ..., 10. The procedure and algorithm was verified using a physical wedge inserted in the beam line during flood-field irradiation. Therefore, a non-uniform beam was used for flood-field calibration and a wedged profile was expected without applying the 2D sensitivity map for an open field delivery. The 2D sensitive map was calculated and applied to a non-uniformed beam acquired with an open field.

Journal of Physics: Conference Series **573** (2015) 012041 doi:10.1088/1742-6596/573/1/012041

3. Results

Figure 1(a) shows an image acquired using a 6 MV beam open field of a 22x22 cm^2. The central circle shows a flattering filter, which was not present in the image of a flattering-filter free beam [6]. After making 4 shifts of the EPID panel, the pixel sensitivity map was calculated as shown in figure 1(b). The variation of detector response ranges from 0.93 - 1.06.

Figure 1(c) shows the corrected image after applying the 2D sensitivity map. The horn effect was less pronounced and the image was more uniform.

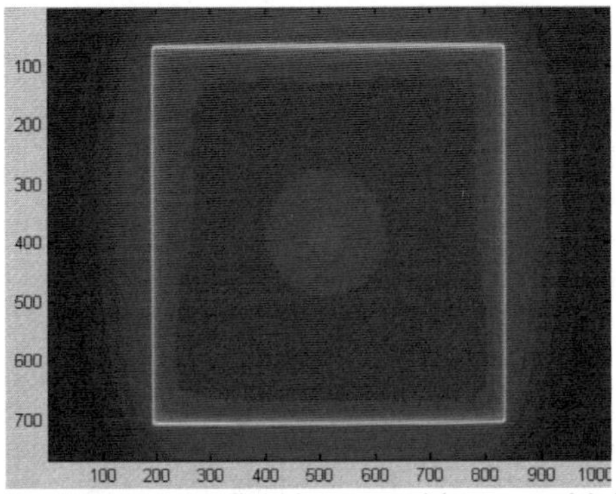

Figure 1(a). Open field images without sensitivity correction.

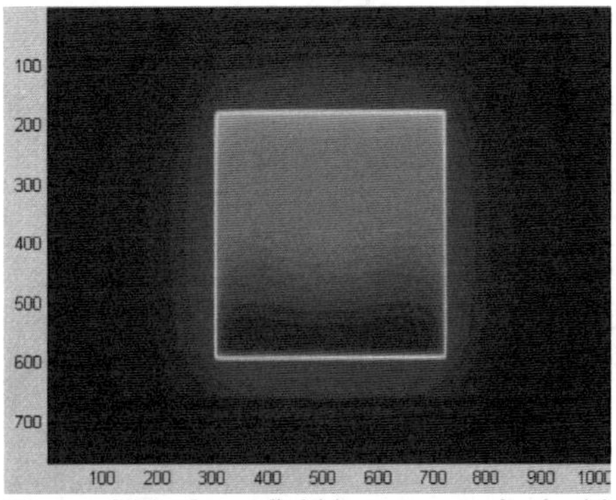

Figure 2(a). Open field images acquired with the flood-field calibrated when a 15 degree physical wedge in the beam line.

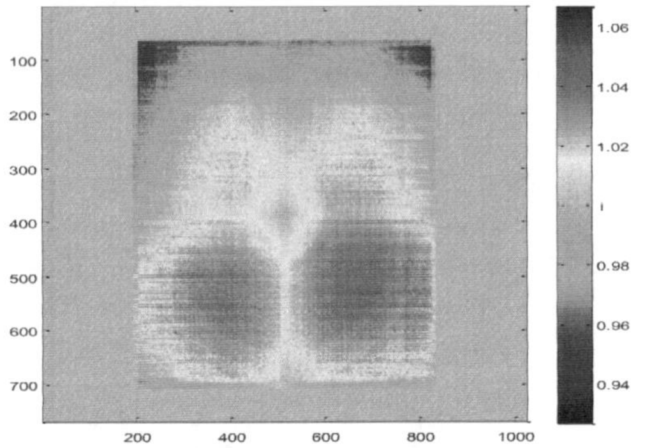

Figure 1(b). Calculated 2D sensitivity map.

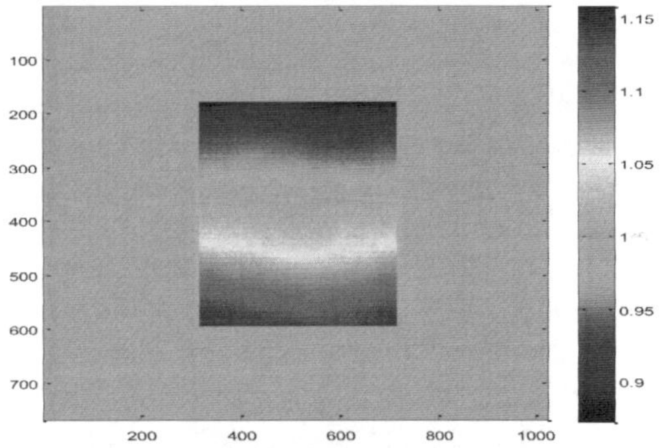

Figure 2(b). Calculated 2D sensitivity map for intentionally modified flooding field calibration.

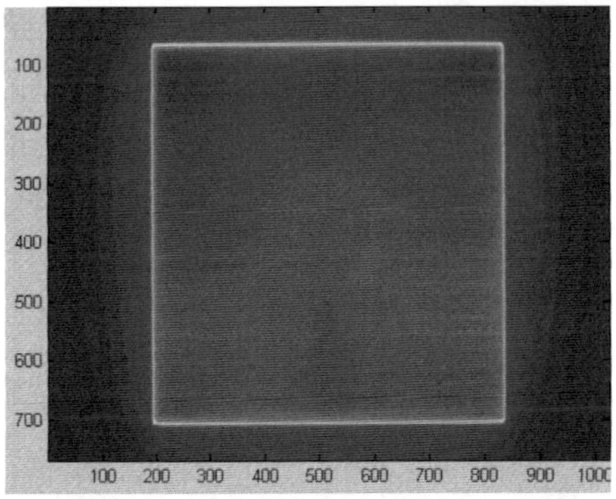

Figure 1(c). Open field images with sensitivity correction.

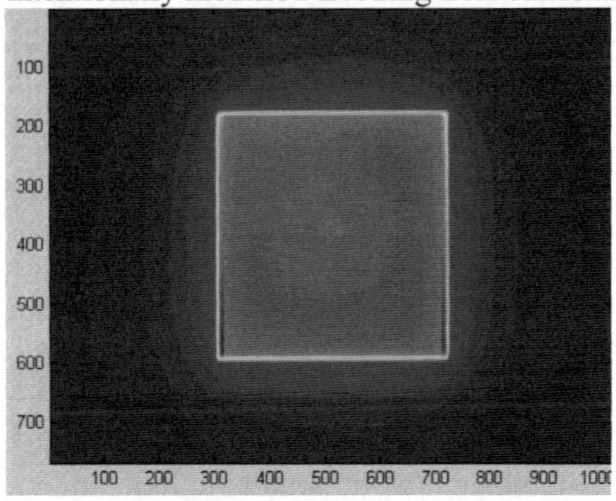

Figure 2(c). EPID image after applying the calculated 2D sensitivity map in figure 2(b).

Figure 2 shows the results of the verification described above. Figure 2(a) shows the image of a 15 x 15 cm^2 open field after intentionally introduced a wedge during the flood-field calibration. The beam

Journal of Physics: Conference Series **573** (2015) 012041 doi:10.1088/1742-6596/573/1/012041

profile presents a wedge beam along the gantry-table direction. Figure 2(b) is the calculated sensitivity map for the verification setup. Figure 2(c) shows the corrected EPID image after applying the calculated gain for each pixel. The coordinates of both x- and y-axes are shown as pixel positions. Figure 3(a) and (b) shows the beam profile in the central axis along the gantry-table direction, before and after the correction of the 2D sensitivity map, respectively. The corrected EPID image and beam profile using our calibration procedure has indicated that the new calibration procedure is independent of the beam profile used in the flood-field irradiation. The corrected beam profile was compared with measurements at D_{max} using a different 2D detector array (Matrixx, IBA Dosimetry GmbH, Germany) and the profile flatness and symmetry were in an agreement within 0.5%.

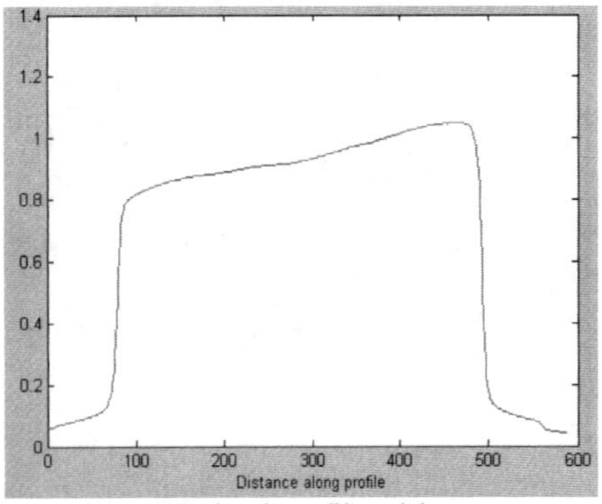

Figure 3(a). Wedged profile without correction of sensitivity map.

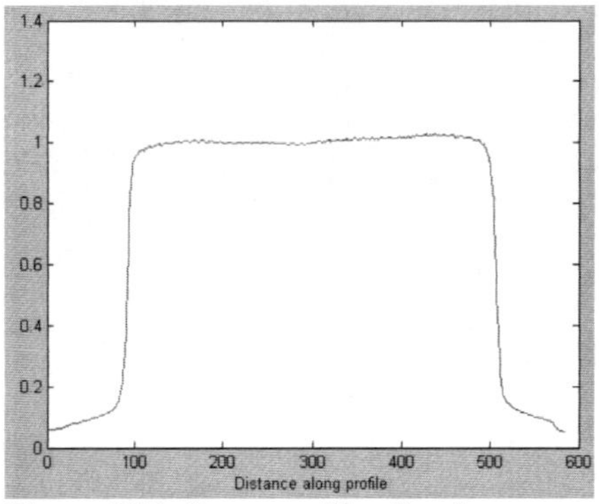

Figure 3(b). Beam profile after the correction.

4. Conclusion
We have developed a self-sufficient procedure and algorithm for calibration of a Varian EPID. This method is easy and fast to implement. It requires only a sequence of five measurements which are then processed with a simple algorithm. This method does not depend on the beam profile required in the flood-field irradiation and is more robust compared to the standard calibration methods recommended by the manufacture's guide. The verification presented in this paper has shown the goodness of the procedure and algorithm.

5. References
[1] Greer P B *et al* 2007 *Med. Phys.* **34** 4389-98
[2] Wang S *et al* 2012 *Int. J. Radiat. Oncol.* **84** S799
[3] Donetti M *et al* 2006 *Phys. Med. Biol.* **51** 485
[4] Simon T A *et al* 2010 *Med. Phys.* **37** 3501
[5] Boriano A *et al* 2013 *J. Appl. Clin. Med. Phys.* **14** 4420
[6] Goddu S *et al* 2013 *Med. Phys.* **40** 159

Implementation of an efficient workflow process for gel dosimetry using 3D Slicer

K M Alexander[1], C Pinter[2], J Andrea[2], G Fichtinger[2] and L J Schreiner[1,3]
[1]Department of Physics, Queen's University, Kingston, Ontario, Canada K7L 3N6
[2]School of Computing, Queen's University, Kingston, Ontario, Canada, K7L 3N6
[3]Department of Medical Physics, Cancer Centre of Southeastern Ontario at Kingston General Hospital, Kingston, Ontario, Canada, K7L 5P9

E-mail: kevin.alexander@krcc.on.ca

Abstract. One challenge in gel dosimetry is the manipulation and analysis of complex data sets from different systems. In this paper, we describe a simple and fast gel dosimetry analysis tool for radiation therapy dose deliveries. Using the open source medical imaging software *3D Slicer*, an extension was designed and implemented for the purpose of importing treatment planning system dose, CT imaging from simulation and at treatment, and optical CT gel dosimeter data. The extension also allows for calibration of gel dosimeter data, registration, and comparison of 3D dose distributions. The development of an open source gel dosimetry processing environment may help adoption of gels in the clinic.

1. Introduction

Recent developments in radiation therapy treatment modalities (intensity modulated radiation therapy, volumetric modulated radiation therapy, stereotactic body radiation therapy, etc.) have increased the complexity of radiation therapy treatments, producing advanced patient dose deliveries. Due to this complexity, verification of dose delivery and quality assurance of new treatment techniques to validate that the dose delivery to the patient is accurate and precise is crucial. While many clinical dosimeters measure point dose or two-dimensional dose distributions, these new treatment modalities benefit from full, three-dimensional verification of dose delivery.

Gel dosimeters have been shown to be promising tools for measuring and verifying radiation dose deliveries [1-4]. Typically, gel dosimeters are read out using modalities such as magnetic resonance imaging, x-ray computed tomography (CT), or optical CT. Optical CT is becoming more frequently used as it is a simple and convenient method of acquiring dose distribution information. One major challenge with gel dosimetry is the extensive post-irradiation data processing time involved in registering and calibrating the gel dosimeter, and making the subsequent comparisons between the measured gel dosimeter dose and the planned dose distribution to confirm accurate dose delivery. A gel dosimetry workflow formerly performed using Matlab (Mathworks, Natick, MA), running the Computational Environment for Radiotherapy Research (CERR, www.cerr.info) was implemented in 3D Slicer (www.slicer.org) using a custom workflow extension called a *slicelet.* 3D Slicer is a powerful, open source, and free computational tool for image analysis and visualization. It features advanced volume rendering, interactive segmentation, rigid and non-rigid registration, slice viewers, and also includes a library of downloadable extensions which are designed and customized by 3D

Slicer users for specific clinical tasks. A set of features in 3D Slicer tailored to radiation therapy specific tasks called *SlicerRT* [5] has recently been created, which enables loading of DICOM-RT data, manipulation of structure set contours, computation and display of dose-volume histograms, quantitative comparison of dose volumes and contours, and dose visualization.

2. Materials and Methods

Building on the SlicerRT extension, a slicelet for gel dosimetry analysis was developed following the general workflow shown in figure 1. In this paper, the main features of the slicelet are presented using the example of a 7 field intensity modulated radiation therapy plan that was delivered to a 1L jar of Fricke xylenol gel, mounted in a Quasar pelvic phantom (Modus Medical Devices Inc., London, Ontario). Gels were imaged using a Vista Optical CT Scanner (Modus Medical Devices Inc.).

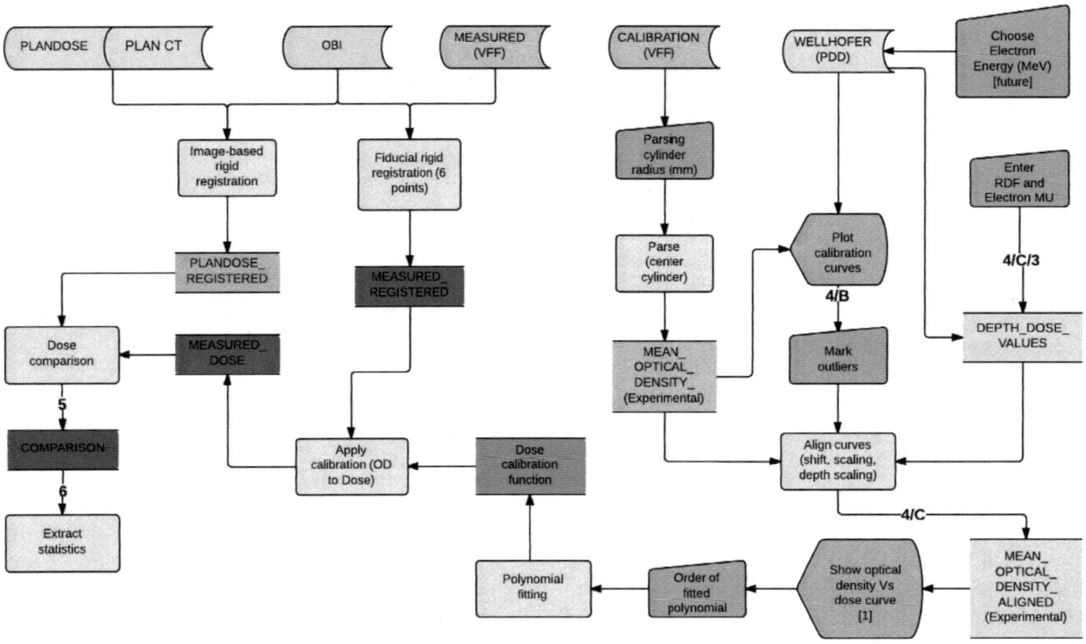

Figure 1. The workflow of the gel dosimetry analysis slicelet showing the incorporation of planning data (upper left), optical CT data (VFF file format), gel calibration (right), and dose comparison (bottom left).

2.1. Import DICOM data and registration to on-board imaging

As shown in the top row of figure 1, the radiation dose, planning CT images (and structure contours), and on-board imager (OBI) CT DICOM files were imported into 3D Slicer (figure 2). Using rigid body intensity based registration (pre-existing in 3D Slicer), the planning CT images, structure contours, and radiation dose were registered to the OBI CT imaging (figure 3).

2.2. Optical CT registration

Prior to dose delivery, 6 pen markings were made on the jar to act as fiducial markers: 2 sets of 3 co-planar points. Small ball bearings were fixed to the jar on top of each pen marking, and a cone beam CT was taken using the OBI to localize the ball bearings. After dose delivery, the ball bearing fiducials were removed, optical CT imaging was performed, and the reconstructed optical CT file ('MEASURED (VFF)' in figure 1) was imported to 3D Slicer using a file importer plugin developed specifically for this purpose. The 6 fiducial points were selected on the optical CT images (figure 4) using the gel analysis slicelet and registered to the OBI fiducial points using a rigid registration.

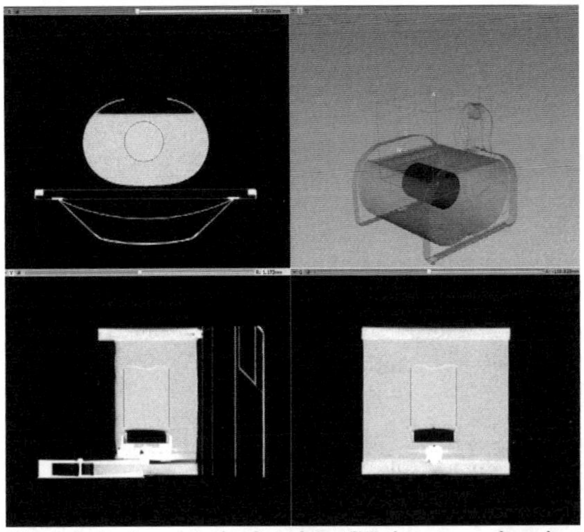

Figure 2. Imported planning CT images, showing gel jar and body structures.

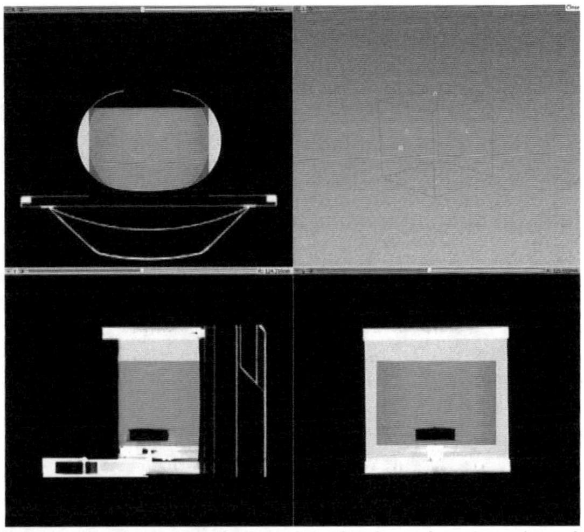

Figure 3. Registered planning CT imaging (grey) and OBI CT imaging (green).

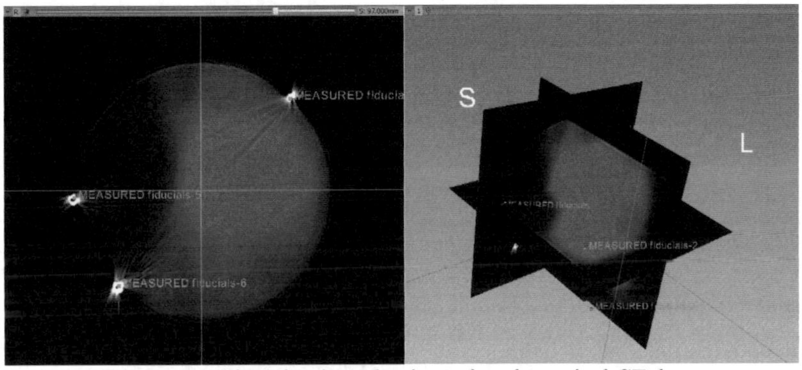

Figure 4. Fiducial point selection using the optical CT data.

2.3. Optical CT calibration

A 12 MeV, 6x6 cm^2 field, 230 MU electron beam was delivered to the surface of a second jar with Fricke xylenol gel from the same batch (SSD=100cm) to calibrate the gel response to dose, and an optical CT of the jar was acquired ('CALIBRATION (VFF)' in figure 1). The optical density values through the central axis of the jar (averaging across a user defined radius (i.e. 10 mm)) were taken, and commissioned ion chamber depth dose data ('WELLHOFER (PDD)' in figure 1) for a 12MeV electron beam was imported to the slicelet. The two curves (optical density vs. depth, and percent depth dose vs. depth) were aligned using a minimization routine in the slicelet. Using the aligned curves, optical density was plotted against dose. The user selected the order of polynomial fit function to be used for the calibration. In this case, the data was fit using a linear function (figure 5). This calibration function was then applied to the first gel jar ('MEASURED_REGISTERED' in figure 1) to convert optical density measurements into dose values (figure 6).

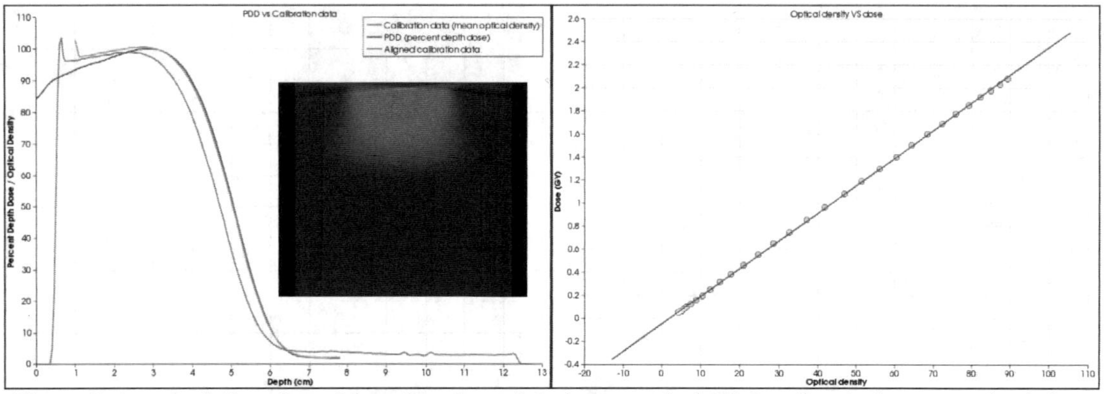

Figure 5. A sagittal slice view of the calibration gel dosimeter optical CT data (inset), the percent depth dose curve alignment (left), and the calibration line data and fit (right).

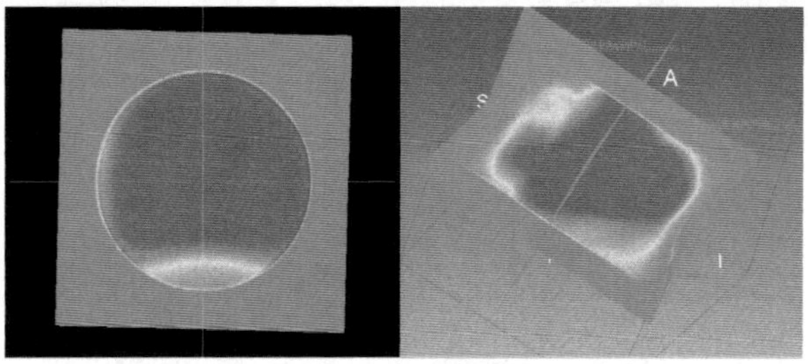

Figure 6. Calibrated optical CT gel data.

3. Results and Conclusions

After calibrating the optical CT gel data, the final step in the gel dosimetry analysis slicelet is to compare the measured optical CT dose data with the planned dose delivery distribution. The slicelet features a gamma comparison tool, and allows the user flexibility in setting the distance to agreement and dose difference, as well as the option to only calculate the gamma values inside the gel jar structure. Examples of the gamma comparison portion of the slicelet can be found in this conference's proceedings [6]. In this work, Olding *et al* used the gel dosimetry analysis slicelet to validate stereotactic ablative body radiation delivery using volumetric modulated arc therapy.

Overall, the gel dosimetry analysis slicelet has been proven to be a simple and efficient way to quickly analyze gel dosimetry data. Previous gel analysis using Matlab and CERR was lengthy and processing time would take several hours. Using this slicelet, processing time is reduced to about 10 minutes, facilitating easier clinical adoption of gel dosimetry. Further improvements are underway to expand the capabilities and scope of the slicelet to allow for more flexibility and slight variations in the typical workflow.

4. References

[1] Baldock C *et al* 2010 *Phys. Med. Biol.* **55** R1-63
[2] Jordan K and Avvakumov N 2009 *Phys. Med. Biol.* **54** 6773-89
[3] Olding T *et al* 2013 *J. Phys.: Conf. Ser.* **444** 012073
[4] Baldock C 2006 *J. Phys.: Conf. Ser.* **56** 14-22
[5] Pinter C *et al* 2012 *Med. Phys.* **39** 6332-7
[6] Olding T *et al* 2014. Proceedings of 8th International Conference on 3D Radiation Dosimetry (IC3DDose) 4-7 September 2014, Ystad, Sweden

The quenching effect in PRESAGE® dosimetry of proton beams: Is an empirical correction feasible?

S J Doran[1,2], T Gorjiara[3], J Adamovics[4,5], Z Kuncic[3], A Kacperek[6] and C Baldock[3]
[1]CRUK Cancer Imaging Centre, Institute of Cancer Research, London, UK
[2]University of Surrey, Guildford, UK
[3]Institute of Medical Physics, School of Physics, University of Sydney, NSW 2006, Australia
[4]Department of Chemistry and Biology, Rider University, Lawrenceville, NJ, USA
[5]Heuris Pharma, LLC, 412 Sunset Rd., Skillman, NJ, 08558 USA
[6]Douglas Cyclotron, Clatterbridge Centre for Oncology, Wirral, UK

E-mail: Simon.Doran@icr.ac.uk

Abstract. Chemical dosimeters, including PRESAGE® as used in optical CT, exhibit significant quenching effects in response to proton irradiation and this may limit their widespread uptake. This study performs careful measurements of the observed quenching of a recently developed variant of PRESAGE® in a 60 MeV proton beam and uses them to attempt an empirical correction of a simple superposition of two unmodulated beams.

1. Introduction

Proton therapy is an advanced radiotherapy technique with the potential for achieving dose distributions superior to those available in x-ray and electron beam therapy. Protons deposit their energy very rapidly at the end of their range in a sharp and well defined region, called the Bragg peak. Furthermore, the range of a proton beam is strongly modulated by the density of the tissue through which it passes. These properties lead to great potential benefits in terms of the conformality of delivered doses to tumours, but make proton therapy an inherently riskier technique, with the possibilities of depositing either significantly too much dose or no dose at all, potentially resulting in either serious damage to healthy tissue or treatment failure and lack of tumour control.

The recommended devices for calibration dosimetry of proton beams are ionisation chambers, but these are not suitable for the high-resolution 3-D mapping needed for whole-system commissioning and the verification of dose plans. A volumetric 3D dosimeter with comparable dose measurement accuracy to that of an ionisation chamber, and dimensions similar to those of the tumour and adjacent organs at risk, is thus highly desirable [1, 2].

But there is a problem: most chemical dosimeters (film, polyacrylamide gels (PAG), Fricke gels, PRESAGE®, alanine, etc.) *under-record dose at depths near the Bragg peak.* This is often known as signal "quenching" [3, 4]. The phenomenon is also referred to as "saturation" and, as we show below, is linked to the high-dose saturation that has long been observed in traditional low-LET measurements, particularly for PAG. Gustavsson *et al* [5] observed quenching effects in polymer gel samples and explained the findings in terms of ion-recombination, whilst Jirasek *et al* [6] interpreted similar observations using the framework of track structure theory [7]. Despite the clear challenge posed by quenching, previous authors [8] have been optimistic that the difficulties could be overcome by appropriate calibration. The aim of this study, therefore, was to perform careful measurements to measure accurately the extent of the quenching phenomenon and to determine whether an empirical correction can be performed.

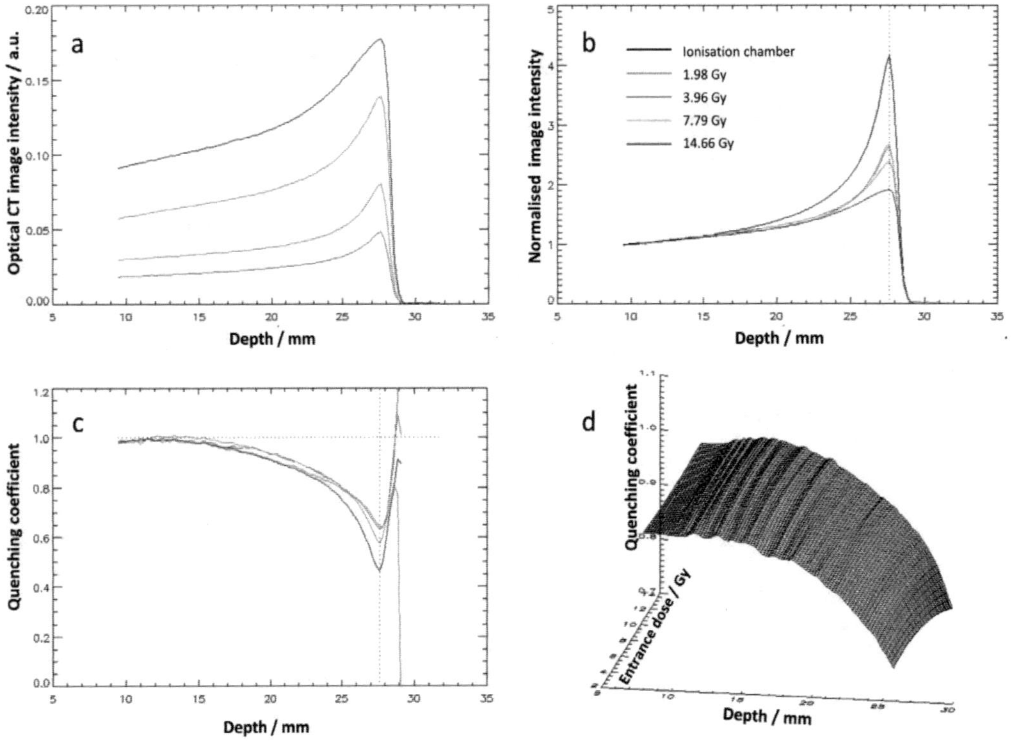

Figure 1. Processed data derived from proton irradiations: (a) Aligned, but un-normalised profiles demonstrate the form of the proton depth dose curve and its scaling with entrance dose, but the profile shapes change subtly with increasing dose. (b) Normalising the curves and comparing with the equivalent ionisation chamber readings demonstrates graphically how quenching is a function of dose. (c) Quenching coefficient; (d) The quenching function appears to be smooth and monotonic along both dose and depth directions and can be interpolated, thus allowing us to calculate the profile for arbitrary doses. N.B. All values plotted are measured, not simulated data.

2. Methods

2.1. Proton irradiations

Five solid cylindrical dosimeters (diameter 61 mm and length 90 mm), made from a PRESAGE® formulation (Heuris Inc, Skillman, NJ) specially designed for use with proton beams [9, 10], were irradiated using an unmodulated 60 MeV proton beam from the Douglas Cyclotron at the Clatterbridge Cancer Centre (CCC), Wirral, UK. The first four samples were each irradiated with a single beam of circular cross-section and with calculated entrance doses of 1.98, 3.96, 7.79 and 14.46 Gy respectively. The fifth sample was irradiated twice in the same place with an unmodulated beam designed to deposit 2 × 3.97 Gy at the entrance, but on the second occasion a 1 cm block of solid water was inserted at the proximal end of the sample to displace the position of the Bragg peak.

2.2. Optical CT imaging

After irradiation, dosimeters were refrigerated at 5°C for between 2 and 4 days before scanning in a CCD-based optical CT scanner [11]. For optical matching, a composition of 93.4% 2-Ethylhexyl Salicylate (Sigma-Aldrich Co. LLC.), 6.6% 4-methoxycinnamic acid 2-ethylhexyl ester (Sigma-Aldrich Co. LLC.) and a small quantity of green dye showed the most similar optical absorption and refractive index to PRESAGE®, with a calculated refractive index of 1.505. 400 raw projection images of the dosimeter, each with 256 × 256 pixels, were recorded, and reconstructed to an isotropic spatial resolution of 0.255 mm.

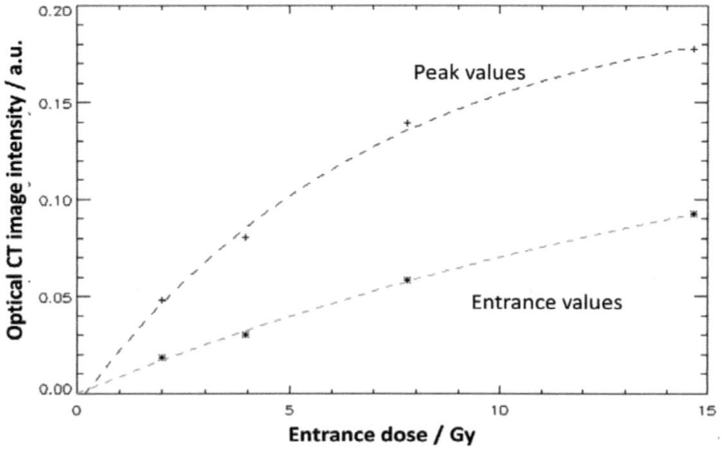

Figure 2. Demonstration of a significant saturation effect in peak region and slight effect in the entrance

2.3. Data processing
Since the Bragg peak covers only a few mm for 60 MeV protons, significant changes in signal occur over a very small number of voxels. Thus, in order to obtain a reliable estimate of the optical CT quenching effect, significant care was required in processing the data. The individual steps were:

- *Averaging over regions-of-interest*: median intensity from a circular region of radius 3.6 mm (approx. 615 pixels), inscribed within the irradiated region and excluding any penumbra;
- *Subtraction of background* using an ROI from the post-irradiation data on a slice-by-slice basis;
- *Alignment of Bragg peaks* to compensate for any inconsistency in manual positioning;
- *Truncation of entrance data*: quantitative values in the first few mm of the data are strongly affected by image artefacts caused by the interface between PRESAGE® and matching liquid;
- *Conversion of ion chamber data* (independent separate calibration of the proton beam in water supplied by AK) recalculated for the equivalent depth in PRESAGE® (see [9] for discussion);
- *Normalisation*: data were divided by the value at the first point on the truncated profile (9.4 mm depth). This point is discussed further below, but is defensible because the quenching coefficient is expected to vary only weakly with distance and dose near this point (see figure 1c).
- *Calculation of quenching coefficient*: we define the quenching coefficient $Q(d, D)$ as the ratio of the normalised dose for the PRESAGE® measurements at depth d in a sample exposed to an entrance dose of D to the normalized ion chamber measurement made at the same depth.
- *Surface fitting*: the variation of the quench profile is a smooth function and thus we can fit an appropriate surface allowing the estimation of the data at an arbitrary intermediate dose.

3. Results
Figure 1a demonstrates the excellent SNR available from the system when an entire ROI is measured. Figures 1b and 1c show the most significant result to come from this study, namely that *signal quenching is a combined function of dose and LET.* Figure 1d is a smooth fitted surface that allows quench profiles to be generated for arbitrary doses.

Figure 2 plots the data from the first point on each profile of figure 1a as a function of dose (lower line) and the Bragg peak (upper line). A significant saturation effect is observed.

Figure 3 shows the results for the sample that was irradiated twice. A clear depiction of the double-Bragg-peak structure is seen in the inset coronal image. The black curve corresponds to the experimental data, whilst the dotted green curve is the same data as in figure 1b. The dotted blue curve represents a first attempt to make an empirical model of the combined irradiation.

4. Discussion and conclusions

The very existence of an LET-dependent quenching phenomenon immediately tells us that we have a

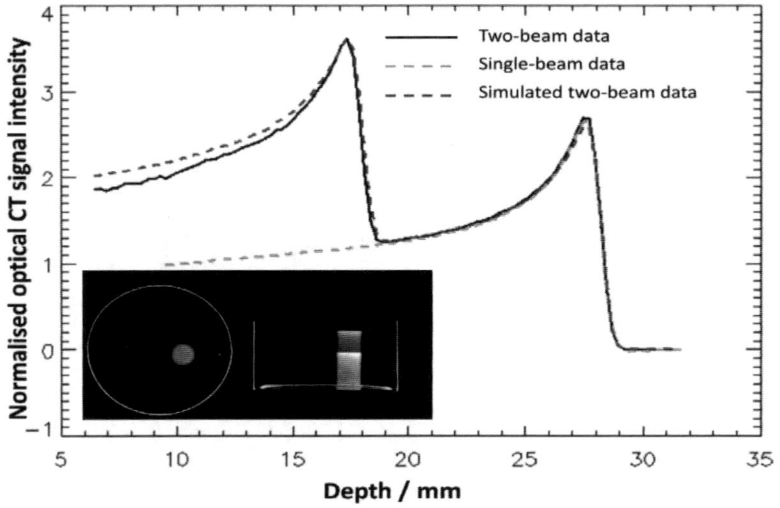

Figure 3. Data from a sample irradiated from one end with two proton beams, one of which has been range-modulated by 1 cm of solid water. The excellent superposition of these two curves between the first and second Bragg peaks shows that the quenching effect is reproducible between samples. The blue dotted line is simulated with a forward model using the known physical dose and the quenching factors previously calculated.

problem that is non-invertible. There is no longer a one-to-one mapping between optical attenuation coefficient (or MRI R2-value) and dose, since a given attenuation may be achieved with different total physical doses via contributions from components with different LET values. Prior to this study, however, grounds for optimism existed that, given a quenching coefficient whose functional form depended only on LET, a simple, linear *forward model* would enable optical effect to be predicted for a known dose pattern. If the quenching effect were indeed dose-independent, we would simply be able to correct each of the physical dose contributions to figure 3 by the relevant quench profile and add them together. The fact that there is a discrepancy tells us that such simple empirical corrections are deficient and further work is necessary to understand the quenching phenomenon theoretically.

5. Acknowledgments

We acknowledge the support received from the CRUK and EPSRC Cancer Imaging Centre in association with the MRC and Department of Health (England) grant C1060/A10334, and NHS funding to the NIHR Biomedical Research Centre.

6. References

[1] Zeidan O A *et al* 2010 *Med. Phys.* **37** 2145-52
[2] Baldock C 2006 *J. Phys.: Conf. Ser.* **56** 14-22
[2] Al-Nowais S *et al* 2009 *Applied Radiation and Isotopes* **67** 415-8
[3] Baldock C *et al* 2010 *Phys. Med. Biol.* **55** R1-63

[4] Gustavsson H *et al* 2004 *Phys. Med. Biol.* **49** 3847-38
[5] Jirasek A *et al* 2002 *Med. Phys.* **29** 569-77
[6] Katz R 1978 *Nuclear Track Detection* **2** 1-28
[7] Zhao L *et al* 2012 *Phys. Med. Biol.* **57** N431-43
[8] Gorjiara T *et al* 2012 *Med. Phys.* **39** 7071-9
[9] Gorjiara T *et al* 2011 *Med. Phys.* **38** 2265-74
[10] Krstajic N *et al* 2007 *Phys. Med. Biol.* **52** 3693

An investigation into the potential influence of oxygen on the efficiency of the PRESAGE® dosimeter

M Alqathami[1], A Blencowe[2] and G Ibbott[1]
[1]Department of Radiation Physics, University of Texas MD Anderson Cancer Center, Houston, TX, 77030, USA
[2]Mawson Institute, Division of Information Technology, Engineering and the Environment, The University of South Australia, SA 5095, Australia

E-mail: malqathami@mdanderson.org

Abstract. The influence of atmospheric oxygen on the efficiency of the PRESAGE® dosimeter was investigated. Batches of PRESAGE® and reporting system solution were deoxygenated using nitrogen and compared to similar batches that were exposed to atmospheric oxygen during fabrication. The overall results show little influence of oxygen on the characteristics of PRESAGE® with the radical initiator oxidizing the leucomalachite green in the presence of oxygen. However, when deoxygenating the reporting system the sensitivity to radiation dose increased by 30% compared to the non-deoxygenated system. A slight improvement in sensitivity (5%) was achieved by deoxygenating the PRESAGE® precursors prior to casting. The results suggest that the solid polyurethane matrix is not permeable to atmospheric oxygen. In addition, there were no observed changes in the dose linearity, absorption spectrum and post-response photofading characteristics of PRESAGE® under the conditions investigated.

1. Introduction

PRESAGE® is an exceptional 3D radiochromic dosimeter with unique characteristics that are desirable for clinical applications [1-4]. Unlike polymer gel dosimeters, PRESAGE® is composed of an optically clear polyurethane matrix that can be fabricated into any shape and requires no supporting container, a reporting system comprising of leucomalachite green (LMG) dye or one of its derivatives [5], and a halocarbon compound as a radical source [6, 7]. Free radicals produced from the radiolysis of the halogen-carbon bond upon irradiation oxidize the LMG dye leading to a change in optical absorbance, which is known to be linear with respect to the absorbed radiation dose [7]. One of the attractive features reported for PRESAGE® dosimeters over gel dosimeters is a lack of sensitivity to oxygen [7]; however, no experimental results have been documented in the literature to support such a claim. As part of a series of PRESAGE® component-specific optimization studies [5, 6, 8], the aim of this study was to investigate the influence of atmospheric oxygen on the efficiency of PRESAGE® as a dosimetric system and independently, on the reporting system (i.e., LMG and radical initiator).

2. Materials and Methods

2.1. PRESAGE® fabrication

The chemical agents used in this study were: Polyurethane resin (Crystal Clear 206, Smooth-On, Easton, PA, USA), which was supplied in two parts (Part A and Part B); the radical initiator, tetrabromoethane; and a commercial LMG dye (Sigma-Aldrich, St Louis, MO, USA). The dosimeters were prepared in poly(methyl methacrylate) spectrophotometer cuvettes with a wall thickness of 1 mm and internal dimensions of 1 × 1 × 4.5 cm. Detailed description of the fabrication procedure is described elsewhere [5, 6, 8]. The chemical composition of the PRESAGE® formula used was tetrabromoethane (1 wt%), LMG (2 wt%) and Crystal Clear 206/catalyst (97 wt%). Deoxygenation of the PRESAGE® was achieved by bubbling nitrogen (N_2) gas through the dosimeter precursor mixture for 10 min prior to casting. The reporting system was prepared by dissolving LMG 2% by mass in tetrabromoethane, which was then transferred into different glass vials fitted with PTFE caps. Deoxygenation of the reporting system was achieved by bubbling N_2 through each vial for 10 min by inserting two needles through the septa of each vial. One needle was connected to the N_2 source and the second needle acted as a vent.

2.2. Absorbance change measurements

Absorbance change measurements were acquired using a dual-beam Thermo scientific GENESYS™ 10S UV-VIS spectrophotometer (Thermo Fisher Scientific Inc, Waltham, MA, USA) over the visible wavelength region (450–750 nm) was measured to determine the absorption maxima of the PRESAGE® dosimeter cuvettes and vials of the reporting system. The optical absorption of each PRESAGE® cuvette was measured pre- and post-irradiation. To investigate the influence of deoxygenation of PRESAGE® on its post-response photo-fading, absorption acquisitions were acquired at different time intervals (1, 3, 6, 24, 48, 92, 120, and 168 hours post-irradiation).

2.3. Irradiation

The PRESAGE® cuvettes were irradiated with a 6 MV Varian True-Beam linear accelerator (Varian Medical Systems, Palo Alto, CA), using a dose rate of 5 Gy/min and a range of radiation doses (0, 1, 5, 10, 20 and 30 Gy). The field size was set to 10 × 10 cm. The dosimeter cuvettes and reporting system glass vials were sandwiched between two solid water phantoms with the top one measuring 1 cm in thickness so that the maximum dose occurred at the middle of each dosimeter. The SSD was set to 100 cm from the solid water surface.

3. Results and Discussion

Initially, the effect of dissolved oxygen on the reporting system, consisting of LMG dissolved in tetrabromoethane, was studied. The absorption spectra of the non-deoxygenated reporting system solution, and an identical solution deoxygenated using N_2, both after irradiation (20 Gy), are shown in figure 1a. For both cases, the absorption maximum (λ_{max}) was found to peak at *ca.* 632 nm. A change in the solution colour was observed upon irradiation even in the non-deoxygenated solution (figure 1b). This suggests that, in contrast to polymer gel dosimeters, the presence of dissolved oxygen in PRESAGE® does not completely inhibit the oxidation reaction and formation of the chromatic cation, which occurs when the methine proton (C-H) of LMG is abstracted by carbon radicals generated by radiolysis of the halocarbons. However, deoxygenating the reporting system improved the sensitivity to radiation dose, with the change in optical density increasing by ~30% compared to the non-deoxygenated solution. This suggests partial oxygen sensitivity, however, this should not be considered a limitation since the colour formation reaction still proceeds in the presence of oxygen.

The measured absorbance changes at 632 nm versus the radiation absorbed doses for the deoxygenated and non-deoxygenated PRESAGE® dosimeters are displayed in figure 2. Absorption values were obtained by subtracting the relevant value of a reference cuvette from the same batch with zero radiation dose from that of the irradiated cuvettes. In both cases very good linearity (correlation coefficient $R^2 > 0.99$) of the dose response was observed for PRESAGE® with/without deoxygenation indicating no influence on the linearity of the dose-response relationship (figure 2). However, it is important to point out that removing dissolved oxygen completely from the PRESAGE® precursors might be challenging because it is expected that oxygen might still diffuse into PRESAGE® during the curing process. Nevertheless, it is expected that potential diffusion of oxygen into the deoxygenated PRESAGE® batch is minimal and not sufficient to make an exact comparison between the two batches in terms of oxygen effects problematic.

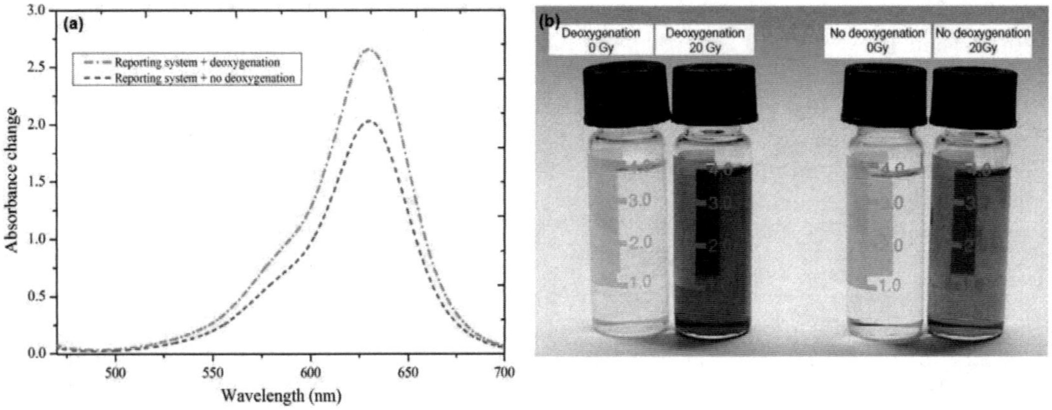

Figure 1. (a) Absorption spectra of the deoxygenated and non-deoxygenated reporting system solutions after irradiation to a dose of 20 Gy. Reference solution (non-irradiated) for each formulation was used as a baseline to establish the zero value. (b) Photograph of the deoxygenated and non-deoxygenated reporting system solutions before and after irradiated to a dose of 20 Gy showing an increase in colour intensity in the deoxygenated solution.

Unlike the reporting system, a slight (~5%) increase in sensitivity was observed for the deoxygenated PRESAGE® compared to the non-deoxygenated batch. Such observation is consistent with the reporting system findings; however, the small variation between the deoxygenated and non-deoxygenated PRESAGE® could be attributed to the polyurethane matrix being solid and thus, minimizing the diffusion of atmospheric gasses.

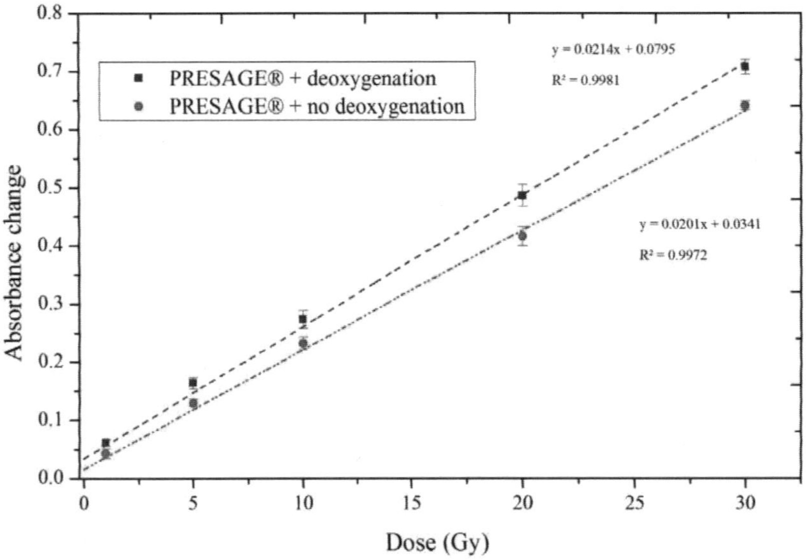

Figure 2. Recorded optical absorbance changes as a function of absorbed radiation dose for the deoxygenated and non-deoxygenated PRESAGE® batches. Error bars represent the standard deviation in the measurement.

4. Conclusion

In this work, the potential influence of atmospheric oxygen on the overall characteristics of the PRESAGE® was investigated. In general, a noticeable oxygen influence on the sensitivity of PRESAGE® and its reporting system has been observed. However, no influence on the linearity, absorption spectra and stability of PRESAGE® has been reported. This confirmed yet another attractive feature of the PRESAGE® dosimeter; its insensitivity to oxygen. An increase in sensitivity to radiation dose could be achieved when deoxygenating the PRESAGE®. However, considering the complexity and time required to deoxygenate the PRESAGE® precursors and the fact that only a slight improvement in sensitivity is achieved, deoxygenation of PRESAGE® precursors is not justifiable and thus not recommended based on our observations.

5. References

[1] Clift C *et al* 2010 *Phys. Med. Biol.* **55** 1279-93
[2] Abdul Rahman AT *et al* 2011 *Phys. Med. Biol.* **56** 4177-99
[3] Baldock C *et al* 2010 *Phys. Med. Biol.* **55** R1-63
[4] Gorjiara T *et al* 2011 *Med. Phys.* **38** 2265-74
[5] Alqathami M *et al* 2013 *Radiat. Phys. Chem.* **85** 204-9
[6] Alqathami, M *et al* 2012 *Radiat. Phys. Chem.* **81** 867-73
[7] Adamovics J *et al* 2006 *Radiat. Prot. Dosim.* **120** 107-12
[8] Alqathami M *et al* 2012 *Radiat. Phys. Chem.* **81** 1688-95

Journal of Physics: Conference Series **573** (2015) 012045 doi:10.1088/1742-6596/573/1/012045

Timing considerations for preclinical MRgRT: effects of ion diffusion, SNR and imaging times on FXG gel calibration

M Welch[1], WD Foltz[2,3] and DA Jaffray[1,2,3,4]

[1]Dept. of Medical Biophysics, University of Toronto, Toronto, ON, Canada
[2]Radiation Medicine Program, Princess Margaret Hospital, Toronto, ON, Canada
[3]Department of Radiation Oncology, University of Toronto, Toronto, ON, Canada
[4]The TECHNA Institute for the Advancement of Technology for Health, Toronto, ON, Canada

E-mail: mattea.welch@rmp.uhn.ca

Abstract. Sub-millimeter resolution images are required for gel dosimeters to be used in preclinical research, which is challenging for MR probed ferrous xylenol-orange (FXG) dosimeters due to ion diffusion and inadequate SNR. A preclinical 7 T MR, small animal irradiator and FXG dosimeters were used in all experiments. Ion diffusion was analyzed using high resolution (0.2 mm/pixel) T_1 MR images collected every 5 minutes, post-irradiation, for an hour. Using Fick's second law, ion diffusion was approximated for the first hour post-irradiation. SNR, T1 map precision and calibration fit were determined for two MR protocols: (1) 10 minute acquisition, 0.35mm/pixel and 3mm slices, (2) 45 minute acquisition, 0.25 mm/pixel and 2 mm slices. SNR and T1 map precision were calculated using a Monte Carlo simulation. Calibration curves were determined by plotting R1 relaxation rates versus depth dose data, and fitting a linear trend line. Ion diffusion was estimated as $0.003mm^2$ in the first hour post-irradiation. For protocols (1) and (2) respectively, Monte Carlo simulation predicted T1 precisions of 3% and 5% within individual voxels using experimental SNRs; the corresponding measured T1 precisions were 8% and 12%. The linear trend lines reported slopes of 27 ± 3 Gy*s (R^2: 0.80 ± 0.04) and 27 ± 4 Gy*s (R^2: 0.90 ± 0.04). Ion diffusion is negligible within the first hour post-irradiation, and an accurate and reproducible calibration can be achieved in a preclinical setting with sub-millimeter resolution.

1. Introduction

Technological advances have revolutionized preclinical radiation studies, resulting in accurate and highly conformal dose delivery with sub-millimeter precision. Novel radiotherapy studies are possible with these systems [1], including MR-guided preclinical radiotherapy (MRgRT) verification.

Sub-millimetre resolution with MR probed FXG gel dosimeters is limited by ion diffusion, inadequate signal-to-noise ratio (SNR), which limits T1 precision, and increased imaging time; high field pre-clinical MRI systems provide an advantageous resolution/SNR/imaging time trade-off. In this paper, high field pre-clinical MRI is used to approximate the rate of ion diffusion and to investigate the feasibility of accurate and reproducible dose calibration with R1 (1/T1).

2. Materials and Methods

2.1 Gel Preparation

FXG dosimeters were used in all experiments and prepared according to the recipe outlined by Babic *et al* [2]; final concentrations were 0.3 mM ferrous ammonium sulphate (Cat.No.203505,

Sigma-Aldrich Ltd, Oakville, Canada), 0.05 mM xylenol orange (Cat.No.398187, Sigma-Aldrich Ltd), 65 mM sulfuric acid (Cat.No.258105, Sigma-Aldrich Ltd) and 6% by mass 300 Bloom Porcine gelatin (Cat.No.G2500, Sigma-Aldrich Ltd) in de-ionized water. This recipe was chosen because of the reduced ion diffusion coefficient resulting from the addition of xylenol-orange [3]. Gelatin was added to a portion of the water and allowed to swell for 10 minutes at room temperature. The mixture was then heated to 50°C, and held at this temperature for approximately 15-20 minutes to dissolve the gelatin. A stock solution containing ferrous ammonium sulphate and xylenol-orange was added to the sulfuric acid and remaining water. The gelatin mixture was cooled to 35°C and the rest of the chemicals were added. The cooled solution was poured into 50 mL centrifuge tubes 3cm in diameter, covered and allowed to set overnight in a refrigerator.

2.2 FXG Gel Irradiations

The FXG gels were irradiated one day post manufacturing using an X-Rad 225Cx small animal irradiator (Precision X-Ray Inc., North Branford, CT, USA). Centrifuge tube caps were removed and the tops of the gel were treated. 1x1 cm square beams (225 kVp, 13 mA) were delivered to a surface dose of 10 Gy (2.85 Gy min^{-1}) for ion diffusion experiments. 1 cm circular beams (225 kVp, 13 mA) were delivered to a surface dose of 4 Gy (2.88 Gy min^{-1}) for calibration experiments.

2.3 MR Imaging

Imaging of the FXG gels were performed pre and post-irradiation using a 7T MR unit (BioSpec 70/30 USR, Bruker Corp., Ettlingen, DE) and 7.2 cm inner diameter cylindrical quadrature RF volume coil provided by the manufacturer; the tubes were aligned in the magnet using pre-existing markings on the centrifuge tubes. The room temperatures were not specified; however, the gels were placed in a temperature controlled 21 °C water bath prior to pre-irradiation imaging for an hour. The variation in dosimeter response due to temperature was considered minimal given that the increase in Fe^{3+} ions in an aqueous solution increases by approximately 0.12% per degree Celsius [4]. The following imaging protocols were used:

2.3.1 High resolution T_1 weighted imaging. To monitor ion diffusion, a 1 minute image acquisition was designed using a spin echo sequence with TR 1000 ms and TE 7 ms to collect twenty, 5 mm thick, axial slices. The field of view (FOV) was 28 mm by 28 mm on a 140 by 140 acquisition matrix.

2.3.2 T1 Mapping. Two variable TR spin echo sequences were designed for R1 dose calibration (TE 6ms; TR 750, 1000, 1500, 2000, 3000 and 5000 ms). Protocol #1: A 10 minute image acquisition with twenty 3 mm axial slices and an in-plane FOV of 32.5 mm by 32.5 mm on a 100 by 100 matrix. Protocol #2: A 45 minute image acquisition consisting of two 22.5 minute image acquisitions with contiguous slice packages providing the same longitudinal volume coverage with 2 mm slice thickness. A FOV of 32.5 mm by 32.5 mm on an acquisition matrix of 128 by 128 was used.

2.4 Ion Diffusion Analysis

Ion diffusion was estimated using T1 weighted high resolution images (Sec. 2.3.1) collected every five minutes, post-irradiation, for an hour. Beam profile noise reduction was achieved by averaging together ten coronal beam profiles (figure 1). As described by Olsson *et al* [5], Fick's second law was used to approximate ion diffusion for the first hour post-irradiation. To encompass the extent of the beam concentration a sigma was chosen to include approximately 95% of the total concentration.

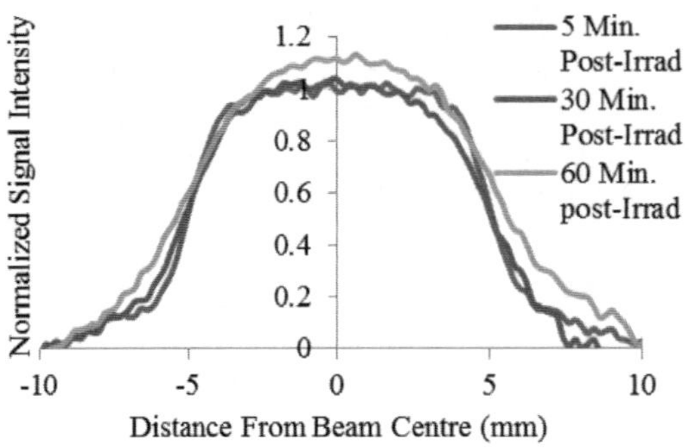

Figure 1. Beam profile for 5, 30 and 60 min. post-irradiation.

2.5 Precision of T1 Estimates

Several factors effect T1 precision, these include the TR sampling strategy, SNR and thermal noise [6]. SNR at the longest TR was found by dividing the average signal amplitude within a ROI in the gel by the standard deviation of the noise, after Rayleigh noise statistics correction. T1 precision for the SNR was then calculated by Monte Carlo simulation using in-house Matlab scripts (250 repetitions of T1 regression with SNR ranging from 20 to 1000). Experimental T1 precision was estimated using pre-irradiation T1 maps and the mean and standard deviation of T1 within a central circular ROI.

2.6 Calibration Curve Fitting Analysis

Calibration curve fitting was analysed with the imaging protocols outlined in Sec. 2.3.2 and 2.3.3. Previously collected dose rate information was assumed linear and interpolated to obtain dose information at depths corresponding to the slice depths of the MR images. Post-irradiation R1 (1/T1) relaxation rates at a given depth were represented by the average of an 11 x 11 pixel area spliced out of the centre of the irradiated column of the gel. The change in relaxation rates was found by subtracting the pre-irradiation R1 relaxation rates from the post-irradiation R1 relaxation rates; these changes were plotted against the depth dose information and a linear trend line was fit to the resulting graphs (figure 2). The R^2 values and slopes of the trend lines were analysed to determine the quality and reproducibility of the fit.

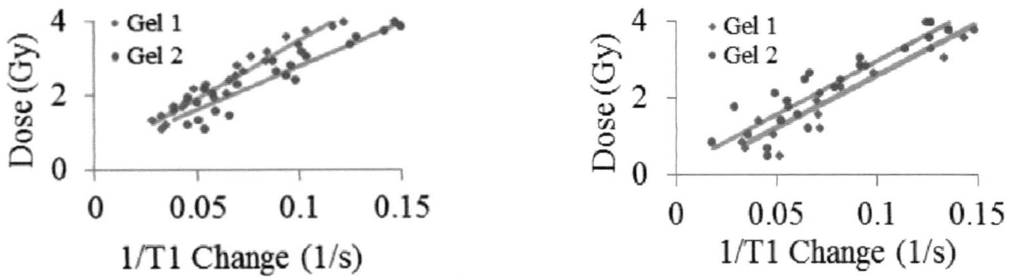

Figure 2. Change in relaxation rates vs dose for Protocol #1: subplot A, and Protocol #2: subplot B.

3. Results

3.1 Ion Diffusion

The ion diffusion coefficient reported when using Fick's second law of diffusion was approximated at 0.003mm^2 in the first hour post-irradiation; a rate undetectable by the image resolution used in these experiments.

3.2 T1 Precision Analysis
SNR at TR 5000 ms was approximately 75 for Protocol #1 and 49 for Protocol #2, as one would expect given 2.5-fold larger voxels for Protocol #1 and two-fold signal averaging for Protocol #2. Based on Monte Carlo analysis, SNR of 75 and 49 should correspond to T1 map precision from thermal noise on the order of 3 and 5%, respectively, and were found to be 8% and 12%, respectively.

3.3 Calibration Fit
Protocol #1 had a mean linear trend line with R^2: 0.80 ± 0.04 and slope: 27 ± 3 Gy*s (figure 2. A). Protocol #2 had a mean linear trend line with R^2: 0.90 ± 0.04 and slope: 27 ± 4 Gy*s (figure 2. B).

4. Discussion

4.1 Ion Diffusion
Olsson *et al* [5] reported an ion diffusion coefficient of 0.19mm^2 hour^{-1} for a standard agarose dosimeter; the use of xylenol-orange as a chelator in these experiments would result in a slower ion diffusion coefficient. The resolution of the images used (0.2mm/pixel) is not high enough to detect ion diffusion in the first hour post-irradiation. Higher resolution and longer wait times post-irradiation would be needed in order to quantify the diffusion coefficient accurately [7]. The diffusion reported in these experiments may be the result of image noise or ion development not associated with diffusion. Furthermore, the changes in beam profile width seen over time in figure 1 may be due to the continued development of Fe^{3+} ions post-irradiation [8].

4.2 T1 Precision Analysis
The measured T1 map precision was higher than predicted from SNR, which may reflect spatial heterogeneity in gel characteristics or systematic biases in the T1 data acquisition. However, the variable TR approach should be robust to RF and static field inhomogeneity; additionally inter-slice interference was reduced by selection of RF pulses with sharp excitation and refocusing profiles. The 95% confidence limits for the T1 maps are 16 and 24% within individual voxels (2 standard deviations), corresponding to 3.4 and 5.7 Gy at a slope of 27 Gy*s. Further improvement to T1 precision can be gained by binning voxels, increasing scan time or using a more sensitive receive coil.

4.3 Calibration Fit
The two imaging protocols used in this study verify that lower resolution T1 maps improved SNR and T1 mapping precision without penalty to calibration; improvements to calibration linearity may be achieved through further T1 mapping optimization. Reproducibility of the fit is also increased when using shorter imaging times, which can be attributed to gel sensitivity.

Gel composition within each imaging protocol is identical; however, it may vary between imaging protocols due to gel batch number. Likely, the varying trend line slopes for Protocol #2 reflect auto-oxidation, because the delay to experimentation for gel 1 was two hours less (1.5 hours in total for imaging, 0.5 hours for set up and irradiation) than for gel 2. This time lapse for Protocol #1 gels would have been less than an hour, giving less time for auto-oxidation. In order to validate these assumptions, further tests need to be conducted regarding auto-oxidation and the effects on sensitivity.

4.4 Clinical MR-Guided RT
MR-guided RT is an emerging field within IGRT. The excellent soft tissue contrast and motion characterization opens new doors in treatment planning, resulting in more aggressive plans with higher tumour conformity that would benefit from 3-Dimentional (3D) dosimetry. MRgRT systems are currently being constructed [9] with minimal preclinical testing. 3D ferrous xylenol-orange gel dosimetry is a proven method utilizing MR [10]. Therefore, it is proposed that a preclinical simulation of the MRgRT workflow using 3D FXG gel dosimeters be performed.

5. Conclusion

The results support potential preclinical MR-guided RT gel dosimetry. The low ion diffusion coefficient and T1 mapping precision provided by imaging Protocol #1 improves the accuracy and reproducibility of the dose calibration. Increased SNR should be considered for dose calibration within individual voxels. Further experiments will establish the effects of auto-oxidation on gel sensitivity, define the ion diffusion coefficient more accurately, and verify preclinical radiation treatment plans.

6. Acknowledgements

The authors thank Shawn Stapleton and Kevin Alexander for valuable discussions regarding project direction and experimental design. This work is supported by the Natural Sciences and Engineering Research Council of Canada, TECHNA, Garron Family and PMCF.

7. References

[1] Kahn J *et al* 2012 *Rad. Onc.* **7** 223
[2] Babic S *et al* 2008 *Phys. Med. Biol.* **53** 1637-50
[3] Bero M A *et al* 2001 *Rad. Phys. Chem* **61** 433-5
[4] Shortt K R 1989 *Phys. Med. Biol.* **34** 1923-6
[5] Olsson LE *et al* 1992, *Phys. Med. Biol.* **37** 2243-52
[6] Nishimura DG 2012 *Principles of Magnetic Resonance Imaging* pp 19, 159-162 and 163-164
[7] Baldock C *et al* 1994. *J. Roy. Soc. Med.* **87** 806-8
[8] Olding T and Schreiner L J 2011 *Phys. Med. Biol.* **56** 1259-79
[9] Jaffray D A *et al* 2010 *SPIE* **7622** 02
[10] Gore J C *et al* 1984 *Magn. Reson. Imaging* **2** 224

A reduction of diffusion in PVA Fricke hydrogels

S T Smith[1], K S Masters[1], K Hosokawa[1], J Blinco[1], S B Crowe[1], T Kairn[1,2] and J V Trapp[1]
[1]School of Chemistry, Physics and Mechanical Engineering,
Queensland University of Technology, Brisbane, Australia
[2]Genesis CancerCare Queensland, Brisbane, Australia

E-mail: s164.smith@qut.edu.au

Abstract. A modification to the PVA-FX hydrogel whereby the chelating agent, xylenol orange, was partially bonded to the gelling agent, poly-vinyl alcohol, resulted in an 8% reduction in the post irradiation Fe^{3+} diffusion, adding approximately 1 hour to the useful timespan between irradiation and readout. This xylenol orange functionalised poly-vinyl alcohol hydrogel had an OD dose sensitivity of 0.014 Gy^{-1} and a diffusion rate of 0.133 $mm^2\,h^{-1}$. As this partial bond yields only incremental improvement, it is proposed that more efficient methods of bonding xylenol orange to poly-vinyl alcohol be investigated to further reduce the diffusion in Fricke gels.

1. Introduction

Fricke gels are based on adding gelling agents to the Fricke solution; an acidic oxygenated aqueous solution of ferrous ion (Fe^{2+}). Upon irradiation, a dose dependant transformation of ferrous (Fe^{2+}) to ferric ions (Fe^{3+}) occurs, which is detectable via MRI or optical CT scanning. Fricke gels are attractive for 3D dosimetry as they are easy to prepare and read out, are tissue equivalent over a very large photon energy range, and give reproducible results.

Unfortunately, the poor spatial stability of Fricke gels due to diffusion of Fe^{3+} ions constrains the permissible time between irradiation and read-out [1, 2]. This introduces an obvious practical inconvenience inhibiting the routine use of gels a clinical environment. Limited success in reducing Fe^{3+} diffusion rates is obtained using different gelling agents (gelatin, agarose, sephadex and poly-vinyl alcohol) and chelating agents such as xylenol orange [3, 4].

Chelators are organic chemicals that form two or more coordination bonds with a central metal ion. In xylenol orange Fricke gels, the oxidation of Fe^{2+} to Fe^{3+} induces an inherent increase in the electropositivity of the ion, allowing the co-ordination of the Fe^{3+} ion by the xylenol orange molecule. Upon this coordination, the Fe^{3+} ion bound to xylenol orange is no longer free to move throughout the gel matrix. As such, any subsequent diffusion of dose information is via movement of the much larger ferric-xylenol orange molecule.

To date, the best Fricke gel diffusion results have been obtained by the poly-vinyl alcohol (PVA) xylenol orange Fricke dosimeters proposed by Chu *et al* [5] Their work reports a diffusion coefficient of 0.14 $mm^2\,h^{-1}$, about one order of magnitude less than that of conventional Fricke gels and about half of that of the agarose and gelatin gels with chelating agents [3, 4]. Here, we present a modification to the PVA xylenol orange Fricke gel that reduces diffusion by partially bonding xylenol orange to the PVA molecule.

2. Methodology

2.1 Gel preparation
All chemicals used for synthesis and gel mixing were purchased from Sigma Aldrich, Sydney. Prior to gel mixing, xylenol orange was chemically linked to PVA (XO-PVA). The final gel formulation was: 20% w/v XO-PVA, 50 mM sulfuric acid and 0.4 mM ferrous sulphate. A 24% w/v XO-PVA solution was made in a flask of 50 mM sulfuric acid while being stirred and heated on a hotplate to 65 °C. Once all XO-PVA was in the reaction vessel, the solution was raised to a temperature of 95 °C with a stopper used to minimize evaporation. Meanwhile, a solution containing 2.4 mM ferrous sulphate in 50 mM sulfuric acid was prepared. The XO-PVA was fully dissolved after 3 hours, indicated by a translucent solution, after which the temperature was dropped to 55 °C. Subsequently, five parts of the XO-PVA solution was mixed with one part of the ferrous sulphate solution to make a final 0.4 mM ferrous ion concentration in a 20% XO-PVA solution. Stirring of this mixture was continued for 30 min before the gel was poured into 1 cm plastic cuvettes and stored at 5 °C. For comparative measurements, the original PVA-FX hydrogel was also made using the method outlined in Chu *et al.* (with 20% PVA, 0.4 mM ferrous sulphate and 0.4 mM xylenol orange) [5].

2.2 Gel Irradiation
Gels were irradiated using a 6 MV photon beam from a Varian EX Clinac. All gel irradiations were performed in a water tank with 9 cm Solid Water and 1 cm Bolus backscatter layers to ensure dose homogeneity. All irradiations were performed with gel samples submersed at a depth of 1 cm with an SSD of 100 cm. For dose response measurements, gel cuvettes were irradiated with 0, 1, 5, 15, and 30 Gy. To perform diffusion measurements, half of each gel cuvette was irradiated with 30 Gy while the other half was shielded by shutting the lower jaw to the central axis and placing a 10 cm lead block 1mm over the light field edge; minimizing the half field penumbra.

2.3 Gel Read-out
To maximise signal to noise ratios and scanning resolution, the optical densities (ODs) of the gels were measured using the green channel of an Epson dual-layer flatbed film scanner. All OD values were obtained through the 1 cm path length defined by the cuvette dimensions with an imaging resolution of 266 dpi.

2.4 Diffusion Measurements
To quantify diffusion rates, the method outlined by Kron *et al.* was employed [3]. OD profiles were collected at varying time intervals between 1.5 and 112.5 hours after irradiation. These diffusion profiles were fit to equation (1):

$$Y = Y_0 + \frac{1}{2}(Y_1 - Y_0)\left[1 + \frac{(X - X_0)}{\sqrt{(X - X_0)^2 + n}}\right] \tag{1}$$

Where Y is the measured OD at the position X, X_0 is the horizontal shift of the inflection point, Y_0 and Y_1 are the minimum and maximum values of Y, and n is the curvature parameter that characterises the slope of the inflection point. As demonstrated by Kron *et al,* a plot of n versus time after irradiation is linear and when the factor 0.212 is multiplied by its slope, the diffusion coefficient, D ($mm^2\ h^{-1}$) is obtained.

3. Results and Discussion
To highlight the reduction in diffusion of XO-PVA gel, figure 1 compares two OD profiles of each gel 72 hours apart. Formation of gel using XO-PVA resulted in increased scatter and consequently more noise was encountered in its OD profiles. Despite this, a clear reduction in the blurring effect across the profiles is present. The diffusion coefficient, measured for the PVA-FX hydrogel matched results

reported by Chu *et al.*, with D = 0.144 ± 0.002 $mm^2\ h^{-1}$. The XO-PVA formulation yielded a diffusion rate of 0.133 ± 0.004 $mm^2\ h^{-1}$, an 8% improvement on the existing PVA-FX gel, and the lowest reported thus far for any Fricke gel.

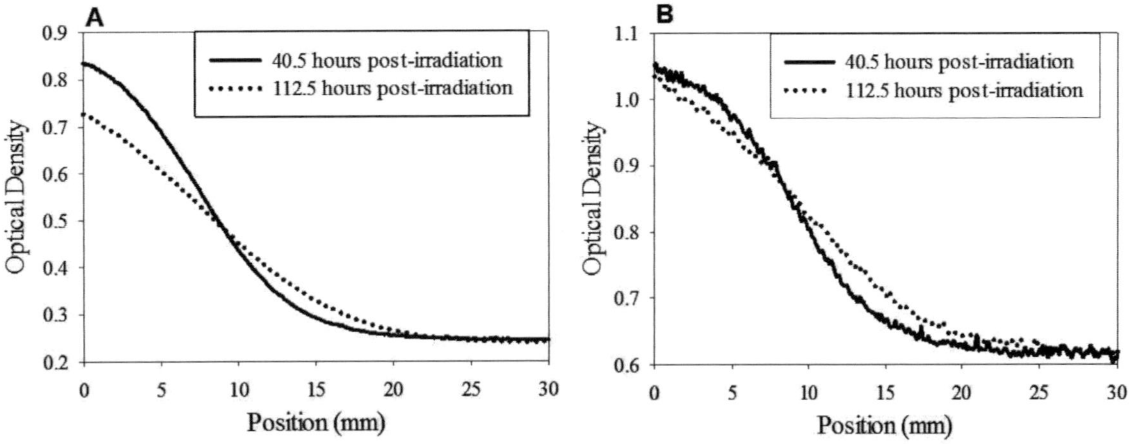

Figure 1. Optical density profiles in half-irradiated cuvettes 40.5 and 112.5 hours post irradiation with air reference for (a) PVA-FX gel and (b) XO-PVA gel.

The mechanism governing the decreased diffusion rate in XO-PVA gel is partial bonding of xylenol orange to PVA. Ferric-xylenol orange molecules bound to the PVA gel matrix are unable to diffuse. Therefore the diffusion measured in this gel can be attributed to the fraction of xylenol orange molecules, which did not successfully bond to PVA during the preparation of XO-PVA. To further improve this result, more efficient methods of bonding xylenol orange to PVA must be explored.

In figure 2, the dose response of both gels is presented. Both show a linear dose response for doses ranging from 0 to 30 Gy. It must be noted that the initial OD of the XO-PVA is almost three times that of the original PVA-FX gel. Our measurements showed a 30% reduction in the dose response of the XO-PVA gel with an OD dose sensitivity of 0.014 Gy^{-1} and a corresponding optical attenuation coefficient of 0.032 $cm^{-1}\ Gy^{-1}$. The uncertainties displayed in figure 2 were calculated by propagating standard deviations of pre- and post-scan pixel values through OD calculations. Referring to this data, it is apparent that these gel dosimeters are capable of resolving single Gy doses, within uncertainty.

4. Conclusions

Preparing gels from xylenol orange functionalised PVA yields a dosimeter with an OD dose sensitivity of 0.014 Gy^{-1} and a diffusion rate of 0.133 $mm^2\ h^{-1}$; an 8% reduction from PVA-FX gels with the same PVA and ferrous ion concentration. Adding approximately 1 hour to the practical time between irradiation and readout, this improvement, although incremental is an important first step in the exploration of functionalised polymers in 3D hydrogel dosimeters. To improve on these results, more efficient methods of synthesising XO-PVA are to be investigated. More broadly, the bonding of reporter molecules to gelling agents may be a general chemical approach to elimination of diffusion in Fricke gels.

5. Acknowledgments

This work was partly funded by Australian Research Council Linkage Project LP110100401 in conjunction with the Wesley Research Institute.

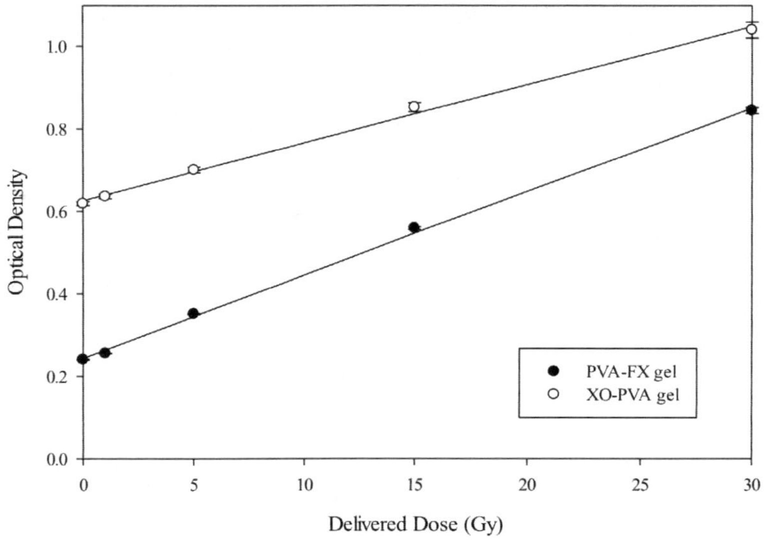

Figure 2. Radiochromic dose response of PVA-FX gel and XO-PVA gel.

6. References

[1] Harris P J *et al* 1996 *Phys. Med. Biol.* **41** 1745-53
[2] Baldock *et al* 2001 *Australas. Phys. Sci. Eng. Med.* **24** 19-30
[3] Schreiner L J 2004 *J. Phys. Conf. Ser.* **3** 9
[4] Kron T *et al* 1977 *Magn. Reson. Im.* **15** 211
[5] Chu K C *et al* 2000 *Phys. Med. Biol.* **45** 955-69.

Verification of micro-beam irradiation

Qiongge Li[1], Titania Juang[1], Rachel Beth[2], Sha Chang[2,3] and Mark Oldham[1]
[1]Duke University Medical Center, Durham, NC, USA
[2]Department of Physics & Astronomy, University of North Carolina at Chapel Hill, Chapel Hill, NC, USA
[3]Department of Radiation Oncology, University of North Carolina at Chapel Hill, Chapel Hill, NC, USA

E-mail: mark.oldham@duke.edu

Abstract. Micro-beam Radiation Therapy (MRT) is an experimental radiation therapy with provocative experimental data indicating potential for improved efficacy in some diseases. Here we demonstrated a comprehensive micro-beam verification method utilizing high resolution (50μm) PRESAGE/Micro-Optical-CT 3D Dosimetry. A small PRESAGE cylindrical dosimeter was irradiated by a novel compact Carbon-Nano-Tube (CNT) field emission based MRT system. The Percentage Depth Dose (PDD), Peak-to-Valley Dose Ratio (PVDR) and beam width (FWHM) data were obtained and analyzed from a three strips radiation experiment. A fast dose drop-off with depth, a preserved beam width with depth (an averaged FWHM across three beams remains constant (405.3um, sigma=13.2um) between depth of 3.0-14.0mm), and a high PVDR value (increases with depth from 6.3 at 3.0mm depth to 8.6 at 14.0mm depth) were discovered during this verification process. Some operating procedures such as precise dosimeter mounting, robust mechanical motions (especially rotation) and stray-light artifact management were optimized and developed to achieve a more accurate and dosimetric verification method.

1. Introduction

Conventional Radiation Therapy (such as Intensity Modulated RT (IMRT) and 3D Conformal RT) can achieve excellent target conformity (with uniform dose) and normal tissue sparing in many sites. However, some diseases such as gliosarcoma require better performance in control rates. Conclusions drawn from previous small animal irradiation experiments confirm the clinical potential and rationale for MRT treatment of deep-seated aggressive brain tumors in developing neurological tissue [3]. A compact MRT system for small animal irradiation has been developed at University of North Carolina (UNC, USA), based on carbon nanotubes (CNT) field emission technology. The CNT MRT system employs linear segmented cathode array and collimating alignment system. Although some preliminary 2D dosimetry measurements have been performed [3, 4], 3D dosimetry characterization has yet to be characterized.

2. Method and Materials

In order to characterize the micro-beam radiation delivered by the CNT MRT system, an ultra-high-resolution prototype 3D dosimetry system was constructed and optimized. This dosimetry system includes two parts: radiochromic dosimetry material (PRESAGE®) and an Optical-CT imaging system (Duke Micro Optical-CT Scanner (DMicrOS)) (shown in figure 1(a) and (b)) for readout purpose. The system can achieve 50μm isotropic high-resolution 3D reconstructed dosimetry data.

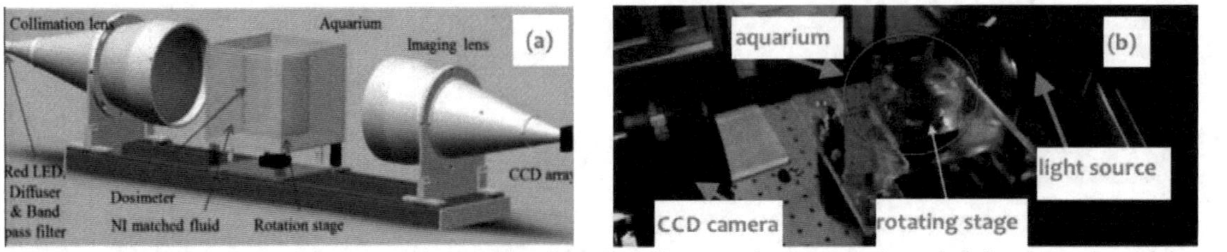

Figure 1. (a): Schematic of the prototype DMicrOS optical-CT scanning system (a scaled down version of the system introduced in [1]) (b): Photograph of the DMicrOS system in the lab – max field-of-view is ~3cm.

PRESAGE® was first introduced in 2006 as a material compatible with 3D dosimetry by optical-CT [2]. It consists of a polyurethane matrix, doped with a halogenated hydrocarbon free radical initiator and the leucodye leucomalachite green (LMG). Free radicals generated from the radiolysis of the halogenated hydrocarbon bond upon irradiation oxidize the LMG dye leading to a change in absorbance. The change in absorbance is linear with respect to the absorbed radiation dose [5].

A small PRESAGE® cylindrical dosimeter was irradiated by CNT MRT with nominal dimension (at the isocenter) of 65mm long x 300μm wide each beam at an energy of 160 kVp (nominal entrance dose was 32 Gy). The count image and OD image of post irradiated dosimeter are shown in figure 2 (a) and (b).

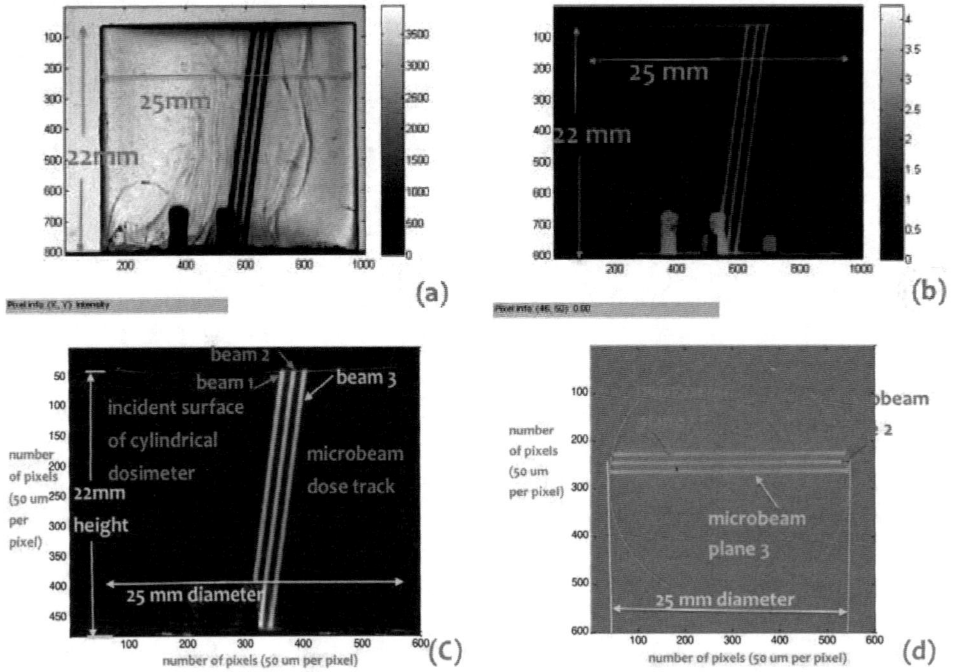

Figure 2. Three micro-planar beam irradiation pattern in the irradiated dosimeter (2.5cm diameter, 2.2 cm height). Top row shows a single projection image as counts (a) and ODs (b). Bottom row shows slices through the reconstructed 3D dose distribution sagittal view (c) and a mid-plane axial view (d).

The dosimeter was affixed and spatially registered to the rotating stage by two holes drilled into the dosimeter bottom that corresponded with metal registration keys on the stage. A weight was placed on top of the dosimeter to prevent dosimeter drift. The optical system was carefully aligned, and the fluid in the tank was matched to the refractive index of the dosimeter. 720 projections were acquired during a 360° rotation as required for a 50μm resolution 3D reconstructed image. A custom, in-house MATLAB-based GUI was employed for 3D reconstruction.

In order to reduce intensity variation brought by the edge of the dosimeter as well as to reduce the noise, a 200 pixel (10 mm) central section of the reconstructed data was averaged and saved as a 2D matrix, which can be used to analyze the data. 50μm (isotropic) 3D dosimetry was performed on all dosimeters using the DMicrOS. The Percentage Depth Dose (PDD), Peak-to-Valley Dose Ratio (PVDR) (taken from averaging three peaks' dose values divided by two valleys' dose values), and beam width (FWHM) were obtained as functions of depth.

3. Results and Discussion

3.1. DMicrOS basic testing

Basic testing results of the DMicrOS system are shown in figure 3, indicating Modulated-Transfer-Function (MTF), dynamic range, spatial resolution limit in object space, and Signal-to-Noise-Ratio (SNR).

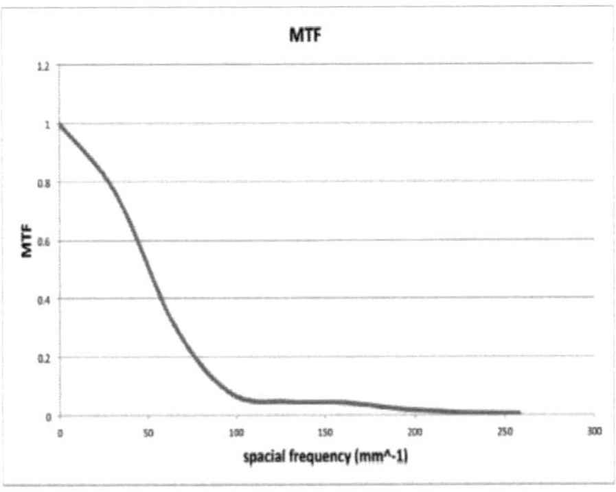

Dynamic range	Pixel size (object space)	Dark noise level	Signal to Noise Ratio	Largest Field- Of-View(FOV)
100~3900 counts	29.6 μm	Mean: 3.8 count σ=0.04μm:	46.6	47.6 mm (HFOV) x 35.8 mm (VFOV)

Figure 3. DMicrOS basic specifications

3.2. 3D dosimetric measurements

3.2.1. PDDs. The PDD measurements were normalized to values at 3mm depth. PDDs of the three beams were plotted as a function of depth (shown in figure 4(a)). Depth was measured along the beam path. A sharp dose drop-off (86.5% at a depth of 14 mm) can be seen in table 2.

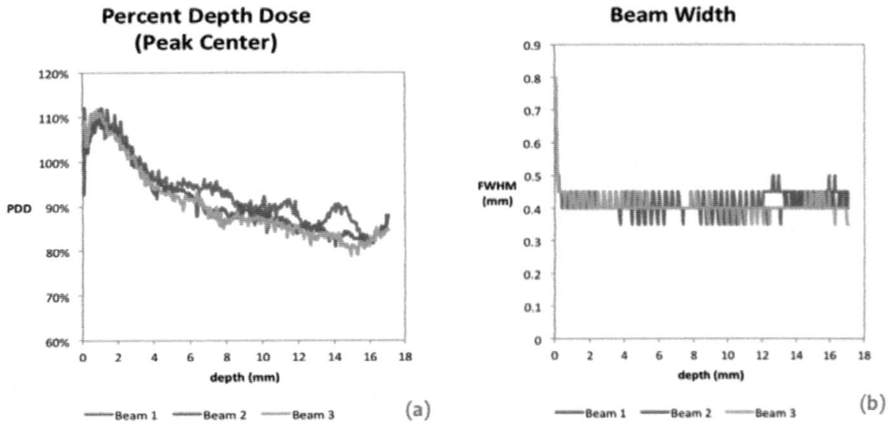

Figure 4. PDDs (a) and FWHMs (b) of three beams.

3.2.2. FWHMs. The FWHMs of the three beams were plotted as a function of depth (shown in figure 4(b)). Depth was measured along the beam path. The variance with different depths is very small (σ~13.2μm). (shown in table 1)

Table 1. FWHMs of three beams.

Entrance dose	Average FWHM (between depth of 3mm~14mm) across three beams	
	mean	Std (σ)
32 Gy	405.3μm	13.2μm

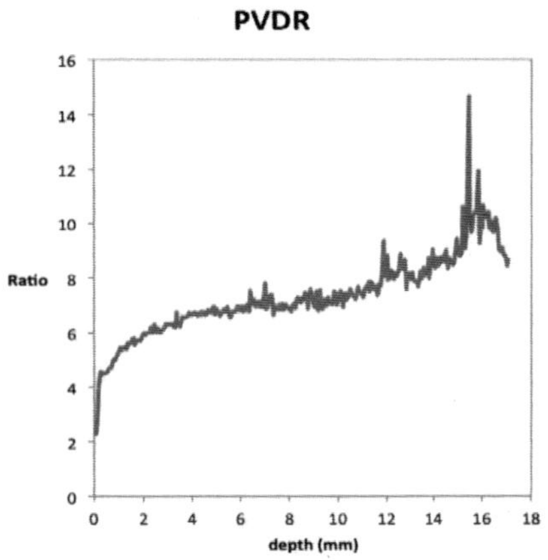

Figure 5. PVDRs of three beams.

Table 2. PVDRs.

Entrance dose	Average PVDR at depth 3mm	Average PVDR at depth 14mm
32 Gy	6.3	8.6

Table 3. Comparison 3D verification with 2D verification.

	Beam width	PVDR
UNC---2D	350	11.5
Duke---3D	350.5	21.9-60.4
%Difference	0.04%	15.6%-34.0%

3.2.3. PVDRs. The PVDR was calculated from averaging of maximum values of each peak divided by average of minimum values of each valley. PVDR was plotted as a function of depth (shown in figure 5). Depth was measured along the beam path. As depth increases, PVDRs also gradually increases from 6.3 (at depth of 3mm) to 8.6 (at depth of 14mm) shown in table 2.

3.3. Comparison with 2D verification

Comparing our results with the independent meawurements from UNC's [3,4] the beam width measurements are very consistent (within 0.04% - table 3). However, a large difference can be seen in PVDR measurements (15.6%-34.0% - table 3). The difference may arise in part from the fact that the 3D measurements were made at depth, while the 2D film measurements were made at the surface. Another potential source of uncertainty is the influence of stray light in this extreme irradiation geometry. According to the report by UNC research group, the PVDR values are consistent between their 1D (with their Nano-crystal dosimeter) and 2D (with EBT2 film) measurements (within ~1% difference). Further work is required to investigate the cause of the differences.

4. Conclusion

MRT dosimetry is extremely challenging due to the utilization of ultra-small field sizes and lower energy beams. The prototype ultra-high resolution optical-CT 3D dosimetry system introduced here showed some characteristics of micro-beams: A fast dose drop off, stabilized beam width and gradually increased PVDRs with the depth. These results are promising but suggest achieving high-resolution 3D dosimetry requires exceptionally careful procedures. Previous efforts to optimize fluid matching, standardize the operating procedures, improving the precision of dosimeter mounting and motion have shown promise. Next step involves developing methods for scanning/artifact correction.

5. References

[1] Thomas A *et al* 2011 *Phys. Med. Biol.* **56** 4433
[2] Adamovics J *et al* 2006 *Radiat. Protect. Dosim.* **120** 107-12
[3] Hadsell M 2013 UNC-Chapel Hill PhD dissertation
[4] Hadsell M *et al* 2013 *Appl. Phys. Lett.* **103** 183505
[5] Juang T *et al* 2013 *J. Phys.: Conf. Ser.* **444** 012029

Verification of motion induced thread effect during tomotherapy using gel dosimetry

Anneli Edvardsson[1], Anna Ljusberg[1], Crister Ceberg[1], Joakim Medin[2], Lee Ambolt[2], Fredrik Nordström[2] and Sofie Ceberg[2]
Medical Radiation Physics, Lund University, Lund, Sweden
Radiation Physics, Skåne University Hospital, Lund, Sweden

E-mail: Anneli.Edvardsson@med.lu.se

Abstract. The purpose of the study was to evaluate how breathing motion during tomotherapy (Accuray, CA, USA) treatment affects the absorbed dose distribution. The experiments were carried out using gel dosimetry and a motion device simulating respiratory-like motion (HexaMotion, ScandiDos, Uppsala, Sweden). Normoxic polyacrylamide gels (nPAG) were irradiated, both during respiratory-like motion and in a static mode. To be able to investigate interplay effects the static absorbed dose distribution was convolved with the motion function and differences between the dynamic and convolved static absorbed dose distributions were interpreted as interplay effects. The expected dose blurring was present and the interplay effects formed a spiral pattern in the lower dose volume. This was expected since the motion induced affects the preset pitch and the theoretically predicted thread effect may emerge. In this study, the motion induced thread effect was experimentally verified for the first time.

1. Introduction

Lung tumors may move due to breathing which can result in differences between the planned and delivered absorbed dose distributions [1]. When respiratory motion is present and not taken into account, for example with increased margins, dose blurring will occur. During advanced radiotherapy such as volumetric modulated arc therapy (VMAT) and intensity modulated radiation therapy (IMRT) breathing induced interplay effects may also occur, which are the differences between planned and delivered absorbed dose that arise from a mismatch between the target and the MLC because of motion [1]. This may cause unpredictable cold and hot spots in the delivered dose distribution. For helical tomotherapy (Accuray, CA, USA) [2] not only the target and MLC move, but also the treatment couch and the radiation source, making the interplay effects even more complex.

In tomotherapy, it is well known that the helical beam junctioning gives rise to a dose ripple described as a "thread effect" [3]. In order to minimize the inherent thread effect, the pitch is set so that the couch travel distance for one (or an integer number of) gantry rotation(s) equals the field width, adjusted for the effects of scattering and beam divergence [3]. In the case of a moving phantom, the combined travel distance of the couch and the phantom may invalidate the pitch setting, and a motion induced thread effect may emerge [4]. This effect can be substantial as long as the frequency of the periodic phantom motion is the same as (or a multiple of) the gantry rotation frequency. However, if the periods mismatch, or if there are random phase shifts, such as in real breathing motions, the effect will decrease [4]. Indeed, while motion induced dose blurring effects have been demonstrated in helical tomotherapy, both by film measurements [5-6] and diode detector arrays [7], thread effects have so far not been experimentally verified.

Interplay effects have previously been investigated for IMRT and VMAT using various methods [8-12]. In this study we take advantage of the very high spatial resolution in three dimensions that the polymer gel detector system provides to measure the motion induced thread effect during tomotherapy. In our previous work it has been shown that polymer gel is a feasible detector for 3D dose verification of dynamic radiotherapy [9, 13-15]. A definition of the interplay effect was proposed as the relative 3D absorbed dose difference between the volumes obtained from the measurement during phantom motion and the measurement to a stationary phantom convolved with the motion function of the moving platform [9].

The aim of this study was to investigate any breathing interplay effects during tomotherapy delivery and identify the expected motion induced thread effect using polymer gel dosimetry.

2. Material and methods

2.1 Polymer gel phantoms

The normoxic polyacrylamide gel (nPAG) used in this study contained 89% w/w ultra-pure deionised water, 3% w/w acrylamide, 3% w/w N,N'-methylenebisacrylamide, 5% w/w gelatine and 10 mM tetrakis(hydroxymethyl)-phosphonium chloride (chemicals from Sigma Aldrich, Germany). Magnetic resonance imaging (MRI) using a 1.5 T unit (Siemens Symphony) was performed about 24 hours post-irradiation. The mixing- and read-out procedures are described elsewhere [14]. MATLAB 7.4.0 was used for image processing and 3D rendering. The R2 data of the irradiated gel phantoms [16] was converted to relative absorbed dose using background subtraction and normalization in a region of homogenous dose [17].

One single batch of gel was used for all experiments. Identical 1.3 liters circular gel phantoms (Ø 10 cm) were used for both measurements during motion and in static mode. An identical un-irradiated gel phantom was used to acquire a background value. Gel vials irradiated to known absorbed doses were used to assure the linearity of the gel dose response.

2.2 Tomotherapy treatment plan and delivery

The treatment plan was calculated with TomoTherapy PlanningStation (Accuray, CA, USA) using 1688 MU (3 Gy) with a beam width of 2.5 cm and a planning modulation factor of 1.4. The total treatment time was 125.2 s with a gantry rotation period of 12 s. The pitch was set to 0.215 to minimize the inherent thread effect. To simulate tumour movement caused by respiratory motion, the gel phantom was positioned on a programmable motion device (Hexamotion, ScandiDos, Uppsala, Sweden) which was set to carry out sinusoidal motion in the superior–inferior (SI) direction with a peak-to-peak distance of 20 mm and a period of 6 seconds. Since the end-to-end distance was traveled in 3 s, this was expected to introduce a motion induced thread effect with an axial advance of 20 mm for a quarter of a complete turn. The delivery was repeated for a second gel phantom with the motion device turned off, i.e. the gel in static mode. The 90% isodose volume for the static gel was 22.6 cm^3.

2.3 Definition of interplay effect

The interplay effect was calculated as the relative 3D absorbed dose difference between the volumes obtained from the measurement during phantom motion and the measurement to a stationary phantom convolved with the motion function of the moving platform. The relative absorbed dose difference between the measured and convolved dose matrices was calculated by voxel-by-voxel subtraction as described earlier [9].

Table 1. The mean value and standard deviation (SD) of the comparison of the different dose distributions.

		Mean value ± 1 SD (%)
	Static gel vs TPS	1.0±1.0
95 % isodose volume	Static gel vs dynamic gel	5.5±3.2
	Dynamic gel vs convolved static gel	-0.5±2.2
	Static gel vs TPS	2.5±1.4
50 % isodose volume	Static gel vs dynamic gel	4.9±5.2
	Dynamic gel vs convolved static gel	0.0±4.6

3. Results and discussion

There was good agreement between the static gel and the TPS while there was a difference in the dose distribution between the dynamic and static gels and the dynamic and convolved static gels (table 1, figure 1). The expected dose blurring was clearly shown comparing the dynamic and static dose distributions. Differences still existed comparing the dynamic and convolved static dose distributions and hence interplay effects were present. These dose differences formed a spiral pattern in the lower dose volume, which was identified as the thread effect with an axial advance of 20 mm for a quarter of a complete turn (figure 2), which was predicted theoretically. The absorbed dose linearity test of the gel batch resulted in R^2= 0.998.

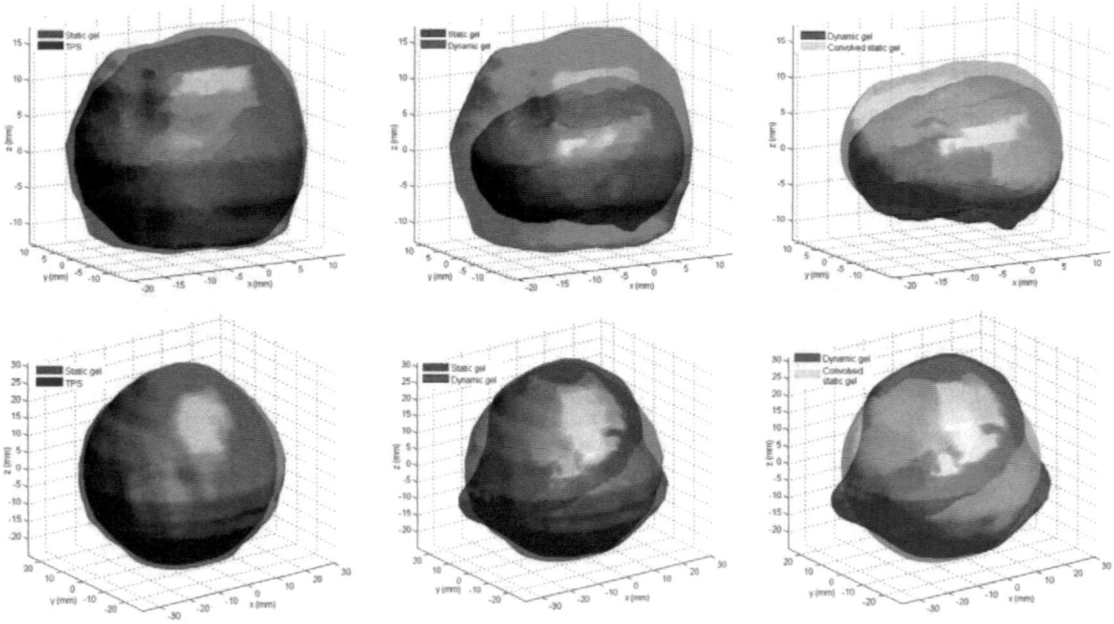

Figure 1. The 95 % isodose surfaces (top) and the 50 % isodose surfaces (bottom) for the static gel and TPS (left), the static gel and dynamic gel (middle) and the dynamic gel and the convolved static gel (right). The motion was induced in the z (SI) direction.

4. Conclusion

In this study, both the expected dose blurring effect but also breathing interplay effects were shown for a tomotherapy treatment of a small target in motion using gel dosimetry and a motion device. For volumes receiving lower absorbed dose over- and under-dosed volumes appeared in a spiral form. Thus the motion induced thread effect was experimentally verified for the first time.

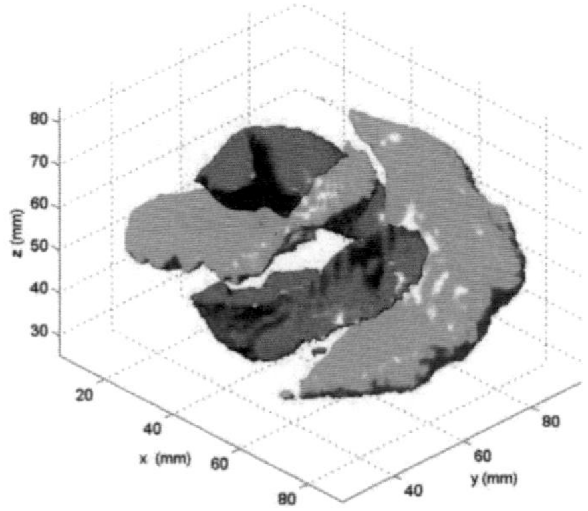

Figure 2. The surface of the voxels enclosed by the 50 % isodose volume with an absorbed dose difference of more than 5% (green) and less than -5% (red) between dynamic gel and convolved static gel, showing a spiral pattern.

5. References

[1] Bortfeld T *et al* 2004 *Semin. Radiat. Oncol.* **14** 41-51
[2] Mackie T R *et al* 1993 *Med. Phys.* **20** 1709-19
[3] Kissick M W *et al* 2005 *Med. Phys.* **32** 1414-23
[4] Kissick M W *et al* 2005 *Med. Phys.* **32** 2346-50
[5] Kanagaki B *et al* 2007 *Phys. Med. Biol.* **52** 243-55
[6] Kim B *et al* 2009 *Phys. Med. Biol.* **54** 5707-34
[7] Klein M *et al* 2013 *JACMP* **14** 52-61
[8] Thomas A *et al* 2013 *J. Phys. Conf. Ser.* **444** 012049
[9] Ceberg S *et al* 2013 *J. Phys. Conf. Ser.* **444** 012098
[10] Duan *et al* 2006 *Med. Phys.* **33** 1380-87
[11] Nelms *et al* 2013 *J. Phys. Conf. Ser.* **444** 012097
[12] Jensen *et al* 2012 *Phys. Med. Biol.* **57** 89-98
[13] Ceberg S *et al* 2008 *Phys. Med. Biol.* **53** 387-96
[14] Ceberg S *et al* 2010 *Phys. Med. Biol.* **55** 4885-98
[15] Ceberg S *et al* 2008 *J. Phys. Conf. Ser.* **250** 012051
[16] Baldock C *et al* 2010 *Phys. Med. Biol.* **55** R1-63
[17] Bjoreland A *et al* 2008 *Acta. Oncol.* **47** 1099-109

Assessment of radiochromic gel dosimeter based on Turnbull Blue dye for relative output factor measurements of the Leksell Gamma Knife® Perfexion™

P Kozubikova[1,3,4], J Solc[2], J Novotny Jr.[1,3,4], K Pilarova[1], J Pipek[1] and J Koncekova[1,3]
[1]CTU in Prague, Faculty of Nuclear Sciences and Physical Engineering, Department of Dosimetry and Application of Ionising Radiation, Prague, Czech Republic
[2]Inspectorate for Ionizing Radiation, Czech Metrology Institute, Radiova 1, Prague, Czech Republic
[3]Na Homolce Hospital, Prague, Czech Republic
[4]Department of Radiation Oncology, University Hospital Motol, Prague, Czech Republic

E-mail: petra.kozubikova@gmail.com

Abstract. The aim of the study was to perform assessment of radiochromic gel dosimeter based on Turnbull blue dye formed by irradiation (TB gel dosimeter) for measurement of ROFs for 4 mm and 8 mm collimators for the Leksell Gamma Knife Perfexion™. All measurements have been carried out using home-made spherical Perspex glass phantom of diameter 160 mm. TB gel dosimeters were scanned using homemade optical CT scanner. The results are compared with vendor recommended Monte Carlo calculated ROFs values of 0.814 and 0.900 for 4 mm and 8 mm collimators, respectively. The comparisons between the gel measurements and the treatment planning system (TPS) calculation are presented in the form of 2D isodoses for the central slices and 1D profile. Measured ROF 0.746 and 0.874 for 4 mm and 8 mm collimators respectively are in a reasonable agreement with vendor recommended values and measured relative dose distribution in a central slice and measured profiles of all shots show excellent correspondence with TPS.

1. Introduction

Radiochromic gel dosimeter based on Turnbull blue dye formed by irradiation (TB gel dosimeter) is an integral chemical dosimeter introduced by Solc *et al* (2007). Basic properties of the TB gel dosimeter are summarized in Solc *et al* (2009). Upon irradiation the gel changes colour from yellow to green and blue because of the formation of Turnbull blue dye $K_3[Fe^{II}Fe^{III}(CN)_6]$, insoluble in water, which has a wide absorption peak with a maximum at a wavelength of approximately 690 nm. It offers several advantages such as inhibited diffusion, linear response from 0 Gy up to at least 400 Gy, easy preparation, non-sensitivity to oxygen and non-toxic composition. The response can be evaluated using both cone- and laser-beam optical CT. The main disadvantage is lower sensitivity and gel ageing caused by the spontaneous interactions of ferric ions with organic compounds resulting in the creation of Turnbull blue not originally initiated by ionizing radiation.

2. Material and Method

A TB gel was prepared according to preparation procedure suggested by Solc *et al* (2009) and is composed of a gel matrix - 0.25 % (w/w) phytagel, 0.5 mM potassium ferricyanide and 0.5 mM ferric chloride compound dissolved in 1mM sulphur acid medium. Phytagel was choose as a gelling agent for TB gel dosimeter since it exhibits the best dosimetric performance as was demonstrated in Solc *et al* (2010). Despite the higher amount of scattered light in phytagel, spontaneous reduction of Fe^{3+} in gelatine makes phytagel the most suitable gelling agent. Prepared gel was poured into four glass cylindrical flasks (0.2 l, with diameter of 6 cm) available in our laboratory (plastic flasks could be also used) – three samples for irradiation and one sample as background. Gel dosimeters were stored in a refrigerator at 5°C. Solidification of TB gel took about 48 hours.

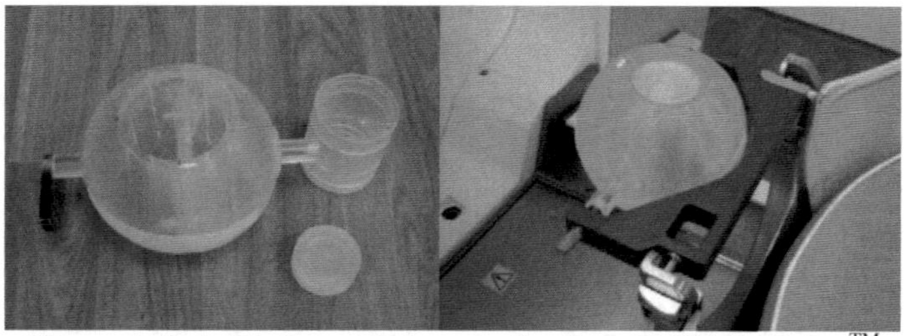

Figure 1. Gel phantom and phantom fixation at LGK Perfexion^{TM}

Irradiation was performed 48 hours after manufacturing on Leksell Gamma Knife (LGK) Perfexion^{TM} (Elekta Instruments AB, Stockholm, Sweden) with the dose rate of 2.216 Gy/min at the date of irradiation. Gel dosimeters were fixed using special phantom designed by Semnicka (2013). The gel phantom made out of Perspex glass is designed in the shape of the standard spherical phantom of diameter 160 mm provided by the manufacturer and enables positioning of gel samples into various well defined positions (figure 1). Its purpose is therefore to map dose distributions that can then be compared with the calculations of the treatment planning system (TPS) (Leksell GammaPlan Version 10.1; Elekta Instruments AB, Stockholm, Sweden). Experiment was designed in the same way as was proposed by Moutsatsos *et al* (2009). Gel dosimeters were subsequently irradiated with one shot, 16 mm, 8 mm and 4 mm, positioned in the isocentre to a prescription dose of 55 Gy to 50% isodose. For planning TPS uses a cubic matrix of 31x31x31 grid points with variable grid spacing Dose distribution was exported with a slice thickness of 0.5 mm.

TB gel dosimeters were scanned using homemade optical CT scanner – 16-bit astronomy CCD camera (G2-0402 type, Moravian Instruments, Czech Republic), stepper motor (65535 micro-steps per one 360° turn), light source (a red diode array emitting light at a peak wavelength of 660 nm). Data analysis was performed with in-house code developed in Matlab® suggested by Solc *et al* (2009). The reconstructed image had resolution of 4.2 px/mm.

The TPS dose distributions were exported from Leksell GammaPlan and imported into Matlab®. Background map was determined from unirradiated gel sample and subtracted from each irradiated dosimeter. For each shot the central value was averaged from 25 pixels for the 16 mm shot, 15 pixels for the 8 mm shot, and 6 pixels for the 4 mm shot. The comparisons between the gel measurements and the TPS calculations are presented in the form of 2D isodoses for the central slices and 1D *x* profiles. Furthermore the relative output factors (ROF) for the 8 mm and 4 mm collimators which were defined in this study as the ratio of the dose rate in the unit centre point (Moutsatsos *et al* (2009)), were calculated using the following equation:

$$ROF_i = \frac{D_i / t_i}{D_{16} / t_{16}} \qquad (1)$$

D_i is the dose delivered to the phantom center, and t_i is the time necessary to deliver the dose D_i. The irradiation time t_i for all collimators was taken from Leksell GammaPlan.

Figure 2. TB gel dosimeters - unirradiated sample and samples irradiated with 4, 8 and 16 mm collimators, respectively.

3. Results

The relative central *x* profiles for all shots are shown in figure 3 and the 2D relative dose distributions are shown in figure 4. The results of the ROF calculations are shown in Table 1. The table also summarizes vendor recommended values (Monte Carlo calculated) of 0.814 and 0.900 for 4 mm and 8 mm collimators, respectively. The discrepancy in the case of the 8 mm collimator was 2.9 % and in the case of the 4 mm collimator 8.4 %.

Collimator	Number of pixels in ROI	Average CCD absorbance in ROI (± σ)	$ROF_{TB\ gel}$	ROF_{LGP}	Difference (%) $(ROF_{LGP} - ROF_{TB\ GEL}) / ROF_{LGP}$
16 mm	25	194.1 (±0.9).10^{-3}	-	1	-
8 mm	15	189.0 (±0.2).10^{-3}	0.874	0.900	2.9
4 mm	6	178.9 (±0.1).10^{-3}	0.746	0.814	8.4

Table 1. Relative output factors measured by the TB gel dosimeter

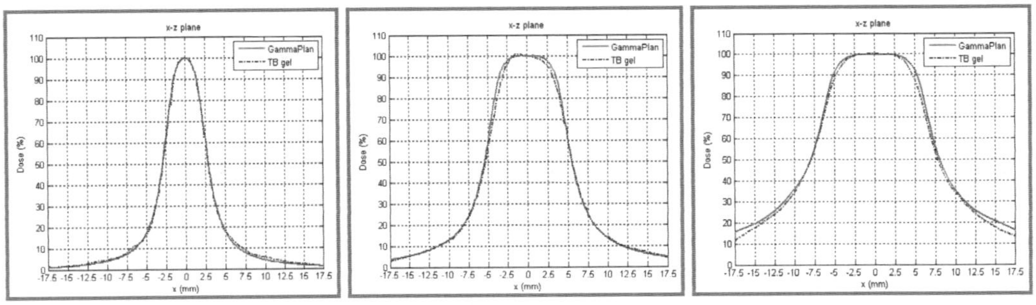

Figure 3. Comparison of relative central x profiles for all shot sizes (a, b, c - 4 mm, 8 mm, 16 mm collimators respectively) measured by the TB gel dosimeter and calculated by Leksell GammaPlan.

4. Discussion

The measured relative dose profiles agreed perfectly with the TPS calculations, larger discrepancies were observed only in low dose region. For 16 mm collimator low dose region (below 20% isodose) was not possible to evaluate due to scatter artefacts from glass wall of flask. For 16 mm collimator using flasks with larger diameter or walls from non-scattering material would allow measurement in low dose region. Since TB gel is not sensitive to oxygen, it can be filled into plastic flasks as well. Replacing glass flasks by Teflon with refractive index close to gel would provide less distorted image, especially near the edge of the projection.

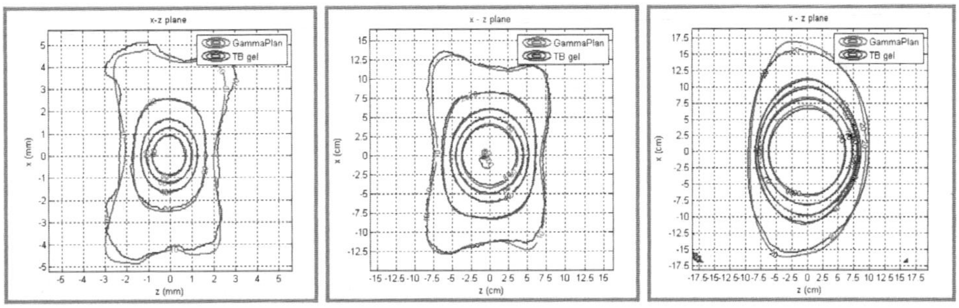

Figure 4. Comparison of 2D relative dose distributions. The red lines are Leksell GammaPlan calculations and the blue lines are gel measurements. Displayed isodoses: 10 (only for 4 and 8 mm collimators), 20, 40, 60, 80 a 100 %.

Measured ROF 0.874 and 0.746 for 8 mm and 4 mm collimator respectively are in a reasonable agreement with the TPS. The discrepancies are caused mainly by underestimated CCD intensities measured in the central part of 16 mm shot. Improved results would be obtained by using yellow light source (yellow diode array, emitting light at a peek wavelength of 590 nm) or delivering smaller dose to dosimeter. In case of 4 mm collimator dose response is averaged from a larger volume than it would be appropriate for the size of the 4 mm shot causing larger measurement error.

5. Conclusion

Measured relative dose distribution in central slice and measured profiles of all shots showed excellent correspondence with treatment planning system and makes TB gel dosimeter suitable for radiotherapy methods with steep dose gradients. Despite low sensitivity of the dosimeter, it appears to be a promising tool for three-dimensional dose measurements in high-dose-rate brachytherapy and stereotactic radiosurgery.

6. Acknowledgement

This work was supported by the Grant Agency of the Czech Technical University in Prague, grant No. SGS13/148/OHK4/2T/14.

7. References

[1] Solc J *et al* 2007 *Book of Abstracts: IEEE Medical Imaging Conf.* (New Jersey: IEEE Engineering in Medicine and Biology Society) 468
[2] Solc J *et al* 2009 *Phys. Med. Biol.* **54** 5095-101
[3] Solc J *et al* 2010 *J. Phys.: Conf. Ser.* **250** 012013
[4] Moutsatsos A *et al* 2009 *Med. Phys.* **36** 4277
[5] Semnicka J 2013 Clinical dosimetry using polymer gels - Application of magnetic resonance imaging *PhD thesis* Czech Technical University in Prague, Faculty of Nuclear Sciences and Physical Engineering, Department of Dosimetry and Application of Ionizing Radiation.

Feasibility on using composite gel-alanine dosimetry on the validation of a multiple brain metastasis radiosurgery VMAT technique

J F Pavoni[1], W F P Neves-Junior[2], M A Silveira[1], P A M M Ramos[2], C M K Haddad[2] and O Baffa[1]
[1]Departamento de Física, FFCLRP, Universidade de São Paulo, Av. Bandeirantes 3900, 14040-901, Monte Alegre, Ribeirão Preto, SP, Brazil
[2]Hospital Sírio Libanês, R. Dona Adma Jafet 91, 01308-000, Bela Vista, São Paulo, SP, Brazil

E-mail: baffa@usp.br

Abstract. This work presents an end-to-end test using a composite Gel-Alanine phantom, in order to validate 3-dimensionally the dose distribution delivered by a single isocenter VMAT technique on the simultaneous treatment of multiple brain metastases. The results obtained with the gel and alanine dosimeters are consistent with the expected by the treatment planning system, showing the potential of this multidosimetric approach and validating dosimetrically the multiple brain metastases treatment using VMAT.

1. Introduction

The addition of Stereotactic Radiosurgery (SRS) to the initial treatment of brain metastasis improves local control [1]. Although these results are well established for up to three lesions, there are promising data for a scenario of more than three and up to ten lesions [2]. Traditionally, multiple brain metastases are treated with SRS each one with its respective isocenter, taking about 20 minutes per lesion. With the advent of Volumetric Modulated Arc Therapy (VMAT) it is now possible to treat multiple metastasis using a single isocenter, in much less time, with more efficiency and comfort to the patient.

Nevertheless, the safe use of the VMAT technique on producing such complex volumetric dose distributions (DD) requires an extensive dosimetric validation [3]. To address this challenge, this work aims to study the feasibility of using a composite Gel-Alanine phantom on an end-to-end test, in order to validate 3-dimensionally [4] the DD delivered by a single isocenter VMAT technique on the treatment of multiple brain metastasis. The phantom is a cylinder filled with dosimetric gel (which plays a role of cortex material) and with alanine pellets (the brain metastasis) distributed along its volume.

2. Materials and Methods

2.1. Phantom preparation

The MAGIC-f gel dosimeter was used, its preparation was performed in a fume hood under normal levels of oxygen following the work of Fernandes *et al* 2008 [5] and applied in the same way as described in papers [6, 7].

Three plastic cylindrical phantoms with approximately 14.5cm of diameter and volume of 1.4l were used. On the first (Gel-Alanine), a PVC holder was used to position 12 alanine dosimeters distributed along the phantom volume. Alanine dosimeters were covered with paraffin film (Parafilm) to avoid contact with gel. The other two phantoms had gel only (Gel 1 and 2). The Gel-Alanine and Gel 1 received the VMAT treatment, and Gel 2 was used as reference in the MRI scanner. Calibration vials of gel were also prepared from the same batch and irradiated with doses of 1 to 12 Gy for the construction of calibration curves.

2.2. Treatment plan process and treatment

The Gel-Alanine phantom was immobilized with a frameless stereotactic thermoplastic head mask (BrainLab AG, Germany) to be attached to a stereotactic fiducial coordinate system. This whole setup was scanned on a Somaton Definition AS CT (Siemens Healthcare AG, Germany) with 1 mm slice thickness and distance, resulting on a final voxel size of 0.70 x 0.70 x 1.00 mm^3.

The CT data was imported on the iPlan treatment planning system (TPS), version 4.1 (BrainLab). The fiducial coordinate system was identified and accounted for. Among the 12 alanine pellets distributed inside the phantom, 5 lesion cores were delineated (term that will be adopted further on the text. Around each lesion core, a PTV was grown and a different dose prescription assigned: "PTV1" (1.5 cm diameter, 1.75 cm^3 volume, around a single pellet and dose of 6 Gy), "PTV2" (irregular C-shape with 2.8 cm and 1.0 cm major and minor axis, 2.9 cm^3 volume, around a single pellet and dose of 5 Gy), "PTV3" (1.0 cm diameter, 0.50 cm^3 volume, around a single pellet and dose of 9 Gy), "PTV4" (3 cm diameter, 14.1 cm^3 volume around three pellets and dose of 7 Gy) and "PTV5" (1.5 cm diameter, 1.75 cm^3 volume, around two pellets and dose of 8 Gy).

The isocenter on iPlan was defined as the geometric center of PTV 4, and the case was exported to Eclipse TPS, version 10 (Varian, USA) and ExacTrac 6D image guidance radiotherapy system (IGRT), version 5 (BrainLab). On Eclipse, a VMAT plan was created for a Novalis Tx linear accelerator equipped with a Millennium High Definition MLC, using 3 arcs: 1 full 360° coplanar arc and 2 partial (155°) arcs at couch angles 45° and 315°, the collimator angle was kept at 15°.

The plan was submitted to the institutional (Hospital Sírio-Libanês) patient specific quality control program, that consists of three separate measurements: (i) composite ionization chamber absolute dose measurement (CC13, 0.125 cm^3 volume, IBA Dosimetry GmbH, Germany), and (ii) composite planar EBT3 Gafchromic (ISP, USA) film dosimetry on a cubic solid water RW3 slabs phantom (IBA Dosimetry) and (iii) per field Portal Dosimetry (PD) technique, version 10.0 (Varian).

Before treatment delivery, the institutional machine pre-treatment quality control program was applied. After that, the phantom was positioned according to the institutional IGRT protocol, that consists of: (i) positioning through ExacTrac 6D correction (translations and rotations), (ii) verification using OBI kV Cone Beam CT technique (kV-CBCT). The doses due to these procedures were considered low enough to be disregarded (< 15 cGy at the phantom surface).

2.3. Magnetic Resonance Imaging and Gel Dosimetry Evaluation

Magnetic resonance images (MRI) of the three phantoms were acquired 1 day after irradiation using a 3T scanner (Phillips, Achieva). These images were centered in the head coil (where the phantoms were positioned) using a GRASE 3D multi spin-echo sequence with 8 echo times multiples of 35 ms, repetition time of 1000 ms and voxel size of 2 x 2 x 2 mm^3. The MRI field inhomogeneity was subtracted from the images of the irradiated phantom. The R2 DD maps were calculated in a software develop in MatLab®.

2.4. Alanine Dosimetry Evaluation

Alanine dosimeter cylinders (2.5 mm radius, 3 height) with 65.5 mg of mass were used. In some cases two and three pellets were stacked to simulate larger metastases. A group of dosimeters was irradiated in the same accelerator with the appropriate build up to produce a calibration curve. They were analysed in a Jeol JES-FA200 X-band Electron Spin Resonance spectrometer at ambient temperature.

3. Results and Discussion

3.1. Plan quality assurance results

(i) Ion chamber: The mean dose on the chamber volume calculated by Eclipse was 644.6 cGy and the measurement was 658.8 cGy (2.2% difference).

(ii) EBT3 composite dosimetry: performed on the coronal plane, at the same level as measured by the chamber (isocenter). The mean dose at a ROI defined on the calculated and on the measured dose matrices were 645.7 cGy and 653.2 cGy, respectively (1.6% difference). The 2%/2mm gamma analysis between them resulted in 99.8% of approved points (doses < 10 % ignored) (figure 1).

(iii) PD: performed per field, on original gantry angles, un-normalized, within jaw aperture. The gamma (2%/2mm) passing rates for each of the 3 arcs where: 97.9%, 98.7% and 98.4% (figure 1).

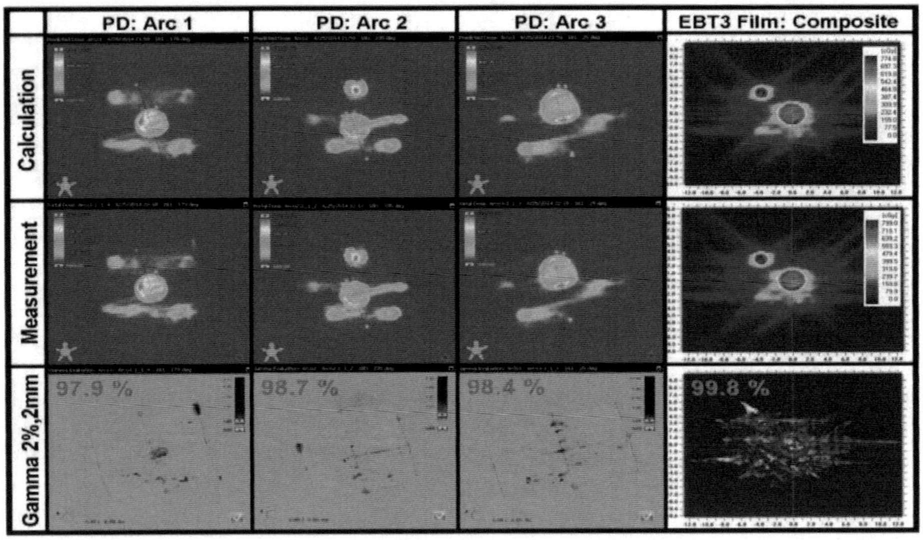

Figure 1. Quality assurance results: calculations (first row), measurements (second row) and gamma evaluations (2 %/2 mm) (third row) for each VMAT arc evaluated by PD (first to third columns) and composite film dosimetry on coronal plane (fourth column).

3.2. Gel and Alanine results

The association of the two detectors in the same measurement enables to evaluate DD 3D (with gel) and point doses (with alanine) inside target volumes and to compare this two methods in order to cross validate them.

For both irradiated gel phantoms, the calculated and measured DD for the 5 PTVs had good passing rates on gamma (figure 2). In the Gel-Alanine, the Parafilm packing trapped some air inside the gel, despite the maximum care on trying to minimize it. Thus, some inhibition of gel response occurred and can be seen especially in PTV5, where some points fails in the gamma map.

The alanine point dose measurements for the lesion cores 1, 2, 3, and 5 are in good agreement with the mean dose expected by Eclipse (TPS) with differences of less than 5% (table 1). For lesion core 4 a leak of the gel into the dosimeter was found that impeded a precise measurement of the dose at this point. Gel mean doses in all lesion cores are also in agreement with the expected value by Eclipse.

The results show good consistency among all the dosimetric methods currently used to verify DD in VMAT and the new methods using Magic-f gel and ESR/alanine

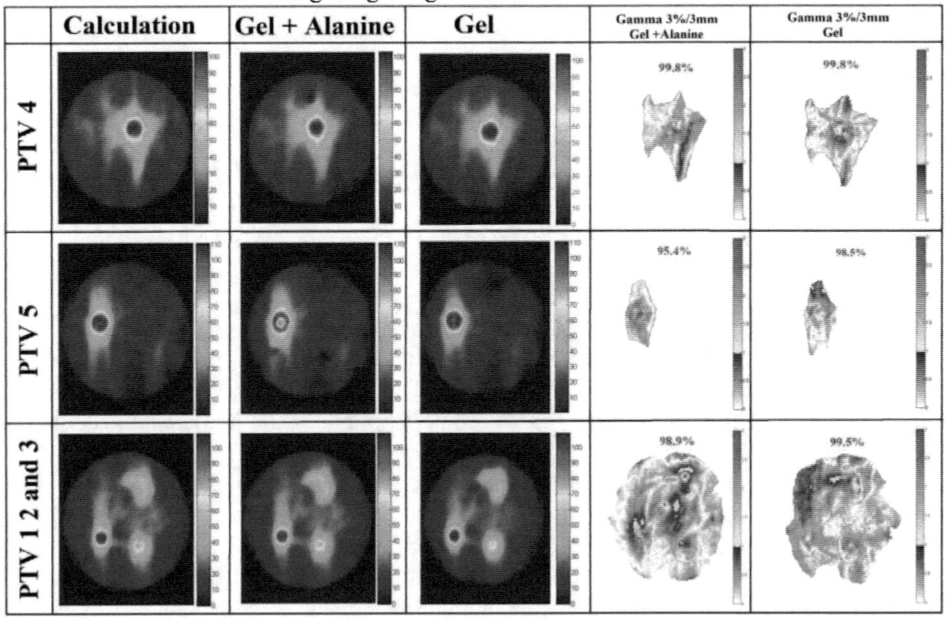

Figure 2. Calculated DD by Eclipse (first column), measured DD by gel+alanine (second column), measured DD by gel (third column) and gamma analysis, 3%/3mm (fourth and fifth columns) for: PTV 4 (first row), PTV 5 (second row) and PTVs 1, 2and 3 (third row). Gamma analysis used threshold of 30%, 25% and 10% for PTVs 4, 5 and (1, 2, 3).

Table 1. Eclipse expected mean dose, alanine doses and gel mean doses as well the percentage difference between the measured and expected values for all lesion cores.

	Lesion 1	Lesion 2	Lesion 3	Lesion 4	Lesion 5
Eclipse Mean dose (Gy)	6.81	5.43	10.78	7.86	9.00
Alanine doses (Gy)	6.55	5.38	10.29	-----	8.64
Difference (%): Alanine-TPS	3.82	0.09	4.54	-----	4.00
Gel Magic-f doses (Gy)	6.93	5.59	10.77	7.73	8.92
Difference (%)	1.7	2.29	0.09	1.65	0.8

3. Conclusions

The Gel-Alanine composite phantom enabled the dosimetric validation of multiple brain metastases treatment using VMAT. Both Alanine and Gel have its own pros and cons, but, when combined, they complement each other and can be considered as an almost ideal tool for this application, featuring: high resolution 3D dose information, reliable absolute dose measurement, dose rate independence, water equivalence, redundant cross validation, flat energy and dose linearity responses.

4. References

[1] Andrews D W *et al* 2004 *Lancet* **363**: 1665
[2] Serizawa T *et al* 2010 *N J Neurooncol* **98** 163
[3] Baldock C 2006 *J. Phys.: Conf. Ser.* **56** 14
[4] Baldock C *et al* 2010 *Phys. Med. Biol.* **55** R1
[5] Fernandes J P *et al* 2008 *Phys. Med. Biol.* **53** N1
[6] Pavoni J F *et al* 2012 *Med. Phys.* **39** 2877
[7] Fernandes J F *et al* 2012 *Radiat. Meas.* **47** 107

Journal of Physics: Conference Series **573** (2015) 012051 doi:10.1088/1742-6596/573/1/012051

A small animal image guided irradiation system study using 3D dosimeters

Xin Qian[1], John Admovics[2,3] and Cheng-Shie Wuu[1]
[1]Department of Radiation Oncology, Columbia University, New York, NY, USA
[2]Department of Chemistry and Biology, Rider University, Lawrenceville, NJ, USA
[4]Heuris Pharma, LLC, 412 Sunset Rd., Skillman, NJ, 08558 USA

E-mail: csw6@columbia.edu

Abstract. In a high resolution image-guided small animal irradiation platform, a cone beam computed tomography (CBCT) is integrated with an irradiation unit for precise targeting. Precise quality assurance is essential for both imaging and irradiation components. The conventional commissioning techniques with films face major challenges due to alignment uncertainty and labour intensive film preparation and scanning. In addition, due to the novel design of this platform the mouse stage rotation for CBCT imaging is perpendicular to the gantry rotation for irradiation. Because these two rotations are associated with different mechanical systems, discrepancy between rotation isocenters exists. In order to deliver x-ray precisely, it is essential to verify coincidence of the imaging and the irradiation isocenters. A 3-D PRESAGE dosimeter can provide an excellent tool for checking dosimetry and verifying coincidence of irradiation and imaging coordinates in one system. Dosimetric measurements were performed to obtain beam profiles and percent depth dose (PDD). Isocentricity and coincidence of the mouse stage and gantry rotations were evaluated with starshots acquired using PRESAGE dosimeters. A single PRESAGE dosimeter can provide 3-D information in both geometric and dosimetric uncertainty, which is crucial for translational studies.

1. Introduction

In order to mimic modern clinical radiotherapy in small animal disease models, several practical conditions impose that a dedicated small animal research platform is required. This means that radiation beams not only need to be downscaled in geometry, but also in energy. Furthermore, the technical precision required to deliver millimetre sized beams to regions within a small animal exceeds the tolerances needed in patient radiotherapy [1]. A few such small animal research platforms have been developed, the one used in this study is SARRP. It is a novel and complete system capable of delivering multidirectional, kilo-voltage radiation beams to targets in small animals under CBCT image guidance [2].

The quality assurance for SARRP includes measurements of absolute dose rates, beam profiles, and central axis percent depth dose (PDD) curves. In this study three dosimetric tools were employed: a farmer type ionization chamber (N30013, PTW), radiochromic EBT2 films, and PRESAGE 3-D [3] dosimeters with optical CT scanner (OCTOPUS, MGS Inc.). The ion-chamber was used to obtain absolute dose rates. Films were used to determine beam profiles and PDD for all fields. Film PDD measurements were challenging because of the difficulty of precise alignment to the small field. In addition, obtaining accurate data using films requires following a strict scanning protocol. Independent measurements were also performed using PRESAGE 3-D dosimeters with optical CT scanner [4-8].

PRESAGE dosimeters are much more efficient compared to film because all parameters of interest can be obtained from a single dosimeter and a single scan, and set-up-errors are negligible because the entire dose distribution is captured in the dosimeter.

2. Materials and Methods

In SARRP the gantry isocenter is at 35 cm from the focal spot. The combination of gantry rotation and mouse stage motion facilitates the dose delivery. The collimator accommodates a range of brass inserts: two square ones 10x10 and 5x5 mm^2, and two circular ones 1 and 0.5 mm in diameter. All irradiations in this work were performed with a 225 kVp and 13 mA beam. For CBCT imaging the gantry and a flat panel detector were set at opposite horizontal positions of 90° and 270° respectively. 360 projection images were obtained by rotating the mouse stage. CBCT imaging was performed with 50-100 kVp and 0.8mA beams.

The output was measured using a calibrated ion-chamber following the AAPM task group report 61 [9]. The field size was $17x17cm^2$ and the depth and source-surface distance (SSD) were 2 and 33 cm, respectively. The output calibration phantom assembly consisted of three plastic water slabs (in order from top to bottom of stack): 20x20x1 cm^3, 20x20x2 cm^3 (with a chamber hole bored out at 1 cm depth), and 20x20x1 cm^3.

EBT2 films were calibrated under the same condition as ion-chamber before it was used for quality assurance measurement. The calibration consisted of 15 dose points spanning 0 to 600 cGy. The exposure times to best approximate the target doses were determined from ion-chamber dose rate measurement. The exposure times were then applied to the ion-chamber measured dose rate to determine the absolute doses to be delivered to films. After film calibration, beam profiles and PDD values for 10x10 mm^2 field were measured using films loaded on the phantom stack at the following depths: 0 (surface dose measurement), 5, 10, 15, 20, 30, 40 and 50 mm. For comparison, PDD curves were also measured from films aligned vertically along the central beam axis using side lasers. Isocentricities of the mouse stage and gantry rotations were evaluated using starshots from films.

The same measurements were performed using four cylindrical PRESAGE dosimeters with 6 cm diameter and 7 cm height. PRESAGE dosimeters were approximately placed at the center on mouse stage with guidance of crossed lasers. The first dosimeter was used to establish a response curve. Dose rate for a given depth and field size can be calculated from the film calibration. Therefore dose delivered to the dosimeter can be calculated for that depth and field size with given exposure time. The linear optical density and dose response of PRESAGE dosimeter was confirmed. The rest of the dosimeters were used to measure beam profiles, PDD, and coincidence of the mouse stage and gantry rotation isocenters. The readouts of the irradiated PRESAGE dosimeters were performed with the optical CT scanner, and data were reconstructed with a resolution of 1 mm^3 voxel size using an in-house program (FBP) written in Matlab. The data readout time and reconstruction time are ~10 hours and ~1.5 minutes respectively.

3. Results and Discussion

From Ion-Chamber measurement, the output at isocenter is 3.9 Gy/min. A calibration curve was obtained from films. Beam profiles for 10x10 mm^2 field were measured at three depths using films orientated perpendicular to the beam, as shown in figure 1 (A). 20%-80% penumbra value is 0.5mm. Full-width-half-maximum (FWHM) of the beam profiles slightly increases over depths due to scattering, the FWHM difference is 0.5mm between depths 0 and 20mm. Figure 1 (B) shows measured PDD for 10x10 mm^2 field. The solid curve is from film vertically aligned with beam central axis. The markers are from the axial films at multi-depths, which are considered as the standard method for measuring PDD. The downside of axial film method is time consuming. The PDD measurements using films highlight the need for the 3-D dosimeters, which avoids the alignment error issues and is very efficient. The isocentricity measurements using films for gantry rotation and mouse stage rotation are summarized in table 1.

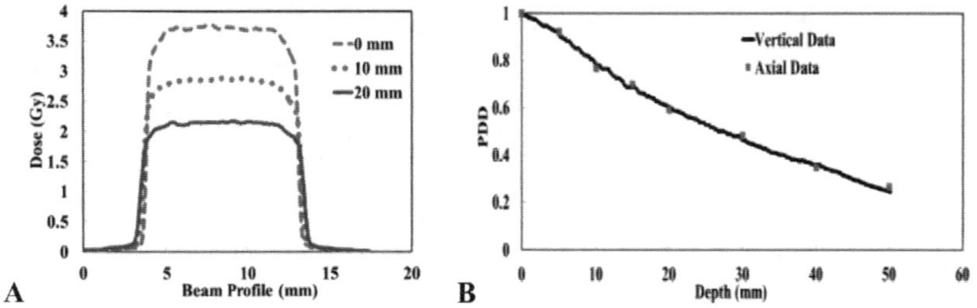

Figure 1. A. Beam profiles at depths of 0, 10 and 20 mm for 10x10 mm^2 square field; B. The PDD curve obtained using films.

Figure 2 shows the linear optical density and dose response at depth of 20 mm in the PRESAGE dosimeter.

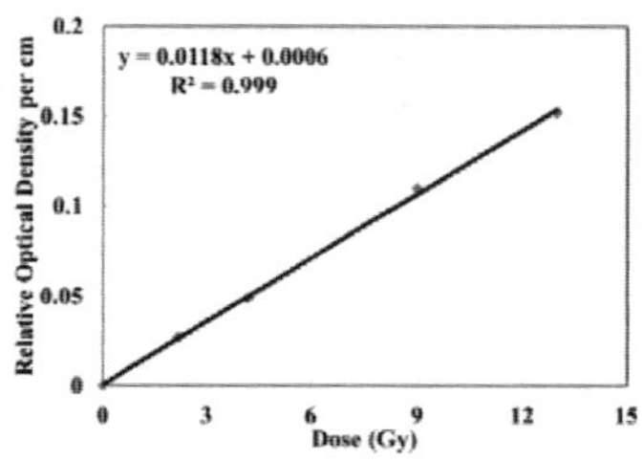

Figure 2. PRESAGE dosimeter calibration curve determined from four dose levels.

Figure 3 (A) shows the central slice of the reconstructed dosimeter image. The dashed lines indicate the locations of two illustrative axial slices at the depths of 10 and 20mm. Figure 3 (B) shows the axial slice at the depth of 20mm. Beam profiles along central cross lines of 10x10 mm^2 field are shown in figure 3 (C). The 20%-80% penumbra value is 1.4mm, which is larger than the one obtained using film. The FWHM difference is 0.25mm between depths 10 and 20mm.

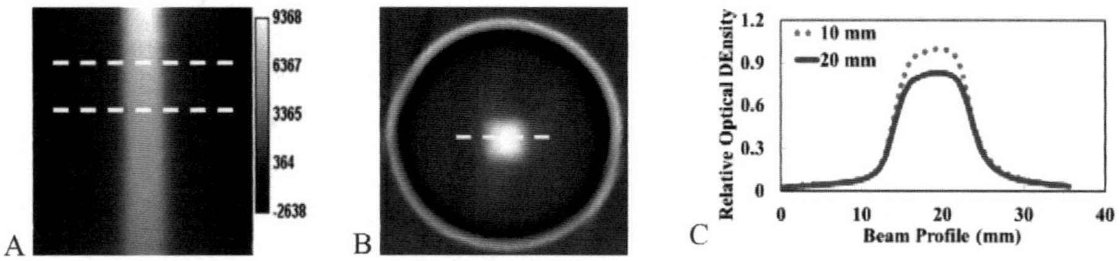

Figure 3. A. Central slice of reconstructed dosimeter image; B. Axial slice at the depth of 20mm; C. Beam profiles along central cross lines of the field at depths of 10 and 20mm.

Figure 4 compares the PDD curves obtained using film and dosimeter irradiated with 10x10 mm^2 field. PDD from film and PRESAGE dosimeter agree within 3% up to the depth of 3cm.

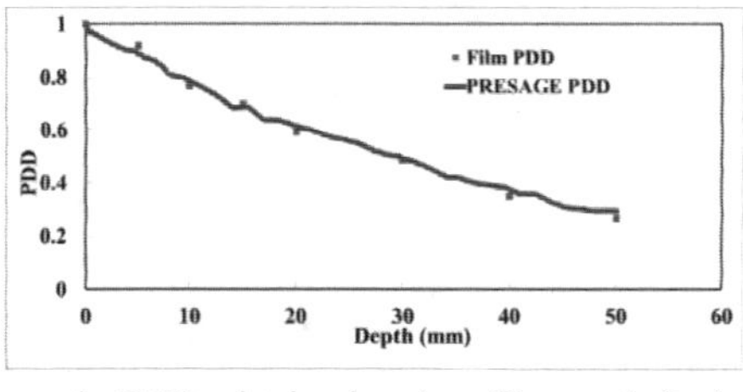

Figure 4. PDD obtained using film and dosimeter irradiated with 10x10 mm^2 field.

Figure 5 (A) shows the irradiated dosimeter for evaluation of isocentricity and coincidence of mouse stage and gantry rotations. Figures 5 (B-C) show beam trajectories delivered at 5 different angles in both irradiation and CBCT imaging coordinates. Table 1 summarizes the obtained results from films and PRESAGE dosimeters.

Figure 5. A. Irradiated dosimeter for both gantry and mouse stage isocentricity measurement; B. Beam trajectories from gantry rotation; C. Beam trajectories from mouse stage rotation.

Table 1. Comparison between films and PRESAGE dosimeters

	Isocentricity of irradiation coordinate (mm)	Isocentricity of CBCT imaging coordinate (mm)	Experiment setup time (mins)	Coincidence of Irradiation and imaging isocentricity
Dosimeters	1	1.02	10	0.8
Films	1.13	1	90	NA

This study presents a comprehensive commissioning method using 3-D PRESAGE dosimeters. Both of film and PRESAGE dosimeter measurements were found to be valuable to ensure accurate and comprehensive commissioning. Good agreement was observed between film and PRESAGE dosimeter measurements. However, film measurements are time consuming due to the difficulty of alignment in small fields, and labour intensive when multi-plane data are required. The PRESAGE dosimeter can be more useful in precise verification of targeting for a small animal irradiation research.

4. References

[1] Tryggestad E *et al* 2009 *Phys. Med. Biol.* **54** 5341-7
[2] Wong J *et al* 2008 *Int. J. Radiat. Oncol. Biol. Phys.* **71** 1591-9
[3] Gorjiara T *et al* 2011 *Med. Phys.* **38** 2265-74
[4] Olding T *et al* 2011 *J. Med. Phys.* **36** 3-14
[5] Clift C *et al* 2010 *Phys. Med. Biol.* **55** 1279-3
[6] Xu Y *et al* 2010 *Med. Phys.* **37** 861-8
[7] Newton J *et al* 2011 *Med. Phys.* **38** 6754-62
[8] Bosi S *et al* 2009 *Phys. Med. Biol.* **54** 275-83
[9] Ma C M *et al* 2001 *Med. Phys.* **28** 868-93

Investigation of a low-cost optical-CT system with minimal refractive index-matching fluid

S Bache[1], J Malcolm[1], J Adamovics[2] and M Oldham[3]
[1]Duke University Medical Physics Graduate Program, 2424 Erwin Rd Suite 101, Durham, NC 27705, USA
[2]Department of Chemistry and Biology, Rider University, Lawrenceville, NJ 08648, USA
[3]Department of Radiation Oncology Physics, Duke University Medical Center, Durham, NC 27710, USA

E-mail: mark.oldham@duke.edu

Abstract. Optical computed tomography (optical-CT) is a method for visualizing 3-dimensional dose distributions in radiochromic dosimeters. Projection images are acquired by collimating a visible light point source into parallel-beam geometry and imaging differential absorption through the sample dosimeter. Practical challenges involved in optical-CT imaging were addressed through the investigation of an in-house Fresnel-based optical-CT system with considerably less refractive index-matching fluid. The "DFOS" (Duke Fresnel-based Optical-CT System) system differed from current optical-CT systems by replacing cumbersome convex telecentric lenses with a lighter and much less expensive Fresnel system. A second major modification was the replacement of the refractive index-matching fluid bath with a solid polyurethane tank. PRESAGE radiochromic dosimeters were irradiated with orthogonal parallel-opposed treatments and dose distributions were readout by the DFOS system and compared to both treatment planning software prediction and two other in-house optical-CT systems. Gamma index passing rate at the 3%/3mm threshold in relation to Eclipse treatment planning software for the treatment was 92.2%%, compared to 96.8% and 95.6% for two other systems featuring a traditional setup. The DFOS system showed promise for 3D dosimetry, but the performance is still substantially inferior at present to the gold-standard systems.

1. Introduction

Optical computed tomography (optical-CT) is a method for reconstructing linear attenuation coefficients of visible light, the visible light analogue of X-ray CT [1]. Optical-CT is used in the imaging of 3D dose distributions [2] given by radiotherapy treatment machines to radiochromic plastic dosimeters such as PRESAGE (Hueris Pharma, Skillman, NJ). In order to image and reconstruct distributions using optical-CT, planar projections must be made of incident visible light attenuation in a sample both before and after irradiation.

One of the distinct challenges in optical-CT is the method for directing incident light through a sample and into a detector or CCD array in a straight path with minimal deviation due to refraction at the air/dosimeter interfaces. Telecentric lenses may be used to achieve parallel-beam geometry for the incident light, while immersing the dosimeter in a fluid of similar refractive index minimizes refraction.

For this work, two questions were addressed: (1) Can the bulky, expensive, telecentric lenses be replaced by a more manageable Fresnel lens system? And (2) can the cumbersome refractive index-matching fluid be minimized utilizing a solid tank concept [3, 4]? This work introduces the Duke Fresnel dry Optical-CT System (DFOS) system that attempts to address both these questions.

2. Methods and Materials

2.1. Modification of traditional Optical-CT systems

The setup and operation of the DLOS optical-CT system (Duke Large Optical-CT Scanner) has been well documented [5, 6]. Here we present a new derivative system called DFOS (Duke Fresnel Optical-CT Scanner). This system has two major differences. First, the convex telecentric lenses are replaced with Fresnel lenses (Light Works, Toledo, OH), resulting in a cost savings of over 90% (~thousands of dollars to hundreds), and a lens weight reduction from ~7kg to 500g. Fresnel lenses have inherent non-uniformity due to the Fresnel grooves, and a moiré effect is noticeable when imaging with a pair of lenses in a telecentric setup. Our rationale was that it may be possible to correct for Fresnel imperfections through the pre and post-irradiation scan information [6]. The second key modification is the replacement of the fluid bath with a solid tank bath which enables a reduction of >90% of the amount of refractive matching fluid. Refractive index matching fluid is used in optical-CT to allow incident light to pass through the dosimeter in a parallel path without refraction at the interface between two media. For the DFOS system, the glass aquarium was replaced with a 17x17x17cm^3 polyurethane cube with a cylindrical bore of diameter 11.5cm and height, allowing for placement of an 11cm diameter x 10cm dosimeter. The 2.5mm air gap surrounding the dosimeter requires just ~90 cm^3 of fluid of match the refractive index between the polyurethane tank (RI = 1.5) to PRESAGE dosimeters (RI = 1.47-1.51), compared to ~13500 cm^3 of fluid for the DLOS system.

For all DFOS image acquisitions shown here, an incident light image or "flood" field was purposefully omitted. This assumes that the LED output, and therefore incident light, remains constant between pre- and post-irradiation scans. In the absence of a dosimeter, the "dry" polyurethane tank essentially becomes an air-gap, refracting all incident light out of the CCD field-of-view and making flood field acquisition impossible. Further investigation into the LED consistency and the validity of flood field omission is required.

2.2. PRESAGE dosimeters and Treatment Plans

To evaluate the DFOS system, a rectangular parallel-opposed treatment was delivered to a cylindrical PRESAGE radiochromic dosimeter (diameter 11cm). The dose distribution was then imaged in the gold standard DLOS, DMOS and prototype DFOS systems immediately sequentially, at 24h post irradiation. Prior in-house work indicated that the optimal dose-readout time for the DEA PRESAGE formulation to be 3-24 hours post-irradiation – this formulation has been found to be stable within 2% over this time period. To generate the treatment plan, CT scans were taken of the dosimeter and imported to the Eclipse Treatment Planning Software (Varian Medical Systems, Palo Alto, CA) for dose prediction. The treatment consisted of two orthogonal sets of parallel-opposed 4cm x 10cm beams, delivered to a cylindrical 11cm diameter x 10cm PRESAGE dosimeter. Dose was configured in order to have 3 distinct dose regions by delivering 200 monitor units each for the right/left lateral and left/right lateral beams (low dose) and 300 monitor units each for the anterior/posterior and posterior/anterior beams (medium dose), leaving a high dose cube in the center of the distribution. Projection images were taken over 360-degrees at 1-degree increments with the Fresnel-lens system (DFOS) as well as two conventional systems (DLOS and DMOS) both before and after irradiation and reconstructed at 2mm resolution with a custom MATLAB GUI (figure 1). For this preliminary study, relative dose was converted to absolute dose by making a point measurement in the Eclipse predicted dose distribution and scaling the DFOS reconstructed dose accordingly. This step was justified as PRESAGE has been shown to exhibit a strong linear optical density change with absorbed dose [7, 8].

DFOS reconstructed dose was compared against Eclipse dose prediction by calculating the gamma index with 3%/3mm criteria.

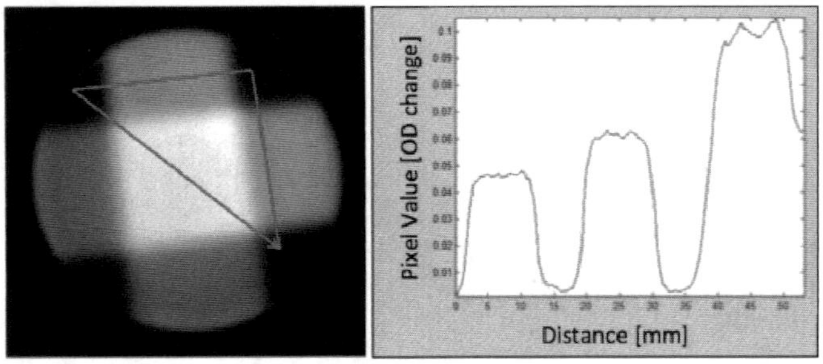

Figure 1. Axial slice from 3D reconstruction of optical density change in treated dosimeter with 2mm resolution (left), with line profiles through the 3 dose regions (right). The dosimeter is 11cm in diameter, and the radiation fields are 4x10cm.

3. Results

The 3D gamma pass rates for the DFOS, DLOS, and DMOS systems, when evaluated against the known Eclipse distribution, were 92.2%, 95.6%, and 96.8%, respectively (figure 2). The DLOS and DMOS distributions were highly consistent (99.0% gamma passing rate when compared to each other). This result is an important demonstration of the reproducibility and reliability of our gold-standard scanning systems. Figure 3 shows comparative line profiles through the dose distributions of each system. The DFOS distribution was observed to be significantly noisier, and contain substantial artifacts that were not fully removed by the pre and post-irradiation correction. A surprise observation was a rounding of the penumbral dose gradients that was observed in all three systems (figure 3). This effect has not been seen at this magnitude in scans of other Presage dosimeters, and may be specific to this experimental formulation. The cause is unknown at this point, and is under investigation. Similarly both DLOS and DMOS distributions showed elevated dose regions at very low dose levels, as seen in the tails of the profiles in figure 3. This effect has also not been seen in other dosimeters, and is possibly another feature of this formulation.

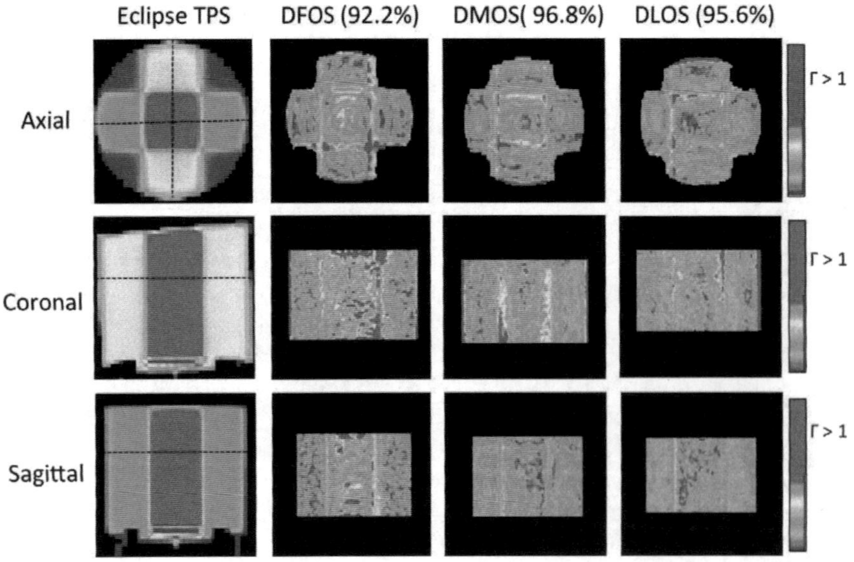

Figure 2. Orthogonal views through 3%/3mm gamma maps evaluated between measured and calculated dose distributions of the same dosimeter scanned by different scanners: DFOS (column 2), DMOS (column 3), and DLOS (right). 3D passing rates are given atop the columns. The eclipse dose is show in 1st column.

Journal of Physics: Conference Series **573** (2015) 012052 doi:10.1088/1742-6596/573/1/012052

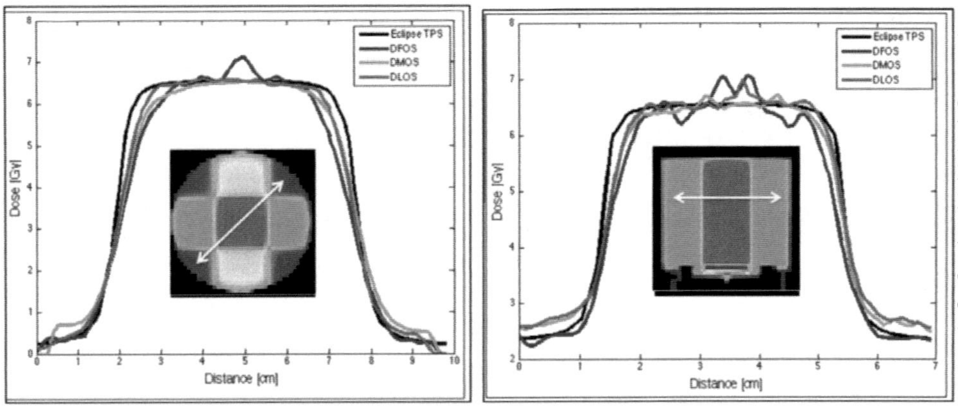

Figure 3. Dose line profiles comparing Eclipse (black), DFOS (blue), DMOS (green), DLOS (red) through the dose distribution of figure 1 and 2.

4. Conclusion

The DFOS scanner incorporates two major design changes from our prior scanners: low-cost Fresnel lenses, and a fluid saving solid tank. Both have dramatic implications for the cost and efficiency of optical-CT 3D dosimetry, provided no significant loss of accuracy is incurred. The Fresnel lenses introduced substantial ring and other flood field artifacts when compared to the DLOS systems. Some correction was possible using pre and post-irradiation correction strategies, but extremely robust apparatus is required to avoid any change in alignment between the two. The huge reduction (>90%) in the amount of required RI fluid associated with the solid-tank, makes fluid matching and general scanning much easier. It may also (in some cases) reduce noise in the floods.

Overall, first attempts for imaging 3D dose distribution with an optical-CT system with minimal fluid and Fresnel lenses is reasonably promising. A 92.6% gamma pass rate was achieved with a simple 4 field irradiation. This value was notably lower than either DLOS or DMOS (both >95%). The reduced agreement is attributed to increased noise of the DFOS and also the presence of other artifacts (e.g. more significant penumbral rounding). Further investigation is warranted to see how close DFOS can approach to DLOS and DMOS.

5. References

[1] Oldham M 2006 *J. Phys.: Conf. Ser.* **56** 58-71
[2] Baldock C *et al* 2010 *Phys. Med. Biol.* **55** R1-63
[3] Oldham M *et al Med. Phys.* **40** 051701
[4] Doran S *et al* 2012 *Phys. Med. Biol.* **57** 665-83
[5] Thomas A *et al* 2010 *J. Phys.: Conf. Ser.* **250** 1-5
[6] Thomas *et al* 2011 *Med. Phys.* **38** 4846
[7] Adamovics J *et al* 2006 *Radiat. Prot. Dosim.* **120** 107-12
[8] Gorjiara T *et al* 2011 *Med. Phys.* **38** 2265-74

8th International Conference on 3D Radiation Dosimetry (IC3DDose) IOP Publishing
Journal of Physics: Conference Series **573** (2015) 012053 doi:10.1088/1742-6596/573/1/012053

Validation of 3DVH estimated DVH metrics for prostate VMAT plans

Sankar Arumugam[1], Aitang Xing[1], Tony Young[1], David Thwaites[2] and Lois Holloway[1,2,3,4]

[1]Liverpool and Macarthur Cancer Therapy Centres and Ingham Institute, NSW, Australia

[2]Institute of Medical Physics, School of Physics, University of Sydney, NSW, Australia

[3]Centre for Medical Radiation Physics, University of Wollongong, NSW, Australia

[4]Uinversity of New South Wales, NSW, Australia

E-mail: sankar.arumugam@sswahs.nsw.gov.au

Abstract. The accuracy of 3DVH (Sun Nuclear Corporation, USA) reported DVH metrics for target volumes and Organs at Risk (OARs) for two Prostate Volumetric Modulated Arc Therapy (VMAT) plans was studied. The accuracy of 3DVH estimated DVH metrics in the presence of Multi Leaf Collimator (MLC) systematic errors was also tested with error introduced plans calculated in Pinnacle. The results of the study show that the DVH metrics estimated by 3DVH for error-free plans agree with the TPS calculation within 3%. The D95 to PTV was shown to be sensitive in detecting studied MLC errors. However the accuracy of 3DVH estimated DVH metrics for Target Volumes and OARs in the presence of MLC errors for VMAT prostate plans has limitations with this small data set. Although for most situations values matched within 3% for small MLC errors, there was up to a 9.8% difference between the TPS and 3DVH in the presence of a simulated 5mm MLC positioning error. Further study with more plans including other treatment sites is required to fully assess the performance of 3DVH in detecting potential clinical delivery errors.

1. Introduction

Measurement based three dimensional (3D) dosimetric verification is highly recommended to ensure dose calculation and delivery accuracy of complex modern radiotherapy techniques [1, 2]. Recently many dosimetric systems have been introduced that provide semi 3D or 3D dose information of the delivered treatment plan in a specific phantom geometry. Many metrics have been recommended to evaluate the agreement between measured and calculated dose matrices quantitatively [3]. Among them gamma (γ) index is widely implemented in many dosimetric systems and used in clinics [4]. However due to lack of correlation between these quantitative metrics and clinically relevant dose parameters there is an increasing demand to evaluate delivered dose to the relevant clinical structures in the patient geometry. Different approaches have been adapted by different dosimetric systems to estimate the dose in patient geometry based on the measurement performed by detectors [5, 6]. Sun Nuclear Corporation, USA recently introduced a patient dosimetric QA software system, 3DVH, that estimates the dose to clinical structures in patient geometry using a planned dose perturbation (PDP) algorithm. In this study we evaluate the accuracy of 3DVH software in estimating dose to patient geometry for Volumetric Modulated Arc Therapy (VMAT) for prostate cancer. To investigate a situation where there may be a delivery error, the accuracy of the software in estimating the dose in the presence of systematic errors in Multi Leaf Collimator (MLC) position has been studied.

2. Materials and Methods

2.1. Planning and delivery system

A 6MV photon beam model for an Elekta –Synergy (Elekta Ltd, Crawley, UK) linac in the Pinnacle (Philips Ltd, USA) treatment planning system (TPS), v9.2, was used to generate all the treatment plans in this study. The Elekta–Synergy linac used has the MLCi head and the VMAT plans were delivered using continuously variable dose rate (CVDR) with Integrity, v1.1, console software.

2.2. Clinical plans and simulated errors

Two prostate VMAT plans were considered to study the accuracy of 3DVH in estimating the delivered dose to patient geometry. Systematic shifts in MLC leaf position, ranging from ±1mm to ±5mm in steps of 1mm, were introduced to the VMAT arc using in-house software [6]. The plans with MLC error were calculated in Pinnacle to study the impact of error on target volume and OAR doses.

2.3. ArcCHECK measurement and verification

The error-free and error-introduced plans were measured using the ArcCHECK dosimetric system. A point dose measurement using a CC13 ion chamber was also performed at the centre of the ArcCHECK phantom for all plans to provide the measured point dose information to the 3DVH, v3.0, software. The PDP algorithm implemented in 3DVH perturbs the planned patient dose voxels based on the ray traced error maps obtained from ArcCHECK measurements and TPS dose matrix calculated in ArcCHECK geometry. More details on the description of the algorithm can be found elsewhere [7].

2.4. Comparison metrics

The TPS and 3DVH estimated DVH parameters including D95 to CTV and PTV and mean dose to PTV, CTV, Rectum and Bladder were compared to verify the dosimetric accuracy of delivered error-free plans. A minimum 3% agreement between the TPS and 3DVH estimated DVH metrics for error free plans are expected to accept the plan. To assess the accuracy of 3DVH reported DVH metrics in the presence of simulated errors, the errors were introduced to the patient plan in the TPS using in-house software and dose calculations were performed. The TPS calculated DVH metrics for the error plans were compared against the 3DVH reported metrics for error introduced plans.

The dose matrices calculated by the TPS and estimated by 3DVH were compared using γ analysis and Dose Volume Histogram (DVH) metrics of the clinical structures. The γ analysis was performed using 2%/2mm and 3%/3mm criteria using global (G) and local (L) dose tolerances. The dose elements within the ArcCHECK-PDP (ACPDP) calculation volume and high dose threshold of 10% of the maximum dose were used in the γ analysis. The change in γ pass rate with increased magnitude of errors was studied to assess the sensitivity of the 3DVH estimated dose matrix in identifying delivery errors.

3. Results

3.1. Gamma comparison

Figure1 shows the γ pass rate results of the 3DVH predicted dose matrix and TPS calculated dose matrix for error-free and error introduced plans with various tolerance criteria. In general Case 1 showed less variation in γ pass rate at all tolerance levels for introduced errors. The pass rate for Case 2 varied as a function of introduced error. 3%/3mm global criteria (3%G/3mm) was observed to be less sensitive in predicting errors. 2%G/2mm and 3%L/3mm tolerance criteria showed similar pass rate variations for all plans. 2%L/2mm criteria showed high sensitivity to error for Case 2, the pass rate was reduced from 96.3%, for the error-free plan, to 83.2% for the plan with 5mm error (figure1).

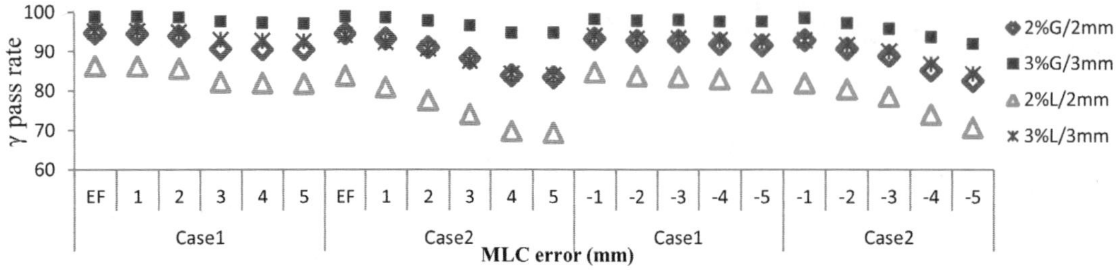

Figure 1. γ pass rate values for γ comparison between error free and +ve and –ve MLC error introduced plans for the studied cases with different γ tolerance criteria.

3.2. Change in DVH metrics calculated by TPS

Figure 2 shows the percentage change in DVH metrics between error-free and +ve MLC error introduced plans calculated by the TPS. Figure 3 shows similar data for –ve MLC shift plans. As expected, the D95 of PTV decreased as the MLC error value increases. A maximum change of -9.8% was observed for the PTV of case2 for the plan with -5mm MLC error. For CTV the D95 decreased for both +ve and –ve errors for case1. For case 2 the D95 improved with +ve errors and decreased with –ve errors. A maximum change of -4.6% decrease in D95 was observed for the CTV with -5mm error. The mean dose to PTV was decreased bya maximum -3.7% with -5mm error. The mean dose to Bladder was increased with +ve MLC error and decreased with –ve MLC errors for both cases. A maximum change of -6.9% was observed for case 1 with -5mm error. The mean rectum dose showed minimum change for both cases with +ve MLC errors whereas with –ve errors the mean dose increased in both cases and case 2 showed an increase of 6.2% with -5mm MLC error.

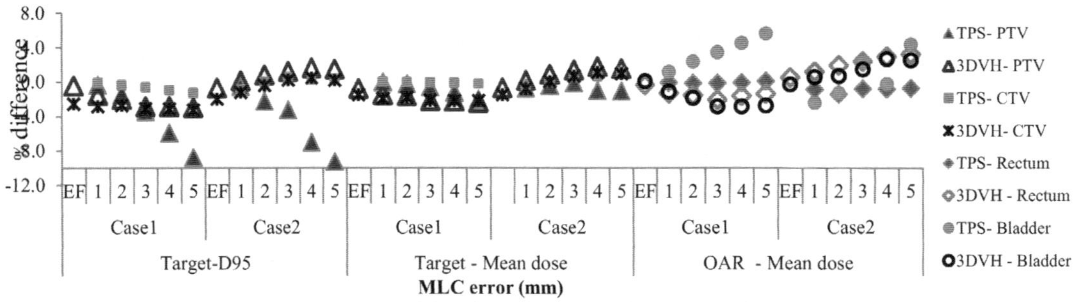

Figure 2. Percentage (%) difference in considered DVH metrics for target volumes and OARs for error-free (EF) and +ve MLC error introduced plans by TPS and 3DVH.

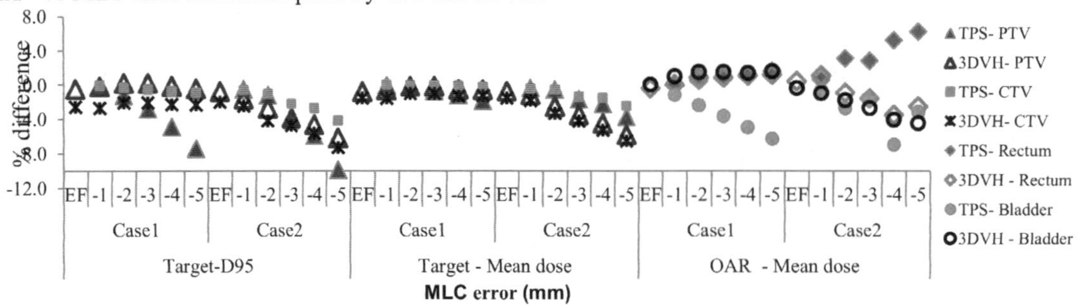

Figure 3. Percentage (%) difference in considered DVH metrics for target volumes and OARs for error-free (EF) and -ve MLC error introduced plans by TPS and 3DVH.

3.3. Accuracy of 3DVH estimated DVH metrics

The DVH metrics estimated by 3DVH for error-free and +ve and –ve MLC error introduced plans are shown in figures 2 and 3. The D95 and mean dose of error-free plans estimated by 3DVH for clinical structures agreed with the TPS within 3%. For MLC error introduced plans the % difference of PTV D95 estimated by 3DVH was consistently less compared to the TPS. A maximum difference of -9.4% was observed between the 3DVH estimated and TPS calculated D95 values for PTV for a plan with 5mm MLC error (figure2). The TPS calculated D95 to CTV showed only small magnitude variation for MLC errors for both cases. The 3DVH estimated D95 values followed the same trend as TPS calculated values (figures2 and 3). The mean dose to target volumes showed only small changes due to +ve MLC errors for studied cases. 3DVH estimation also followed the same trend. In the case of –ve MLC errors for case 2 the 3DVH estimated mean dose change was consistently higher compared to TPS calculated changes. A maximum difference of 4.5% was observed between TPS and 3DVH mean dose values at -5 mm MLC error. The % difference in mean dose to OARs, estimated by 3DVH, showed inconsistent results compared to TPS calculation. The % difference in bladder dose calculated by TPS for Case 1 and the dose to rectum and bladder estimated by 3DVH showed contrary results compared to TPS calculations. For +ve MLC errors the TPS calculated mean dose to Rectum showed minor changes (maximum -1.4% for 5mm error for case 1).The 3DVH estimated % differences for Case 2 were observed to have consistently higher values compared to TPS calculations. In the case of –ve MLC errors the 3DVH estimated % difference was consistently low compared to TPS calculation for Case 2. The % difference in mean dose to bladder for Case 1 estimated by 3DVH showed an inverse trend to that with TPS calculation. The 3DVH and TPS calculated mean dose for Rectum in Case 2 showed relatively close agreement, with a maximum difference of 3% observed between TPS calculated and 3DVH estimated mean dose for the plan with -4 mm error.

4. Discussion

The DVH metrics estimated by 3DVH for error-free plans agreed with the TPS calculation within a generally accepted tolerance of 3%. Clinically significant changes (>5%) in D95 to PTV and mean dose to OARs were observed with the simulated MLC errors. 3DVH estimated D95% difference for error simulated plans agreed with TPS calculation only for –ve MLC error plan scenario of Case 2 (figures 2 and 3). The mean dose to Rectum estimated by 3DVH in the presence of MLC error did not agree with the TPS calculation for Case2. Similarly the mean dose to Bladder estimated by 3DVH in the presence of MLC error did not agree with the TPS calculation for Case 1.The ACPDP has been shown to have limitations in predicting changes in DVH metrics due to the studied MLC shifts. However the γ pass rate with 2%L/2mm tolerance criteria showed a consistent decrease in pass rates with MLC error for at least one of the studied cases (Case 2 results in figure 1).

5. Conclusion

The accuracy of 3DVH estimated DVH metrics for target volumes and OARs in MLC error introduced VMAT prostate plans has been shown to have limitations in this small data set with up to a 9.8% difference between the TPS and 3DVH in the presence of a simulated 5mm MLC positioning error. Further study with more plans, including other treatment sites, is required to fully assess the performance of 3DVH in detecting potential clinical delivery errors.

6. References

[1] Baldock C *et al* 2010 *Phys. Med. Biol.* **55** R1-63
[2] Venning A J *et al 2004 J. Phys.: Conf. Ser.* **3** 155-8
[3] Thomas S J *et al* 2012 *Int. J. Radiat. Oncol. Biol. Phys.* **82** 717-23
[4] Low D A *et al* 2003 *Med. Phys.* **30** 2455
[5] Van Elmpt W *et al* 2008 *Radiother. Oncol.* **86** 86-92
[6] Nelms B E *et al* 2001 US Patent US 7945022 B2
[7] Arumugam S *et al* 2013 *J. Phys. Conf. Ser.* **444** 012019

Evaluation of 3D Gamma index calculation implemented in two commercial dosimetry systems

Aitang Xing[1], Sankar Arumugam[1], Shrikant Deshpande[1], Armia George[1], Philip Vial[1], Lois Holloway[1,2,3] and Gary Goozee[1]
[1]Liverpool and Macarthur cancer therapy centres and Ingham Institute, Liverpool Hospital, New South Wales, Australia
[2]Institute of Medical Physics, School of Physics, University of Sydney, New South Wales, Australia
[3]Centre for Medical Radiation Physics, University of Wollongong, Wollongong, New South Wales, Australia

E-mail: aitang.xing@sswahs.nsw.gov.au

Abstract. 3D Gamma index is one of the metrics which have been widely used for clinical routine patient specific quality assurance for IMRT, Tomotherapy and VMAT. The algorithms for calculating the 3D Gamma index using global and local methods implemented in two software tools: PTW- VeriSoft® as a part of OCTIVIUS 4D dosimeter systems and 3DVH™ from Sun Nuclear were assessed. The Gamma index calculated by the two systems was compared with manual calculated for one data set. The Gamma pass rate calculated by the two systems was compared using 3%/3mm, 2%/2mm, 3%/2mm and 2%/3mm for two additional data sets. The Gamma indexes calculated by the two systems were accurate, but Gamma pass rates calculated by the two software tools for same data set with the same dose threshold were different due to the different interpolation of raw dose data by the two systems and different implementation of Gamma index calculation and other modules in the two software tools. The mean difference was -1.3%±3.38 (1SD) with a maximum difference of 11.7%.

1. Introduction

With increasing complexity of radiotherapy delivery from 3D-CRT to IMRT, VMAT and Tomotherapy, patient-specific quality assurance (QA) for patient plans is gradually moving away from verifying the planar dose and the dose at one point to verifying the dose distribution in three dimensional (3D) space [1]. The basic task for patient-specific QA based on 3D dosimetry is to compare two data sets of points of known spatial position and an associated absorbed dose value. One data set is usually 3D dose calculated in patient or phantom geometry by the treatment planning system (TPS) and another is the measured dose in the phantom or reconstructed dose in patient from 2D detector measured dose or fluence. To quantify the agreement between these two data sets, the Gamma index proposed by Low [2] has been widely used for clinical routine work and research.

The algorithm to calculate the 3D Gamma index was implemented in two software tools: VeriSoft® from PWT as an integrated component of the OCTIVIUS 4D dosimeter system [3] and 3DVH™ software from Sun Nuclear Corporation, Melbourne, and FL. [4]. The purpose of this paper is to evaluate the accuracy of the 3D Gamma index calculation engine implemented independently in these systems by two different vendors.

Journal of Physics: Conference Series **573** (2015) 012054 doi:10.1088/1742-6596/573/1/012054

2. Materials and Methods

2.1. PTW-VeriSoft and 3DVH[TM]

VeriSoft® (version 5.1) is the patient plan verification software for PTW OCTIVIUS 4D detector [3], whereas 3DVH[TM] (version 2.21.28208) is a software tool that is able to estimate the 3D dose distribution in the patient using the measured dose in the phantom from ArcCHECK®, MapCHECK ® or EPIDose[TM] [4]. Both of them are able to load DICOM-RT dose files and calculate a 3D Gamma map using their built-in 3D Gamma calculation engine. Depending on the normalization value of the dose difference between measured and calculated dose points, there are two schemes for calculating Gamma index: local calculation method and global calculation method [4]. In the local normalized method the TPS calculated dose corresponding to the point measured is used to determine the acceptable dose difference criteria when determining percentage error. In the global calculation method, the acceptable dose difference criteria use the maximum dose found in the patient plan dose.

2.2. Preparation of the Data set

To compare the accuracy of the 3D Gamma calculators in VeriSoft® and 3DVH[TM], three reference/evaluation pairs were used for this study (summarized in Table 1). All data sets were in the standard DICOM RT dose format. For Sample A, Reference and Evaluated dose are dose cubes with 127 slices, each of which is a 144-by-201 matrix. The voxel resolution along X, Y and Z (slice direction) is 1mm for reference and the Evaluated 3D dose. The dose at the point (0, 0, 0) in reference 3D dose is 200cGy and 0cGy at all other points. In evaluated 3D dose set, the doses are 195cGy at points (1,1,1) and (-1,-1,-1) , 198cGy at points (-2,2,-2) and (2,-2,-2) and 0cGy at all other points.

Table 1. A summary of the datasets compared in this study.

Data set name	Reference 3D dose	Evaluated 3D dose	Case study
Sample A	Reference dose	Evaluated dose	
Sample B	Original plan dose	Plan dose with 1mm MLC errors	VMAT-Prostate
Sample C	Original plan dose	Plan dose with 3mm MLC errors	VMAT-Prostate
Sample D	Original plan dose	Plan dose with 3mm MLC errors	IMRT-prostate

The original plan dose in Samples B and C were exported as DICOM-RT dose files from two clinical patient plans. These two plans were generated in the Pinnacle treatment planning system (Philips, Version 9.6). IMRT plan was a typical 9 beam plan and contains 95 control points. The VMAT plan was a single arc plan and has 89 control points. The calculation grid was 2.5 mm for the X, Y and Z direction. A 1 mm systematic error was introduced into the multileaf collimator (MLC) position of each control point for the VMAT plan while the remaining plan parameters were unchanged. The plan with MLC errors was recalculated and DICOM RT dose was exported and used as the evaluated 3D dose. For the IMRT plan, 3 mm MLC errors were introduced instead of 1 mm errors to simulate a large dose variation across the whole dose volume. Here use of the TPS-calculated dose distribution as the evaluated dose cube instead of the measured dose cubes from different detectors was used to avoid uncertainties that may be introduced due to phantom setup, machine output variation or the resolution of detector array.

2.3. Calculation of 3D Gamma map in VeriSoft® and 3DVH[TM]

The user interface for the 3D dose map calculation in VeriSoft® is different from 3DVH[TM]'s, but they both provide options for users to choose the Gamma calculation method and change the Gamma criteria such as distance to agreement (DTA), dose difference and dose threshold. The search distance cannot be changed by the user in either software. It was noticed that the DICOM-RT dose loaded into 3DVH[TM] was automatically interpolated into 1mm resolution along three directions. VeriSoft® interpolated the DICOM-RT dose into 1mm resolution but kept the slice direction resolution unchanged. The end-user has no control or option on interpolation. The 3D Gamma indices were calculated for the whole dose volume for the three data sets using both software tools. The Gamma pass rates were calculated and compared using global and local methods with Gamma criteria of 3%/3mm, 2%/2mm, 3%/2mm, and 2%/3 mm, respectively.

3. Results and Discussion

For sample A, the Gamma index at point (0,0,0) can be manually calculated. Within 3mm search distance around this point and using 3 % dose criteria, the following four values can be calculated: $\sqrt{(2/6)\hat{}2+(2/3)\hat{}2} = 0.7454$, $\sqrt{(200/6)^2+(2/3)^2} = 33.34$, $\sqrt{(5/6)\hat{}2+(\square\square\square\square(3)/3)\hat{}2} = 1.0138$ and $\sqrt{(200/6)^2+(1/3)^2} = 33.34$. The Gamma value at this point is 0.7454 and $\square = \sqrt{(0/6)\hat{}2+(1/3)\hat{}2} = 0.3333$ at all other points. Table 2 also listed the Gamma indices calculated using the two software tools. The 3DVH calculated Gamma index was closer to the manually calculated value than the VeriSoft calculated value at point (0,0,0). At other points, both VeriSoft and 3DVH give the same values as the manually calculated. The difference is negligible for both software options due to the uncertainty in the numerical calculation.

Table 2. Comparison of 3D Gamma index calculated by VeriSoft, 3DVH and manually for Sample A.

Software	VeriSoft	3DVH	Manual
Gamma index	0.746	0.745	0.7454

Tables 3-5 presents the Gamma pass rate over the 3D dose volume for sample B, C and D using the global and local calculation methods. The dose threshold of 10% of maximum dose in reference dose cube was used. The mean Gamma pass rate difference calculated by two software tools was -1.3%±3.38 (1SD) with a maximum difference of 11.7%. The Gamma pass rate calculated by the two software options showed the same trend when the Gamma criteria changed from 3mm/3% to 2mm/2% or from global calculation method to local calculation method. In both situations, the Gamma pass rate decreased, which agrees with observation by Brescian et al [5]. As expected, the Gamma pass rates calculated by two systems also decreased for cases with larger MLC errors due to the larger dose difference and distance to agreement.

Table 3. Comparison of 3D Gamma pass rates for Sample B using global and local calculation.

Gamma Criteria	3%/3mm		2%/2mm		3%/2mm		2%/3mm	
	Global	Local	Global	Local	Global	Local	Global	Local
3DVH (%)	99.7	94.7	96	72.8	99.3	79.6	98.2	91
VeriSoft (%)	99.9	97.9	98.6	80.3	99.9	91.3	99.9	93.8
Difference (%)	-0.2	-3.2	-2.6	-7.5	-0.6	-11.7	-1.7	-2.8

Table 4. Comparison of 3D Gamma pass rates for Sample C using global and local calculation.

Gamma Criteria	3%/3mm		2%/2mm		3%/2mm		2%/3mm	
	Global	Local	Global	Local	Global	Local	Global	Local
3DVH (%)	94.1	90.4	91.6	78.2	93.2	82.3	93.4	88.9
VeriSoft (%)	93.4	89.4	89.19	77.2	92.1	79.19	92.1	88.9
Difference (%)	0.7	1	2.41	1	1.1	3.11	1.3	0

Table 5. Comparison of 3D Gamma pass rates for Sample D using the global and local calculation.

Gamma Criteria	3%/3mm		2%/2mm		3%/2mm		2%/3mm	
	Global	Local	Global	Local	Global	Local	Global	Local
3DVH (%)	60.3	52.1	48.2	40.6	55.6	42.5	55.4	51
VeriSoft (%)	64.7	52.1	51.6	38.9	59.8	42.6	58.2	49.4
Difference (%)	-4.4	0	-3.4	1.7	-4.2	-0.1	-2.8	1.6

Although the results from simple dataset A indicated that the basic Gamma index calculation is accurate in both software tools, the Gamma pass rates for clinical dose set varied depending not only on Gamma criteria, Gamma calculation method but also the data set. For example, a larger difference as 11.7% was observed for data set B using 3%/2mm and local method. The reported Gamma pass rate difference by 3DVH™and VeriSoft® is contributed to the combination of several factors: (1) Interpolation of original dose into fine resolution before Gamma calculation; (2) DTA search algorithm implementation for both local and global calculation method; (3) reporting of Gamma pass rate. Both software tools are black boxes to the end user as the user does not know what interpolation method was used, which specific DTA search algorithm was adopted and the total number of evaluated voxel for same threshold. For comparison studies or routine patient-specific quality assurance using different dosimeter systems, it is important to be aware that the Gamma pass rate can be different due to different software implementation.

4. Conclusion

The Gamma pass rates reported by two different software systems for the same data set may differ by a few percent due to the different implementation by different vendors.

5. References

[1] Zhen H *et al* 2011 *Med. Phys.* **38** 5477-89
[2] Low D A *et al* 1998 *Med Phys.* **25** 656-61
[3] McGarry C *et al* 2013 *Med Phys.* **40** 091707
[4] Olch A 2012 *Med. Phys.* **39** 81-6
[5] Brescian S *et al* 2013 *Med. Phys.* **40** 121711

Comparison of 2D and 3D Gamma calculations for an IMRT QA phantom

R Lafratta[1,2], G Ibbott[1,2], J Adamovics[3,4] and D Followill[1,2]
[1]Department of Radiation Physics, University of Texas MD Anderson Cancer Center, 1515 Holcombe Blvd., Houston, TX 77030 USA
[2]University of Texas at Houston Graduate School of Biomedical Sciences, Houston, TX 77030 USA
[3]Rider University, Lawerenceville, NJ, USA
[4]Heuris Pharma, LLC, 412 Sunset Rd., Skillman, NJ, 08558 USA

E-mail: gibbott@mdanderson.org

Abstract. Gamma index pass rates were evaluated for an anthropomorphic phantom using both 2D and 3D calculations. The phantom was irradiated with the traditional dosimetry insert loaded with radiochromic film and TLD, and then with a 3D dosimetry insert. A comparison with the calculated dose distribution showed that both the PRESAGE® dosimeter and the film and TLD system agreed with the plan to within 5% using 2D gamma index criteria. The 3D gamma index showed a slightly higher pass rate than the 2D gamma index at 3%/3mm, and comparable pass rates using more generous constraints. The acceptable number of pixels passing the tighter constraints (3%/3mm) might be dependent upon the choice between a 2D calculation versus a 3D calculation.

1. Background

The ability for 3D dosimetry to obtain a large amount of data in a single irradiation is advantageous [1]. The benefits have led us to consider outfitting the Imaging and Radiation Oncology Core (IROC - formerly the RPC) anthropomorphic phantoms with 3D dosimeters. The phantoms currently use film and TLD; we propose to use PRESAGE® dosimeters [2]. In doing so, we must also look at how we will evaluate the data. Currently, 2D gamma calculations from film measurements are the IROC standard for evaluating an irradiation. Moving to a 3D dosimeter would allow for a 3D gamma calculation and true volumetric quality assurance.

The extra dimension while searching for agreement in a gamma calculation should produce a higher pass rate when compared to a 2D gamma calculation using the same criteria [3, 4]. This study investigated gamma pass rates for identification of appropriate constraints for quality assurance evaluation.

2. Methods

An IROC head and neck (H&N) phantom was fitted with a cylindrical insert to hold a 500 g cylinder of PRESAGE®. The original IROC insert contained radiochromic film in the axial and sagittal planes and 8 TLD capsules. The PRESAGE® insert covered the same volume as the structures and dosimeters in the original phantom. Both inserts are shown in figure 1.

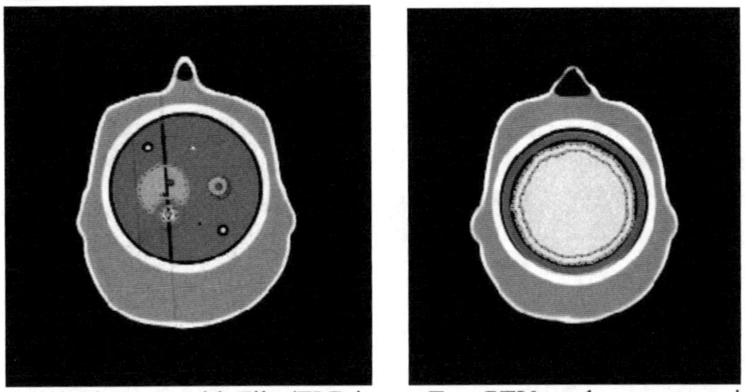

Figure 1. (A) Phantom with Film/TLD insert. Two PTVs and an organ at risk are contoured. (B) Phantom with PRESAGE® insert.

A 9-beam IMRT plan was developed with a Pinnacle planning system (Philips Radiation Oncology Systems, Fitchburg, WI) for delivery with a Varian linear accelerator (Varian Medical Systems, Palo Alto, CA). The phantom containing the film/TLD insert was irradiated 3 times, then the phantom containing PRESAGE® was irradiated an additional 3 times. The film and TLD were evaluated according to the IROC standard procedures for the H&N phantom described by Molineu *et. al.* [5]. The PRESAGE® dosimeters were scanned using the Duke Midsized Optical Scanner for IROC (DMOS-IROC) using a 1 mm voxel edge for reconstruction. Change in optical density (OD) was measured by subtracting pre-irradiation scans from post-irradiation scans [6]. The dose was imported into the Computational Environment for Radiotherapy Research (CERR) for evaluation [7].

The film and PRESAGE® measured doses were scaled to the mean TLD dose at relevant locations and 2D gamma calculations were made in the axial and sagittal planes bisecting the primary target for the film and the PRESAGE®. 3D gamma measurements excluded the outermost 5 mm of each PRESAGE® dosimeter to eliminate edge artifacts. Gamma constraints of 3%/3mm distance to agreement (DTA), 5%/3mm DTA and 7%/4mm DTA were used in the study. The gamma-index analyses for the 3 irradiations for each insert were averaged together for comparison.

3. Results

The average pass rate from the 2D gamma calculations for each trial is shown in table 1. The results from a 3D gamma over the full volume are shown in table 2.

Table 1. Average 2D gamma pass rates

	Constraints	Film	PRESAGE®
Axial Plane	3%/3mm	85%	85%
	5%/3mm	93%	94%
	7%/4mm	99%	97%
Sagittal Plane	3%/3mm	85%	89%
	5%/3mm	90%	95%
	7%/4mm	99%	99%

Table 2. Average 3D gamma pass rates

Constraints	PRESAGE®
3%/3mm	90%
5%/3mm	95%
7%/4mm	99%

4. Discussion and Conclusions

The film and PRESAGE® 2D pass rates showed agreement within 5% for all gamma index criteria. Random uncertainties as well as small differences in registration of the separate dosimeters could account for the difference. The 3D gamma showed a slightly higher pass rate at 3%/3mm and comparable pass rates at the more generous constraints. This agrees with literature and studies predicting a higher pass rate; however, to our knowledge, the higher pass rate has not previously been demonstrated experimentally. Once the constraints were loosened the discrepancy in calculation space did not cause a large difference in passing pixels. However, when passing rates are high, as they are for the looser constraints, identifying a difference between dosimetry methods is not possible.

The gamma constraints useful for quality assurance do appear to differ between 2D and 3D gamma calculations. The number of passing pixels for the tighter constraint (3%/3mm) might need to differ when using 2D calculation versus a 3D calculation. More studies need to be performed to define an appropriate criterion. At this time, the best option might be to use both: an overall 3D gamma analysis for initial QA with select planar analysis at critical structures.

5. References

[1] Baldock C *et al* 2010 *Phys. Med. Biol.* **55** R1-63
[2] Adamovics J *et al* 2006 *Radiat. Protect. Dosim.* **120** 107-12
[3] Persoon L C *et al* 2011 *Med. Phys.* **38** 4032-5
[4] Wendling M *et al* 2007 *Med. Phys.* **34** 1647-54
[5] Molineu A *et al* 2005 *Int. J. Radiat. Oncol.* **63** 577-83
[6] Thomas A *et al* 2011 *Med. Phys.* **38** 4846-57
[7] Deasy J O *et al* 2003 *Med. Phys.* **30** 979-85

New method for estimation of fluence complexity in IMRT fields and correlation with gamma analysis

T Hanušová[1,2], V Vondráček[3], K Badraoui-Čuprová[1,3], I Horáková[4] and I Koniarová[1,4]

[1]Department of Dosimetry and Application of Ionizing Radiation, Faculty of Nuclear Sciences and Physical Engineering, Czech Technical University in Prague, Břehová 7, 115 19 Praha 1, Czech Republic
[2]Hospital Na Bulovce, Budínova 67/2, 180 81 Praha 8, Czech Republic
[3]Proton Therapy Center Czech, Budínova 2437/1a, 180 00 Praha 8, Czech Republic
[4]National Radiation Protection Institute, Bartoškova 28, 140 00 Praha 4, Czech Republic

E-mail: terezahanusova@centrum.cz

Abstract. A new method for estimation of *fluence complexity* in Intensity Modulated Radiation Therapy (IMRT) fields is proposed. Unlike other previously published works, it is based on portal images calculated by the Portal Dose Calculation algorithm in Eclipse (version 8.6, Varian Medical Systems) in the plane of the EPID aS500 detector (Varian Medical Systems). *Fluence complexity* is given by the number and the amplitudes of dose gradients in these matrices. Our method is validated using a set of clinical plans where fluence has been smoothed manually so that each plan has a different level of complexity. *Fluence complexity* calculated with our tool is in accordance with the different levels of smoothing as well as results of gamma analysis, when calculated and measured dose matrices are compared. Thus, it is possible to estimate *plan complexity* before carrying out the measurement. If appropriate thresholds are determined which would distinguish between acceptably and overly modulated plans, this might save time in the re-planning and re-measuring process.

1. Introduction

Intensity Modulated Radiation Therapy (IMRT) helps to avoid organs at risk in proximity of the tumor and thus allows to increase prescribed dose compared to 3D conformal radiotherapy. Because the technique is quite complex, clinics that are not confident with their implementation of IMRT still verify each patient's plan prior to treatment. As IMRT is indicated in an increasing number of cases, this process might soon be inapplicable. The question is how to reduce time needed for verification without compromising patient safety. One solution is to avoid production of overly modulated fields by recognizing excess *plan complexity* on the planning stage, and thus avoid complete replanning and remeasurement of the plan. Here we propose a method for *fluence complexity* calculation that can help to distinguish between adequately and inadequately complex IMRT fields or plans. Moreover, our method might be suitable to calculate *dose complexity* in all three dimensions, unlike previously published works.

2. Fluence complexity calculation

Works that have been published so far calculate *fluence complexity* in an IMRT field using plan parameters such as the optimal fluence estimated by the treatment planning system (TPS) or the number of monitor units (MU). The most frequently used parameters are probably the *Modulation Index* [1] and the *Modulation Complexity Score* [2, 3]. Other parameters are described in [4, 5, 6]. Recently, another promising solution has been proposed by Nauta et al. [7] using fractal analysis. Here we define *fluence complexity* using the number and the amplitude of dose gradients in matrices of dose distribution calculated by the Portal Dose Calculation algorithm (PDC) in Eclipse (version 8.6, Varian Medical Systems, Palo Alto, USA) in the plane of the Electronic Portal Imaging Device (EPID) when creating a verification plan. Mathematically, the quantity can be defined as

$$P_q = \sum_{i=1}^{m}\sum_{j=1}^{n}\sum_{\varphi=1}^{8} k_{ij\varphi} \qquad k_{ij\varphi} = \begin{cases} 1 & c_{ij\varphi} > q \\ 0 & c_{ij\varphi} \le q \end{cases}$$

$$V_q = \sum_{i=1}^{m}\sum_{j=1}^{n}\sum_{\varphi=1}^{8} l_{ij\varphi} \qquad l_{ij\varphi} = \begin{cases} c_{ij\varphi} & c_{ij\varphi} > q \\ 0 & c_{ij\varphi} \le q \end{cases},$$

where P_q is the number and V_q is the sum of amplitudes of dose gradients in the matrix mentioned above which are greater than a certain limit q, m and n are the matrix dimensions and $c_{ij\varphi}$ is the amplitude of the dose gradient on the position [i, j] in the direction φ. There are 8 directions for each point of the matrix[5] as illustrated in figure 1. The amplitude $c_{ij\varphi}$ is calculated as the difference between values in adjacent pixels divided by the real distance between measuring points of the detector. The value of q was chosen to be 400 arbitrary units[6], based on histograms showing the number of different

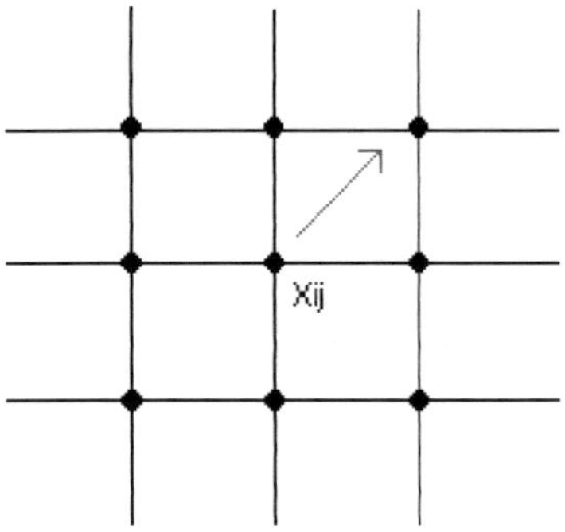

Figure 1. Distribution of the EPID detector measuring points.

dose gradient amplitudes in the IMRT fields considered. This is a way to exclude small dose gradients that are present outside the actual treatment field, because the PDC algorithm calculates dose distribution in the entire sensitive area of the EPID detector.

Even if dose distribution is used for calculation, we use the term *fluence complexity*, supposing that these two physical quantities are correlated. If *fluence complexity* in an IMRT field is changed at the planning stage, *dose complexity* in the distribution perpendicular to the beam central axis should change accordingly. However, when speaking of the third dimension, it would be more appropriate to use the term *dose complexity*.

Matrices of dose distribution predicted in the plane of the detector EPID were exported in DICOM format from TPS Eclipse. MATLAB was used to calculate *fluence complexity*. The experiment was

[5] Implementation of the algorithm in MATLAB takes account of the matrix edges.

[6] Values in the predicted matrix of dose distribution which is then exported in DICOM format from the TPS Eclipse are relatively proportional to dose. However, no physical quantity can be assigned to them.

carried out with EPID aS500 and Varian CLINAC 600C/D as well as Varian CLINAC 2100C/D (Varian Medical Systems, Palo Alto, USA).

3. Evaluation of the method

The method was evaluated using clinical IMRT plans that were further modified. Six patients with different diagnoses (head and neck, pelvis, pancreas) and different photon energies (6 MV, 18 MV) were included. Plan 1 was the original plan with no smoothing performed after optimization. The optimization parameters X smooth and Y smooth were set to 40 and 30, respectively. Plan 2 was a modification of Plan 1 – manual smoothing of field fluences was applied, using the Smooth Transmission Factors tool in Eclipse Fluence Editor, with brush size equal to one third of the first field X dimension, as is normally done during the planning process at our institute. Fluences were further modified in Plan 3 and Plan 4 in the same way, reducing the brush size to one third of its previous size in each iteration, unless the brush size dropped below 1 cm. Finally, for each patient we obtained 4 plans with increasing level of smoothing and decreasing number of MU.

Verification plans to be measured with the EPID detector were then created for all plans and *fluence complexity* was calculated using our method. For each plan we estimated the *average sum of dose gradient amplitudes in a field* (average V_q) and the *average number of dose gradients in a field* (average P_q) averaging out all fields in each plan. These two parameters were used as the measure of *plan complexity*. The calculated *plan complexity* agreed with the manual level of smoothing in all cases. An example for one patient is shown in figure 2.

All verification plans were delivered on the respective linear accelerator and dose distributions were measured with the EPID aS500 detector. We looked for a correlation between our measure of *plan complexity* and results of gamma analysis using the criteria 3 %/ 3 mm. Due to the small number of plans for each patient, the correlation could only be determined qualitatively. An example is shown in figure 3. Similar results were obtained for all the six patients studied, both for the average V_q in a field and the average P_q in a field (not shown).

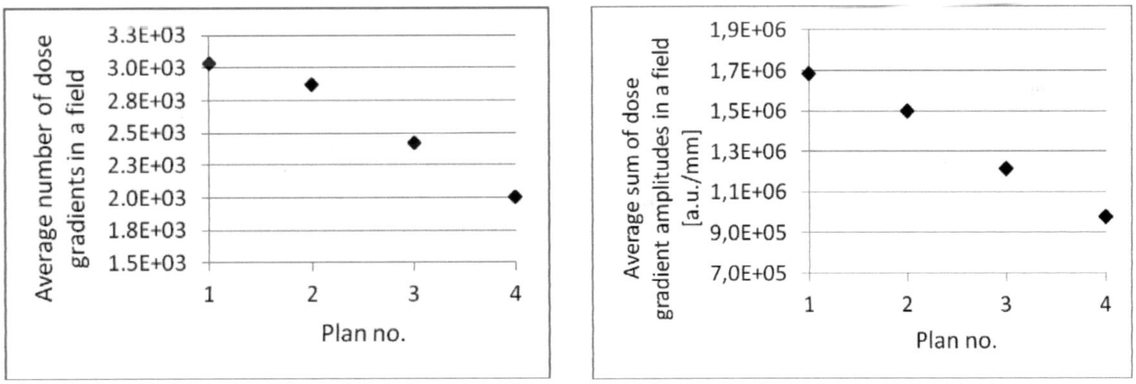

(a) (b)

Figure 2. Correlation between *plan complexity* as defined with our method (average number of dose gradients (a) and average sum of dose gradient amplitudes in a field (b), averaged over all fields in a plan) and the level of manual smoothing of field fluences. Plan 1 is not smoothed, Plan 4 is smoothed the most.

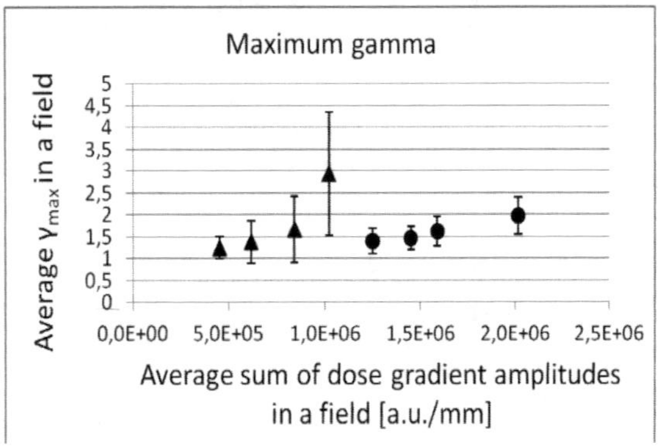

(a)

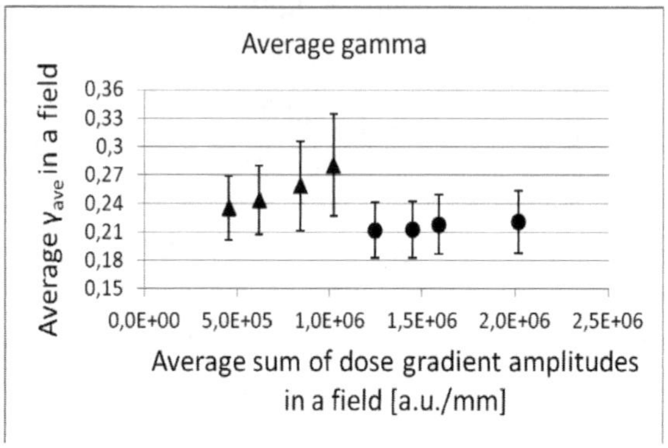

(b)

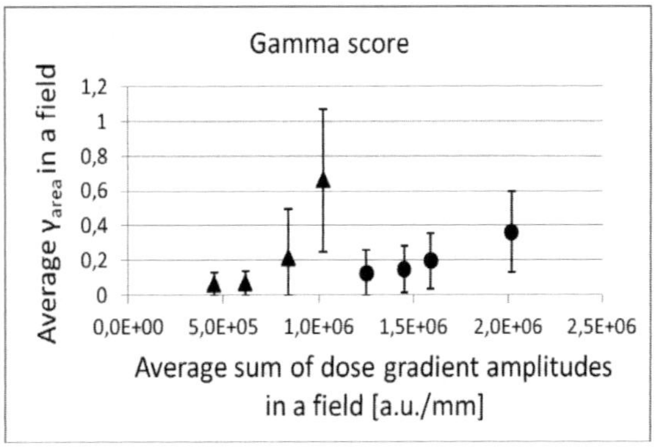

(c)

Figure 3. Correlation between *plan complexity* as calculated with our method and results of gamma analysis for two patients (▲ - patient no. 1, ● - patient no. 2). Each point represents one plan with a different level of complexity. Error bars show one standard deviation. (a) maximum gamma index, (b) average gamma index, (c) gamma score (percentage of points in the gamma distribution with $\gamma > 1$), averaged over all fields in a plan.

4. Conclusions

A new method for *fluence complexity* calculation in IMRT fields has been proposed. It differs in principle from other methods published so far. Evaluation of our tool with six IMRT cases showed agreement between the level of manual smoothing of field fluences, *plan complexity* calculation with our tool and plan verification with the EPID detector using gamma analysis. Estimation of *fluence complexity* prior to plan verification could help to recognize overly modulated fields or plans and save time avoiding reoptimization and remeasurement. However, *fluence complexity* is systematically different for different patients. This might be due to many factors, such as the site of treatment, shape of tumor, position of organs at risk etc. Therefore, further investigations need to be done to estimate appropriate thresholds in order to recognize overmodulation. Individual fields must be taken into account, rather than average value for a plan. Owing to the different design and relatively short computation time of our method, it might allow to determine *plan complexity* in all three dimensions, unlike previously published works. This has not been tested yet but will be done in the near future.

5. References

[1] Webb S 2003 *Phys. Med. Biol.* **48** 2051-62
[2] Nicolini G *et al* 2007 *Radiat. Oncol.* **42** 42-54
[3] McNiven A L *et al* 2010 *Med. Phys.* **37** 505-15
[4] Mohan R *et al* 2000 *Med. Phys.* **27** 1226-37
[5] Llacer J *et al* 2001 *Phys. Med. Biol.* **46** 2637-63
[6] Coselmon M M *et al* 2005 *Med. Phys.* **32** 1234-45
[7] Nauta M *et al* 2011 *Med. Phys.* **38** 5385-93

Feasibility of a dual wavelength laser optical CT scanner with in-air gel readout

D Ramm[1,2] and T P Rutten[1]
[1]Department of Medical Physics, Royal Adelaide Hospital Cancer Centre, South Australia, Australia
[2]School of Chemistry and Physics, University of Adelaide, South Australia, Australia

E-mail: daniel.ramm@health.sa.gov.au

Abstract. Net optical attenuation in optical CT scanning is usually determined by pre and post-irradiation scans. Replacement of the pre-irradiation scan by a scan of different wavelength, acquired concurrently with the post irradiation scan is proposed. This would result in greater practicality of gel dosimetry and potentially improved image quality. This study indicates that the approach may be viable, however experimental investigation is required for analysis of the prospective benefits of removing inter-scan variations.

1. Introduction

Contemporary optical CT scanners typically require the use of refractive index matching fluid and pre-irradiation scans [1-3][1-4]. Removing both of these requirements simplifies the scanning process and may help make optical CT scanning a more attractive option for dosimetry in the radiotherapy clinic. In this work, the viability of a dual wavelength scanner is investigated for the case of a previously developed fluid-less optical CT at the Royal Adelaide Hospital (RAH) [4][5]. A laser beam of different wavelength scanned sequentially to the measurement laser could provide more relevant reference data than pre-irradiation scanning as dust, fingerprints and other scattering particles are sampled at the time of readout.

2. Dual wavelength concepts and methods

Differential absorption measurements require two wavelengths, λ_1 and λ_2 giving significantly different responses to dose. As refractive index is wavelength dependant, the difference in λ_1 and λ_2 is minimized to mitigate the effects of refractive index change. Therefore the gel dose response absorption spectrum was searched for a steep gradient where λ_1 gives near maximum response to dose, while λ_2 gives a substantially reduced response. Ferrous, xylenol orange gel (FXG) is scanned with the existing RAH scanner using a He-Ne 594 nm laser, providing a suitable λ_1. At increased wavelengths the dose response decreases. For example, at 633 nm the response is approx. 1/3 of that at 594 nm, and drops further to 1/6 at 650 nm. A λ_2 of 633 nm (He-Ne laser) is considered in this work however laser diodes can provide other wavelength options.

The second laser is directed at the rotating mirror as shown in figure 1, resulting in a λ_1 scan across the cylinder followed by λ_2. The measured net transmission for each data point of a projection as a

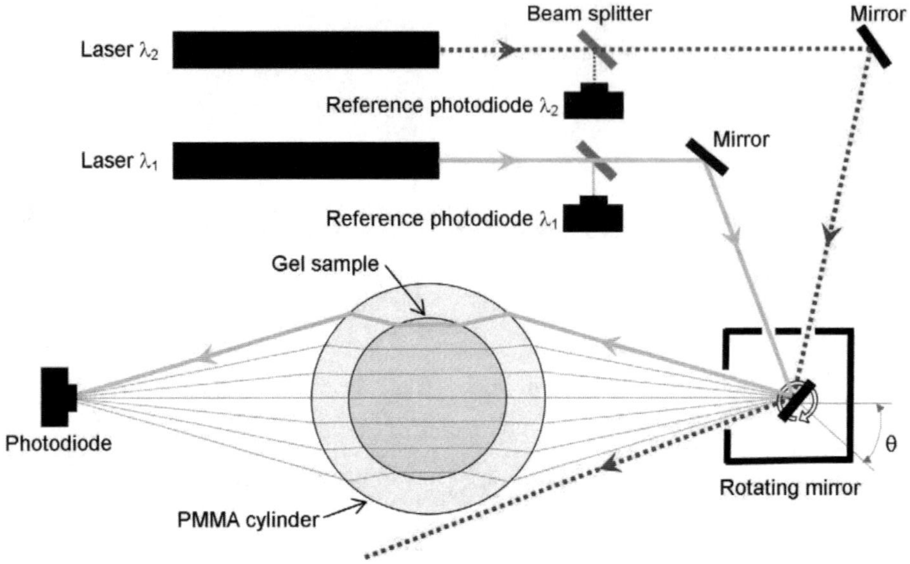

Figure 1. Schematic diagram of the RAH in-air optical CT with the addition of a second laser and reference detector.

function of dose, D and mirror angle, θ (refer figure 1), is:

$$T_{(\lambda_1-\lambda_2)}(\theta,D)=\frac{R_1(\theta,D_1)/R_1^{\mathrm{ref}}}{R_2(\theta,D_2)/R_2^{\mathrm{ref}}}\times\frac{R_2^{\mathrm{air}}/R_2^{\mathrm{ref}}}{R_1^{\mathrm{air}}/R_1^{\mathrm{ref}}} \tag{1}$$

where subscripts 1 and 2 correspond to λ_1 and λ_2, R and R^{ref} are the measurement and reference photodiode readings, and R^{air} is the measurement photodiode reading with the gel cylinder removed from the beam path. The measured net transmission is converted to attenuation and processed as usual, with conversion to dose via gel cuvette calibration samples, scanned by the optical CT.

Conceptually the net transmission comprises a number of elements that may be significant for this scanner configuration:

$$T_{(\lambda_1-\lambda_2)}(\theta,D)=\frac{T_1^{\mathrm{gel}}(\theta,D_1)}{T_2^{\mathrm{gel}}(\theta,D_2)}\times\frac{T_1^{\mathrm{refl}}(\theta)\times T_1^{\mathrm{contam}}(\theta)\times T_1^{\mathrm{defect}}(\theta)\times T_1^{\mathrm{int}}(\theta)}{T_2^{\mathrm{refl}}(\theta)\times T_2^{\mathrm{contam}}(\theta)\times T_2^{\mathrm{defect}}(\theta)\times T_2^{\mathrm{int}}(\theta)} \tag{2}$$

where there is a dosimetric component, T^{gel} for attenuation by the gel sample, and non-dosimetric components, T^{refl} for losses due to reflection at material interfaces, T^{contam} for losses due to contaminant particles in the gel or on the cylinder surfaces, T^{defect} for losses due to defects of gel cylinder, T^{int} for losses/gains due to optical interference. Since the intention is that the non-dosimetric terms will cancel, they need to be examined as a function of wavelength.

Change in refractive indices due to wavelength gives altered ray paths, which may affect all components of equation (2). T^{gel} changes if rays traverse gel differently. T^{refl} depends on ray angle and refractive indices. T^{contam} and T^{defect} change if rays do not intersect the particles or defects in the same way. T^{int} is affected by ray path change and wavelength. Because of these issues, the intent is to minimize change in ray paths rather than apply corrections. Ray paths and reflection losses were calculated using methods similar to previous [4][5], for λ_1 = 594 nm and λ_2 = 633 nm. Refractive indices used in calculations were n_1^{gel} = 1.342, n_2^{gel} = 1.340, n_1^{PMMA} = 1.491 and n_2^{PMMA} = 1.489. The gel PMMA cylinder dimensions were 76 mm outer diameter and 51 mm inner diameter, according to the existing scanner configuration [4][5].

3. Results and discussion

The existing RAH in-air scanner could incorporate an additional laser of different wavelength, while retaining a scan acquisition speed of 5 sec per slice due to the continuous 360° rotation of the scanning mirror. Alignment of the two lasers would be a fundamental aspect of the experimental setup, together with matching of polarization and spot sizes. Two sets of projection data would be acquired in a volumetric scan, one at the measurement wavelength of 594 nm and another at 633 nm, giving net transmission using equation (1). This has the advantage of removing inter-scan variability that may occur when the dosimeter is scanned on two separate occasions. It may be particularly advantageous where positioning reproducibility is limited. A guaranteed benefit is the convenience and time saving of a single scan.

A disadvantage of the proposed dual wavelength method is the loss of around 1/3 of the FXG dose sensitivity (T^{gel}). Some of this loss could be recovered by use of a higher wavelength than 633 nm, at the expense of greater ray deviations.

Calculated ray paths are similar for 594 and 633 nm as shown in figure 2. The differences between red and yellow ray locations at each of the four interfaces of the gel cylinder are given in figure 2(c). As expected, deviations are greater at the exit interfaces, reaching 0.13 mm at the gel-PMMA exit

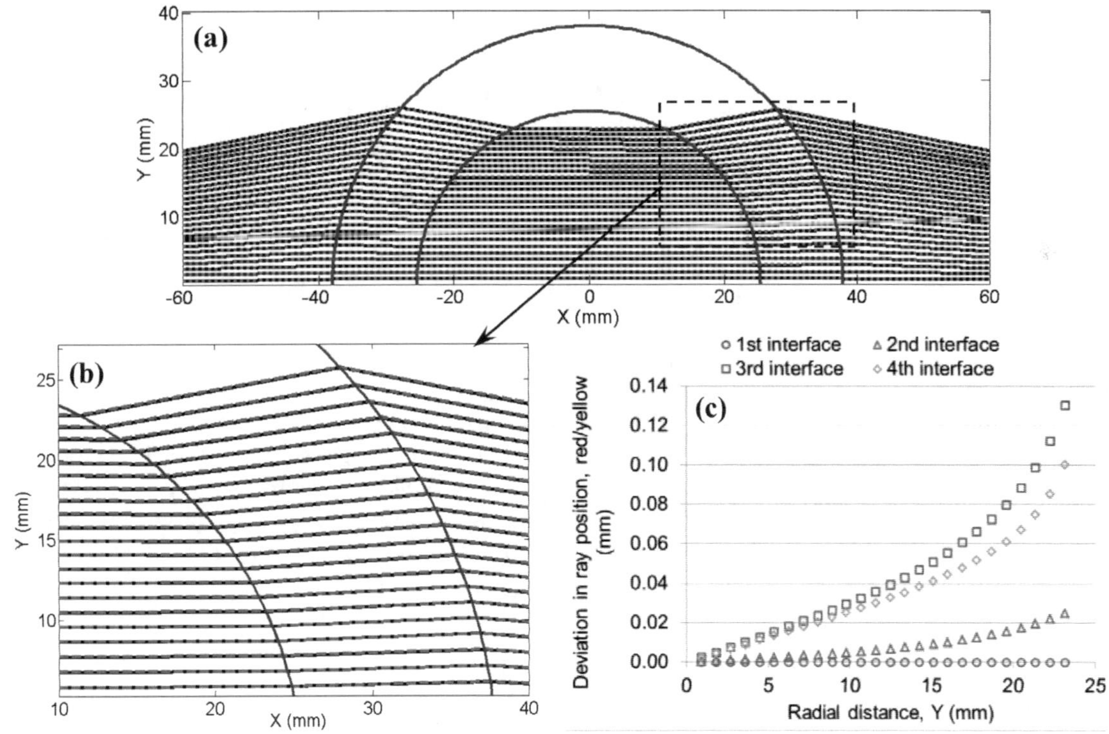

Figure 2. (a) Ray tracing through the gel cylinder from left to right for 594 nm (black lines) and 633 nm (dashed red lines). **(b)** Zoomed view of figure 2 (a) in the region of greatest path deviations between 633 and 594 nm (3rd and 4th interfaces). **(c)** Calculated differences in ray positions at each interface of the PMMA cylinder. Radial distance, Y is effectively proportional to the mirror angle, θ. The deviations are < 0.1 mm for the majority of rays and interfaces. Considering a laser beam FWHM of 0.4 mm, the largest deviations while not large, are not completely insignificant. However, their comparison to inter-scan spatial precision may be favourable. If so, both $T_1^{contam}/T_2^{contam}$ and $T_1^{defect}/T_2^{defect}$ could be closer to unity than that of pre-irradiation scanning.

Reflection losses (T^{refl}) occurring as the beam scans across the cylinder must be accounted for as there is variation of at least several percent. A 633 nm beam may be used as a reference since there are small differences compared to 594 nm as shown in figure 3. The difference is 0.1 – 0.2 % at 90 % of the gel radius (where the data is usually truncated during processing). Therefore T_1^{refl}/T_2^{refl} should be close to unity.

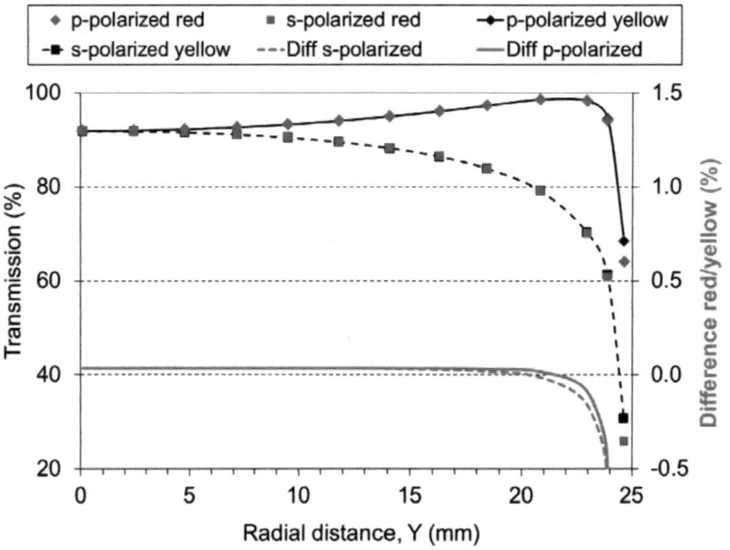

Figure 3. Calculated transmission through the gel cylinder, assuming 100 % transmission for the gel. The differences between 633 nm and 594 nm are given by the right y-axis.

Optical interference must be avoided since T_1^{int}/T_2^{int} is unlikely to be unity. The gel cylinder hard coating has been previously noted to cause optical interference [5][6]. To avoid this, the cylinder should have a refractive index matched hard coating or be uncoated.

4. Conclusions

A dual wavelength optical CT scanner has been proposed, allowing both measurement and reference scans to be obtained in one scan, giving improved practicality for optical CT in gel dosimetry. There is the prospect of improved scan quality by better spatial registration of artifact generating flaws in reference and measurement projection data. The tradeoff between loss of gel measured sensitivity and potential improvement of artefacts is to be examined by experimental investigation. While this proposal is specific to the RAH in-air optical CT, it may also be suitable for other scanner configurations and gels/solids.

5. References

[1] Thomas A *et al* 2011 *Med. Phys.* **38** 4846-57
[2] Olding T *et al* 2011 *Phys. Med. Biol.* **56** 1259-79
[3] Xu Y *et al* 2013 *Phys. Med. Biol.* **58** 479-95
[4] Baldock C *et al* 2010 *Phys. Med. Biol.* **55** R1-63
[5] Ramm D *et al* 2012 *Phys. Med. Biol.* **57** 3853-68
[6] Ramm D 2012 *J. Phys.: Conf. Ser.* **444** 012078

Dual wavelength optical CT scanning of anthropomorphic shaped 3D dosimeters

Yves De Deene
Department of Engineering, Faculty of Science, Macquarie University, Sydney, Australia

E-mail: yves.dedeene@mq.edu.au

Abstract. To create an optical density map of 3D dosimeter phantoms, the ratio of the transmission profile (either a line or planar) acquired after irradiation of the dosimeter and a pre-irradiation reference scan of the same dosimeter phantom is taken. Any uncertainty in repositioning of the phantom may result in an uncertainty in the optical density map and finally also in the derived dose maps. Correct repositioning is paramount when scanning non-cylindrical dosimeter phantoms as any repositioning error will give rise to severe imaging artifacts. We hereby propose a different scanning technique that does not require any repositioning of the dosimeter phantom. In this method, no pre-irradiation san is recorded but the dosimeter phantom is scanned twice with light at two different wavelengths. It is demonstrated that this method is accurate in scanning non-cylindrical anthropomorphic shaped phantoms.

1. Introduction

Three-dimensional radiation dosimetry has received growing interest with the implementation of highly conformal radiotherapy treatments [1]. Much research has been conducted towards both the development of different radiation sensitive three-dimensional dosimeters and adequate scanning techniques. Although an acceptable precision and accuracy has been achieved with polymer gel dosimeters and MRI scanning [2-4], the procedure is relatively labor intensive and some expertise is needed to acquire reliable quantitative MR images. As an alternative to MR scanning, optical CT scanning has been proposed as an alternative [5]. Different staining dosimeters have been proposed as a better alternative to polymer gel dosimeters including polyurethane based dosimeters, PRESAGE™ [6] and micelle gel dosimeters [7, 8] and another system will be proposed at this conference.

In 3D radiation dosimetry with optical CT scanning readout, projections of transmitted light are recorded in either lines or planar by use of an optical CT laser scanner or cone beam CT scanner using a CCD camera respectively. Dose maps are acquired by taking a blank (reference) scan of the dosimeter phantom prior to irradiation and a scan of the dosimeter phantom after irradiation. The change in optical density that is a measure of the absorbed radiation dose can then be easily obtained by taking the logarithm of the ratio of the dark current corrected reference scan and data scan. Absolute dose maps can be derived from the optical density maps by calibration of the optical absorption coefficients to dose values. The method using a pre-irradiation reference scan and post-irradiation scan relies on the repositioning and scanning of the dosimeter in exactly the same position.

Accurate repositioning becomes even more crucial when scanning anthropomorphic shaped and deformable dosimeter phantoms.

2. Methods and Materials

The principle behind dual wavelength scanning is that the radiation sensitive dye in the dosimeter has a selective absorption for light of a particular wavelength (figure 1a). As a result, only light with that specific wavelength will be absorbed by the dye while any impurities such as impurities or Schlieren will cause attenuation in the transmitted light projection scan at both wavelengths (figure 1b).

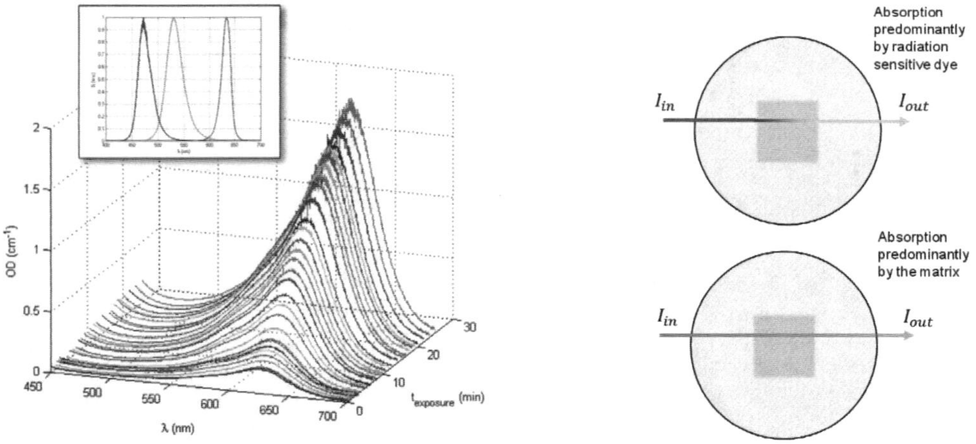

(a) (b)

Figure 1. Absorption spectrum of Flexydos3D, a new staining dosimeter (a) with an absorption peak at red light wavelengths (a) and basic principle of dual wavelength scanning (b). The inset of (a) shows the measured spectrum of the different cone beam optical CT scanner light sources.

Two different scanner types have been modified for scanning at two different wavelengths. The first type of scanner is an in house constructed CT laser scanner (figure 2a) where a second green laser is added and is directed onto the galvano-mirror by use of broadband half-reflecting mirrors. The second type of scanner is rebuilt from a cone beam Modus Vista™ scanner (figure 2b).

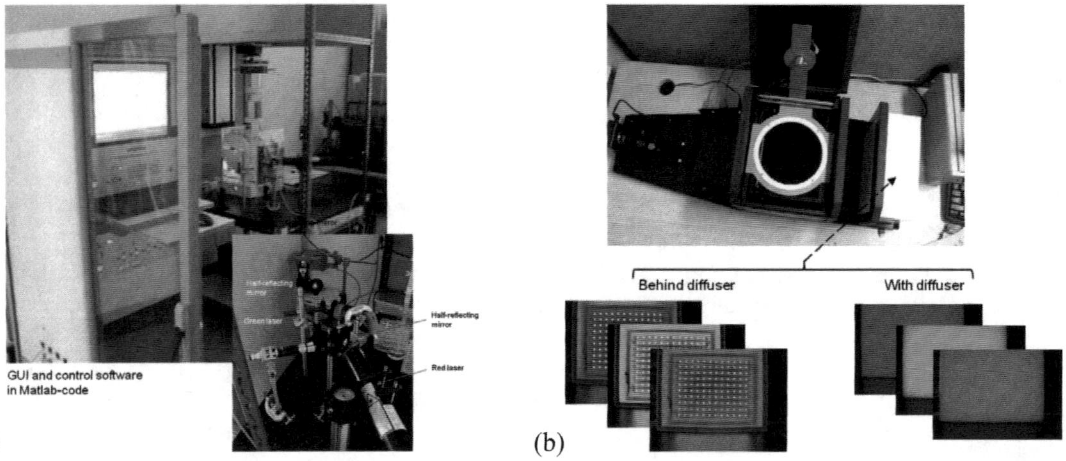

(a) (b)

Figure 2. Optical laser CT scanner equipped with a second green laser that is focused on the same galvano-mirror as the red laser (a) and an optical cone beam scanner equipped with new light sources at three different wavelengths (b).

The CCD camera in the Modus scanner is recognized as DragonFly2 DR2-HIBW. The light source of the scanner has been replaced by a flat bed diffuse light source panel in which the principle colours (red, green, blue) can be selected. The driver for the LED light source was developed in house and is based on a simple bipolar transistor circuit. The light intensity of each of the colours can be adjusted to optimise the dynamic range without adjusting the exposure time or gain of the CCD camera. All reconstruction software for both scanners is developed in house in Matlab code and takes into account the difference in light intensities and wavelength dependent sensitivity of the detector.

Projection maps of optical density difference between the red and green light projections are calculated prior to the cone beam reconstruction. As shown in equation 1, the blank projections drop out of the equation:

$$\Delta OD\lambda=530\ nm\lambda=630\ nm=OD\lambda=630\ nm-OD\lambda=530\ nm=log10I\ \lambda=530\ nm-IdI\ \lambda=630\ nm-Id \quad (1)$$

where $I\lambda$ is the measured intensity in a pixel of the projection image recorded with central wavelength λ and Id is the dark current. It is assumed that the blank current is the same for both wavelengths, which is easily obtained by scaling the intensity of the projection images on the basis of the background signal intensity (i.e. next to the phantom). Then a cone beam reconstruction is performed on the optical density difference projection images. From figure 1, it can be seen that there is nearly no light absorption by the leucodye, which implies that $I\ \lambda=530\ nm \cong I0$. This is also apparent from figure 4a where the square beam is invisible. It is imperative that this is not a necessary condition for this method to work.

3. Results and Discussion

3.1. Spatial accuracy QA phantom

In order to check the reliability of scanning a non-cylindrical phantom, a spherical phantom (figure 3) and other non-cylindrical test phantoms have been scanned. A grid pattern was created in these phantoms by use of syringes arranged in a regular pattern.

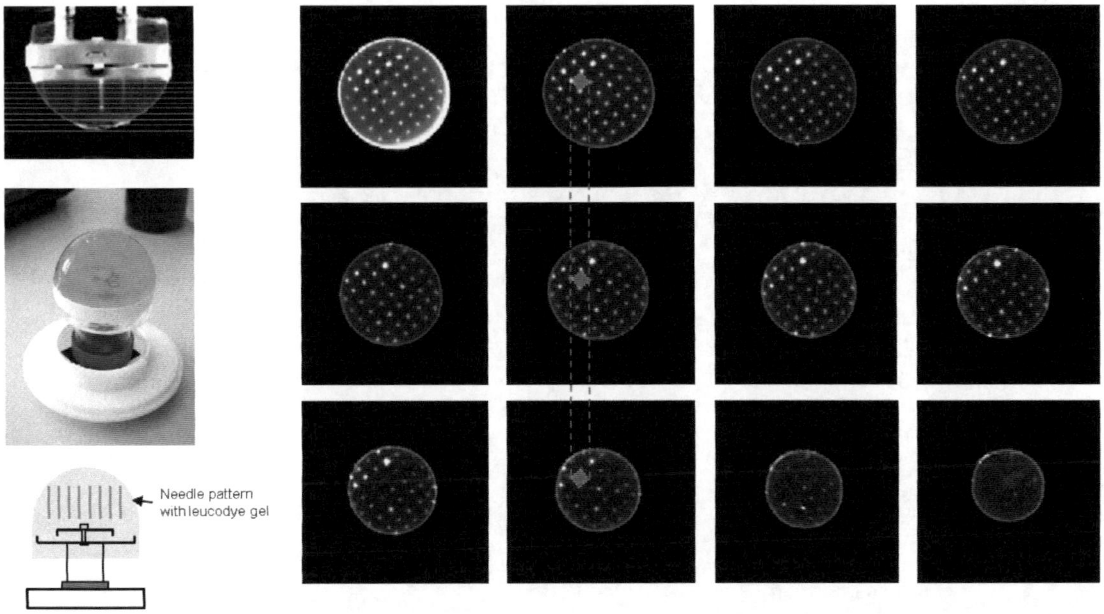

Figure 3. A grid pattern created in a demi-spherical phantom is preserved in the scans independent of the cross section diameter. The scans were obtained with the laser scanner but comparable results were obtained with the cone beam scanner and in a non-spherical prostate shaped phantom.

After removal of the syringes, the cavities were filled with a colored gel. Intensity differences in the grid pattern are related to deviations in the diameter of the needle cavities and are not attributed to optical effects.

3.2. Square photon beam

Figure 4 demonstrates the difference in absorption of red and green laser light by the leucodye after exposure of a micelle gel dosimeter to a square photon beam.

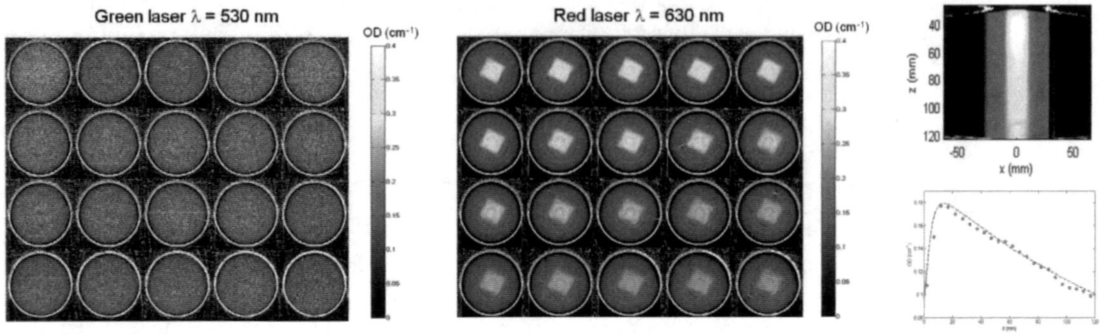

(a) (b) (c)

Figure 4. Axial OD maps scanned with a 530 nm green laser (a) and with a 630 nm red laser (b) and reconstructed longitudinal dose map along the beam axis with depth dose profile (c).

4. Conclusion

Dual wavelength scanning avoids the need for a blank pre-irradiation scan, hence diminishing any imaging artifacts related to a non-accurate repositioning of the phantom in the optical scanner. This is particular helpful in scanning anthropomorphic shaped dosimeter phantoms. In principle, it is possible to perform dual wavelength scanning by use of a broadband light source and using optical filters at the detector side or a color CCD camera with significant dynamic range at each of the principle colors. This would speed up the optical scanning process with a factor of two.

5. Acknowledgements

Dr. Robin Hill is thanked for assisting in the irradiation of the cylindrical phantom with a square photon beam (figure 4) at Chris O'Brien Lifehouse.

6. References

[1] Baldock C *et al* 2010 *Phys. Med. Biol.* **55** R1-63
[2] Vandecasteele J and De Deene Y 2013 *Phys. Med. Biol.* **58** 19-42
[3] Vandecasteele J and De Deene Y 2013 *Phys. Med. Biol.* **58** 43-61
[4] Vandecasteele J and De Deene Y 2013 *Phys. Med. Biol.* **58** 63-85
[5] Gore J C *et al Phys. Med. Biol.* **41** 2695-704
[6] Adamovics J A and Maryanski M 2004 *Med. Phys.* **30** 1349
[7] Jordan K and Avvakumov N 2009 *Phys. Med. Biol.* **54** 6773-89
[8] Vandecasteele J and De Deene Y 2013 *Phys. Med. Biol.* **58** 6241-62

Radiochromic film thickness correction with convergent cone-beam optical CT scanner

Kevin Jordan
London Regional Cancer Program, London Health Sciences Centre, Department Medical Biophysics, Western University, London, Ontario, Canada

E-mail: kevin.jordan@lhsc.on.ca

Abstract. A cone-beam optical computed tomography (CT) scanner was modified by replacing the diffuse planar yellow light emitting diode (LED) source with violet and red LEDs and a large Fresnel lens. The narrow band sources provided transmission images of radiochromic EBT2 film at 420 and 633 nm, with air as a reference. The dose image was not detectable with the violet source. This demonstrated spectral independence of the two images. Assuming attenuation at 420 nm was dominated by absorption from yellow dye in the active film layer allowed a relative thickness image to be calculated. By scaling the 633 nm optical density image for relative thickness, non-uniformities in the recorded dose distribution due to film thickness variations, were removed.

1. Introduction

Self-developing radiochromic film is a versatile, tissue-equivalent dosimeter. The commercially available Gafchromic® EBT2 film from Ashland Chemicals has acceptable spatial resolution and dose sensitivity for measuring most radiotherapy dose distributions. Bending the film sheet allows sampling a surface within a 3D dose distribution. The main problem with this dosimeter is the non-uniform thickness of the active layer. Until this manufacturing limitation is resolved, methods to measure and correct for thickness variations will be required. A yellow dye is included in the active layer to provide strong absorption of violet and ultraviolet light to reduce film sensitivity to light at these wavelengths. This allows the film to be handled in typical indoor lighting conditions without initiating a photochromic response. By measuring the absorption in the violet or blue spectrum a relative film thickness map can be determined, assuming the dye doping is uniform throughout the film. White papers on the manufacturer's website provide information for scanning the film for accurate dosimetry [1]. Micke *et al* have also developed a commercial program for analyzing EBT2 transmission data from Epson flatbed document scanners using the red, green and blue channels to correct for film uniformity and optimize dynamic range of the combined film-scanner system [2]. These document scanners have limitations for quantitative spectral imaging. One of the largest issues is the cross-talk between the blue and red channels which limits accurate measurement of the recorded dose distribution. In this study a prototype light source for a convergent, cone-beam, optical CT scanner is evaluated for imaging of EBT2 radiochromic film.

By choosing a violet light source, in the spectral range of 400 to 430 nm, it is expected that EBT2 film attenuation will be primarily due to absorption from the yellow dye. It is also anticipated that violet light absorption by the dose-induced chromophores will be minimal. The red source, 633 nm, fwhm = 10 nm is near the dose product absorption peak at 635 nm. It is expected that spectrally independent violet and red transmission images obtained with this light source will provide higher quality images for dosimetry compared to the Epson document scanner.

2. Instrumentation and methods

A schematic of the modified Vista10, Modus Medical Devices Inc., cone-beam optical CT scanner is shown in figure 1. The red source consisted of a Luxeon III, 3W LED operating at 0.8 mA (note maximum recommended current is 1.3 A). The LED light was pseudo-collimated with a composite optic to a beam divergence of 20 degrees, filtered by a 10 nm bandpass filter with centre wavelength of 633 nm and then focused to a 2 mm spot at a white diffuser. The spot source was aligned with the optic axis of the Vista scanner. A Fresnel lens, 25 cm diameter, was located ~ 30 cm from the rotation axis of the scanner and the source position was adjusted to provide a cone of light focused at the camera position. The violet source was realized by substituting a 3W, 420 nm LED, operating at 0.2 mA (StevesLEDs, note maximum recommended current is 1 A) and a dark blue glass filter. The optimum source position was closer to the Fresnel lens for the violet source because of dispersion.

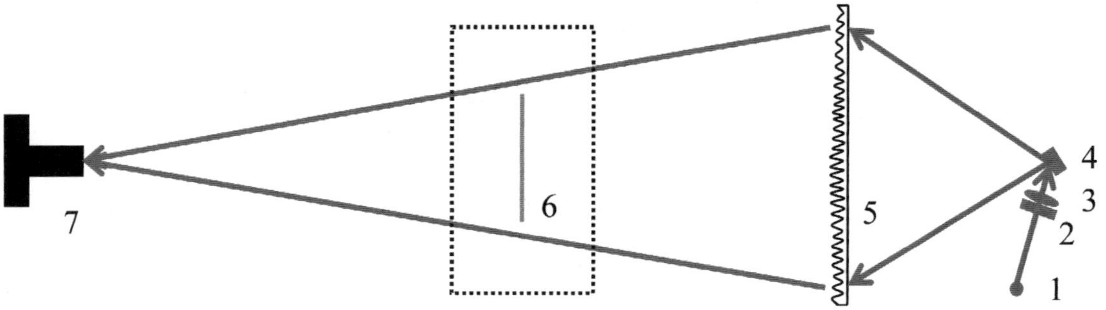

Figure 1. Schematic top view of modified Vista10 convergent cone-beam optical CT scanner with beam: 1) LED, 2) bandpass filter, 3) lens, 4) diffuser, 5) Fresnel lens, 6) film, 7) digital camera.

A reference air image for the red source was acquired with an exposure time of 3 msec (note the equivalent exposure with the planar, diffuser source required ~60 msec). The camera lens was focused to the scanner's rotation axis plane and an f4 aperture was used for all images. An EBT2 film sample that had been previously exposed for another measurement, was attached to a frame and placed in the rotation axis plane and centred on the optic axis. The sequence was repeated with the violet source. Dark images were acquired by blocking the light at the LED. Images were processed in Matlab. Optical density images were calculated with air as reference, since the film had been exposed.

3. Results

Power measurements, with a calibrated silicon photodiode, determined blue and red LED light temporal stabilities to within 1% during the experiment. Figure 2a, shows the violet, optical density image. The mean optical density is 2.5. No filtering was applied. By comparison the zero-dose, blue-channel OD with the Epson scanner was 0.48, see figure 2 of reference 2. No image of the dose distribution is evident in the violet image. In contrast, the blue channel has ~25% the dose sensitivity of the red channel with the Epson scanner see figure 2 of reference 2. The relative thickness correction image was generated by spatially filtering the normalized, violet, OD image with a 5x5 median filter, see figure 2b. Note the horizontal features are similar in magnitude (+1% to -1% of the mean), to those reported with the Epson scanner, see figure 3 of reference 2.

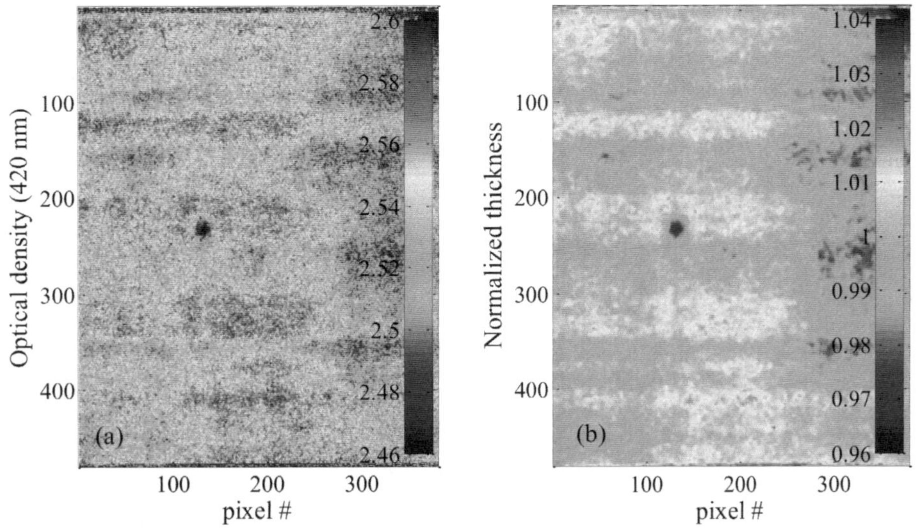

Figure 2. Optical density image of EBT2 film at 420 nm with air reference (a), relative thickness image (b), pixel size 0.235 x 0.235 mm.

The red OD image shows a 6 MV dose distribution with an average maximum dose of 2.2 Gy, see figure 3a. At this dose, the maximum optical density is approximately 2.0 compared to 0.28 for the Epson scanner, see figure 2 of reference 2. Features present in the relative thickness map are also observed in red OD image. Correction for relative thickness was performed by dividing the red OD image by the relative thickness image, pixel by pixel. This ratio removed non-uniformity features, see figure 3 (right panel).

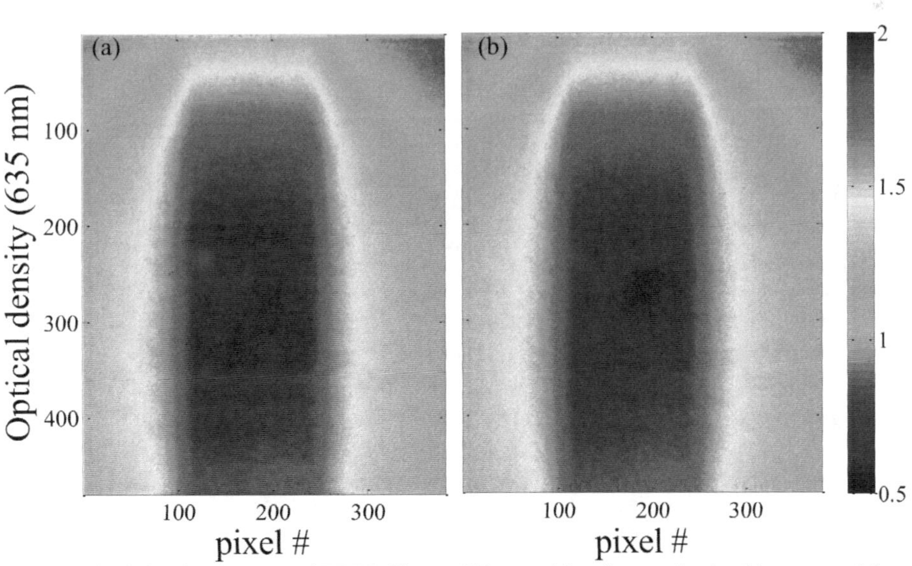

Figure 3. Optical density images of EBT2 film at 633 nm: (a) reference is air, (b) corrected for non-uniform film thickness, pixel size 0.235 x 0.235 mm.

4. Discussion

This sevenfold signal increase, compared to Epson scanner, is similar to EBT film data from 633 nm scanning laser film densitometers [3]. The impact of film scatter requires further investigation.

However, point measurements with a 633nm laser (polarization at 45 degree, to approximate non-polarized light) are similar. This indicates that scatter is minimal in these images. This result is consistent with the fact that the irradiance is approximately 100 times lower in the film plane compared to the diffuse planar source. Imaging efficiency is much greater with the small source-lens combination compared to the previous planar diffuser source again indicating the primary to stray light ratio has increased with this more collimated source. Multiple slot array images will be used to quantify magnitude and spatial distribution of stray light in this system [4]. Since the violet image is insensitive to dose, previously irradiated EBT2 films can be rescanned to provide high-resolution thickness corrections to the dose images. Acquiring multiple images and averaging would decrease the noise. Polarization effects with EBT2 film are large and will be reported with ongoing paired measurements for EBT3 film.

5. Acknowledgements

Disclosure: The author has a research agreement with Modus Medical Devices Inc concerning optical CT scanners and dosimetry applications.

6. References

[1] Lewis D 2010, Using radiochromic film tips and techniques, technical paper www.ashland.com
[2] Micke A *et al* 2011 *Med. Phys.* **38** 2523-34
[3] Soares G 2006 *Radiat. Prot. Dosim.* **120** 100-6
[4] Babic S *et al* 2012 *Phys. Med. Biol.* **57** N377-89

Development of CCD-based optical computed tomography and comparison with single-beam optical CT scanner

YJ Chang[1,2]
[1]Institute of Biomedical Engineering and Materials Science, Central Taiwan University of Science and Technology, Taichung, Taiwan (R.O.C.)
[2]Department of Management Information Systems, Central Taiwan University of Science and Technology, Taichung, Taiwan (R.O.C.)

E-mail: ronchang@ctust.edu.tw

Abstract. This study reports on the development of CCD-based optical computed tomography (CT) CT-s2. A commercially available 10× fast optical computed tomography scanner (OCTOPUSTM-10X, MGS Research, Inc., Madison, CT, USA) was used for comparison. NIPAM polymer gel dosimeter was used to validate the performance of CT-s2. The gamma pass rate can reach 96.00% when using a 3% dose difference and 3 mm dose-to-agreement criteria. The results of CT-s2 are as good as those of the single-beam optical-CT scanner, but the scanning time of CT-s2 is only one-tenth of that of the single-beam optical-CT scanner.

1. Introduction

As a readout tool, optical computed tomography (optical CT) is superior to MRI owing to such advantages as low-cost, high-speed scanning time, and high resolution [1]. The introduction of the radiochromic plastic dosimeter has also gained considerable attraction in the field of optical CT. Many geometries of the optical CT scanner have been revealed in the application of gel dosimetry [2]. The first-generation single-beam optical CT scanner was developed by Gore *et al* [3]. The IMRT measurement result obtained by using the first-generation single-beam optical CT is still considered as the gold standard in the field [4-6]. However, this technology is hindered by long scanning time. To reduce the scanning time, the fast single-beam optical CT (OCTOPUS^{TM}-10× optical CT scanner, MGS Research Inc., Madison, CT) based on rotating mirror and Fresnel lens was revealed by MGS to be 10 times faster than the first-generation optical CT scanner [7, 8]. Another fast imaging modality utilizes a diffuser and CCD to acquire a complete 2D projection image at each rotating angle, tremendously reducing the scanning time to approximately several minutes [9-12]. However, some challenges associated with the use of CCD-based optical CT have to be addressed. These challenges include spatial uniformity of light source, FOV size, scattering artifacts, low-noise CCD, and best-fit matching liquid [9].

Although many geometries of CCD-based optical CT have been presented previously, stable, low-cost, high accuracy, high precision and fast optical-CT was rarely reported. Based on the NIPAM gel, on which our research group has conducted detailed investigation [14-16], this study uses prior knowledge on optical CT to implement an improved CCD-based parallel beam optical CT (CT-s2). This device is based on a specially designed positioning mechanism involving a uniform light plane that uses a spatial filter, a microscope objective lens, a pinhole and one rotating diffuser, two large-diameter lenses, and an automated acquisition system. Regarding to the NIPAM gel dosimeter, the scanning result of CT-s2 is comparable with that by OCTOPUS^{TM}-10×.

2. Materials and Methods

Figure 1 shows a photograph of the CT-s2 optical CT scanner implemented in this study. CT-s2 includes a solid-state laser (120 mW power and 660 nm wavelength, Power Technology Inc., Little Rock, AR, USA), spatial filter, a microscope objective lens (40X), pinhole (diameter 5 μm), rotating diffuser, two collimating lenses, and a CCD (1296 pixels × 966 pixels, scA1300-32fm, BASLER AG, Ahrensburg, Germany). The laser was used because it can generate a uniform light source combined with a spatial filter and pinhole. The gel container with 10 cm diameter and 10 cm height is fixed on a rotating plate in an aquarium with dimensions of 32 cm × 32 cm × 32 cm. The rotating plate is mounted on a stepper motor fixed under the aquarium. All components are mounted on a stable anti-vibration optical table (Newport Corporation, Irvine, CA, USA). An in-house control program written by Labview (NI Inc.) rotates the gel phantom while capturing the image. The rotating speed is 1 degree per second, and 360 images are acquired at 360 degrees. The gel phantom filled with NIPAM [5% gelatin (300 Bloom Type A, Sigma-Aldrich], 5% NIPAM (97% pure; Sigma–Aldrich), 3% N, N'-methylene bisacrylamide (Bis), and 5 mM Tetrakis (hydroxymethyl phosphonium chloride) was prepared and placed in an acrylic phantom. The gel phantom was irradiated using a 6 MV Varian Clinac IX linear accelerator (Varian Corporation, Palo Alto, CA, USA). The treatment plan consists of five fields of IMRT irradiation with gantries aligned at 120°, 155°, 180°, 215°, and 245°. The maximum prescribed dose is 6.8 Gy at the central region.

After irradiation, the gel phantom was scanned by the CT-s2 and OCTOPUS^{TM}-10× optical CT scanner. The preparation process was based on the steps outlined in previous research [13-15]. The matching liquid was 1.346 following the best fit RI of NIPAM [16].

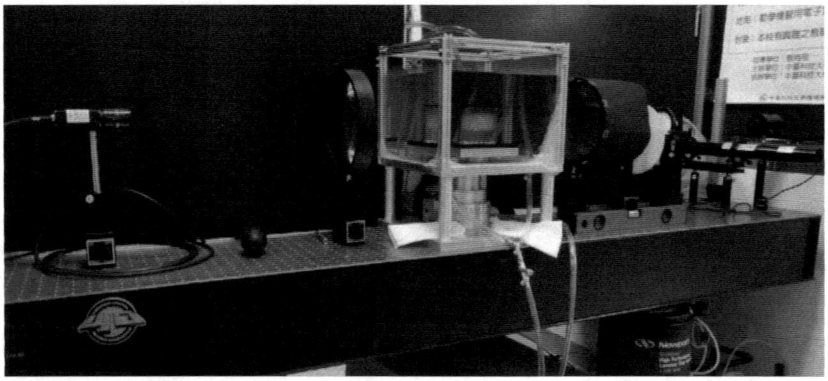

Figure 1. Photograph of CT-s2 optical CT.

3. Results and discussion

3.1. Projection images and sinogram

Before acquiring an image, optical alignment was carefully conducted to ensure a uniform light field. Repeated adjustment is required to achieve a high-quality light plane. The uniformity is 7.8% [(maximum grey value-minimum grey value)/ (mean grey value)/2] for the central 10 cm × 10 cm region of the FOV. Figure 2 shows the projection image and sinogram of non-irradiated gel phantom acquired by the CCD camera, whereas figure 3 shows the projection image and sinogram of irradiated gel phantom acquired by the CCD camera.

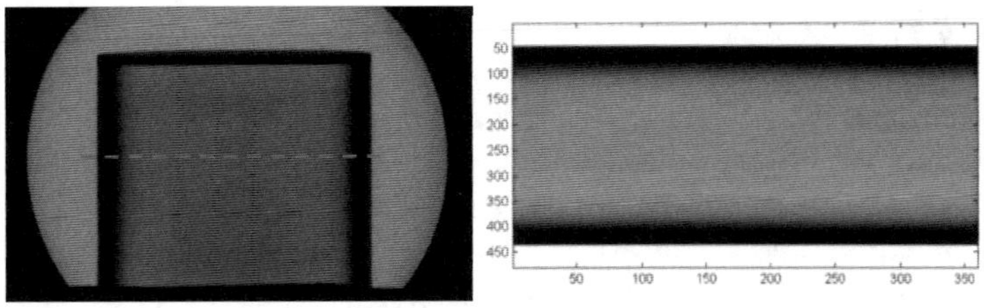

Figure 2. Left image is the projection image of non-irradiated gel phantom acquired by a CCD camera. Right image is the sinogram of 360 projections at 40 mm depth (from the top of gel container).

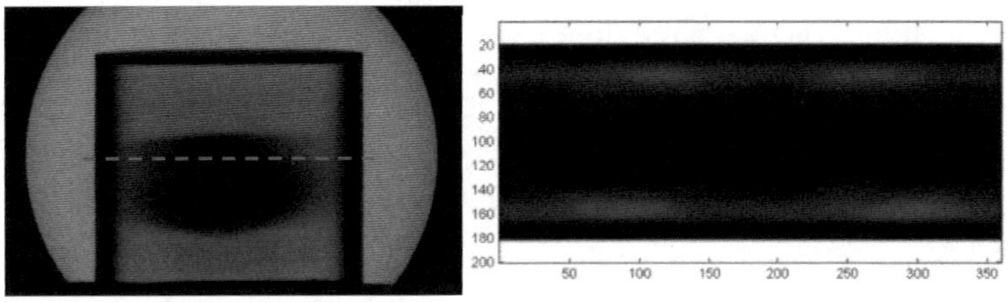

Figure 3. Left image is the projection image of irradiated gel phantom acquired by a CCD camera. Right image is the sinogram of 360 projections at 40 mm depth (from the top of gel container).

3.2. Reconstruction

After acquiring 360 projections, an inverse-radon transform algorithm was implemented in MATLAB (The Math-Works, Inc.) to reconstruct the image. Figure 4 illustrates the reconstruction images of the CT-s2 (upper row) and OCTOPUS™-10× (lower row) optical CT scanner. The gel phantoms are the same for both scanners but the FOV is different because the design of both scanner is different. The image processing technique used for the results in this study is wiener2, a 2D noise-removal filtering function built in MATLAB.

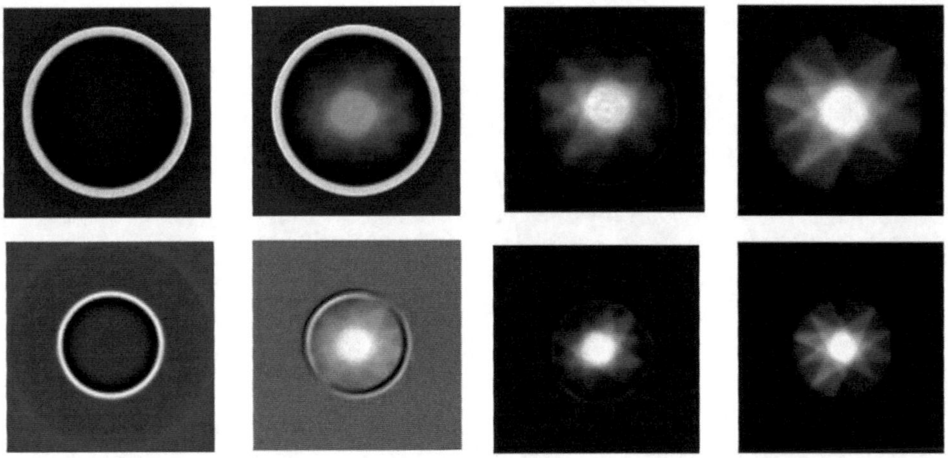

Figure 4. Reconstruction images of transverse slice at 40 mm depth. The images above are the results of CCD-based optical CT. The images below are the results of OCTOPUS™-10× optical CT scanner. The images from left to right: non-irradiated gel, irradiated gel, subtraction of irradiated gel and non-irradiated gel, the TPS image.

3.3. Gamma evaluation

A quantitative evaluation method was adopted in this study. The gamma evaluation technique [17] compares the measured and reference doses by using composite of dose-difference and distance-to-agreement (DTA) comparison criteria. The TPS image shown in figure 4 was used as reference for gamma passing rate calculations for both CT-s2 and OCTOPUS^{TM}-10× optical CT scanner. The criteria are 3% dose-difference and 3 mm DTA. The pass rate is 96.00% and 95.13% for CT-s2 and OCTOPUS^{TM}-10× optical CT scanner, respectively.

4. Conclusion

This study showed that a CCD-based parallel optical CT used as gel dosimeter readout tool performs as well as a single-beam optical CT. Better optical alignment, best fit matching liquid, and mechanical positioning design enhanced the acquired images. Hence, less image processing techniques are needed. However, the adaptive noise-removal filter for parallel-beam CCD-based optical CT scanner improves image reconstruction without performing an average technique for each projection. The results indicated that a fast, accurate, and high-resolution dosimeter can be achieved by improving the positioning accuracy, uniform light plan, and image processing technique.

5. Acknowledgements

This work was funded by the National Science Council, Executive Yuan, Taipei, Taiwan under Grants NSC 102–2314–B–166-003, NSC 101–2314–B–166–005, and NSC 99–2632–B–166–001–MY3.

6. References

[1] Baldock C *et al* 2010 *Phys. Med. Biol.* **55** R1-63
[2] Bosi S *et al* 2009 *Phys. Med. Biol.* **54** 275-83
[3] Gore J C *et al* 1996 *Phys. Med. Biol.* **41** 2695-704
[4] Wuu C S *et al* 2006 *J. Phys.: Conf. Ser.* **56** 239-95
[5] Xu Y *et al* 2004 *Med. Phys.* **31** 3024-30
[6] Baldock C 2006 *J. Phys.: Conf. Ser.* **56** 14-22
[7] Xu Y *et al* 2013 *Phys. Med. Biol.* 58 479-95
[8] Hsieh B T *et al* 2013 *IEEE Trans. Nucl. Sci.* **60** 560-5
[9] Doran S J *et al* 2001 *Phys. Med. Biol.* **46** 3191-213
[10] Sakhalkar H S *et al* 2008 *Med. Phys.* **35** 101-11
[11] Papadakis A *et al* 2010 *IEEE Trans. Med. Imag.* **29** 1204-12
[12] Thomas A *et al* 2011 *Med Phys.* **38** 4846-57
[13] Senden R J 2006 *Phys. Med. Biol.* **51** 3301-14
[14] Chang Y J 2011 *Nucl. Instrum. Meth. A* **652** 78-5
[15] Hsieh B T 2011 *J. Radioanal Nucl. Chem.* **290** 141-8
[16] Chen C H 2014 *Radiat. Phys. Chem.* (In press)
[17] Low D A *et al* 1998 *Med. Phys.* **25** 656-61

PAGAT gel dosimeters for dose distribution measurements in the vicinity of high-density implants: A preliminary study

A Asena[1], T Kairn[1,2], S B Crowe[1], S T Smith[1] and J V Trapp[1]
[1]School of Chemistry, Physics and Mechanical Engineering, Queensland University of Technology, Brisbane, Australia
[2]Genesis CancerCare Queensland, Brisbane, Australia

E-mail: a.asena@qut.edu.au

Abstract. This work examined the suitability of the PAGAT gel dosimeter for use in dose distribution measurements around high-density implants. An assessment of the gels reactivity with various metals was performed and no corrosive effects were observed. An artefact reduction technique was also investigated in order to minimise scattering of the laser light in the optical CT scans. The potential for attenuation and backscatter measurements using this gel dosimeter were examined for a temporary tissue expander's internal magnetic port.

1. Introduction

The success or failure of radiation therapy treatments is largely dependent on the ability to deliver the prescribed dose to the patient within a narrow tolerance. However, the human body consists of many components that vary in density which have been shown to cause complications in treatment planning and hence delivery [1, 2]. These can include naturally occurring internal inhomogeneities, as well as man-made biomaterials such as implants and prostheses. While these may be necessary for the survival or quality of life of radiotherapy patients, they also have the potential to seriously compromise the accuracy of radiotherapy treatment planning, and hence delivery. This work examines the feasibility of using the PAGAT gel dosimeter [3] for dose distribution measurements around high-density implants.

Gel dosimeters have the capacity to record radiation dose distributions in three-dimensions (3D) making them a very promising dosimetric tool for external beam radiotherapy [4-13]. It has been shown that the PAGAT gel formulation is approximately radiologically water equivalent for electron and photon beams [14] and is therefore an ideal 3D dosimeter for studying the effects of implants on patient dosimetry. The main advantage of using a gel dosimeter for this work is that complex shaped implants can be investigated due to their complete immersion in the gel thus alleviating the requirement of simplified geometry. Using PAGAT gel also means the conditions of the implant placement in the patient are matched. Here we present an examination of the suitability of using PAGAT gel dosimeters for dose distribution measurements in the vicinity of a temporary tissue expander's internal magnetic port.

2. Methodology

2.1 System optimisation
Before the PAGAT gel dosimeter was used for dose distribution measurements around high-density implants, an assessment of the reactivity with various metals was performed. Objects composed of stainless steel, brass, and zinc were suspended in the mixture and observed over a period of 7 days to examine any potential corrosive effects. An artefact reduction technique was also investigated whereby metal objects suspended in the gel formulation were imaged in the optical CT scanner with, and without a coat of black paint to determine the extent of scattering of the laser light in the imaging system.

2.2 Manufacturing the PAGAT gel
As per the recipe recommended by Venning *et. al.* [3] and Khoei *et. al.*[15], the PAGAT polymer gel formulation by % mass consisted of 4.5% N,N'-methylenbis-acrylamide (bis), 4.5% acrylamide (AA), 5% gelatine, 7 mM tetrakis (hydroxymethyl) phosphonium chloride (THPC), 0.01 mM hydroquinone (HQ) and 86% HPLC (Water). When the preparation of the final polymer gel solution was completed, it was transferred into phantom along with the implant and allowed to set by storage in a refrigerator.

2.3 Gel Irradiation
The internal magnetic port (IMP) of a Mentor temporary tissue expander was extracted from the implant and suspended in a container of gel. The magnetic disk (1.5cm diameter, 0.5cm thickness) allows the position of the valve inside the patient's body to be determined. However, this IMP is composed of a high-density neodymium magnet cased in titanium, which has the potential to compromise the accuracy of radiotherapy treatment planning, illustrating the need for 3D dose distribution measurements. The gel container was irradiated with a 5x5cm^2, 6MV photon beam from a Varian 21iX Clinac. The post-manufacturing time was selected to be 24 hours.

2.4 Imaging
The container was imaged in a MGS Research IQScan optical CT scanner specifically designed for readout of irradiated gel dosimeters. The imaging parameters included a field of view (FOV) of 220mm, 720 projections, a 360° rotation range, allowing a resolution of 0.5mm pixels. This pre- and post-scan data was reconstructed using filtered-back projection to create a 3D model of the dose distribution surrounding the high-density implant.

3. Results and Discussion
The results displayed in figure 1 demonstrate that the application of a non-reflective coating to the suspended metallic object has substantially reduced imaging artefacts in the optical CT scanner. Additionally, no corrosive effects were observed for the various metal objects that were submerged in the PAGAT gel formulation.

The radiation-induced polymerisation of the PAGAT gel formulation displayed in figure 2 shows a visible dose depletion downstream of the magnetic disk. However, the suspension angle of the internal magnetic port prevented the optical CT scanner from detecting the under-dosed region. This issue is resolved by simply implementing a horizontal, and therefore symmetrical, suspension method for the implant that allows the scanning laser to traverse this under-dosed volume.

Figure 3 shows a single reconstructed slice produced by the optical CT scanner with a resolution of 0.5mm. Also included is a normalised profile of the slice; thus employing internal calibration for optimal dose to gel [16]. This image slice illustrates that dose measurements in the proximity of the container wall are susceptible to edge artefacts and therefore should be avoided. Similarly, the cylindrical shape of the metallic implant makes it difficult to discern whether the profile peaks visible in figure 3 are the result of the production of secondary particles in the metal causing increased polymerisation, or more likely an increase in the laser light scatter.

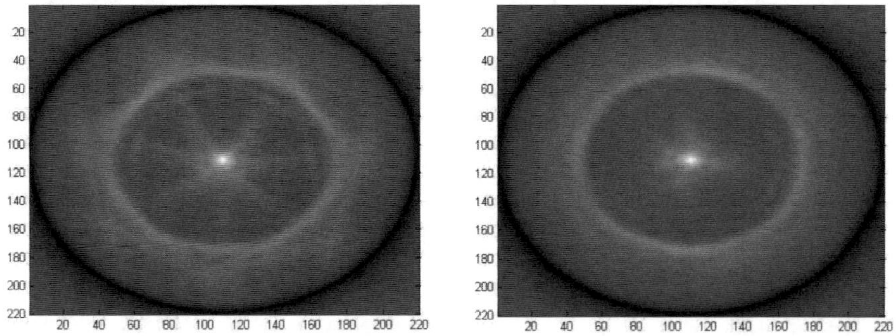

Figure 1. Image slices of stainless steel bolt uncoated (left) and coated (right) with black paint submerged in PAGAT gel. These were obtained using the optical CT scanner at a resolution of 1mm.

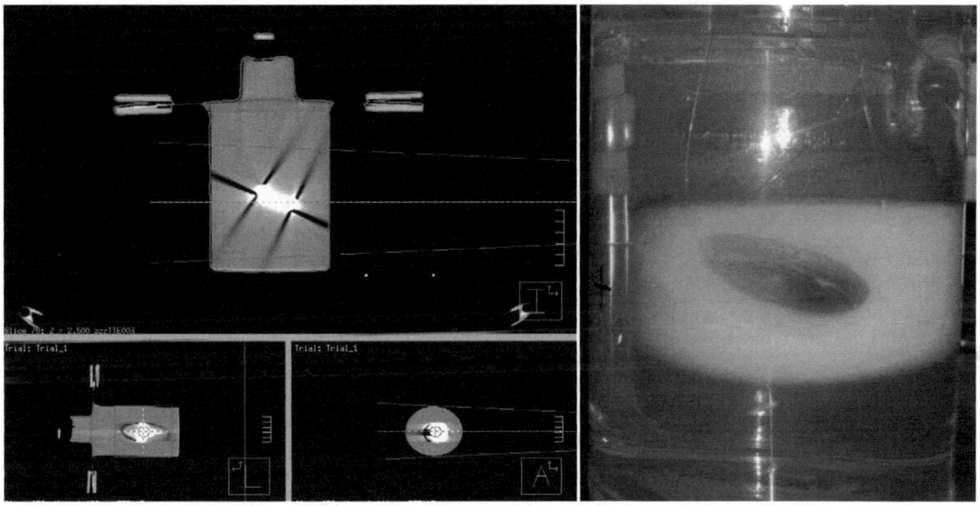

Figure 2. An example treatment being delivered to the gel container is shown (left) alongside a photo of the PAGAT gel container with the suspended internal magnetic port post-irradiation (right).

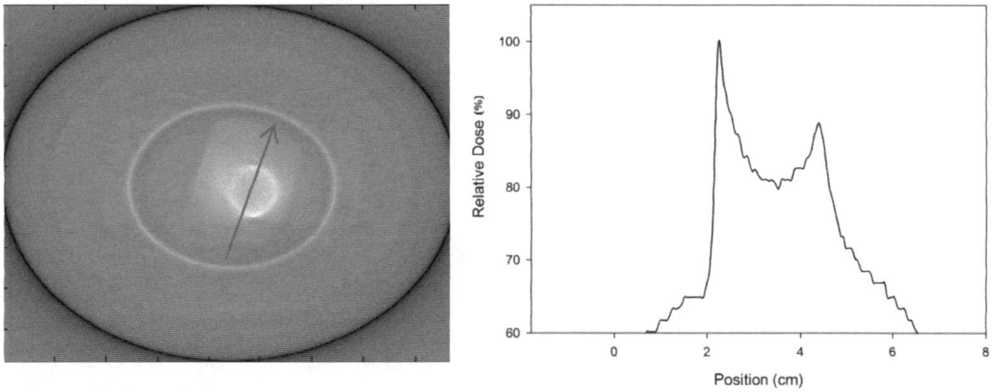

Figure 3. Optical CT image slice of gel container with 0.5mm resolution with an indicator of profile location and beam direction (left) with corresponding profile plot (right).

Despite the absence of qualitative data regarding attenuation effects of the implant, this study demonstrated the viability of using the PAGAT gel dosimeter for attenuation and backscatter measurements of a temporary tissue expander's internal magnetic port.

4. Conclusions

The suitability of the PAGAT gel dosimeter for use in dose distribution measurements around high-density implants was evaluated. This work demonstrated the potential that with an optimized gel dosimetry system, various implants can be placed into phantoms and followed through the clinical scenario of CT scanning, treatment planning, irradiation, and evaluation.

5. Acknowledgments

This work was partly funded by Australian Research Council Linkage Project LP110100401 in conjunction with the Wesley Research Institute.

6. References

[1] Kairn T *et al* 2013 *Australas. Phys. Eng. Sci. Med.* **36** 209-17
[2] Kairn T *et al* 2013 *J. Phys.: Conf. Ser.* **444** 012108
[3] Venning A J *et al* 2005 *Phys. Med. Biol.* **50** 3875-88
[4] Trapp J V *et al* 2001 *Phys. Med. Biol.* **46** 2939-51
[5] Trapp J V *et al* 2002 *Phys. Med. Biol.* **47** 4247-58
[6] Lepage M *et al* 2002 *Phys. Med. Biol.* **47** 1881-90
[7] Mather M L *et al* 2003 *Ultrasonics* **41** 551-9
[8] Trapp J V *et al* 2004 *Phys. Med. Biol.* **49** N139-46
[9] Venning A J *et al 2004 J. Phys.: Conf. Ser.* **3** 155-8
[10] Baldock C 2006 *J. Phys.: Conf. Ser.* **56** 14-22
[11] Bosi S *et al* 2009 *Phys. Med. Biol.* **54** 275-83
[12] Gorjiara T *et al* 2011 *Med. Phys.* **38** 2265-74
[13] Baldock C *et al* 2010 *Phys. Med. Biol.* **55** R1-63
[14] Brown S *et al* 2008 *Appl. Radiat. Isot.* **66** 1970-4
[15] Khoei S *et al J. Phys.: Conf. Ser.* **250** 012019
[16] Taylor M L *et al* 2007 *Phys. Med. Biol.* **52** 3991-4005

Photon beam dose distributions for patients with implanted temporary tissue expanders

A Asena[1], T Kairn[1,2], S B Crowe[1] and J V Trapp[1]
[1]School of Chemistry, Physics and Mechanical Engineering, Queensland University of Technology, Brisbane, Australia
[2]Genesis CancerCare Queensland, Brisbane, Australia

E-mail: a.asena@qut.edu.au

Abstract. This study examines the effects of temporary tissue expanders (TTEs) on the dose distributions of photon beams in breast cancer radiotherapy treatments. EBT2 radiochromic film and ion chamber measurements were taken to quantify the attenuation and backscatter effects of the inhomogeneity. Results illustrate that the internal magnetic port present in a tissue expander causes a dose reduction of approximately 25% in photon tangent fields immediately downstream of the implant. It was also shown that the silicone elastomer shell of the tissue expander reduced the dose to the target volume by as much as 8%. This work demonstrates the importance for an accurately modelled high-density implant in the treatment planning system for post-mastectomy breast cancer patients.

1. Introduction

High-density implants and prostheses are known to cause local increases in dose due to lateral-scatter and backscatter of radiation, in addition to depletion of downstream dose due to attenuation [1]. The size of these dose perturbations due to the high-density media depends on the design and composition of the implant and characteristics of the incident radiation. It is currently recommended to avoid treatments where the fields pass through the high-Z media before they reach the planning target volume (PTV) to reduce uncertainty in the treatment planning stages [2] and hence the delivered dose [3]. However, this is sometimes not possible and often results in a sub-optimal treatment being delivered to the patient to avoid uncertainties in the delivered dose. This work aims to establish the dosimetric effects of a temporary tissue expander (TTE), which is a device that cannot be avoided during postmastectomy radiotherapy treatments.

It is well established that postmastectomy radiation therapy improves survival for selected breast cancer patients [4]. However, many of these patients also desire a cosmetic breast reconstruction after mastectomy, which involves the subcutaneous implantation of a TTE. The shell of the TTE is composed of a silicone elastomer membrane, which is a chemically and mechanically resistant material and is reported to be radiologically water equivalent. Additionally, a magnetic disk allows the position of the implant's valve to be determined inside the patient's body. However, this high-density disk has the potential to compromise the accuracy of radiotherapy treatment planning, and hence delivery. Here we present an investigation into the impact of the TTE on photon beam dose distributions using EBT2 film and ion chamber measurements.

2. Methodology

2.1 EBT2 film measurements

The Mentor TTE was first investigated in a radiotherapy treatment using EBT2 radiochromic film in a tissue and lung phantom. The implant was placed on 10cm planar phantom composed of an arrangement of water and lung equivalent materials with a 1cm layer of bolus positioned over the top. Then using a 6MV photon beam at an incidence perpendicular to the heterogeneity, the TTE was irradiated. The radiochromic film was placed 0cm, 2cm and 8cm downstream of the implant as well as one piece of film immediately upstream of the implant to measure any backscatter caused by the high-z material. The implant was filled with 250cc of a 0.9% saline solution before being irradiated isocentrically with $15x15cm^2$ fields at a gantry angle of 0°.

Film measurements were also taken using a CIRS IMRT thorax phantom with segmented, tangent photon fields being delivered at angles of 70° and 250° this time with the TTE filled with 400cc of the saline solution. This beam arrangement was adopted from a clinical plan and used a forward planned IMRT (field-in-field) technique which has been shown to provide more homogeneous dose distributions in the PTV and reduced doses in the organs at risk [5]. Each piece of film was scanned and evaluated as per the protocol outlined in Kairn [6] and Aland [7] in order to minimise the effects of film heterogeneity and scanner output variations. The dose reduction values reported throughout this study were calculated as the percentage difference between the average doses outside the shadow of the implant, relative to the doses in the region directly under the magnetic inhomogeneity. Measurement uncertainties were taken as the standard deviation of doses in each region.

2.2 Ion chamber measurements

The internal magnetic port (IMP) was removed from its silicone housing and depth dose and profile measurements were taken underneath with a Roos ionization chamber (Type 34001) and a water tank in 6 and 10 MV photon fields. These measurements were repeated in water with no IMP present for comparison. A field size of $15x15cm^2$ was used. The Varian 21iX linac was continuously delivering dose at a rate of 600MU/min while the ionisation chamber recorded a series of cross-profiles and percentage depth doses. Photon profile depths were at 2, 5, 10, 20 and 30 cm for both the 6 and 10 MV beams. Percentage depth doses were taken from 2cm onwards due to a minimum scanning depth caused by the presence of the IMP.

3. Results and Discussion

3.1 EBT2 film measurements

Figure 1 (a) shows the dose profile of the implant at three different depths, downstream of the breast implant in the planar phantom for a 6MV photon beam at 0° gantry. At 0cm, in the region directly beneath the magnetic valve, the dose is reduced by as much as 15 ± 3 %. This figure also reports a 12 ± 2 % dose reduction at 2cm and an 11 ± 2 % dose reduction as far as 8cm downstream of the magnetic disk. No backscatter dose enhancements were reported in the radiochromic film and were therefore not included.

Figure 1 (b) describes the measured attenuation effects of each component of the implant taken in the CIRS IMRT thorax phantom. It was determined that profiles downstream of the silicone elastomer/saline interface averaged doses around 8% lower when compared to profiles taken outside the implant's shadow. Differences for profiles taken under the titanium case, titanium ring and neodymium magnet were approximately 12%, 15% and 19% respectively.

Photon tangent fields were also delivered to the thorax phantom and the results in figure 1 (c) illustrate a 25% dose reduction caused by the high-density magnet in both treatment directions. Given that post-mastectomy treatments are typically delivered as per this experimental setup, the impact of the TTE on photon dose distributions would be significant if it's not accurately accounted for in planning. The insets indicate in figure 1 (b) indicate the approximate profile locations as well as pictographically demonstrating the perturbation caused by the TTE's internal magnet and silicone elastomer shell.

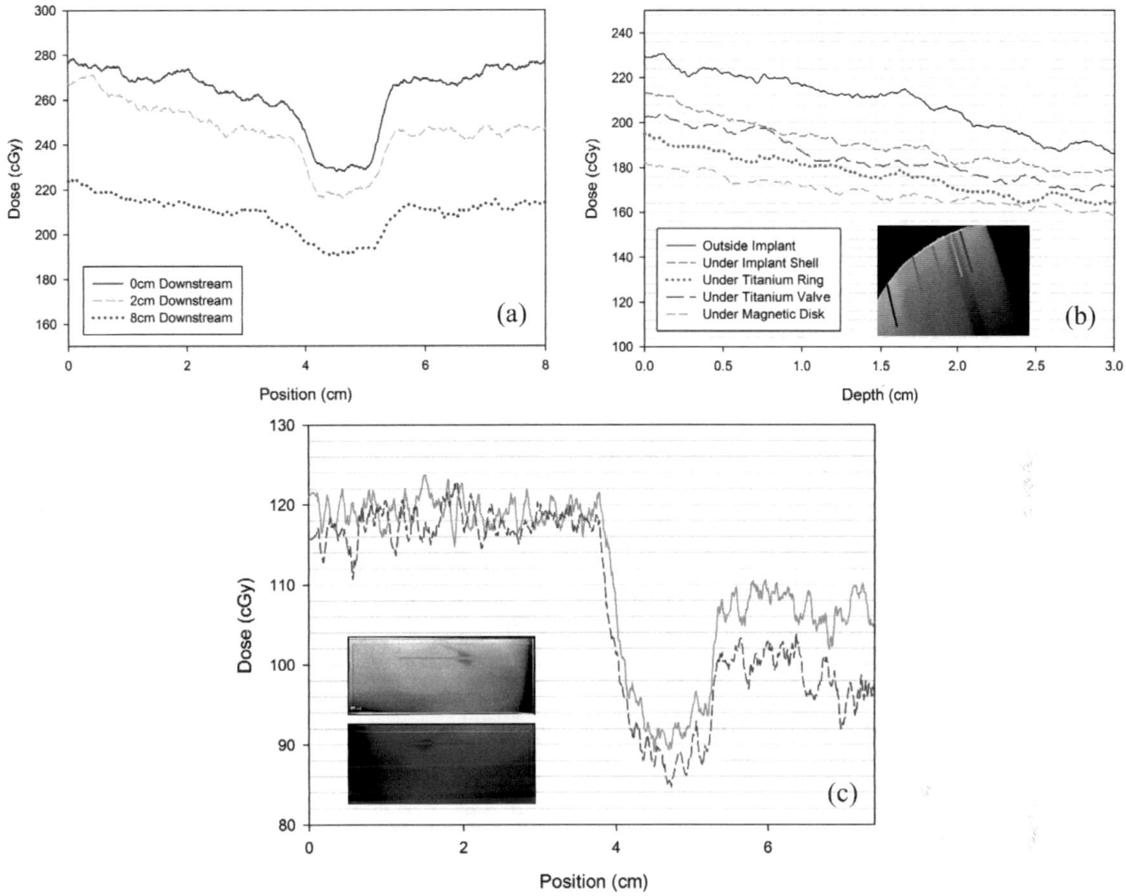

Figure 1. Film dose profiles in; (a) Planar tissue-lung phantom at 0cm, 2cm and 8cm downstream of the implant, (b) CIRS thorax phantom downstream of different components in the implant, and (c) CIRS thorax phantom for two photon beam tangents at gantry angles of 70° and 250°.

3.2 Ion chamber measurements

Figures 2 and 3 show dose reductions of 8% and 7% at a depth of 2cm for the 6MV and 10MV beams, respectively which are similar to our results obtained using radiochromic film. Reductions of 2% and 3% were present for both energies as far down as 30cm. While these results don't offer a direct clinical perspective on dose depletions caused by the implant, they do illustrate the effects that inhomogeneities have on photon dose distributions. In addition to the cold spots beyond the inhomogeneity caused by the greater mass scattering power in the IMP, hot spots lateral to the inhomogeneity are also produced and are most evident at 2cm in the 10MV profile in figure 2.

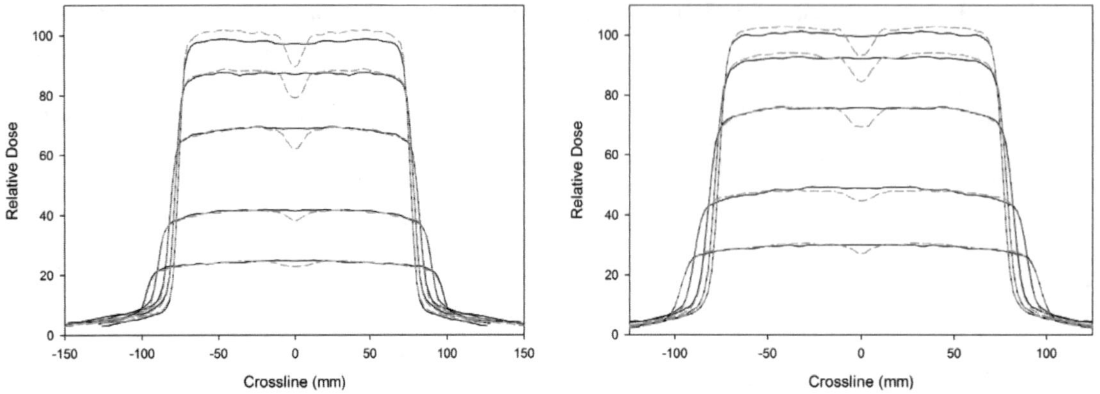

Figure 2. Cross profiles with (green) and without (black) the IMP at depths of 2, 5, 10, 20 and 30 cm for (a) 6MV and (b) 10MV photons.

4. Conclusions

This work indicates that the magnetic disk present in a tissue expander causes a dose reduction of approximately 25% in photon tangent fields immediately downstream of the implant. The silicone elastomer shell of the Mentor implant has also been shown to reduce the dose to the target volume by as much as 8%, which in turn reduces the probability of tumour control. Each component of the TTE attenuates the radiation beam to different degrees. This highlights the importance for an accurately modelled high-density implant in the treatment planning system for post-mastectomy patients.

5. Acknowledgments

This work was partly funded by Australian Research Council Linkage Project LP110100401 in conjunction with the Wesley research institute.

6. References

[1] Kairn T *et al* 2013 *Australas. Phys. Eng. Sci. Med.* **36** 209-17
[2] Kairn T *et al* Trapp JV 2013 *J. Phys. Conf. Ser.* **444** 012108
[3] Reft C *et al* 2003 *Med. Phys.* **30** 1162-82.
[4] Chen S A *et al* 2013 *Rad. Oncol.* **8** 45
[5] Al-Rahbi Z S *et al* 2013 *J. Med. Phys.* **38** 22-9
[6] Kairn T *et al* 2010 *Phys. Med. Biol.* **55** 37-42
[7] Aland T *et al* 2011 *Australas. Phys. Eng. Sci. Med.* **34** 251-60

Journal of Physics: Conference Series **573** (2015) 012063 doi:10.1088/1742-6596/573/1/012063

Artifacts suppression in optical CT for gel dosimeters by iterative reconstruction

Du Yi, Wang Xiangang and Xiang Xincheng
Institute of Nuclear and New Energy Technology, Tsinghua University, Beijing, 100084, China

E-mail: wangxiangang@tsinghua.edu.cn

Abstract. Optical CT has been considered as an important and promising readout method for 3D gel dosimetry. However, tomographic image qualities are often corrupted by artifacts such as streaks and rings, which are induced by projection discontinuities and magnified by FBP reconstruction. These artifacts will surely deteriorate the accuracy and precision of dose measurement. In this paper, we performed a preliminary study on our in-house optical CT scanner using an iterative algorithm instead of the commonly used FBP for image reconstruction. Comparative analysis of the results validates the proposed method in artifacts suppression and image quality improvement when the convergent process is properly controlled.

1. Introduction

Gel dosimeters such as PAGs and FXGs have potential advantages over conventional dosimeters especially for high resolution and tissue equivalent 3D dosimetry, which serves suitable for 3D dose verification in radiation therapy quality control and assurance (QC&A) [1-4]. After exposure to ionizing radiation, the optical densities (ODs) of gel dosimeters change quantitatively to absorbed doses, and thus 3D dose information is stored in the form of OD distribution, which can be read out by optical CT. Several prototype and two commercial optical CT scanners have been proposed [5]. Key components of a typical optical CT scanner consist of a visible source, a water tank with matching liquid, a sample gel dosimeter and a detector. Since incident light rays pass through several interfaces before finally reaching the detector, interface imperfections such as surface damages and stains will introduce discontinuities into projection signals. Unfortunately, these errors will be manifested in the form of image artifacts such as streaks and rings after filtered back-projection (FBP) reconstruction, which will magnify discontinuities in the projection filtering [6]. In order to suppress streak and ring artifacts in optical CT, a lot of work has been done in projection preprocessing. To achieve the same goal, we utilized an iterative reconstruction algorithm instead of the classic FBP in this paper.

2. Method

2.1. Optical CT scanner & polymer gel dosimeter

We have built an in-house prototype optical CT scanner showed in figure 1, with a Hg light as the source and a CCD camera as the detector. Incident light rays are monochromized by a 580nm filter, and are collimated by an aperture [7]. The sample container is suspended and rotated from its top. 400 frames of projection signals were obtained at equal intervals over 360 degrees, and each frame consisted of 50 rows. The source-to-iso distance is so large as 900mm that the scanning geometry can be approximated to the parallel beam geometry.

A polymer gel dosimeter known as MAGIC was in-house made in strict accordance with [8] and used in this study. The MAGIC sample was held in a PMMA container, and a square field at the center was irradiated uniformly from the top by a medical linear acceleration for radiation therapy. Figure 2 shows one frame of the raw projections of the irradiated sample with imperfection of two visible seams on the container.

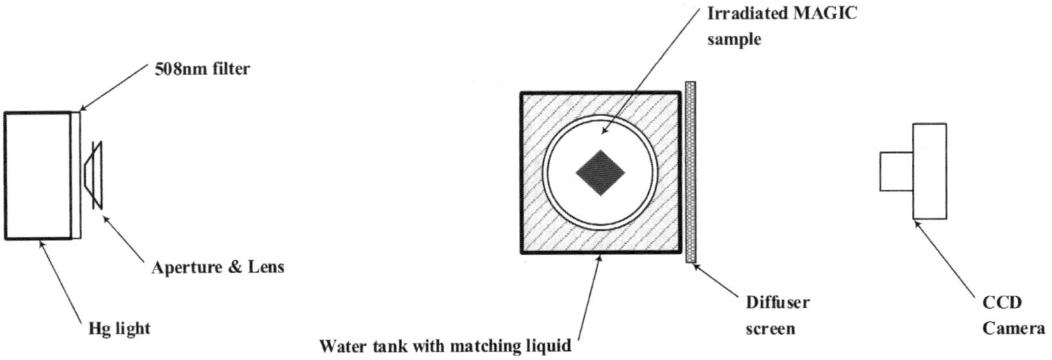

Figure 1. Schematic of the prototype optical CT scanner.

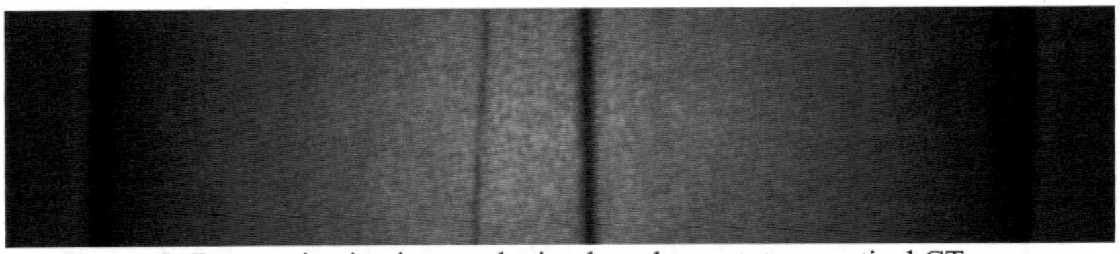

Figure 2. Raw projection image obtained on the prototype optical CT scanner.

2.2. Projection preprocessing

The sample gel dosimeter was scanned twice in this study: before the irradiation (P_{before}) and after the irradiation (P_{after}). Background noise were also obtained in the process. The logarithmic projections (g) used for reconstruction were calculated by the following equation (1) [9]:

$$g_\theta = -\ln\left(\frac{P_{after} - Noise_{after}}{P_{before} - Noise_{before}}\right)_\theta \quad (1)$$

Since the cylindrical gel container interfaces reflected and refracted the incident rays, and the ray path deflection got larger near the edges of the container, the bilateral outer parts of the projections were discarded to eliminate edge reflection and refraction related errors.

2.3. Simultaneous iterative reconstruction technique (SIRT) with total variance (TV) minimization

Several iterative CT reconstruction algorithms such as ART, SART, SIRT, ML-EM have been under research for long and applied in certain areas. Compared to the classic FBP algorithm that is analytic in nature, iterative algorithms can be more exact in physical modelling and capable of incorporating prior-knowledge in reconstruction process.

In this study, we used SIRT for optical CT reconstruction, which stands out for great stability, noise suppression and convergence. The algorithm can be expressed as equation 2 [10]:

$$\begin{cases} f^{(k+1)} = f^{(k)} + \lambda R^T (g - Rf^{(k)}) \\ \quad s.t. Rf^{(*)} = g \end{cases} \tag{2}$$

where R is the system matrix, f is the target tomographic image to be reconstructed, g is the logarithmic projections, and λ is the relaxation factor.

Besides, total variance (TV, defined in equation 3) [11] minimization was also added into the iteration. TV minimization is a piece-wise based optimization rule, and has shown its high performance in edge-preservation and noise reduction. It's introduced here for further artifact reduction.

$$TV(f) = \int_x \int_y \sqrt{\left(\frac{\partial f}{\partial x}\right)^2 + \left(\frac{\partial f}{\partial y}\right)^2} \, dx \, dy \approx \sum_{i,j} \sqrt{(\Delta f_i)^2 + (\Delta f_j)^2} \tag{3}$$

where (i, j) is the pixel location index of the tomographic image.

3. Results & discussion

For comparison, tomographic images were reconstructed respectively by FBP and SIRT (TV). The projection preprocessing and reconstruction software was in-house developed in C++ language on the *Visual Studio 2012*^{TM} platform. All images were of 208*208 pixels, and each pixel was 0.18*0.18 mm^2. For illustration, figure 3 shows the middle plane (Slice 25th) results by FBP (a) and SIRT (TV) at different iteration times (10, 30, 50, 100, 200) with zero initials.

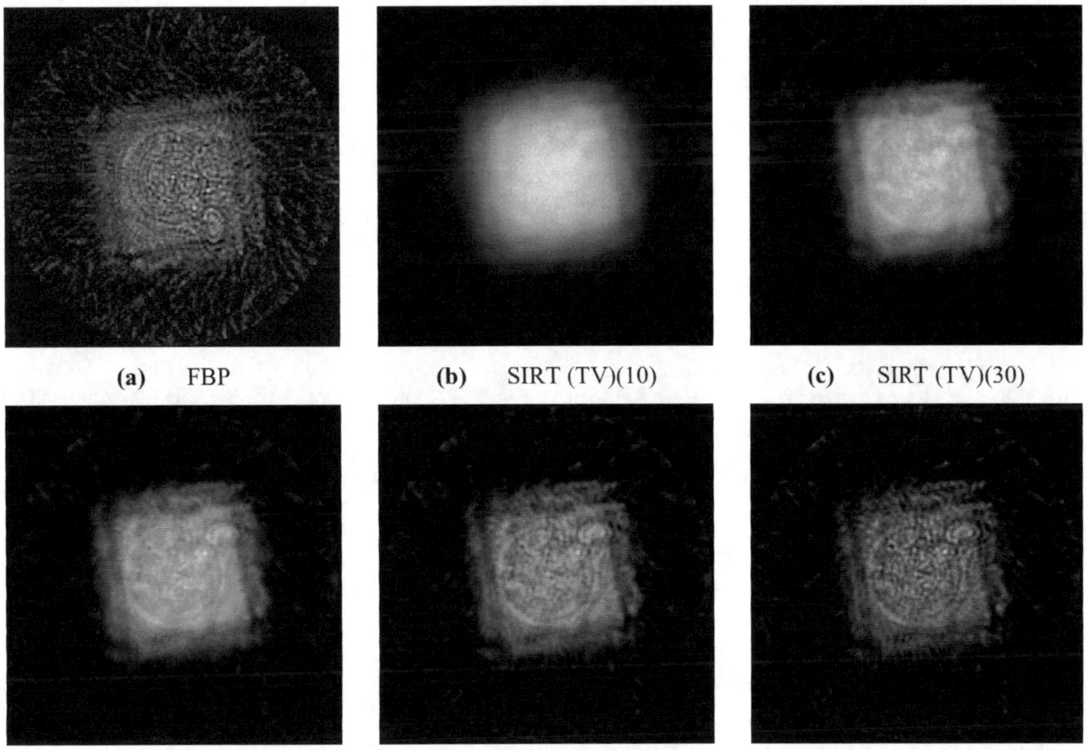

(a) FBP **(b)** SIRT (TV)(10) **(c)** SIRT (TV)(30)

(d) SIRT (TV)(50) **(e)** SIRT (TV)(100) **(f)** SIRT (TV)(200)

Figure 3. Reconstruction results of the middle plane by FBP and SIRT (TV)

In figure 3(a), streaks and rings are easy to identify as well as high noise, which are probably related to the imperfect surfaces of the water tank and sample container. Figure 3(b)~(f) shows the convergent process of SIRT (TV). It's found that the image qualities didn't get improved all the way as the iteration continued. At the start, low frequent information of the image, such as the shape of the irradiated area, was reconstructed. Then, the high frequent details, say the edges, got restored, while noise and artifacts also became visible and began to deteriorate at the same time (this could be attributed to the poor quality projection data which the iteration algorithm used as the convergent target). Therefore, the iteration process should not go too far, in case of serious artifacts as well as long calculation time.

Figure 3(c) was selected as the optimal SIRT (TV) result, and its middle profile was drawn in figure 3 as well as that of figure 3(a). From figure 4, we can clearly see that the large vibration, which represents sharp artifacts, has been well suppressed as we desire.

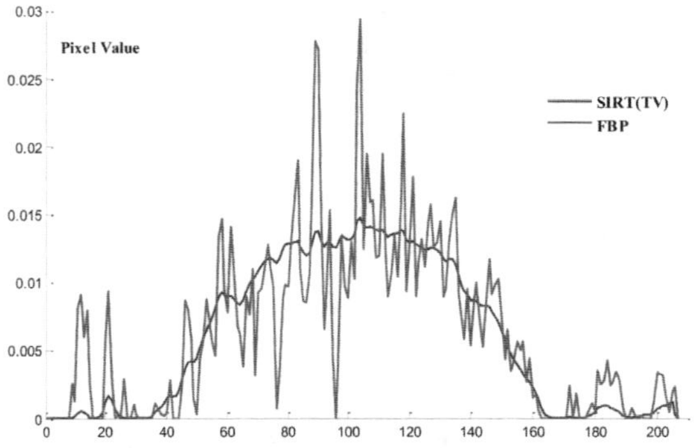

Figure 4. Profiles of reconstructed images by different algorithms

4. Conclusion

We have performed a preliminary study on artifact suppression by using iterative algorithm instead of the widely used FBP for optical CT reconstruction on an in-house optical CT scanner with MAGIC gel dosimeters. The results by SIRT (TV) showed much fewer artifacts and less noise than that of FBP, when the iteration time was properly selected. The profile comparison demonstrated the effectiveness and efficiency of the iterative reconstruction algorithm in optical CT image quality improvement with streak and ring artifact suppression.

5. Acknowledgements

Work supported in part by the National Natural Science Foundation of China (NSFC) under Grant 61171115 and the Beijing Higher Education Young Elite Teacher Project under Grant YETP0158.

6. References

[1] Schreiner L J 2009 *J. Phys.: Conf. Ser.* **164** 012001
[2] Bosi *et al* 2007 *Phys. Med. Biol.* **52** 2893-903
[3] Bosi et al 2009 *Phys. Med. Biol.* **54** 275-283
[4] Baldock C 2006 *J. Phys.: Conf. Ser.* **56** 14-22
[5] Baldock C *et al* 2010 *Phys. Med. Biol.* **55** R1-63
[6] Hsieh J 2009 *Computed tomography: principles, design, artifacts, and recent advances* (Bellingham: SPIE)
[7] Wang P *et al* 2014 *CT Theory and Applications* **23** 37-43
[8] Fong P M *et al* 2001 *Phys. Med. Biol.* **46** 3105-13
[9] Krstajić N *et al* 2007 *Phys. Med. Biol.* **52** 3693-713
[10] Zeng G L 2010 *Medical Image Reconstruction* (Heidelberg: Springer)
[11] Sidky E Y *et al* 2008 *Phys. Med. Biol.* **53** 4777-807

Particle size analysis of PAGAT gel dosimetry

E J J Samuel, P Sathiyaraj, D Titus and D S Kumar
Photonics, Nuclear and Medical Physics Division, School of Advanced Sciences, VIT University, Vellore, India

E-mail: ejames@vit.ac.in

Abstract. This work represents the response of PAGAT gel dosimeter using UV-Visible Spectrophotometry. The particle size and optimal wavelength of the gel sample were analyzed from the obtained spectrum. In addition, the compressibility was estimated using Ultrasonic Interferometer. The results showed that the particle size of the PAGAT varied appreciably with respect to the dose applied but did not vary significantly with the post irradiation time.

1. Introduction

Understanding the chemistry of polymer gel dosimetry is worthy of attention. Jirasek *et al* [1] discussed the consequence of chemistry on gel dosimetry. Baldock *et al* [2] reviewed the basic radiation chemistry of the polymer gel dosimetry. Poly Acrylamide Gelatin and Tetrakis hydroxyl phosphonium chloride (PAGAT) is one of the members of polymer gel family. Radiation induced polymerization in PAGAT [3] is the witness of absorbed dose and the dose can be extracted from different techniques such as Magnetic Resonance Imaging (MRI) [4-6], Optical Computed Tomography (Optical CT) [7, 8], X-ray CT [9, 10], Ultrasound CT (UCT) [11-13] and vibrational spectroscopy [14-16]. Chemistry of radiation induced polymerization in PAGAT contains the processes such as water decomposition, radical formation, chain growth, cross linking and termination. Each process depends on different factors including concentration of ingredients, concentration of radicals, temperature, pH, dose, dose rate, post irradiation time, energy of interacting radiation and their types (charged particle, uncharged particle and photons). Among these, we considered the effects of dose and post irradiation time for analysing particle size of PAGAT. This study may handout one of the faces of the PAGAT gel dosimeter via particle size variation. Heather *et al* [17] measured the particle size for MAGIC-2 gel dosimetry for different doses and dose rates. In this study, spectrophotometer acts as an analytical tool to measure the particle diameter for PAGAT gel at different doses at various time intervals post irradiation. .

2. Materials and Method

2.1. Gel Preparation

We chose PAGAT gel due to its tissue equivalence property and stability [18]. PAGAT gel consisting of 3.5% (w/w) BIS ($C_7 H_{10} N_2 O_2$), 3.5% (w/w) acrylamide ($C_3 H_5 NO$), 5% (w/w) of gelatin ($C_{17} H_{32} N_5 O_6$) n, 10Mm of THPC ($C_4 H_{12} CLO_4 P$) and 89% of distilled water (mille pore) were prepared in normal atmospheric condition. The gel preparation is reported elsewhere [19]. After preparation, the solution was filled in plastic cuvettes of 4.5 ml volume and the cuvettes openings were sealed with transparent adhesive tapes to prevent oxygen entry. All the samples were refrigerated overnight for perfect gellation.

2.2. Gel Irradiation
Gel samples were irradiated with cobalt-60 (Theratron 780c). Source to surface distance (SSD) was maintained at 80 cm with the field size of 10 x 10 cm^2. Samples were irradiated from 3 Gy to 18 Gy in the range of 3 Gy intervals. After irradiation, the samples were maintained at 4 degree Celsius for 24 hours to attain complete polymerization.

2.3. Functional group identification
The functional group of PAGAT has been confirmed by Fourier Transform Infrared (FT-IR) spectrometer. 1654.92cm^{-1} and 1629.85 cm^{-1} represents the carbonyl group(C=O) associated with Acrylamide and Bisacrylamide. 3485.37 cm^{-1} represents the OH functional group associated with THPC and Gelatin.1550.77 cm^{-1} denotes the bending vibration of N-H in amide group. After polymerization the band 3400 cm^{-1} was shifted to 3450 cm^{-1} [20]. The shift of stretching N-H group from Acrylamide to Polyacrylamide indicated the presence of polymerization action. Figure 1 shows the IR-spectrum of the samples.

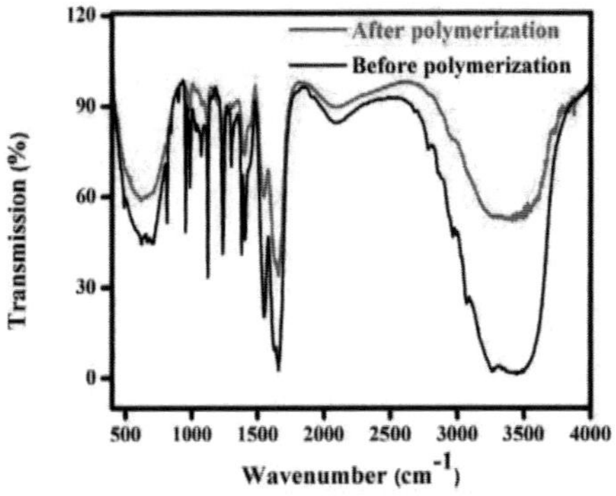

Figure 1. FTIR Spectrum of PAGAT.

2.4. UV Spectrophotometry for PAGAT
U2800-Hitachi was used for UV Spectrophotometry on PAGAT. The un- irradiated gel (0 Gy) acted as a control sample for irradiated gel, and it was used for baseline correction in spectrophotometer analysis. The absorption wavelength was selected from 350 to 800 nm with 1 nm interval.

The samples were read after one day of irradiation. The particle size was determined from the relation between maximum absorption and Mie-Debye efficiency factor. It is given by the following equation,

$$D = \frac{(Ka)\lambda_{\max}}{n\pi} \quad (1)$$

where, D is the diameter of the particle, Ka is the Mie-Debye efficiency factor which is found to be 4.34 [20], n is the refractive index of the sample which is calculated as 1.5.

2.5. Linearity and Particle size measurement
Linearity of the PAGAT was obtained by choosing the different absorption values such as 460nm, 480nm, 500nm and 550 nm. Samples were read at different time intervals of 24, 96 and 240 hours.

Journal of Physics: Conference Series **573** (2015) 012064 doi:10.1088/1742-6596/573/1/012064

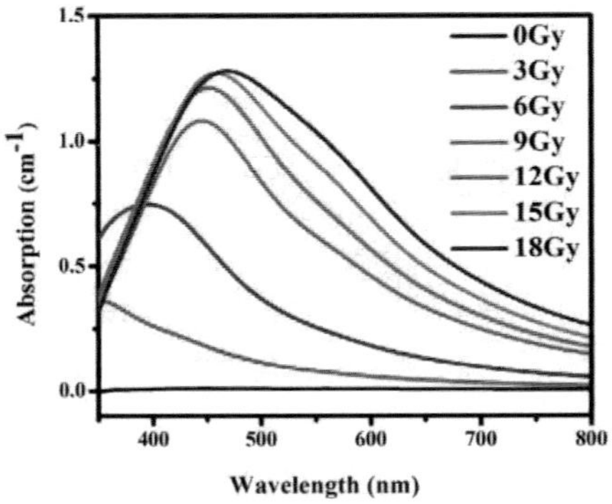

Figure 2. UV-Visible spectrum of PAGAT

Figure 3. Irradiated PAGAT gel

2.6. Refractive index and compressibility of PAGAT

The travelling microscope method was used for measuring the refractive index (RI) of non- irradiated gel. The same method was not used to determine the RI of exposed gel due to its opacity to light. Instead, Abbe's refractometer was used for measuring the RI on irradiated gel. The Ultrasonic interferometer was used for measuring the velocity of ultrasound in PAGAT at 27^{0}C.The compressibility of irradiated PAGAT gels was calculated by the following relation,

$$\beta = \frac{1}{\rho \upsilon^2} \tag{2}$$

where β is the compressibility, v^2 is the velocity of sound, ρ is density of the gel. The density of PAGAT was 1026 Kg/m^3 [21].

3. Result and Discussion

3.1. UV Spectrophotometry Analysis of PAGAT

Absorption of the PAGAT increased with respect to dose it is clearly shown in the absorption spectrum (figure 2). No other absorption peaks were seen in the spectrum up to 18 Gy. Particle size at different doses and time intervals as shown in figure 5.It shows that the particle size increased with increasing dose. Heather M *et al* [17] Maryanski *et al* [22] and Oldham *et al* [23] demonstrated particle size increased with increasing dose. But our study found that time intervals of post irradiation did not change the particle size significantly.

3.2. Linearity and particle size measurement

Figure 4 shows the linearity of PAGAT for different absorption wavelengths, It shows good correlation (R^2 = 0.9677) than other wavelengths. Figure 5 shows the particle size variation of PAGAT as a function of dose and post irradiation time. Post irradiation time was not influence in particle sizes of PAGAT.

3.3. Refractive index and compressibility of PAGAT

The travelling microscope method measured the refractive index of non- irradiated gel as 1.4899. Abbe's refractometer measured the RI of irradiated gel as 1.4998, 1.5101 and 1.5123 for 1 Gy, 2 Gy and 3 Gy respectively. The average value of 1.5 was considered as the RI at 26° C. The velocity of ultrasound in irradiated PAGAT was found to be 1520.49 ms^{-1}, 1555.5 ms^{-1} and 1570 ms^{-1} for 1 Gy, 2 Gy and 3 Gy respectively. The corresponding compressibility of the irradiated PAGAT gels was 4.2, 4.02 and 3.95 x 10^{-10} m^2/N for 1 Gy, 2 Gy and 3 Gy respectively. These values show that the compressibility of gel reduced with respect to dose which is due to the stiffness of gel after irradiation.

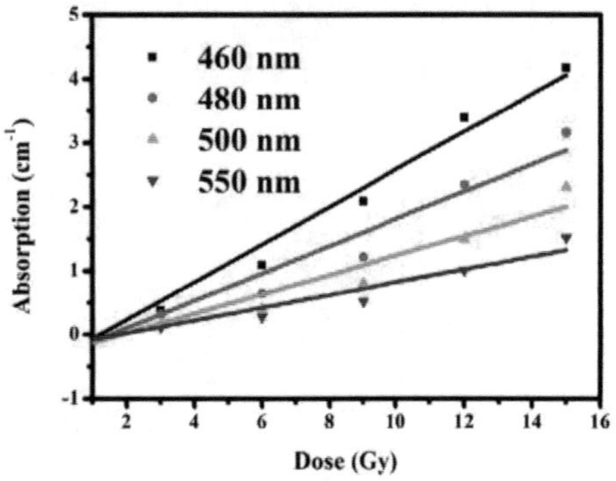

Figure 4. Absorption for different wavelength.

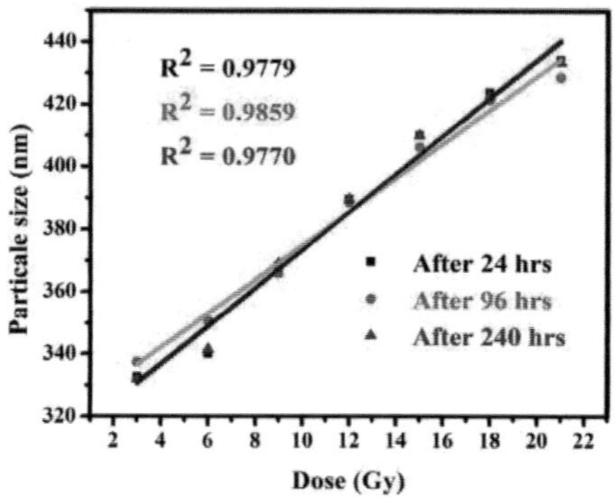

Figure 5. Particle size for diff. post-irradiation time.

4. Conclusion
The average particle size of PAGAT was found to be 385 nm and was not affected by the post irradiation time. The particle size analysis will be extended to several types of gel dosimeters with different composition and irradiation. The linearity of PAGAT was excellent at 460 nm, and this is the suggested optimal wavelength for PAGAT. The extended study will correlate the ultrasound velocity and molecular weight of PAGAT with respect to different dose rates.

5. Acknowledgements
The authors are grateful to the DST-SERB, Government of India for their financial aid and one of the author gratefully acknowledge the Research Associateship provided by the VIT University.

6. References
[1] Jirasek A *et al* 2009 *J. Phys.: Conf. Ser.* **164** 012003
[2] Baldock C *et al* 2010 *Phys. Med. Biol.* **55** R1-63
[3] Venning A J *et al* 2005 *Phys. Med. Biol.* **50** 3875-88
[4] Lepage M *et al* 2002 *Phys. Med. Biol.* **47** 1881-90
[5] Venning A J *et al* 2004 *J. Phys.: Conf. Ser.* **3** 155-8
[6] Murry P *et al* 2000 *Australas. Phys. Eng. Sci.* **23** 44-51
[7] Bosi S *et al* 2007 *Phys. Med. Biol.* **52** 2893-903
[8] Bosi S *et al* 2009 *Phys. Med. Biol.* **54** 275-83
[9] Hill B *et al* 2005 *Brit. J. Radol.* **78** 623-30
[10] Hill B *et al* 2005 *Med. Phys.* **32** 1589-97
[11] Mather M L *et al* 2002 *Phys. Med. Biol.* **47** 4397-409
[12] Mather M L *et al* 2003 *Ultrasonics* **41** 551-9
[13] Mather M L *et al* 2003 *Phys. Med. Biol.* **48** N269-75
[14] Baldock C *et al* 1998 *Phys. Med. Biol.* **43** 3617-27
[15] Lepage M *et al* 2001 *J. Appl. Polym. Sci.* **79** 1572-81
[16] Rintoul L *et al* 2003 *Appl. Spectrosc.* **57** 51-7
[17] Heather M *et al* 2006 *J. Phys.: Conf. Ser.* **56** 160-3
[18] Taylor M L *et al* 2008 *Australas. Phy.Eng. Sci. Med.* **31** 131-8
[19] Subramanian B *et al* 2006 *J. Med. Phys.* **31** 72-77
[20] Ozeroglu *et al* 2007 *eXpress Polymer Letter* **1** 132-41
[21] Brown S *et al* 2008 *Appl. Radiat. Isotope* **66** 1970-1974
[22] Maryanski M J *et al* 1996 *Phys. Med. Biol.* **41** 2705-17
[23] Oldham M *et al* 2003 *Med. Phys.* **30** 623-34

Antioxidant effect of green tea on polymer gel dosimeter

E J J Samuel[1], P Sathiyaraj[1], T Deena[1] and D S Kumar[1]
[1]Photonics, Nuclear and Medical Physics Division, School of Advanced Sciences, VIT University, Vellore, India

Email: ejames@vit.ac.in

Abstract. Extract from Green Tea (GTE) acts as an antioxidant in acrylamide based polymer gel dosimeter. In this work, PAGAT gel was used for investigation of antioxidant effect of GTE.PAGAT was called PAGTEG (Polyacrylamide green tea extract gel dosimeter) after adding GTE. Free radicals in water cause pre polymerization of polymer gel before irradiation. Polyphenols from GTE are highly effective to absorb the free radicals in water. THPC is used as an antioxidant in polymer gel dosimeter but here we were replaced it by GTE and investigated its effect by spectrophotometer. GTE added PAGAT samples response was lower compared to THPC added sample. To increase the sensitivity of the PAGTEG, sugar was added. This study confirmed that THPC was a good antioxidant for polymer gel dosimeter. However, GTE also can be used as an antioxidant in polymer gel if use less quantity (GTE) and add sugar as sensitivity enhancer.

1. Introduction

Advanced radiotherapy requires highly effective 3 dimensional dose verification systems for patient specific quality assurance. Polymer gel dosimeter is an emerging system and serves as good tool for dose verification in radiotherapy [1, 2]. Absorbed dose and the dose can be extracted from different techniques such as Magnetic Resonance Imaging (MRI) [3-5], Optical Computed Tomography (Optical CT) [6, 7], X-ray CT [8, 9], Ultrasound CT (UCT) [10-12] and vibrational spectroscopy [13-15].

Polymer gel consists of toxic ingredients such as acrylamide, Bis-acrylamide and THPC. Utilization of these ingredients is very harmful to individual and the environment. In this study, we replaced one of the ingredients (THPC) from pre-existing PAGAT recipe and analysed their performance for adaptability in clinical applications. Many natural products having good antioxidant activity such as Acai berry, red berry, green tea, turmeric etc [16-18] have good antioxidant activity. Among these, green tea easily available, it does not change in the colour of gel and is easy to prepare its extract. For these reasons, it was chosen for this study. Most of GTE polyphenols are flavonols known as catechin it contains (+) catechin, (-) E picatechin, (-) E pigalloctechin, (-) Epicatechin gallate, (-) Gallocatechin gallate, and (-) Epigallocatechin gallate etc. This catechin consumes the free radicals in water and protects the prepolymerization of gel. Figure 1 shows some major catechin from GTE [19]. Xuping Zhu *et al* [20] and Kwon *et al* [21] reported glucose and urea increase the sensitivity of polymer gel. Since common sugar is the disaccharide of glucose and fructose hypothesis was made that sugar may increase the sensitivity of polymer gel.

Journal of Physics: Conference Series **573** (2015) 012065 doi:10.1088/1742-6596/573/1/012065

2. Materials and Methods

2.1. Green extract preparation
Dried green tea leaves were commercially purchased (Tetley) from Allmart shop -VIT campus - Vellore (India). 1gm of leaves was soaked in 50 ml of Millipore water for 4 hours. The solution was stirred in very low RPM with a magnetic stirrer for 5 minutes. Then the solution was heated to 30 degree Celsius for 30 minutes. After the solution cooled down (Hereafter called Green tea extract) it was filtered and kept in room temperature for PAGTEG preparation.

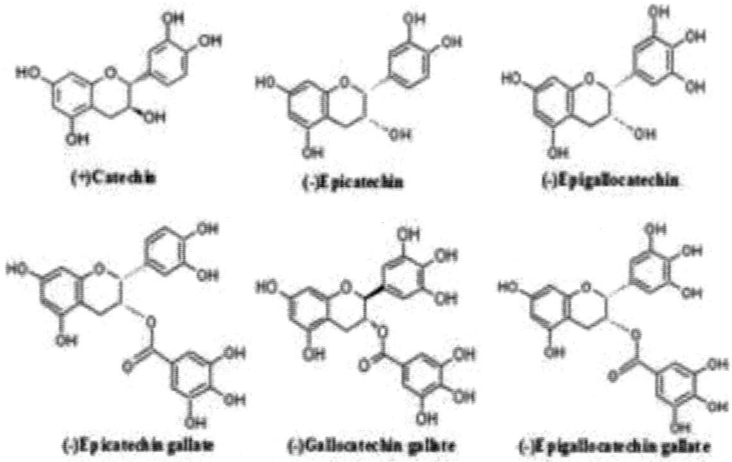

Figure 1. Polyphenol from GTE.

2.2. Gel preparation
PAGTEG was prepared from PAGAT recipe except replacing THPC by GTE. PAGAT preparation is reported in literature [22]. PAGTEG was prepared by adding green tea extract with different concentrations of 0.6%, 1% and 0.23%+2.5 gm of sugar. Gel solution was filled in the plastic cuvettes and kept in a refrigerator for gelation purpose.

2.3. Gel irradiation
Gel samples were irradiated with cobalt-60 (Theratron 780c). Source to surface distance (SSD) was maintained at 80 cm and field size opened for 10×10 cm^2. Samples were irradiated up to 18 Gy with 3Gy interval. After irradiation, the samples were kept at 4 degree for 24 hours to enable it to attain complete polymerization.

3. Results and Discussion

3.1 Spectrophotometry of gel samples
After one day of refrigeration of samples (PAGTEG) they were analysed by double beam spectrophotometer. Un- irradiated gel samples acted as baseline correction samples (Control samples). Samples were read by spectrophotometer before irradiation. All spectrums (Not shown) super imposed on each other it denotes that there is no difference in absorption among them and no pre polymerization occurred. Figures 2 and 3 show the UV-Visible absorption spectrum of PAGTEG-1 & PAGTEG-2 (After irradiation). Both spectrums did not show any significant absorption because polyphenols from GTE absorbed most of the free radicals which were formed during irradiation. Due to insufficient quantity of free radicals, polymerization process was almost inhibited by higher concentration of polyphenols from GTE.

Figure 4 shows appreciable absorption spectrum of PAGAT. It contains no GTE. In figure 5 PAGTEG-3 shows very little absorption because PAGTEG-3 contains very less GTE. In addition sugar was also added in that recipe. Combination of Less concentration of GTE and sugar slightly increased the sensitivity of PAGTEG-3 compared with PAGTEG-2&3

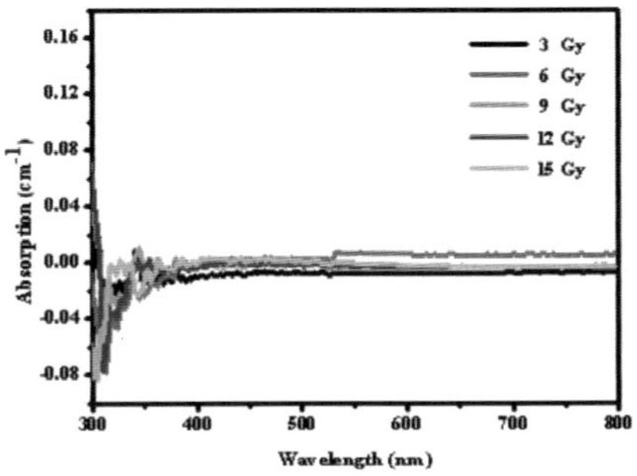

Figure 2. UV-Visible spectrum of PAGTEG-1.

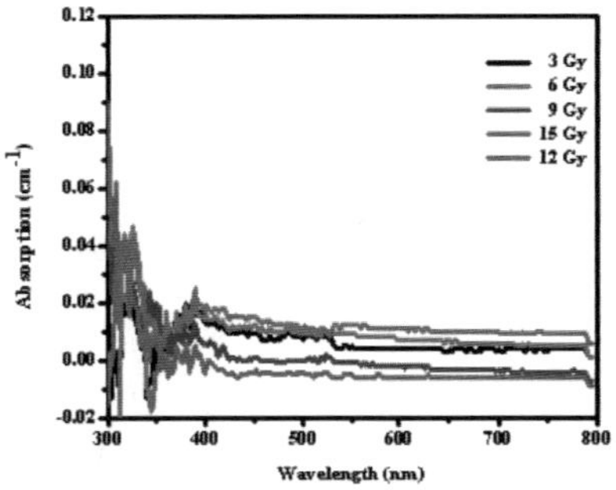

Figure 3. UV-Visible spectrum of PAGTEG-2.

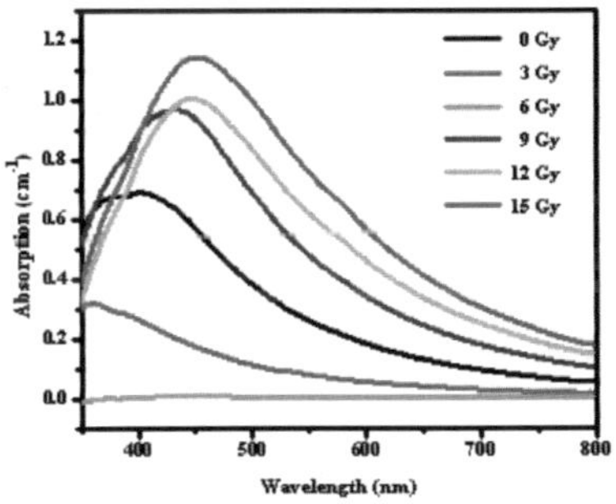

Figure 4. UV-Visible spectrum of PAGAT.

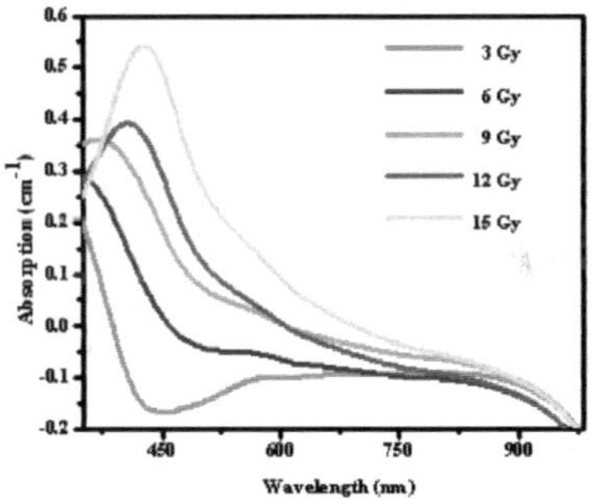

Figure 5. UV-Visible spectrum of PAGTEG-3.

3.2. Particle size measurement of PAGTEG

Particle size was determined for the PAGTEG gel dosimeter from the following equation:

$$2\square = \frac{(\square\square)\square\square\square\square}{\square\square}$$

where a is the maximum particle radius, (Ka) is the Mie Debye efficiency factor, λ_{max} is the wavelength at which the maximum turbidity is observed and n is the refractive index of the gel sample. (Ka) is 4.3 [23] Particle size was increased with respect to dose and the Correlation value R^2 =0.98986 showed that a good relation existed between dose and particle size. Figure 6 shows the relation between the particle size and dose.

4. Conclusion

From this study we suggest THPC as the appropriate antioxidant agent for polymer gel dosimetry.GTE acts as a high radiation protector and so inhibit the polymerization process during irradiation by absorbing free radicals. But when GTE is used in lesser quantity (0.23%) with sugar it may allow a little polymerization and we can protect the pre irradiated gel from auto polymerization by adding GTE with less quantity. This study will be extend using the same procedure for different time interval of irradiation and optimize the GTE concentration levels. The tail of study will be continued to find out the nature products for polymer gel dosimetry and eliminate toxic ingredients such as THPC, Acrylamide and Bis- Acrylamide.

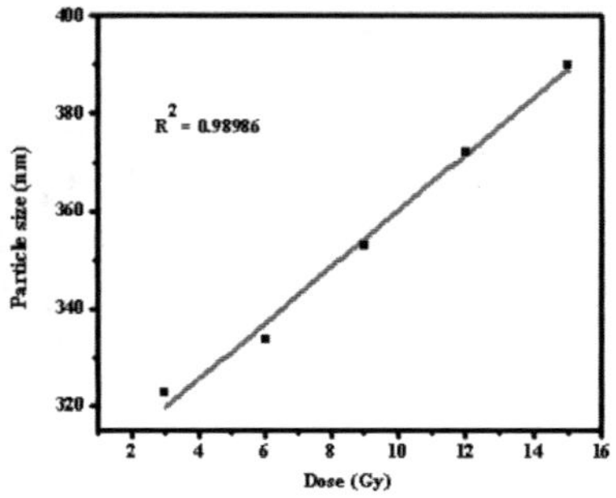

Figure 6. Linear relation between particle size and dose of PAGTEG.

5. Acknowledgements

The authors are grateful to the DST, Government of India for their financial support and to the Management of VIT University for their continuous encouragement.

6. References

[1] Dae-Hong Kim *et al* 2012 *J. Korean Phys. Soc.* **60** 149-54
[2] Baldock C *et al* 2010 *Phys. Med. Biol.* **55** R1-63
[3] Lepage M *et al* 2002 *Phys. Med. Biol.* **47** 1881-90
[4] Venning A J *et al* 2004 *J. Phys.: Conf. Ser.* **3** 155-8
[5] Murry P *et al* 2000 *Australas. Phys. Eng. Sci.* **23** 44-51
[6] Bosi S *et al* 2007 *Phys. Med. Biol.* **52** 2893-903
[7] Bosi S *et al* 2009 *Phys. Med. Biol.* **54** 275-83
[8] Hill B *et al* 2005 *Brit. J. Radol.* **78** 623-30
[9] Hill B *et al* 2005 *Med. Phys.* **32** 1589-97
[10] Mather M L *et al* 2002 *Phys. Med. Biol.* **47** 4397-409
[11] Mather M L *et al* 2003 *Ultrasonics* **41** 551-9
[12] Mather M L *et al* 2003 *Phys. Med. Biol.* **48** N269-75
[13] Baldock C *et al* 1998 *Phys. Med. Biol.* **43** 3617-27
[14] Lepage M *et al* 2001 *J. Appl. Polym. Sci.* **79** 1572-81
[15] Rintoul L *et al* 2003 *Appl. Spectrosc.* **57** 51-7
[16] Kahkonen M P *et al* 2001 *J. Agric. Food. Chem.* **49** 4076-82
[17] Weni Ying Huang *et al* 2013 *The Scientific World Journal* 368350, 6
[18] Durgaprasad S *et al* 2005 *Indian. J. Med. Res.* 122 315-8
[19] Magdy Sayed Aly *et al* 2013 *J. Intech.* Chapter 12

[20] Xuping Zhu *et al* 2010 *Med. Phys.* **37**(1) 183-8
[21] Kwon *et al* 2013 *New Physics: Sae Mulli* **63** 684-8
[22] Subrimanian B *et al* 2006 *J. Med. Phys.* **31** 72-7
[23] Heather M *et al* 2006 *J. Phys. Conf. Ser* **56** 160-3

Effect of bloom strength on radiochromic gel dosimeters

S Ebenezer Suman Babu and B Paul Ravindran
Department of Radiation Oncology, Christian Medical College, Vellore 632004, India

E-mail: emailstoebi@gmail.com

Abstract. The Fricke gel dosimeter has been the widely used dosimeter among the gel dosimeters because of its dose response characteristics and easy preparation. The ferrous to ferric conversion that happens in this gel dosimeter on irradiation, corresponds to the absorbed dose of radiation. Gel dosimetry in India is not moving forward because of the import restrictions on the commercially available high bloom strength gelatin (imported 300 bloom). The feasibility of using Fricke gel dosimeter prepared with the locally available gelatin of 240 bloom and 200 bloom were compared with the 300 bloom gelatin taken as standard. The gel samples were prepared with 5% gelatin by weight and irradiated with ^{60}Co gamma radiation for a dose range from 0-3 Gy used clinically. The optical absorption of gel samples were analyzed using spectrophotometer at 585 nm and dose response curves were generated. The results indicate that Fricke gels prepared with 240 bloom have linear dose response and comparable with those prepared with 300 bloom but the use of gels prepared with 200 bloom was found to be limited because of its poor optical transmittance.

1. Introduction

Fricke and Polymer gel dosimeters are manufactured from radiation sensitive chemicals, which upon irradiation polymerize as a function of the absorbed radiation dose [1]. These gel dosimeters which record the radiation dose distribution in three-dimensions (3D) have specific advantages when compared to one-dimensional dosimeters and two-dimensional dosimeters [2]. These 3D dosimeters are radiologically soft-tissue equivalent [3, 4] with properties that may be modified depending on the application. The 3D radiation dose distribution in polymer gel dosimeters may be imaged using magnetic resonance imaging (MRI) [5, 6], optical-computerized tomography (optical-CT) [7, 8], x-ray CT [9, 10], ultrasound [11, 12] or vibrational spectroscopy [13, 14].

Gelatin is the key ingredient in the gel dosimeter that holds the radiation induced change by holding it in its matrix for determination of dose [15-19]. The strength of the gel is often given in terms of "bloom strength". Bloom strength is defined as the ability of the gelatin to swell when dissolved in water to form a matrix generally termed as gel strength. Though there are various bloom strengths of gelatin commercially available, the restrictions imposed by the government due to misuse of gelatin by anti-social elements have created an artificial non-availability of higher bloom strength gelatin in India. Hence the need arises to investigate the use of different bloom strengths of gelatin and their limitations when radiochomic gels are prepared with them.

The dose response of Fricke gel prepared with 270 bloom strength of gelatin has been reported [20] as good performance when compared with 300 bloom gelatin that is usually recommended for preparation of gels. However 270 bloom is also considered to be a higher bloom strength when looking for gelatin available in the local market. Further the non-availability of Type A gelatin at this bloom strength requires the use of Type B gelatin as an alternative. Studies on gelatin obtained from manufacturers in India (Rama industries, Mumbai, India) have showed that the source of gelatin whether it is from Porcine (Type A) or Bovine (Type B) did not alter the quality of gel prepared but bloom strength did affect the outcome [21]. Various researchers have investigated Fricke gel dosimeter, the role of various combinations of components and their influence on the optical properties [22-25].

The purpose of this study is to analyze the spectrophotometric response of Fricke gel dosimeter prepared with gelatin (300 bloom) and to compare the outcome with the cost effective 240 and 200 bloom gelatin locally available, belonging to Type B category.

2. Materials and Methods

Three different bloom strengths of gelatin 300 (Sigma), 240 (SD-fine), 200 (Raymon) were chosen for the experiments.

2.1. Preparation of Fricke Gel Dosimeter

Fricke gel dosimeter was prepared using 5% gelatin by weight. Gelatin was added to triple distilled water and stirred well for 1 hour inside a water bath maintained at 45^0 C. After 1 hour the gelatin solution was allowed to cool down to 25^0 C and required quantities of Xylenol Orange, Ferrous Ammonium Sulphate and 1 M H_2SO_4 were added. The final gel consisted of 50 mM H_2SO_4, 0.05 mM Xylenol Orange and 0.3 mM Ferrous Ammonium Sulphate. The gel solution was then transferred to cuvettes of 3.5 ml capacity and path length 1cm. The samples as gel solution were kept in the refrigerator for 12 hours maintained at 4^0 C.

2.2. Irradiation setup

The gel samples were irradiated on a ^{60}Co gamma irradiation unit (Theratron-780C) for clinical range of doses ranging from 0-3 Gy for Fricke gel. The gel samples were placed in a water (23^0 C) bath kept on acrylic sheets for sufficient backscatter. The irradiation setup is shown in figure1.

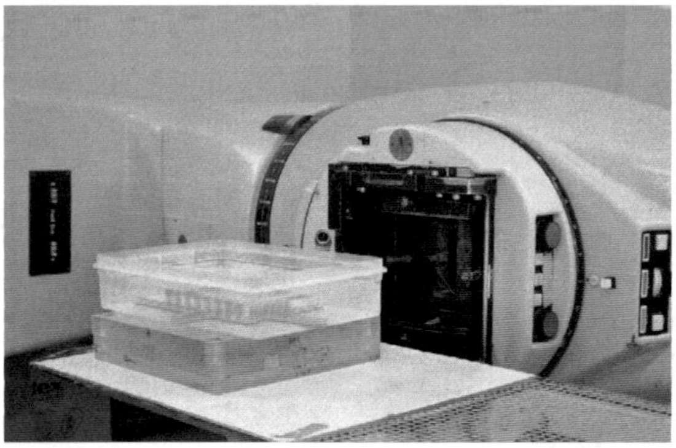

Figure 1. Irradiation setup of Fricke gel samples in a water bath on ^{60}Co Tele gamma unit.

2.3. Evaluation of Gel samples

The samples were evaluated after 1 hour of irradiation using a UV-1800 spectrophotometer (Shimadzu, Japan) in the optical range of 200 – 900 nm. The wavelength for photometric measurement was fixed at 585 nm. Un-irradiated gel sample readings were subtracted from all measurement to account for pre-irradiation data from gel dosimeter.

2.4. Parameters studied
The cuvettes with the Fricke gel were analyzed for the optical absorption per path length of the gel (1 cm). The room temperature was at 28^0 C when the samples were measured. The physical condition of the gel samples at this temperature was also noted.

3. Results and Discussions
Dose response curves were generated from arithmetic mean of data from five samples of the gel dosimeter. The temperature during the preparation, irradiation and measurement of data using spectrophotometer was noted.

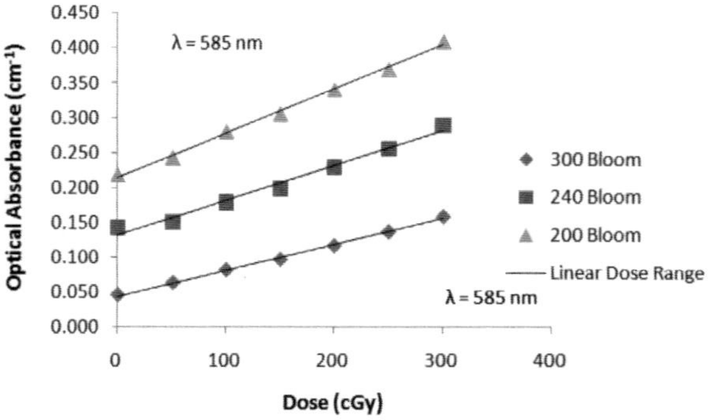

Figure 2. Dose response of Fricke gel dosimeter prepared with 300, 240, 200 bloom gelatin evaluated at a wavelength of 585 nm.

3.1. Fricke Gel – 300 Bloom
The Fricke gel dosimeter prepared with 300 bloom gelatine was taken as the reference. The measurements were performed at a wavelength of 585 nm (figure 2). Considering the uncertainties associated with the gel samples, the evaluation of R^2 value from the graph showed 99.77% reliability level for the gel dosimeter.

3.2. Fricke Gel -240 Bloom
The gel samples were stable at room temperature (28^0 C) and were similar to the physical condition of the samples of 300 bloom gelatin during irradiation and measurements. The transparency of the gel prepared with the 240 bloom gelatin was little lower than the 300 bloom gelatin when prepared as a bulk gel (~1 L) but did not seem to affect the dose response as it was similar to the linear dose response of gels prepared with 300 bloom gelatin.

3.3. Fricke Gel – 200 Bloom
Fricke gel dosimeters prepared with 200 bloom gelatin were less transparent than the 300 and 240 bloom gelatin. Moreover the gel samples were not affected at the temperature (28^0 C) but showed signs of melting when the room temperature increased more than 30^0 C. Though there were limitations, the gel samples prepared with 200 bloom showed linearity in dose response but poor optical transmittance when a larger quantity (~ 1 L) was prepared.

4. Conclusions
The obtained results indicate that the gels prepared with 240 bloom gelatin exhibit similar linearity in dose response when compared with 300 bloom gelatin and a potential substitute that is locally available at a low cost if the bloom strength can be improved further. Though the 200 bloom gelatin showed good dose response its use as a bulk gel is limited because of its lower optical transmittance.

Further studies on improving the gel strength of locally available 240 bloom gelatin should be performed to remove the limitations of its use.

5. References

[1] Baldock C 2006 *J. Phys.: Conf. Ser.* **56** 14-22
[2] Baldock C *et al* 2010 *Phys. Med. Biol.* **55** R1-63
[3] Brown S *et al* 2008 *Appl. Radiat. Isotopes* **66** 1970-4
[4] Gorjiara T *et al* 2011 *Med. Phys.* **38** 2265-74
[5] Baldock C *et al* 1998 *Phys. Med. Biol.* **43** 695-702
[6] Venning A J *et al* 2004 *J. Phys.: Conf. Ser.* **3** 155-8
[7] Bosi S *et al* 2009 *Phys. Med. Biol.* **54** 275-83
[8] Bosi S *et al* 2007 *Phys. Med. Biol.* **52** 2893-903
[9] Hill B *et al* 2005 *Brit. J. Radol.* **78** 623-30
[10] Brindha S *et al* 2004 *Phys. Med. Biol.* **49** N353-61
[11] Mather M L *et al* 2003 *Ultrasonics* **41** 551-9
[12] Mather M L *et al* 2003 *Phys. Med. Biol.* **48** N269-75
[13] Baldock C *et al* 1998 *Phys. Med. Biol.* **43** 3617-27
[14] Rintoul L *et al* 2003 *Appl. Spectrosc.* **57** 51-7
[15] Lepage M *et al* 2002 *Phys. Med. Biol.* **47** 1881-90
[16] Hill B *et al* 2005 *Med. Phys.* **32** 1589-97
[17] Murry P *et al* 2000 *Australas. Phys. Eng. Sci.* **23** 44-51
[18] Gustavsson H *et al* 2004 *Phys. Med. Biol.* **49** 227-41
[19] Hurley C *et al* 2005 *Appl. Radiat. Isotopes* **63** 443-56
[20] Cavinato C C *et al* 2010 *J. Phys.: Conf. Ser.* **249** 012064
[21] Rama Rao K V and Singh S 2002 *Pharmaceutical Technology* 42-46 (Dec 2002)
[22] Gambarini G *et al* 1999 *Nuclear Inst. and Methods in Physics Research, A* **422** 643-8
[23] Scharf K *et al* 1962 *Radiat. Res.* **16** 115-24
[24] Davies J B *et al* 2010 *Radiat. Phys. Chem.* **79** 660-2
[25] Schreiner L J 2004 *J. Phys.: Conf. Ser.* **3** 9

A new dosimeter formulation for deformable 3D dose verification

E M Høye[1], P S Skyt[1], E S Yates[1], L P Muren[1], J B B Petersen[1] and P Balling[2]
[1] Dept of Medical Physics, Aarhus University /Aarhus University Hospital, Denmark
[2] Dept of Physics and Astronomy, Aarhus University, Denmark

E-mail: elhoey@rm.dk

Abstract. We present the characteristics of a new silicone-based radiochromic dosimeter containing the leuco-malachite green (LMG) dye. The dose response as well as the dose-rate and photon-energy dependence of the dosimeter were characterized. To optimise the dose response, different concentrations of the chemical components were investigated. The dose response was found to decrease exponentially as a function of time after irradiation. A cylindrical dosimeter was produced and irradiated with a volumetric modulated arc therapy plan; the standard deviation between measured and calculated dose was 5% of the total dose.

1. Introduction

The increasing complexity of radiotherapy demands dose verification in three dimensions with high spatial resolution [1]. Gel dosimeters capable of performing these measurements include polymerizing and radiochromic dosimetry [2, 3]. Radiochromic dosimeters have been shown to be well suited for readout with optical computed tomography (CT) [4]. The dose response in radiochromic micelle dosimeters is caused by the dye leuco-malachite green (LMG) [5]. LMG is a nonpolar molecule, which restricts its use with hydrogels for forming a 3D dosimeter. Micelle dosimeters successfully avoid this restriction by encapsulating the dye in micelles which can be dissolved in the hydrogel. The dose response of the micelle dosimeter has been restricted by its chemistry and has been shown to be dose-rate dependent [5, 6]. LMG is also used in the commercial Presage® dosimeters, in this case embedded in a polyurethane polymer. In the current study, silicone has been investigated as a host matrix for LMG dissolved in chloroform, in order to create a 3D dosimeter with new clinical applications. The dose response, the dose-rate and energy dependence have been investigated in photon beams. The characteristics of the host material open potentials for the dosimeter to be molded into anthropomorphic phantoms and for subsequent mechanical manipulation (compression, bending etc.) during irradiation.

2. Materials and methods

2.1. Fabrication

The new radiochromic dosimeter that was investigated in this study uses the commercially available Dow Corning SYLGARD® 184 Silicone Elastomer Kit as a host material for the active component LMG, which was dissolved in chloroform. The curing agent and the silicone elastomer were used in a 1:10 weight ratio. LMG dissolved in chloroform was first mixed with the curing agent and subsequently with the silicone elastomer. Air bubbles were removed from the dosimeter by use of a vacuum desiccator in two steps; before and after being poured into the containers. The containers were standard PMMA cuvettes ($1 \times 1 \times 4.5$ cm^3) and a cylindrical container of 15 cm height and 15 cm diameter. The dosimeters were covered with aluminium foil to avoid light pollution and left to cure at room temperature.

The cylindrical dosimeter and the batch which was investigated for dose-rate and energy dependence contained 0.4 mM LMG and 85 mM chloroform, and were allowed to cure for 48 hours. Batches of varying LMG and chloroform concentrations were produced to investigate the dose response. Four batches with chloroform concentration of 100 mM and LMG concentrations of 0.2 mM, 0.4 mM, 0.6 mM and 1.0 mM were produced. Three batches with LMG concentration of 0.4 mM and chloroform concentrations of 80 mM, 150 mM and 200 mM were also produced. These batches were allowed to cure for 24 hours.

2.2. *Irradiation*

The dosimeters were irradiated with a Varian linear accelerator at Aarhus University Hospital. The cuvettes were irradiated with 10×10 cm^2 photon fields of 6 MV and 15 MV beam quality, at a dose-rate of 600 MU/min, as well as at dose-rates of 100, 200 and 400 MU/min with 6 MV. The cuvettes were placed between two 5 cm slabs of solid water in SSD 94.5 cm.

A cylindrical dosimeter was irradiated with a volumetric modulated arc therapy (VMAT) plan, which delivered a dose distribution of 3 Gy in the shape of a cone in the inferior end of the dosimeter. From the middle to the superior end, the dosimeter was irradiated with a cylindrically-shaped dose gradient from 5 Gy to 1 Gy in steps of 1 Gy, and finally 0 Gy and 1 Gy in the superior end. The irradiation consisted of three fractionations with a maximum of 5 Gy in each, giving a total of 15 Gy.

2.3. *Readout and data analysis*

The optical densities of the cuvettes were measured with a spectrophotometer (Helios Alpha, Thermo Spectronic) at 624 nm. A pre-scan was performed a couple of hours before irradiation and repeated post-scans were performed during the days following irradiation. The change in optical density caused by the irradiation was found by subtracting the pre-scan from the post-scans. The optical response (i.e. the optical absorptivity *versus* the irradiated dose) was fitted to a linear equation, and the dose response was found as the slope of the fit. The change in dose response with time was fitted to an exponentially decreasing function with an offset.

The cylindrical dosimeter was scanned with an OCTOPUS IQ optical-CT scanner (MGS Research, Inc., Madison, CT, USA) approximately four hours before irradiation and post-scanned three hours after irradiation. The data analysis was performed as in [7] but employing a direct voxel-to-voxel correlation between the VMAT plan and the CT measurement.

3. Results

The optical response was highest for the high-energy beam: At 10 Gy the optical response was approximately 2% higher for the cuvette irradiated with a 15 MV beam compared to that irradiated with a 6 MV beam. The energy dependence increased linearly with dose, reaching 9% at 60 Gy. A dose-rate dependence was also found; from 600 MU/min to 100 MU/min the dose response increased by 32 ± 1 % relative to that at 600 MU/min.

The optical responses approximately one hour after irradiation for the batches with different LMG concentrations are shown in figure 1. The dose response was fitted by a linear function, and the slopes were higher for the batches with high LMG concentration. However, the dose response did not increase regularly with concentration, as seen in figure 2. The same trend was observed for the batches with increasing chloroform concentration. The response was linear up to 60 Gy at all post-scans.

The post-scans performed over six days revealed that the dose response of all batches decreased exponentially in time, as shown in figure 3. The decay time did not depend on the concentration of either LMG or chloroform. The zero-dose optical response (i.e. the auto oxidation) increased with time.

Journal of Physics: Conference Series **573** (2015) 012067 doi:10.1088/1742-6596/573/1/012067

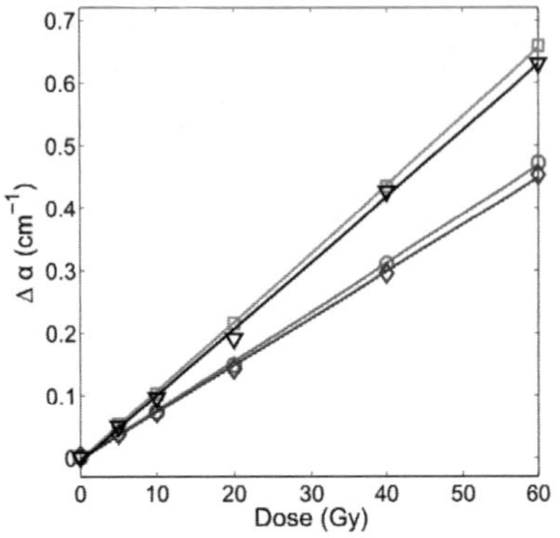

Figure 1. The optical response, $\Delta\alpha$, for batches with 100 mM chloroform and LMG concentrations of 0.2 mM (red circles), 0.4 mM (blue diamonds), 0.6 mM (black triangles) and 1.0 mM (green squares).

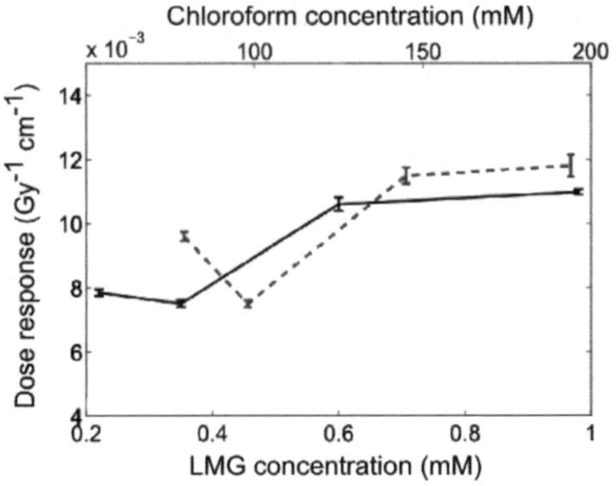

Figure 2. The dose response one hour after irradiation as a function of LMG (full) and chloroform (dashed) concentrations. Error bars from the linear fits to the optical response are shown.

The VMAT dose distribution measured in the cylindrical dosimeter reproduced the features of the plan successfully, as seen in figure 4. In a voxel to voxel analysis of the full dosimeter, the standard deviation between measured and calculated dose was 5% of the total dose.

4. Discussion

The new silicone-based dosimeter showed a significant dose-rate dependence, as well as a small energy dependence. Further investigations at different concentrations of LMG and chloroform are needed to determine if these effects can be minimized. Dose-rate dependencies have for other dosimeters to some extent been minimized by pre-irradiating the dosimeter with a low dose [5]. The high dose response makes it possible to optimize the dosimeter for the lowest dose-rate and energy dependence.

Measurements at higher concentrations are needed to verify whether a saturation of the dose response is observed at the highest LMG and chloroform concentrations in figure 2. The dose responses faded quickly in time and fitted exponentially decaying functions very well. This facilitates determining the dose response at any time, provided that the decay function is well known. The dose response was significantly higher than in micelle dosimeters containing the same dye and solvent concentrations, reported to be 3.5×10^{-3} Gy^{-1} cm^{-1} in [6] with 0.37 mM LMG and 80 mM chloroform. The highest dose response observed in this study was 1.2×10^{-2} Gy^{-1} cm^{-1} with 0.4 mM LMG and 200 mM chloroform. This is comparable to the dose response of Presage®, which we have measured to

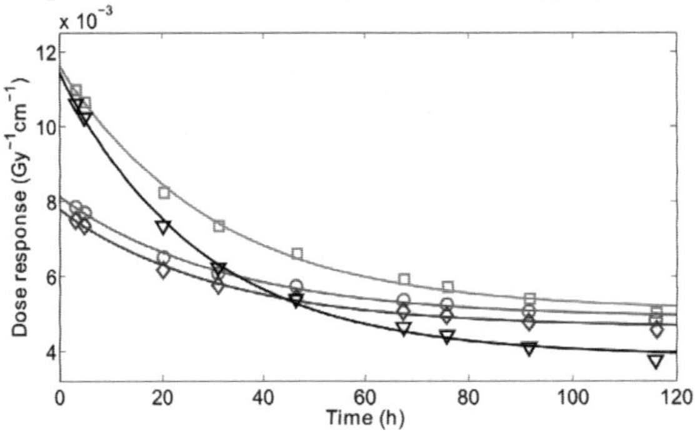

Figure 3. The dose response decayed exponentially with time after irradiation for all LMG concentrations (symbols as in figure 1).

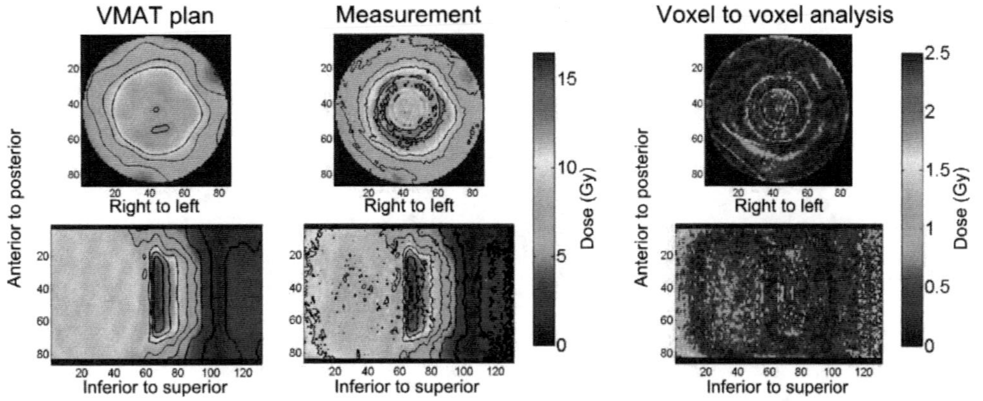

Figure 4. First 3D results with the new silicone-based dosimeter. The upper row shows the transversal plane through the high-dose region (slice 76) in the middle of the dosimeter, and the lower row shows the coronal plane through the central part of the dosimeter. Isocurves are plotted in the calculated VMAT plan and in the results from the optical CT measurement (1 mm^3 voxels) from 2 Gy to 14 Gy with 2 Gy intervals. The final column shows a voxel to voxel analysis of the dose difference between plan and measurement.

2.3×10^{-2} Gy^{-1} cm^{-1} [8]. The data seem to indicate that the dose response of the silicone dosimeter could be increased even further by increasing the dye and chloroform concentrations. Unfortunately, significant colouring of the dosimeters before irradiation was observed, which eventually will restrict the dynamical range of the dosimeter-scanner system.

The dosimeter reproduced the 3D distribution of a challenging VMAT plan with high dose gradients, which shows that it has great potential for use as a clinical dosimeter. The production of the dosimeters is straightforward and fast and requires few utilities. The consistency of the dosimeter was such that it could be used as a compressible dosimeter. This would be useful for studying, e.g., the impact of heart and lung movements on a delivered dose distribution. The dosimeter is solid enough to be handled without a container, and the dosimeter could therefore be moulded into the shape of organs. This would also significantly reduce the image artefacts close to the walls of the dosimeter, caused by light refraction at the boundary between container and dosimeter. A further analysis of the 3D properties of the dosimeter will not be provided here, as problems with the refractive-index-matching liquid created artefacts in the measurement that can easily be avoided in future studies.

5. Conclusion

This first study of the LMG/silicone dosimeter shows that it has significant potential for use as a radiochromic dosimeter in clinical practise, but that further investigations and development is necessary. A high dose response is conceivable, presumably by applying higher LMG and chloroform concentrations. Silicone elastomer has physical characteristics that introduce the possibility of moulding the dosimeter into the desired shape and manipulating the dosimeter mechanically during irradiation to mimic clinically relevant deformation of organs.

6. References

[1] Baldock C 2006 *J. Phys.: Conf. Ser.* **56** 14-22
[2] Doran S J 2009 *Appl. Radiat. Isotopes* **67** 393
[3] Baldock C *et al* 2010 *Phys. Med. Biol.* **55** R1–63
[4] Jordan K 2004 *J. Phys.: Conf. Ser.* **3** 115
[5] Vandecasteele J *et al* 2011 *Phys. Med. Biol.* **56** 627-51
[6] Skyt P S *et al* 2013a *J. Phys.: Conf. Ser.* **444** 012036
[7] Skyt *et al* 2013 *Acta Oncologica* **52** 1445-50
[8] Skyt P S *et al* 2012 *Med. Phys.* **39** 7232

Polymer gel dosimetry for neutron beam in the Neutron Exposure Accelerator System for Biological Effect Experiments (NASBEE)

H Kawamura[1], H Sato[1], T Hamano[2], M Suda[3] and H Yoshii[4]
[1]Department of Radiological Sciences, Ibaraki Prefectural University of Health Sciences, 4669-2, Ami, Ami-machi, Inashiki-gun, Ibaraki, Japan
[2]Research Center for Charged Particle Therapy, National Institute of Radiological Sciences, 4-9-1 Anagawa, Inage-ku, Chiba, Japan
[3]Research, Development and Support Center, National Institute of Radiological Sciences, 4-9-1 Anagawa, Inage-ku, Chiba, Japan
[4]Research Center for Radiation Emergency Medicine, National Institute of Radiological Sciences, 4-9-1 Anagawa, Inage-ku, Chiba, Japan

E-mail: kawamura@ipu.ac.jp

Abstract. This study aimed to investigate whether gel dosimetry could be used to measure neutron beams. We irradiated a BANG3-type polymer gel dosimeter using neutron beams in the Neutron exposure Accelerator System for Biological Effect Experiments (NASBEE) at the National Institute of Radiological Sciences (NIRS) in Japan. First, the polymer gels were irradiated from 0 to 7.0 Gy to investigate the dose-R_2 responses. Irradiated gels were evaluated using 1.5-T magnetic resonance R_2 images. Second, the polymer gels were irradiated to 1.0, 3.0, and 5.0 Gy to acquire a depth-R_2 response curve. The dose-R_2 response curve was linear up to approximately 7 Gy, with a slope of 1.25 $Gy^{-1}\cdot s^{-1}$. Additionally, compared with the photon-irradiated gels, the neutron-irradiated gels had lower R_2 values. The acquired depth-R_2 curves of the central axis from the 3.0- and 5.0-Gy neutron dose-irradiated gels exhibited an initial build-up. Although, a detailed investigation is needed, polymer gel dosimetry is effective for measuring the dose-related R_2 linearity and depth-R_2 relationships of neutron beams.

1. Introduction

Radiotherapy techniques have been applied to intensity-modulated radiotherapy (IMRT), stereotactic radiotherapy (SRT), and particle therapy. Although these therapies produce better dose distributions than conventional techniques, in particle therapies such as proton therapy and carbon therapy, energetic protons and carbon ions generate neutrons upon interaction with collimators, and the generation of secondly neutron radiation cannot be avoid. Patients receiving these therapies as well as technicians and surrounding staff are exposed to secondly neutron radiation from the collimators and patients. Younger patients who are exposed to such radiation and thus to extra radiation treatment doses cannot be ignored.

Polymer gels are 3-dimensional (3D) dosimetric tools used to evaluate radiation dose-related radiation-induced polymerization [1-4]. Therefore, polymer gel dosimetry is desirable for radiation therapy. Additionally, this technique features the advantages of water and soft-tissue equivalent material dosimeters. When using a polymer gel as a phantom, it is possible to measure the radiation dose without disturbing the radiation fields and surrounding areas.

Polymer gel dosimetry of radiation beams with contributions from neutron [5-7] and carbon ion [8] have been previously reported. Further work dose need to be undertaken to calibrate the response of the gel to neutron beams and high linear energy transfer (LET) particles.

The purpose of the present study was to investigate whether polymer gels were available to measure the generated neutron-beam radiation doses in radiotherapy.

2. Materials and Methods

2.1 Preparation of polymer gel

A BANG-3 type polymer gel was prepared. The BANG kit (MGS Research Inc. Guilford, CT, USA) was melted in a hot water bath at 55°C, and a modifying buffer and an antioxidant agent were added. Prepared gels were poured into polyethyleneterepthalate (PET) containers of 30 mm × 30 mm × 65 mm for dose-R_2 calibration and 40 mm × 40 mm × 150 mm for depth-R_2 response curve acquisition. These containers had been filled with an ascorbic acid solution 2 days before gel preparation for the purpose of oxygen scavenging [9]. The gels in the containers were wrapped in aluminium foil and stored in a refrigerator at 4°C until irradiation.

2.2 Irradiation field of neutron beams and set up

The prepared gels were irradiated with a neutron beam in Neutron-exposure Accelerator System for Biological Effect Experiments (NASBEE) at the National Institute of Radiological Sciences (NIRS) in Japan. This facility was developed to study the biological effects of neutrons [10]. The neutron was produced by 4 MeV deuterons bombarding into the beryllium target. The neutron energy spectrum revealed a mean energy of 2.0 MeV. In this facility, the beam gantry was set in the vertical direction.

2.3 Irradiation of prepared gels for calibration and for stereotactic irradiation

The prepared gels were placed upside-down 240 mm below the surface with the longitudinal axis placed in the beam axis for the dose-R_2 response measurements and depth-R_2 measurements. For the dose-R_2 response measurements, the polymer gels were set in the central axis, three plastic 10-mm-thick phantoms were placed on the prepared gels, and plastic phantoms surrounded the prepared gels. The prepared gels were irradiated to calibrate the dose versus R_2 from 0 to 7.0 Gy. For the depth-R_2 measurements, a 150-mm-length polymer gel was inserted parallel to the beam axis in a plastic phantom. The three 10-mm-thick plastic phantoms on the prepared gels were removed from the beam line [figure 1 (a)]. The prepared gels were irradiated to 1.0, 3.0, and 5.0 Gy for the depth-R_2 examination. After irradiation, these gels were stored in a refrigerator at 4°C until 1.5-T magnetic resonance imaging (MRI).

2.4 Irradiation for photon beams and set up

For comparison of the sensitivity of the gel for neutron, prepared gels were irradiated with a 10 MV X-ray (photon) beam on a linear accelerator (EXL-15SP, MITSUBISHI, Japan) at Ibaraki Prefectural University of Health Sciences. The gels in PET container were irradiated with 1.0, 3.0, and 5.0 Gy at the isocenter at a depth of 5 cm in a 300×300×300 mm^3 water tank. The dose rate was 4 Gy/min.

2.5 MRI measurements of irradiated gels and Analysis

MRI measurements were performed on a 1.5-T Toshiba Excelart Vantage scanner (Toshiba Medical Systems, Tochigi, Japan) 4 days after irradiation. The irradiated gels were scanned with a quadrature (QD) Torso SPEEDER coil (Toshiba Medical Systems). The gels were set in the coil and imaged using a multi-spin echo pulse sequence [with a first effective echo time (TE1) =30 ms and second effective TE (TE2) = 60 ms]. For each scan, a repetition time of 4000 ms was used to reduce the influence of spin-lattice relaxation along with a 1-mm^2 resolution, and 5-mm-thick planes. The R_2

Journal of Physics: Conference Series **573** (2015) 012068 doi:10.1088/1742-6596/573/1/012068

image was calculated from two different echo time images using an in-house program [11]. For the dose-R_2 response measurements, regions of interest (ROI) were drawn in the R_2 images, and the mean values and standard deviations of ROI were measured. The depth-R_2 profiles were evaluated from the R_2 images in the central axis profile.

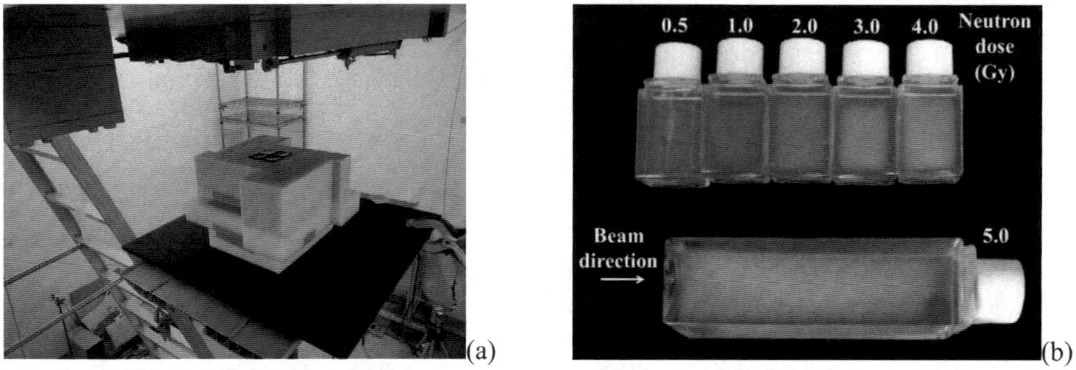

Figure 1. (a) Photograph showing the gel phantom set-up for the depth-R_2 distribution experiments with a neutron beam in NASBEE at NIRS. (b) Upper: 0.5, 1.0, 2.0, 3.0 and 4.0 Gy-irradiated gels (left to right) for the dose-R_2 response measurements; lower: 5.0 Gy-irradiated gel for the depth-R2 measurements. The gel for the depth-R2 measurement was exposed from left to right direction (arrow).

3. Results and Discussion

3.1 Neutron-beam dose-response curve of the polymer gel and R_2 versus dose linearity

A photograph of the neutron beam-irradiated gels is shown in figure.1 (b). The results indicated visual dose-related change in polymerization. The R_2 response curve as a function of the neutron dose from 0 to 7 Gy was linear related to the dose [figure 2 (a)]. The data points were obtained by averaging the R_2 values in the polymer gel ROI at the same dose. The fitted straight line had a slope of 1.25 $Gy^{-1} \cdot s^{-1}$, intercept of 6.29 s^{-1}, and correlation coefficient of 0.984.

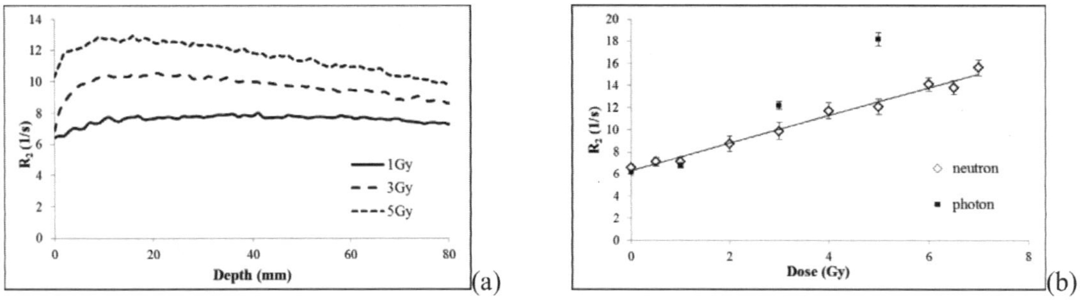

Figure 2. (a) Dose-R_2 response curves for neutrons and photons. The neutron curve was linear from 0 to 7 Gy. (b) Depth-R_2 dose distributions for 1-, 3-, 5-Gy neutron irradiation were shown.

For comparison, the photon-irradiated gels; the R_2 value of irradiated gels with 3- and 5-Gy photon beams were higher than those of gels irradiated with identical neutron beams [figure 2 (a)]. It was difficult to measure the dose from neutron alone because photons and other particles developed when the neutrons impinged on the gels and plastic phantoms. Additionally, neutron have high LET and relative biological effectiveness; therefore, the sensitivity of polymer gel to neutrons is possibly lower than the sensitivity to photon, as same as proton and carbon [12]. The result in figure 2 (a) will require further investigation.

3.2 Depth-R_2 profiles between gel measurements

Figure 2 (b) shows the depth-R_2 profiles in the central axis line of the 1.0-, 3.0- and 5.0-Gy- irradiated gels. The curves obtained at 3.0 and 5.0 Gy revealed an initial R_2 build-up at a shallow depth. This build-up was caused by the increasing gamma rays resulting from the interactions of the neutrons with the gels and surrounding plastic phantoms. This curve decreased with the gel depth. The R_2 values at identical depths were increased with higher neutron doses.

4. Conclusions

This study revealed a linear neutron dose-R_2 response curve. Therefore the polymer gel dosimetry can be used for neutron measurements. The polymer gel depth-R_2 curve in response to neutron beam irradiation indicated an initial build-up and reduction in depth-R_2 curves with increasing depth. The neutron beam depth-dose curve can be more effectively acquired with polymer gels than with other dosimetry tools. Further investigation will be needed to evaluate the effects of neutrons, photons and the other particles.

5. Acknowledgement

The authors would like to acknowledge the staff of NIRS for help with the experiments using neutron beams. This work was supported in part by JSPS KAKENHI Grant Number 24601012.

6. References

[1] Baldock C *et al* 2010 *Phys. Med. Biol.* **55** R1-63
[2] Mather M L *et al* 2003 *Ultrasonics* 41 551-9
[3] Venning A J *et al* 2004 *J. Phys.: Conf. Ser.* **3** 155-8
[4] Baldock C 2006 *J. Phys.: Conf. Ser.* **56** 14-22
[5] Uusi-Simola J et al 2003 *Phys. Med. Biol.* **48** 2895-906
[6] Farajollahi A R *et al* 2000 *Phys. Med. Biol.* **45** N9-14
[7] Vanossi E *et al* 2008 *Radiat. Meas.* **43** 442-5
[8] Ramm U *et al* 2000 *Phys. Med. Biol.* **45** R95-102
[9] Murakami Y *et al* 2007 *Phys. Med. Biol.* **52** 2301-11
[10] Takada M *et al* 2011 *AIP Conf. Proc.* **401** 1336
[11] Kawamura H *et al* 2013 *Appl. Radiat. Isot.* **77** 94-102
[12] Heufelder J *et al* 2003 *Med Phys.* **30** 1235-40

Energy dependence of Fricke-xylenol orange gel and gel based on Turnbull blue for low-energy photons

Jaroslav Šolc[1], Vladimír Sochor[1] and Petra Kozubíková[2]
[1] Czech Metrology Institute, Radiova 1a, Prague, Czech Republic
[2] CTU in Prague, Faculty of Nuclear Sciences and Physical Engineering, Department of Dosimetry and Application of Ionising Radiation, Prague, Czech Republic

E-mail: jsolc@cmi.cz

Abstract. The paper describes the energy dependence of two types of radiochromic gels, a Fricke-xylenol orange gel (FX gel) and a gel based on Turnbull blue (TB gel), on low energy photons between 14 and 145 keV. Gel samples were irradiated at reference photon fields at the Czech Metrology Institute and evaluated by two independent optical methods. Measurements revealed that the response of the TB gel is independent on photon energy down to at least 14 keV photons, while the FX gel is energy dependent for photons below roughly 50 keV.

1. Introduction
One of the aims of the Joint Research Project "Metrology for radiotherapy using complex radiation fields" (MetrExtRT) [1], a collaboration of European metrological institutes, is to determine the absorbed dose-to-water distribution around miniature electronic brachytherapy X-ray sources INTRABEAM® (Carl Zeiss, Germany) and AXXENT® (XOFT, USA). The dose distribution should be determined by radiochromic gels therefore information about gels energy dependence is necessary. Although there are papers describing the energy dependence of the FX gel (e.g., [2]), no detailed study related to the energy dependence has been done on the TB gel.

2. Materials and Methods

2.1. Cuvettes
Classic spectrophotometric cuvettes with inner dimensions of $1.0 \times 1.0 \times 4.5$ cm^3 and 0.1 cm thick polystyrene walls were used. In order to avoid photon attenuation one wall was removed and replaced by (6.03 ± 0.14) µm thick transparent Mylar foil. The cuvettes were faced towards the beam by this wall during irradiation.

2.2. Fricke-xylenol orange gel
The Fricke-infused xylenol orange ion indicator gelatine-based gel (FX gel) was firstly described in [3]. It is the most often used radiochromic gel because of its high dose sensitivity. A drawback is the diffusion of dose patterns [4] due to which the evaluation of dose response should be done shortly after irradiation. Composition of the FX gel was as follows: deionized water (96% w/w), edible gelatin (Natura, Czech Republic; 4% w/w), ferrous ammonium sulphate (0.5 mM), xylenol orange (0.05 mM), and sulphuric acid (50 mM). Finished gel in liquid form was poured into the cuvettes and put into a refrigerator to solidify.

2.3. *Gel based on Turnbull blue dye*
The radiochromic gel based on Turnbull blue dye (TB gel) uses phytagel instead of more common gelatin. It has lower dose sensitivity than the FX gel but it does not suffer from diffusion of dose patterns [4]. Composition of the TB gel was as follows: deionized water (99.75% w/w), phytagel (Sigma Aldrich®; 0.25% w/w), potassium ferricyanide (0.5 mM), ferric chloride (0.5 mM), and sulphuric acid (1.0 mM). Finished gel in liquid form was poured into the cuvettes and put for the whole night into a refrigerator to solidify.

2.4. *Irradiation*
High-dose-rate X-ray spectra [5] and ^{60}Co photon beam [6] were utilized for the experiment due to their high air kerma rate necessary for delivering sufficient dose in reasonable time. Parameters of photon beams and the reference values of air kerma at the reference point are presented in Table 1. The reference point was the front wall (Mylar foil) of a cuvette. The targeted air kerma was 4 Gy and 15 Gy in case of FX gel and TB gel, respectively.

Table 1. Parameters of photon beams and air kerma rate at the reference point.

Beam quality	Mean energy (keV) [5]	Air kerma rate (Gy/h)[c]
H020[a]	14	16.2
H030[a]	20.1	15.7
H060[a]	38.1	11.6
H100[a]	57.4	14.9
H200[a]	99.4	15.1
H280[a]	145.1	22.2
^{60}Co[b]	982.4	78.4

[a] High-dose-rate X-ray spectra [5]. The number means bias voltage in kilovolts.
[b] Chisobalt radiotherapeutic unit. For this energy only, a 4 mm thick PMMA sheet was placed just in front of the cuvettes. It is the reference photon energy.
[c] At the front wall of cuvettes with gel.

2.5. *Monte Carlo simulations*
The aim of the Monte Carlo (MC) simulations was to obtain, for each gel and each photon beam, a conversion from reference air kerma at the reference point to absorbed dose-to-gel averaged over the whole volume of the gel in a cuvette. The MC simulations were performed using a general-purpose code MCNPX™ in version 2.7E [7]. A simple model of experimental geometry included a cuvette with the gel, a realistically defined photon beam and, in case of ^{60}Co beam, a 4 mm thick PMMA plate in front of the cuvette to achieve electron equilibrium.

2.6. *Evaluation of gel response*
Gel response was evaluated by two independent optical methods: by measurement of spectrophotometric absorbance with a UV-VIS spectrophotometer and by determination of a CCD absorbance from photographs taken with a sensitive CCD camera [8].

3. Results
MC simulated dependences of conversion factors from reference air kerma to mean absorbed dose-to-gel averaged over the whole volume of gel inside a cuvette are presented in figure 1. The mean absorbed dose inside the gel per unit air kerma at the cuvette front wall significantly decreases for photon energies below 50 keV due to increasing photon self-absorption with decreasing energy.

Figures 2 and 3 show measured energy dependence of both radiochromic gels. Data are normalized to the ^{60}Co beam gel response. The results are in agreement with previous investigations done for the limited range of photon energies [9]. The TB gel does not show the energy dependence for low energy photons. The assumption is that this is caused by very high weight fraction of water in the gel reaching 99.75%. However, the measurement uncertainties are higher compared to the FX gel due to lower dose response of the TB gel. Although irradiated to lower dose the FX gel responses were higher and the uncertainties of FX data are lower. This gel is energy dependent for low energy photons where the response decreases by 8% and 20% for high-dose-rate X-ray spectra with the mean photon energy of 20 keV and 14 keV, respectively.

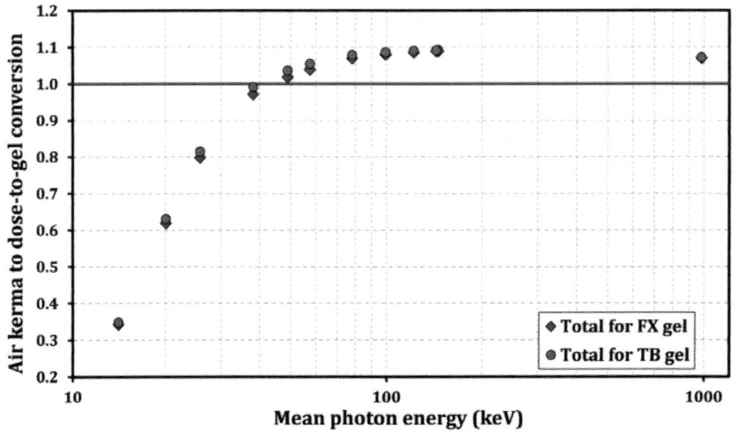

Figure 1. Dependence of the conversion factor from reference air kerma to dose-to-gel as a function of mean photon energy for FX gel (blue diamonds) and TB gel (red circles). The plot includes more high-dose-rate X-ray beams [5] than used in the experiment. Data point around 1000 keV corresponds to ^{60}Co beam with 4 mm PMMA placed in front of a cuvette.

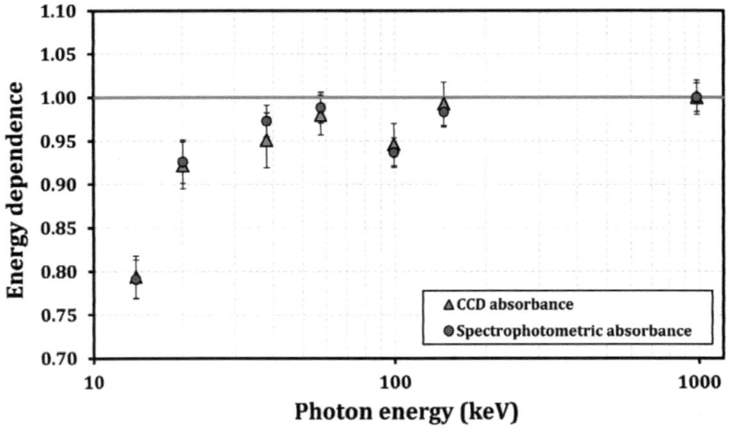

Figure 2. Energy dependence of the FX gel as determined by measurement of CCD absorbance (blue triangles) and spectrophotometric absorbance (red circles).

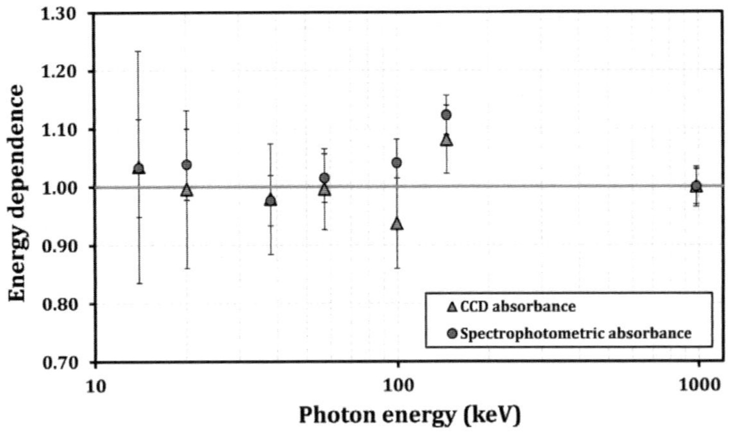

Figure 3. Energy dependence of the TB gel as determined by measurement of CCD absorbance (blue triangles, larger uncertainties) and spectrophotometric absorbance (red circles, smaller uncertainties).

It is noted that the same behavior of gels energy response is obtained by both evaluation methods – optical spectrophotometry and photographing by a CCD camera.

4. Conclusions

The energy dependence of two types of radiochromic gel dosimeters was determined for X-ray photon beams with mean energy from 14 keV to 145 keV, and for ^{60}Co photons (mean energy of photon spectrum 982 keV). The TB gel does not show energy dependence, while sensitivity of the FX gel decreases at photon energies below 40 keV. The energy dependence of the FX gel should be taken into account in planned measurements of spatial dose distribution around electronic brachytherapy X-ray sources using this radiochromic gel.

5. Acknowledgement

This work was supported by the EMRP joint research project "Metrology for radiotherapy using complex radiation fields" (MetrExtRT) which has received funding from the European Union on the basis of Decision No 912/2009/EC. The EMRP is jointly funded by the EMRP participating countries within EURAMET and the European Union.

6. References

[1] MetrExtRT. http://radiotherapy-emrp.eu/
[2] Cavinato C C *et al* 2008 Proceedings of IRPA 12. http://irpa12.org.ar/fullpapers/FP1044.pdf
[3] Bero M A *et al* 2000 *Nucl. Instrum. Meth. A* **166** 820-5
[4] Šolc J *et al*. 2009 *Phys. Med. Biol.* **54** 5095-5
[5] Ankerhold U 2000 PTB-Dos-34 Report
[6] Šolc J 2010 *Nucl. Instrum. Meth. A* 580 141-4
[7] Pelowitz D B *et al* 2011 Los Alamos National Laboratory Report LA-UR-11-01502
[8] Šolc J *et al* 2009 *Nucl. Instrum. Meth. A* **619** 163-16
[9] Šolc *et al* 2012 *Metrologia* **49** S231

The polyGeVero® software for fast and easy computation of 3D radiotherapy dosimetry data

Marek Kozicki and Piotr Maras
GeVero Co., Tansmana 2/11, 92-548 Lodz, Poland

E-mail: gevero@polygevero.com

Abstract. The polyGeVero® software package was elaborated for calculations of 3D dosimetry data such as the polymer gel dosimetry. It comprises four workspaces designed for: *i)* calculating calibrations, *ii)* storing calibrations in a database, *iii)* calculating dose distribution 3D cubes, *iv)* comparing two datasets e.g. a measured one with a 3D dosimetry with a calculated one with the aid of a treatment planning system. To accomplish calculations the software was equipped with a number of tools such as the brachytherapy isotopes database, brachytherapy dose versus distance calculation based on the line approximation approach, automatic spatial alignment of two 3D dose cubes for comparison purposes, 3D gamma index, 3D gamma angle, 3D dose difference, Pearson's coefficient, histograms calculations, isodoses superimposition for two datasets, and profiles calculations in any desired direction. This communication is to briefly present the main functions of the software and report on the speed of calculations performed by polyGeVero®.

1. Introduction

Advanced radiotherapy techniques of tumors irradiation require corresponding 3D dosimetry [1]. As such, polymer gel dosimetry and radiochromic gel/plastic dosimetry attracted a great attention [2, 3]. The dosimetric systems after irradiation are measured with 3D scanning techniques such as magnetic resonance imaging, computed tomography, ultrasonography and optical computed tomography. Afterwards, the scanning results, in form of a set of 2D planes or a 3D cube, are processed usually with the aid of Matlab® software package (MathWorks, Inc.). This in turn requires the ability of writing codes for any new experiment that may be laborious, time consuming and discouraging both for 3D dosimetry false beginners and advanced users.

A solution to this problem can be a dedicated software package capable of processing 3D dosimetry data with ease and a high speed. A proposition of such one was described previously [4, 5]. This communication, however, is to briefly report on the new polyGeVero® software package that was elaborated for the 3D dosimetry data calculations.

2. A brief review of the software's functions

The polyGeVero® software operates on DICOM (modality: MR, CT, US) and VFF type files; its structure can be expanded for other file formats and is to be further enriched in the new functionalities. All operations on the files obtained after scanning a 3D dosimeter are performed in four workspaces that were projected for the preparation of calibration equations, storing the calibration characteristics of the 3D dosimeters, calculating dose distribution cubes and comparing two datasets e.g. a measured dose distribution with the aid of a 3D dosimeter with the calculated dose distribution with the aid of a

Journal of Physics: Conference Series **573** (2015) 012070 doi:10.1088/1742-6596/573/1/012070

treatment planning system. The calculations' results can be exported as TXT and BMP files (1D and 2D graphs: signal and dose maps as well as profiles and other results). They can also be printed or saved as BMP images or in the form of ready-made reports.

In calibration workspace, linear, exponential and polynomial calibration equations are allowed for a few generally known calibration options that employs specific calibration phantoms (e.g. GeVero Co. phantoms): multi vials (a few phantoms are exposed to different doses such that each phantom comprises 2-3 irradiation bands), depth dose (a build-up and decrease of dose is observed for a depth dose phantom), cross beam (beams of radiation are crossed such that the crossing areas are taken for calculating calibration data points), and brachytherapy (a dosimeter is irradiated with an isotope usually at a single dwell position). These calibration methods give different number of calibration points that is the highest for the depth dose and brachytherapy and are characterised by different complexity which is the lowest for the cross beam calibration. Once the calibration equations are calculated they are stored in a calibration database with corresponding characteristics of a dosimeter.

In gel dosimetry dose distribution workspace the user calculates 3D dose cube (*.vec file) after application a of calibration equation and selection of an isocentre point using dedicated tools. This cube, which corresponds to the measured dose distribution, is further used in gel dosimetry vs. treatment planning system (TPS) workspace for comparison with a TPS calculated dose distribution. In general, any two datasets can be compared in this workspace on 2D-2D or 3D-3D basis and calculations of gamma index, gamma angle, dose difference, Pearson's correlation, histograms as well as isodoses superimposition and profiles analysis is allowed. The polyGeVero® is also equipped with tools for displaying a single 2D plane and multiple 2D planes of dose distribution in 3D. In figure 1 an example print screen of the polyGeVero® gel dosimetry vs. TPS workspace with two datasets uploaded and calculated is presented.

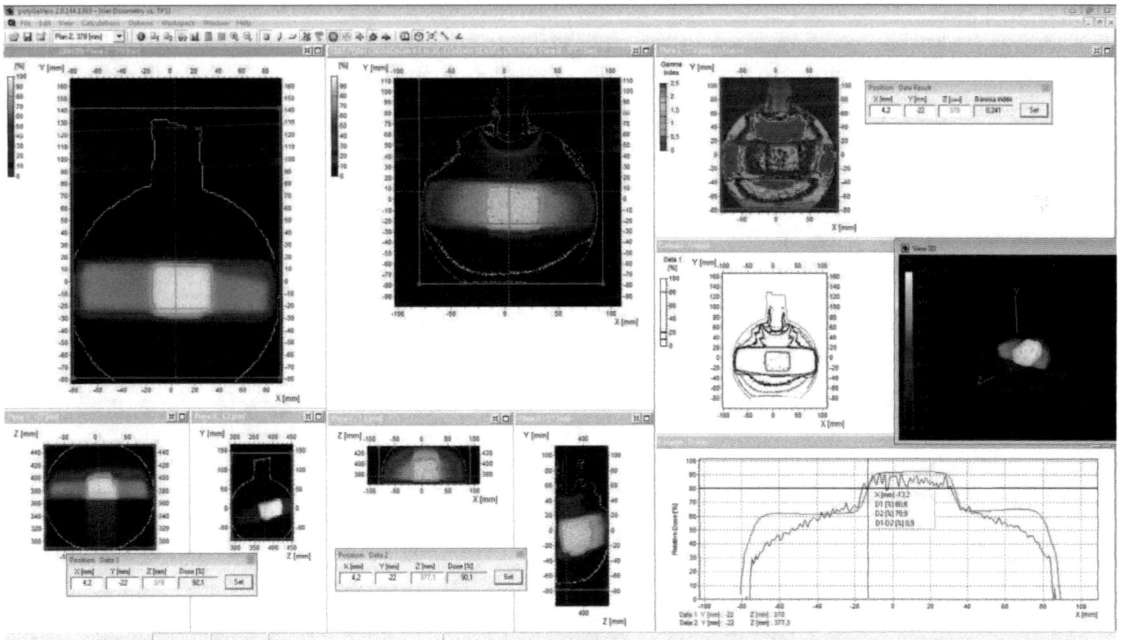

Figure 1. An example view of one of four workspaces of polyGeVero®: gel dosimetry vs. TPS, with a polymer gel and TPS data compared. A window with a 3D view of dose distribution is shown as well.

3. How fast does it calculate dosimetry data

It is believed that the easiness of calculations in polyGeVero® does not require thorough discussion. Due to its structure the software allows for most important for routine clinical dosimetry calculations without the need for writing new codes as in case of e.g. Matlab® calculations. For the routine dosimetry purposes those calculations should also be quick. Therefore in this section we present the results of a test of calculations speed. This test included the following operations: *i)* calculations of the calibrations equations based on four calibration approaches: the multi vials, cross beam, brachytherapy and depth dose, *ii)* storage of the calibration characteristic in a calibration database, *iii)* calculations of four dose distributions 3D cubes after application of four calibration equations, *iv)* comparison of each 3D dose cube measured with a TPS calculated one. The test was performed on a computer of the following parameters: processor AMD Phenom™ II X4 965 3.4 GHz, RAM: 4 GB, operating system: 64 bit Win 7. The results are presented in table 1. It should be noted that the total time of calculations is equal to less than thirty minutes. The calculations, even complex ones such as the 3D gamma index, take just seconds for a parameter. The most time consuming during the test were the other operations such as indication of a folder with raw data or typing additional information for the data being uploaded.

Table 1. Results of the test of calculations speed for the polyGeVero® software.

Workspace/Operations		Time of operation [min]
Calibration 1. Searching for DICOMs 2. Building images series 3. Recognising layers 4. Calculating R2 5. Preparation of complete R2 vs. dose curve, linear calibration equation and export of calibration characteristics to the calibration database	Brachytherapy calibration	2.8
	Depth dose calibration	7.6
	Cross beam calibration	12.0
	Multi vials calibration	19.8
Gel dosimetry dose distribution 1. Searching for DICOMs 2. Building images series 3. Recognising layers 4. Calculating R2 5. Calculating dose distribution for brachytherapy, depth dose, cross beam and multi vial calibration equations 6. Calculating 3D dose cube (*.vec file) for the dose distributions mentioned above	*.vec file for cross beam calibration	23.2
	*.vec file for depth dose calibration	24.0
	*.vec file for brachytherapy calibration	25.0
	*.vec file for multi vial calibration	25.8
Gel dosimetry vs. TPS 1. Import TPS generated plan of irradiation (head size phantom: number of pixels of 1 layer: 5460; 100 layers) 2. Import gel dosimetry 3D dose cubes as calculated in gel dosimetry dose distribution workspace (head size phantom: number of pixels of 1 layer: 16384; 62 layers) 3. Calculation gamma index in 3D mode for all planes for each set of data: TPS vs. a dose distribution 3D cube of gel dosimetry	Gamma index for TPS vs. gel dosimetry 3D cube (brachytherapy calibration)	27.4
	Gamma index for TPS vs. gel dosimetry 3D cube (cross beam calibration)	28.2
	Gamma index for TPS vs. gel dosimetry 3D cube (depth dose calibration)	29.0
	Gamma index for TPS vs. gel dosimetry 3D cube (multi vial calibration)	29.8
Total time of calculations		29.8

Journal of Physics: Conference Series **573** (2015) 012070 doi:10.1088/1742-6596/573/1/012070

4. Conclusions

In this communication the polyGeVero® software package, which was prepared for 3D dosimetry data calculations, was briefly presented. The software calculates data with a high speed. It is also easy to perform calculations typical for routine dosimetry with the software since, *inter alia,* it is not necessary to write separate protocols for any new experiment performed. This software package is subject to further development. For instance, presently DICOM (modality MR, CT, US) and VFF files (of VISTA optical scanner, Modus Medical Devices, Canada) can be processed, however, the software's structure is ready for expansion to comply with the other file formats as well.

More information on polyGeVero® including short movies presenting its main options can be viewed on the GeVero Co. web site: http://polygevero.com as well as elsewhere [6].

5. References

[1] Baldock C 2006 *J. Phys.: Conf. Ser.* **56** 14-22

[2] Adamovics J *et al* 2006 *Rad. Prot. Dosimetry* **120** 107

[3] Baldock C *et al* 2010 *Phys. Med. Biol.* **55** R1

[4] Kozicki M *et al* 2009 *J. Phys.: Conf. Ser.* **164** 12062

[5] Kozicki M *et al* 2009 *J. Phys.: Conf. Ser.* **164** 12061

[6] Kozicki M *et al* 2014 *Phys. Med. Biol.* **59** 4111

Feasibility of PET/CT 3-D dosimetry for proton-activated PRESAGE® dosimeters

M Carroll[1,2] G Ibbott[1] and J Adamovics[3,4]
[1]Department of Radiation Physics, University of Texas MD Anderson Cancer Center, 1515 Holcombe Blvd., Houston, TX 77030 USA
[2]University of Texas at Houston Graduate School of Biomedical Sciences, Houston, TX 77030 USA
[3]Rider University, Lawerenceville, NJ, USA
[4]Heuris Pharma, LLC, 412 Sunset Rd., Skillman, NJ, 08558 USA

E-mail: mcarroll1@mdanderson.org

Abstract: A feasibility study is considered using PRESAGE® dosimeters as a direct dosimetry comparison tool to correlate proton activation to dose in a clinical beam. Proton activation measured by PET/CT has been studied as a future tool method for in vivo dosimetry. PRESAGE® is a 3D offline dosimeter that has a material composition giving it a longer activation decay time than similar dosimeters allowing easier measurements. In this study, the PRESAGE® positron emissions were measured by PET and directly compared to dose measurements made by optical-CT to make spatial comparisons of dose and proton activation. Profiles along the central axis of the beam found a shift in the distal-80% resulting in the proton activation profile being 1 cm shallower than the dose profile. Cross beam profiles demonstrated little discrepancy between long (3 hour) and short (30 minute) PET scan times.

1. Background

Previous studies have investigated the potential of proton positron activation as a method of noninvasive in vivo dosimetry using positron emission tomography (PET) [1]. Early work has theorized that positron activity distributions could be used to reconstruct dose distributions [2], but prior to clinical use, a better correlation must be developed between these two distributions. The cross-section for activation of positron emitters is dependent on the energy of the proton beam that changes with depth in tissue, leading to a shift in areas of maximum positron activity away from the Bragg Peak. Using a separate device for relative dosimetry can lead to dose mapping uncertainties [3]. A 3D dosimeter [4] can allow mapping of proton dose distribution to the PET activity measurement within a single detector and reduce these uncertainties.

Previous studies have tested polymer gels to map positron-emitter activation. Gels rely primarily on ^{15}O emitters which have half-lives of ~2 minutes making rapid readout essential [5]. We have investigated use of the radiochromic polyurethane dosimeter PRESAGE® [6] to correlate the proton dose distribution to the PET activity measurement within a single detector. The polyurethane in PRESAGE® is primarily composed of carbon (approx. 62% w/o) which has a more manageable half-life of 20 minutes. The PRESAGE® formulation used was developed to minimize the sensitivity to proton LET and has similar density and RLSP to tissue [7].

2. Methods

Cylindrical PRESAGE® dosimeters approximately 6 cm in height and 6 cm in diameter were used in this study. PRESAGE® is homogeneously radiation sensitive without the necessity of a container which allows dosimetric measurements through its volume. The dosimeters used in this study were manufactured by Heuris Pharma, LLC. Irradiations were performed at the M. D. Anderson Proton Therapy Center (PTC).

Figure 1. PRESAGE® dosimeters in the water phantom prior to proton irradiation.

Three cylindrical dosimeters were irradiated to 500 cGy with 180-MeV energies in a single, wide-beam shot with passive double scattered protons. Irradiations were done in a water phantom, shown in figure 1, with the dosimeters placed approximately 13.8 cm deep. An 18x18 cm^2 field size was used to irradiate all three dosimeters at once with the beam directed along the dosimeter's central axis to deliver a 2 cm spread-out Bragg peak (SOBP) 2 cm deep in the dosimeter. The lower halves of the dosimeters were left unirradiated to measure PET activity in steep dose gradients across the full dose profile.

After irradiation, the dosimeters were rushed to a nearby PET/CT for the first stage of imaging on a GE Discovery 690 FX. Between irradiation end and scan beginning was just under 15 minutes which is less than one half-life of the activated ^{11}C. Pixel spacing of 1.0x1.0x1.0 mm^3 was used for the CT. The PET scan was then run for 3 hours allowing acquisition of the full decay of the activated isotopes. The CT was used for attenuation correction of the PRESAGE®. Immediately after, the dosimeters were analyzed using the Duke Mid field-of-view Optical CT Scanner (DMOS) at a nearby facility to matching pixel sizes. The PET image was compared with the calculated dose distribution using the Computational Environment for Radiotherapy Research (CERR) software platform.

3. Results

PET reconstructions of the first 30 minutes, 1 hour, 2 hours and 3 hours were correlated to the dose measurement of the optical-CT to determine time sensitivity of the activation decay. Measurements of the PRESAGE® dosimeter taken with PET/CT and optical-CT are shown in figure 2 and figure respectively. The PET/CT shows the effect of dose as positron emission while the optical-CT measures dose by a change in the optical density in the dosimeter.

A comparison of the two imaging modalities demonstrates the proton activation geometrically with respect to the beam profile. A visual comparison shows that the proximal region of the SOBP displayed the highest PET activity, meaning the positron hotspots shifted towards the beam entrance compared to the measured dose. This correlation was consistent with previous proton activation studies [5]. A profile along one irradiated PRESAGE® central axis, along with ion chamber data of the beam in a water phantom and corrected for RLSP, reflects this shift in the activity peak and is shown in figure 4. At the distal 80%, an approximately 1 cm discrepancy between normalized dose and PET activity was found.

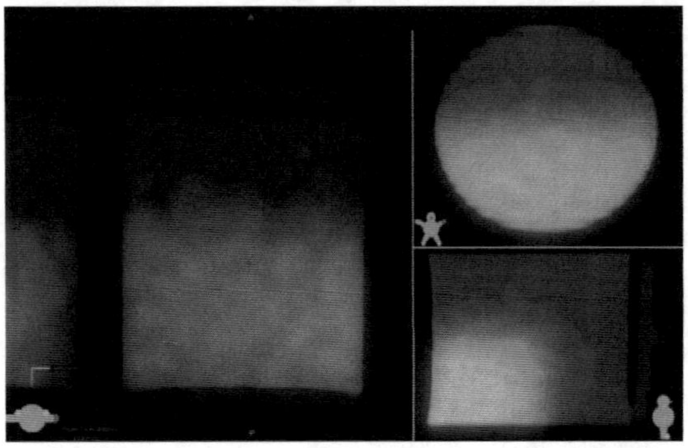

Figure 2. Co-registered PET/CT cross-section. Three hour reconstruction

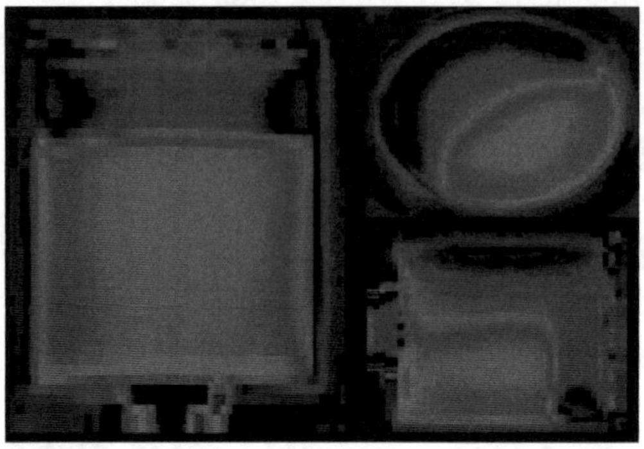

Figure 3. PRESAGE® optical-CT image cross-section.

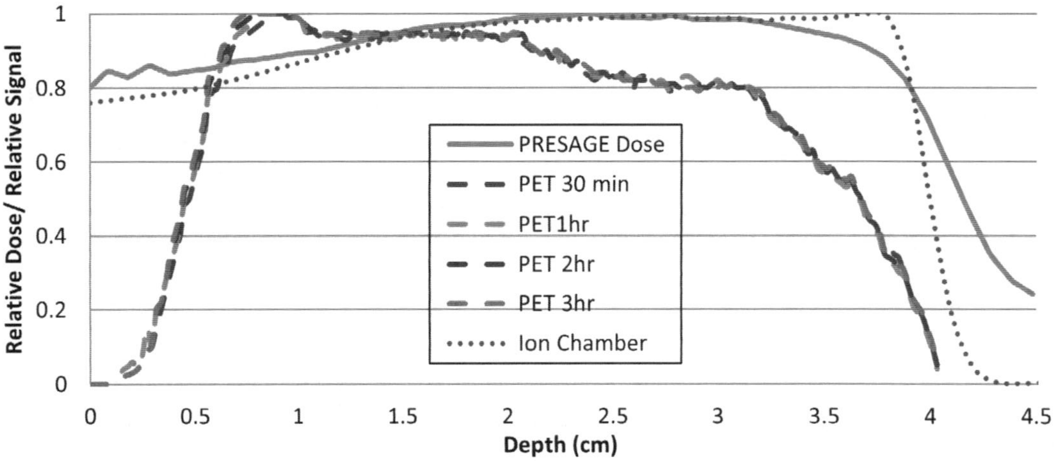

Figure 4. Dose profile and PET activity along the PRESAGE® central axis.

Spatial distributions between the measured dose and PET signal are perhaps the most relevant for in vivo verification. The proximal region of the dose is the region with the highest activity while less signal is detected in the distal side of the SOBP. The falloff region following the SOBP in which the beam energy has significantly decreased shows no signal as protons energies have fallen below the activation thresholds. A lateral axis profile comparison shows another cross-section of relatively uniform dose that demonstrates the positron activity across a monoenergetic profile and is shown in figure 5. PET measurements showed minor improvements in stability for scans beyond 30 minutes. Profile discrepancy in the central axis data showed up to a 5% discrepancy comparing scans of 30 minutes and 3 hours which fell to less than 2% comparing scans of 1 hour and 3 hours. Similar trends were observed in the lateral profiles.

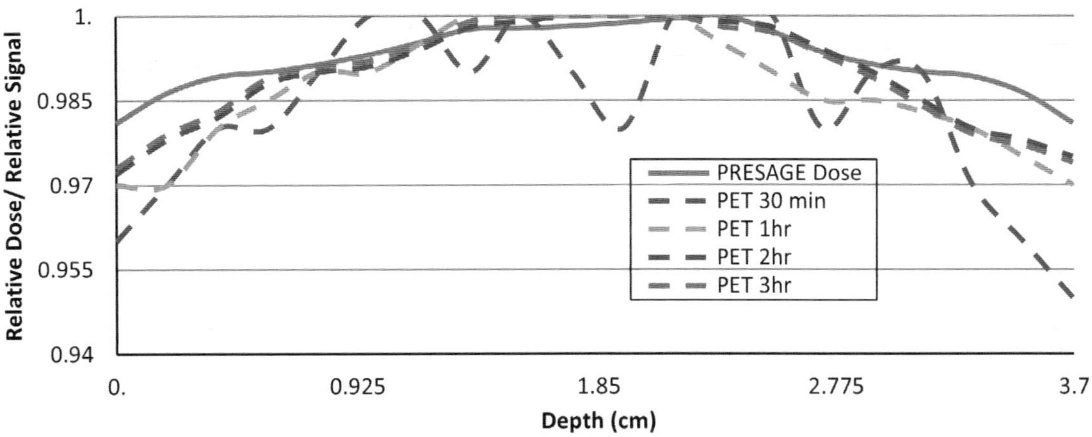

Figure 5. Dose profile and PET activity comparison along the PRESAGE® lateral axis.

4. Discussion and Conclusions

PRESAGE® dosimeters offer a unique potential to accurately correlate dosimetric and PET activation information. Implementation in an anthropomorphic phantom could be used to study representative treatment plans.

5. References

[1] Parodi K *et al* 2007 *Med. Phys.* **34** 419-35
[2] Parodi K *et al* 2007 *Int. J. Radiat. Oncol. Biol. Phys.* **68** 920-34
[3] Schaffner B and Pedroni E 1998 *Phys. Med. Biol.* **43** 1579-92
[4] Baldock C *et al* 2010 *Phys. Med. Biol.* **55** R1-63
[5] Zeidan O A *et al* 2010 *J. Phys.: Conf. Ser.* **250** 012032
[6] Gorjiara T *et al* 2011 *Med. Phys.* **38** 2265-74
[7] Adamovics J and Maryanski M J 2006 *Rad. Prot. Dosim.* **120** 107-12

Dosimetry of Strontium eye applicator: Comparison of Monte Carlo calculations and radiochromic film measurements

M Laoues[1,2], R Khelifi[2] and A S Moussa[1,2]
[1]Laboratory of Nuclear Science and Radiation -Matter Interactions (USTHB), El Alia 16111 Bab Ezzouar, Algiers, Algeria
[2]Laboratory of Theoretical Physics and Radiation-Matter Interactions (USDB), Somaa 09000 Blida, Algeria

E-mail: laouesmus192@gmail.com

Abstract. Strontium-90 eye applicators are a beta-ray emitter with a relatively high-energy (maximum energy about 2.28 MeV and average energy about 0.9 MeV). These applicators come in different shapes and dimensions; they are used for the treatment of eye diseases. Whenever, radiation is used in treatment, dosimetry is essential. However, knowledge of the exact dose distribution is a critical decision-making to the outcome of the treatment. The main aim of our study is to simulate the dosimetry of the SIA.20 eye applicator with Monte Carlo GATE 6.1 platform and to compare the calculated results with those measured with EBT2 films. This means that GATE and EBT2 were used to quantify the surface and depths dose-rate, the relative dose profile and the dosimetric parameters in according to international recommendations. Calculated and measured results are in good agreement and they are consistent with the ICRU and NCS recommendations.

1. Introduction

Due to fast and worldwide introduction of beta sources, Strontium-90 eye plaque brachytherapy is an established technique for the treatment of ocular melanoma. However, self-absorption and scattering cause difficulty to measure beta-isotope dose accurately [1]. The specific shape of this applicator, size of tumor and type of particles makes the calculation of dose distribution complicated. Indeed, the high dose gradient over short distances (~12 mm), requires using fast and accurate tools to estimate the dose. In the first method, radiochromic films are the suitable detectors for their capability of such spatial resolution and in the other way Monte Carlo methods, seem to be the best solution to obtain an accurate determination of dose distributions.

In this work, GATE (Geant4 Application for Tomographic Emission) Monte Carlo simulations v6.1 platform based on GEANT4 code was used to calculate dose distribution of Strontium ophthalmic applicator in water phantom and compared with measurements obtained with EBT2 radiochromic films [GafChromicTM, ISP Technologies Inc., Wayne, NJ, USA].

Finally; surface dose-rate, relative central axis depths dose-rate, relative central axis dose profile and dosimetric parameters calculated with GATE are compared with EBT2 radiochromic films measurements according to Report 14 of the Netherlands Commission on Radiation Dosimetry [Nederlandse Commissie voor Stralingsdosimetrie] (NCS) and report 72 of the International Commission on Radiation Units and measurements (ICRU) recommendations.

Journal of Physics: Conference Series **573** (2015) 012072 doi:10.1088/1742-6596/573/1/012072

2. Materials and Methods

2.1 Source description
Strontium eye applicator SIA.20 manufactured by (Amersham Healthcare, Arlington Heights, IL) details are given in figure 1. For more details the reader is referred to Reft *et al* [1].

2.2 Monte Carlo calculations
MC code employed in this study is GATE 6.1 running on Lunix-based operating system. The first step in the process was to determine and verify the 90Sr/90Y spectrum (figure 2) by simulating a combined 90Sr and 90Y spectrums. It was run in a spherical water-phantom with a radius of 20 cm by 4x106 photons-electrons combination mode. The source was simulated as an isotropic point source centred in a water-phantom using an energy range of 0.00 MeV to 2.28 MeV [2].

GATE was used to simulate a SIA.20 eye applicator surrounded by a spherical water-phantom of 30 mm diameter and cylindrical detectors as shown in figure 3(a). Relative dose along the central axis of applicator (figure 3(b)) was calculated by summing deposited doses in each detector (1 mm3).

Six million primary electrons were generated depending on the 90Sr/90Y combined spectrum.

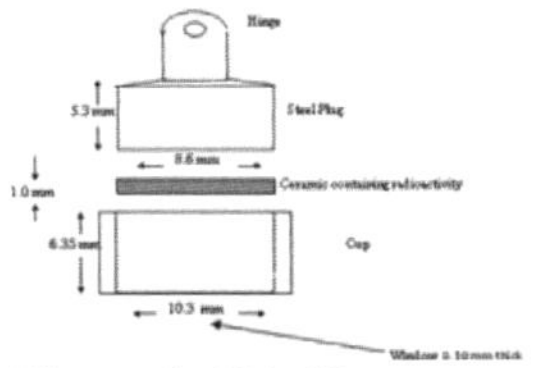

Figure 1. SIA.20 ophthalmic applicator.

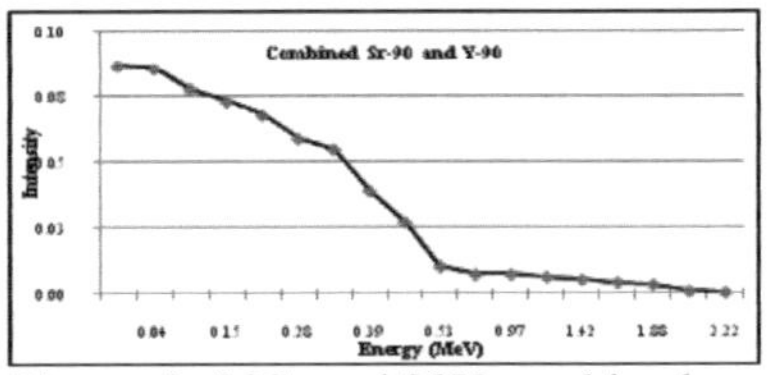

Figure 2. 90Sr and 90Y combined spectrum.

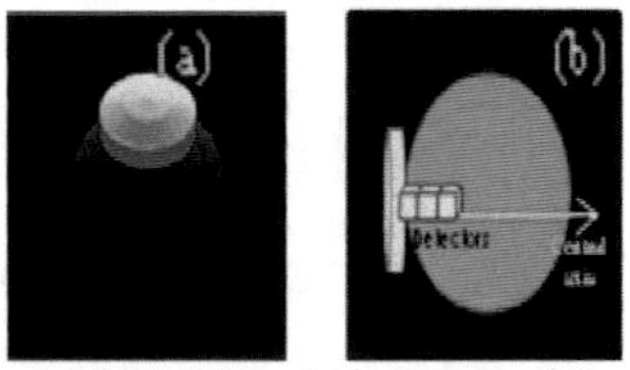

Figure 3. SIA.20 and phantom with GATE (a) and (b) detectors.

2.3 EBT2 Radiochromic film measurements

2.3.1 Film calibration
Seven EBT2 films (lot # A08301202; Expiration: August 2014) of 2x2 cm2 were calibrated with a 6 MeV electron beam using a 3x3 cm2 field (seven dose levels of 0.5 Gy to 10 Gy). Each film was positioned at a 108 mm source axis distance; to eliminate backscatter radiations a PMMA plate of 50 mm was placed underneath the film and another plate of 13 mm was placed in above to obtain a source phantom distance of 95 mm. Unexposed film was used as a zero point. Irradiated films were scanned in transmission mode 24 hours after exposure with an EPSON perfection V750-M scanner (Seiko Epson Corporation, Nagano, Japan). Obtained images (72 dpi and 48 bit RGB) were analyzed with QA FilmPro measuring average optical density (OD) derived from the red channel. Calibration curve of the dose depending on the OD is given in figure 4.

2.3.2 Irradiation procedure of EBT2 films with the eye applicator and dose evaluation
Eight films of 2x2 cm2 were used for SIA.20 doses measuring. EBT2 films were sandwiched in our specific phantom made of polystyrene (~1.05 g/cm3, thickness: 1 mm) and irradiated (Surface dose-rate was 0.53 Gy/s as on 15th Jun 1999) as shown in figure 5, for 30 seconds in the 03rd Feb 2014. The same process of scanning and analysis of films was followed. OD values of EBT2 films irradiated with the SIA.20 eye applicator were translated into dose values by extrapolation points using a third degree polynomial using the dose response curve [4].

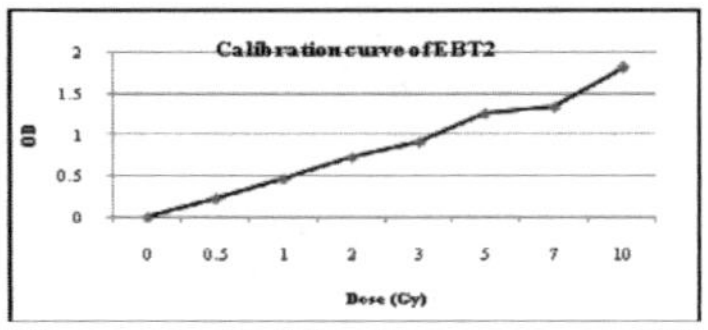

Figure 4. Calibration curve of the EBT2 film.

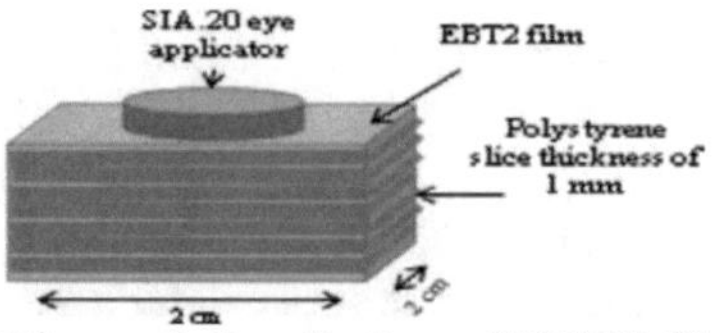

Figure 5. Irradiation of EBT2 films with the SIA.20 eye applicator.

2.4 Quality control (QC) of eye applicator

NCS and ICRU were established a task group to provide recommendations on detectors and procedures for the QC of eye applicators dosimetric parameters. These parameters had the same definition except reference depth, z0 = 1 mm (ICRU) and z0 = 2 mm (NCS). For more details the reader is referred to NCS Report 14 [4] and ICRU Report 72 [5].

3. Results and Discussion

3.1 Comparison of calculated and measured relative depth-dose along the central axis the SIA.20

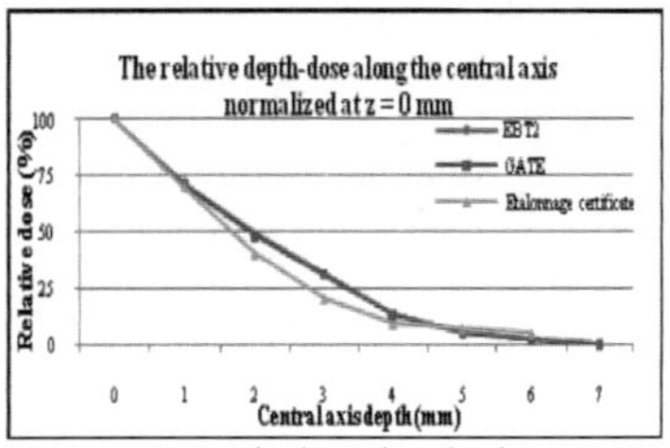

Figure 6. Relative depth-dose rate normalized at z = 0 mm.

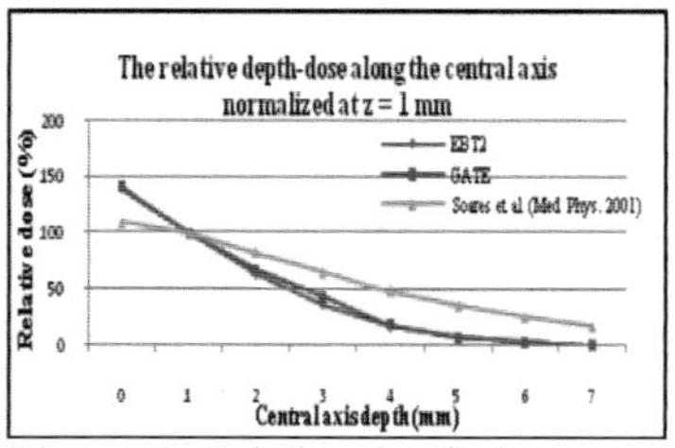

Figure 7. Relative depth-dose rate normalized at z = 1 mm.

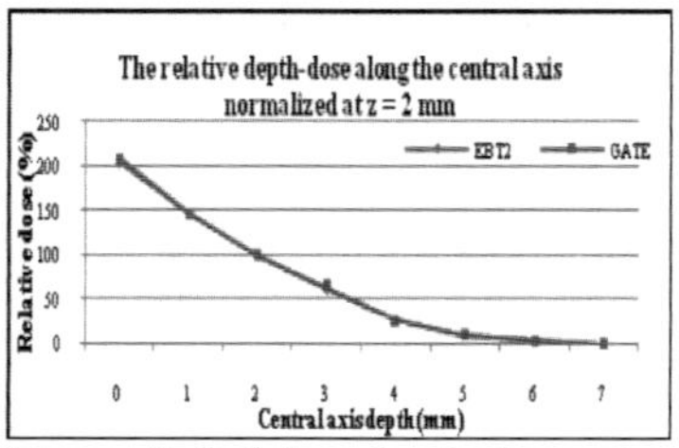

Figure 8. Relative depth-dose rate normalized at z = 2 mm.

Calculated and measured relative depth-dose rate at depth z = 0 mm as shown in figure 6, are in perfect agreement, but show poor agreement with the results provided by the manufacturer [6], with maximum deviations of the order of 63%. Based on experimental measurements the surface dose-rate of the applicator (the day of measurement) equal to 0.42 Gy/s, not 0.38 Gy/s calculated from the calibration certificate; therefore at reception of the applicator the surface dose-rate was 0.6 Gy/s, not 0.534 Gy/s which was provided by the manufacturer.

Figure 7 shows our SIA.20 calculated and measured relative depth-dose rate compared with Soares *et al* [7]. Depth-dose rate was normalized at depth 1 mm. Relative discrepancy between our calculated and measured values is very satisfactory, it doesn't exceed 7%, except 16% for the depth z = 3 mm which may be due to the bad placement of the detector in the simulation. However, relative discrepancies increase with depth between our calculated and measured values on one side and Soares *et al* [7] results other side until almost 100%.

Figure 8 shows the SIA.20 relative central axis depth-dose rate normalized at depth 2 mm for both the calculated and measured. Maximum relative discrepancy at each point doesn't exceed 7 %, which validates our MC simulation code.

3.2 Comparison of calculated and measured relative dose profile of the SIA.20 applicator

Calculated relative dose profiles along the central axis of applicator are in good agreement with measurements. However, relative discrepancies augment with increasing of depth and of perpendicular distance from the applicator central axis, but they do not exceed 17% in the worst case. This comparison demonstrates the good distribution of the applicators' radioactive material.

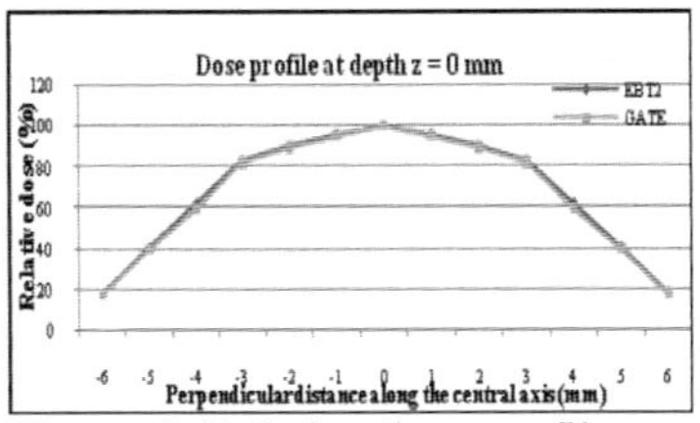

Figure 9. Relative dose-profile normalized at z = 0 mm.

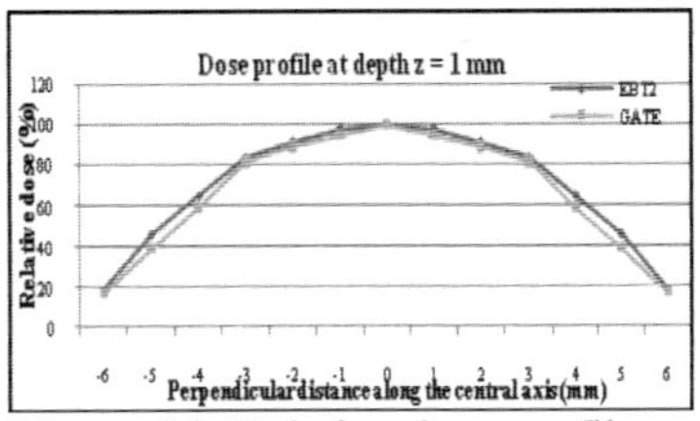

Figure 10. Relative dose-profile normalized at z = 1 mm.

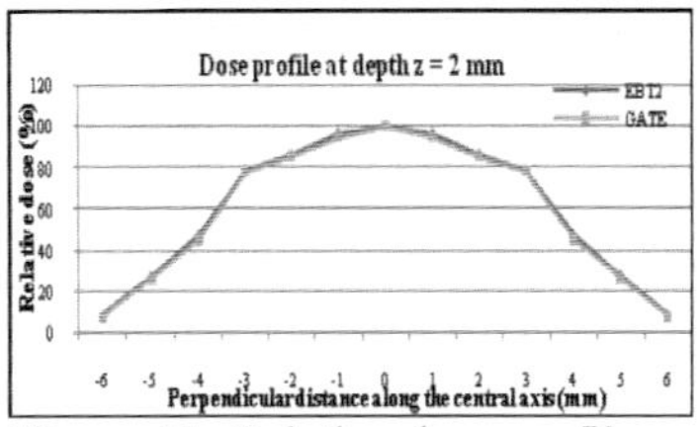

Figure 11. Relative dose-profile normalized at z = 2 mm.

3.3 Quality control of the of the eye applicator: comparison of calculated and measured

Table 1. Calculated and measured dosimetric parameters of the SIA.20 eye applicator.

	EBT2 films		GATE v6.1		Tolerance	
	NCS 14	ICRU 72	NCS 14	ICRU 72	NCS 14	ICRU 72
Average radius (R50)	03.60	04.25	03.70	04.30	/	/
Source non-uniformity (%)	19.00	18.00	18.00	17.00	30.00 ± 1	20.00 ± 1
Source asymmetry (%)	20.99	19.78	19.78	18.57	20.00 ± 1	20.00 ± 1

Calculated and measured dosimetric parameters are of the same order of magnitude. They comply with international recommendations seen previously.

4. Conclusion

GATE platform is an easy and powerful simulation tool for modeling accurate geometries; it is powered directly through command lines, and gives very accurate results. Relative central axis depth-dose rate and relative dose profile calculated with GATE are in good agreement with experimental results obtained with EBT2 films. These results demonstrate a good distribution of the applicators' radioactive material; also that maximum dose is on the applicator central axis. The QA of dosimetric parameters shows that SIA.20 non-uniformity and asymmetry are consistent with ICRU and NCS recommendations.

GATE offers to physicists a good alternative to experimentation i.e., good radioprotection, but the problem with this method is the large simulation time, which requires the use of a calculating grid.

5. References

[1] Reft C S *et al* 1990 *Med. Phys.* **17** 641-6
[2] Briesmeister J F 1997 MCNP - A general Monte Carlo N-particle transport code. Los Alamos National Laboratory Report LA-12625-M Version 4B. Los Alamos, New Mexico.
[3] Menon G *et al* 2000 *Med. Dosim.* **25** 171-7
[4] NCS 2004 Quality control of sealed beta sources in brachytherapy: Recommendations on detectors, measurement procedures and quality control of beta sources. Netherlands Commission on Radiation Dosimetry Report 14. Delft.
[5] ICRU 2004 Dosimetry of beta rays and low-energy photons for brachytherapy with sealed sources. International Commission on Radiation Units and Measurements. Oxford University Press 2004
[6] Amersham International 1999, SIA.20 calibration certificate, S/N 1147ML eye applicator, UK, (unpublished)
[7] Soares C G *et al* 2001 *Med. Phys.* **28** 1373-84
[8] IAEA 2002 Calibration of photon and beta ray sources used in brachytherapy. IAEATECDOC-1274, International Atomic Energy Agency, Vienna (ISSN 1011–4289)

Characterisation of the half-field beam penumbra for a variety of blocking set-ups

S T Smith[1], T Kairn[1,2], S B Crowe[1], A Asena[1] and J V Trapp[1]
[1]School of Chemistry, Physics and Mechanical Engineering, Queensland University of Technology, Brisbane, Australia
[2]Genesis CancerCare Queensland, Brisbane, Australia

E-mail: s164.smith@qut.edu.au

Abstract. Measurements of half-field beam penumbra were taken using EBT2 film for a variety of blocking techniques. It was shown that minimizing the SSD reduces the penumbra as the effects of beam divergence are diminished. The addition of a lead block directly on the surface provides optimal results with a 10-90% penumbra of 0.53 ± 0.02 cm. To resolve the uncertainties encountered in film measurements, future Monte Carlo measurements of half-field penumbras are to be conducted.

1. Introduction

In the development of new gel dosimeters, several characterisation studies are generally undertaken, including dose response and stability, manufacturing conditions, calibration effects and spatial stability [1-7]. To characterise the rate of diffusion in gel dosimeter candidates, dose gradients are measured across the edge of a half-blocked field at varying, post-irradiation intervals. In other words, the experimentally determined diffusion coefficient [8] is a calculation of the loss of edge sharpness as a function of time after irradiation [9].

However, when conducting half-field irradiations, it is apparent that scatter from shielding lead blocks together with beam divergence could contribute to an inherent beam penumbra. In turn, this penumbra may appear to be diffusion of the dose distribution recorded in a gel. Thus, to ensure accurate measurements, the width of the penumbra should be minimized. Across the numerous diffusion studies that have been performed, no consistent irradiation technique has been used to maximize the sharpness of the half-field edge [8-15]. And, although a protocol has been developed which minimizes the effects of initial penumbra [10], it may still be useful to determine how best to produce a half-blocked field using a 6 MV linear accelerator. This work aims to achieve this by establishing the inherent penumbra shape and width, for various field-blocking techniques.

2. Methodology

2.1 Half Field Blocking Set-ups

This work involved measurements of the 10-90% beam penumbra from a Varian Clinac 600C (Varian Medical Systems, Palo Alto, USA) linear accelerator using 20 pieces of EBT2 film. All irradiations used a 6 MV photon beam with an 11 cm solid water backscatter layer. The remaining film pieces were evenly separated into two groups to take measurements at 85 cm and 100 cm SSDs. At each

SSD, 100MU were applied for 5 different field-blocking set-ups with film placed on the surface and at a depth of 4 cm (figure 1). The studied techniques of half-blocking fields were:

1. Upper jaw on the central axis (CAX)
2. Lower jaw on CAX, to provide a reduction in beam divergence from setup 1
3. Lower jaw on CAX, with 10cm lead block placed 1 mm behind light field edge, to clean up the jaw penumbra
4. 10 cm Lead shield only, as utilised in a number of published diffusion studies [11-13]
5. Lower jaw on CAX, with 10 cm lead block placed 1 mm over light field edge, so lead collimates the field and jaw reduces lead transmission and scatter from setup 4.

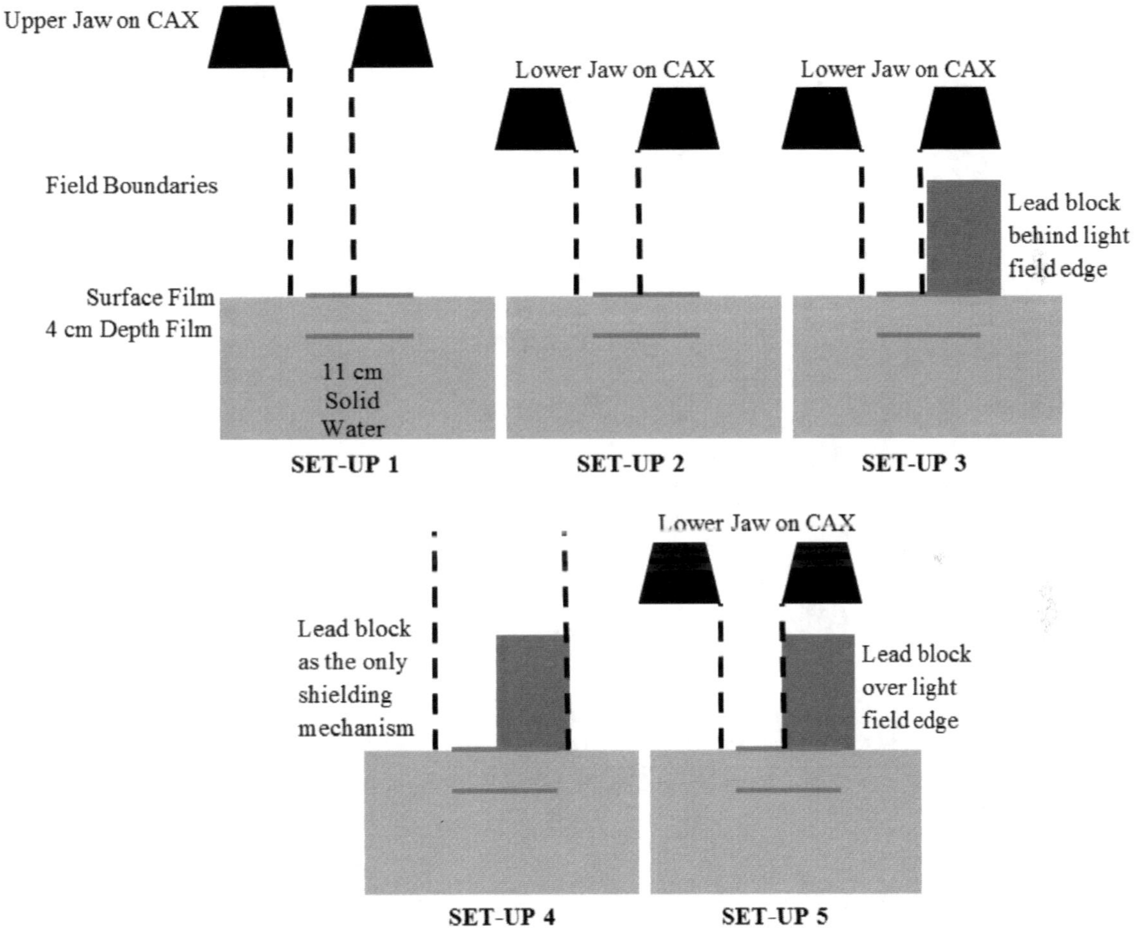

Figure 1. A schematic of the five half-field blocking techniques, which were investigated.

2.2 Penumbra Measurements

The relative dose to EBT2 film was determined according to the protocol outlined in Kairn *et al* [16] in order to minimize the effects of film heterogeneity and scanner output variations. Each piece of irradiated film was scanned before and after irradiation to correct for any film heterogeneities. To account for overall scanner output variation from one scan to the next, the ratio of the mean pixel value in an unexposed piece of film was calculated, before and after the other pieces of film were irradiated. Red channel images for the pre-scan and post- scan data were obtained and the net optical density change due to dose deposition was calculated.

From these corrected images, OD profiles were obtained across the half field dose gradient using ImageJ. To eliminate noise, a 3-point moving average was applied to each profile. The penumbra was taken as being 10% from either extent of the range of optical densities covered by the profile as demonstrated in figure 2.

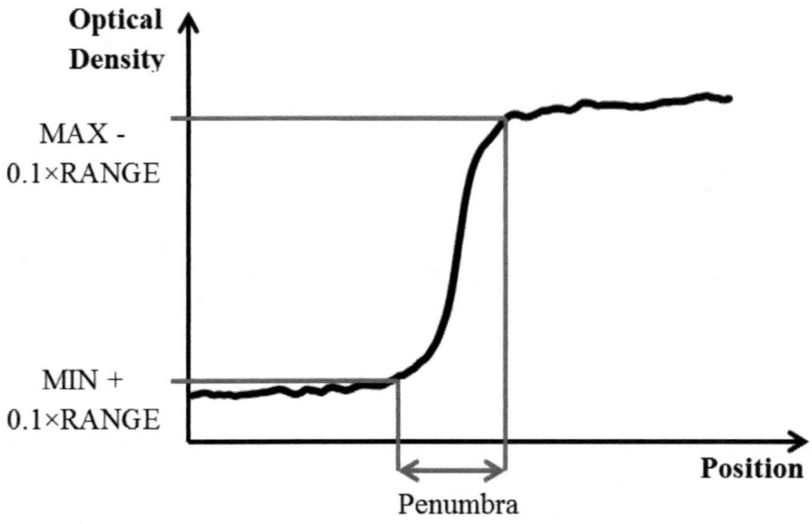

Figure 2. A typical noise reduced half-field profile obtained from the EBT2 film. The penumbra is obtained between points, 10% of the range of OD values from the maximum and minimum.

3. Results and Discussion

Table 1 shows the measured 10-90% penumbra (in cm) for the 5 different half-field setups for SSDs of 85 and 100 cm at both surface and depth. Note that the uncertainty due to limitations of scanner resolution is ± 0.02 cm for all values.

Table 1. 10-90% penumbra measurements (cm) for the various half-field blocking set-ups investigated.

SSD (cm)	Set-up 1		Set-up 2		Set-up 3		Set-up 4		Set-up 5	
	Surface	Depth	Surface	Depth	Surface	Depth	Surface	Depth	Surface	Depth
85	0.95	0.67	1.24	0.67	0.56	0.56	0.53	0.53	0.64	0.56
100	1.31	0.71	1.24	0.74	0.99	0.64	0.92	0.53	0.88	0.56

For almost all arrangements, the penumbra at depth is less than that on the surface. This broadening effect at the surface may be attributed to electron contamination and low-energy, obliquely-scattered photons which are filtered out within the first couple of cm of water. Additionally, measurements at the lower SSD of 85 cm all yield lower penumbra widths than those at the 100 cm SSD. This is clearly a result of beam divergence, widening the field as distance to the source increases. As almost all measurements involving half-fields are taken at depth, it is prudent to compare the penumbra widths at depth to evaluate which set-up is best in practice.

Accordingly, observation of the highlighted cells shows that the blocking systems involving the lead block (i.e. 3,4, and 5) had clearly lower penumbra widths than those that did not (1 and 2). As beam divergence evidently provides a significant contribution to the penumbra, these lead blocks act to collimate the divergent photons. Beyond this, set-up 4 provides the best results for penumbra reduction using lead only with a width of 0.53 ± 0.02 cm. This is somewhat counterintuitive, as it would seem that the addition of the lower jaw as in system 5, would reduce lead transmission and scatter therefore minimizing the penumbra. However, the difference in penumbra width between set-ups 3, 4, and 5 is within the range of the combined uncertainty. Furthermore, this uncertainty cannot account for the inherent noise in OD profiles of the EBT2 film. Therefore, to consolidate these results, it is suggested that Monte Carlo measurements should be made to accurately characterize the penumbra width for these set-ups.

4. Acknowledgments
This work was partly funded by Australian Research Council Linkage Project LP110100401 in conjunction with the Wesley Research Institute.

5. References
[1] Baldock *et al* 2010 *Phys. Med. Biol.* **55** R1-63
[2] Baldock *et al* 2001 *Australas. Phys. Sci. Eng. Med.* **24** 19-30
[3] Deene Y D *et al* 2000 *Phys. Med. Biol.* **45** 859-78
[4] Deene Y D *et al* 2002 *Phys. Med. Biol.* **47** 3441-63
[5] Taylor M L *et al* 2007 *Phys. Med. Biol.* **52** 3991-4005
[6] Taylor M L *et al* 2009 *Trans. Nucl. Sci.* **56** 429
[7] Venning A J *et al 2004 J. Phys.: Conf. Ser.* **3** 155-8
[8] Harris P J *et al* 1996 *Phys. Med. Biol.* **41** 1745-53
[9] Schulz R J *et al* 1990 *Phys. Med. Biol.* **35** 1611-2
[10] Kron T *et al* 1997 *Magn. Reson. Imaging* **15** 211
[11] Chu K C *et al* 2000 *Phys. Med. Biol.* **45** 955-69
[12] Penev K *et al* 2013 *J. Phys. Conf. Ser.* **444**
[13] Babic S *et al* 2009 *Phys. Med. Biol.* **54** 6791
[14] Rae W I D *et al* 1996 *Med. Phys.* **23** 15
[15] Skyt P S *et al* 2013 *J. Phys. Conf. Ser.* **444**
[16] Kairn T *et al* 2010 *Phys. Med. Biol.* **55** 37

Preliminary study of MAGAT polymer gel dosimetry for boron-neutron capture therapy

Shin-ichiro Hayashi[1], Yoshinori Sakurai[2], Ryohei Uchida[3], Minoru Suzuki[2], Shuji Usui[1] and Takahiro Tominaga[1]
[1]Department of Clinical Radiology, Faculty of Health Sciences, Hiroshima International University, Higashi-Hiroshima, Hiroshima 739-2695, Japan
[2]Research Reactor Institute, Kyoto University, Kumatori-cho, Sennan-gun, Osaka 590-0494, Japan
[3]Graduate School of Engineering, Kyoto University, Nishikyo-ku, Kyoto 615-8530, Japan

E-mail: rin@hs.hirokoku-u.ac.jp

Abstract. MAGAT gel dosimeter with boron is irradiated in Heavy Water Neutron Irradiation Facility (HWNIF) of Kyoto University Research Reactor (KUR). The cylindrical gel phantoms are exposed to neutron beams of three different energy spectra (thermal neutron rich, epi-thermal and fast neutron rich and the mixed modes) in air. Preliminary results corresponding to depth-dose responses are obtained as the transverse relaxation rate ($R_2=1/T_2$) from magnetic resonance imaging data. As the results MAGAT gel dosimeter has the higher sensitivity on thermal neutron than on epi-thermal and fast neutron, and the gel with boron showed an enhancement and a change in the depth-R_2 response explicitly. From these results, it is suggested that MAGAT gel dosimeter can be an effective tool in BNCT dosimetry.

1. Introduction

Boron-neutron capture therapy (BNCT) is a promising cancer therapy using the nuclear reaction of ^{10}B with thermal neutron (< 0.5 eV) [1, 2]. ^{10}B is delivered selectively to tumor-cells by boron carriers such as BPA (*p*-boronophenylalanin) and BSH (sodium borocaptate). In the tumor-cell, the boron nucleus captured thermal neutron splits into an alpha particle and a lithium nucleus [^{10}B(n, α)^{7}Li]. They are high LET particles and almost entirely deposit their energy in the small area comparable with a single cell. As the result, only tumor-cells can be killed selectively while the damage on normal cells is minimized. So far, nuclear reactors have been mainly used as the neutron source. Recently, accelerator-based neutron source facilities are under development, and it is expected to spread into the clinical use.

However, neutron delivered to tissue reacts with not only ^{10}B but also other nuclei depending on the energy. Fast and epi-thermal neutrons spin off recoil protons due to elastic scattering with hydrogen nuclei [^{1}H(n, n)^{1}H]. Thermal neutron leads to the nitrogen capture [^{14}N(n, p)^{14}C] and the hydrogen capture [^{1}H(n, γ)^{2}H]. These secondary radiations result to the background absorbed dose in the normal tissue. Therefore quality assurance and quality control (QA/QC) for BNCT to evaluate each dose contribution from the complicated radioactive components is required.

On the other hand, polymer gel dosimeters have been investigated for the 3D dose measurement of the complex conformal dose distributions in the clinical applications [3-5]. These devices utilize radiation-induced polymerization reactions of vinyl monomer in the gel to preserve information about the radiation dose. The 3D absorbed dose distribution is deduced from the polymer distribution measured by imaging modalities, such as MRI, optical or x-ray CT [6, 7]. Applications of polymer gel dosimeters to neutron irradiation have been investigated by several authors, and the potential as a 3D dosimeter has been suggested [8-12]. Polymer gel dosimeter is also regarded as tissue equivalent to neutron beam because the components are mainly water and a small amount of other chemicals consisting of carbon, nitrogen and oxygen. As a further advantage of the polymer gels, the interaction of neutron with the gels could be controlled by addition of some compounds with neutron-capture-nuclei such as nitrogen or boron. By the variety of elemental composition, each dose component can be distinguished from complex dose due to various primary and secondary radiations. In the previous studies, only Farajollahi *et al* [8] has reported the application of gel with ^{10}B to the dose measurements in boron neutron capture using a size of test tube gel dosimeter exposed to accelerator beam. In this work, the NMR response of MAGAT gel with boron is examined its availability to measure the depth-dose responses in the irradiation of neutron beams with different energy spectra from nuclear reactor.

2. Materials and methods

The composition of MAGAT gels with and without boron is shown in table 1. Boric acid, $B(OH)_3$, containing ^{10}B of 20% naturally was added, and the concentration in the gel is the same order (approximately 50 ppm) as the clinical use. The resulting solution was subdivided by pouring into quartz tall beakers (65 mm diameter and 135 mm length, 400 mL). After allowed to gel at room temperature, all samples were stored in a refrigerator at a fixed temperature (4 °C) for three days.

The neutron irradiations were performed using Heavy Water Neutron Irradiation Facility (HWNIF) of Kyoto University Research Reactor (KUR, power of 1 MW) [13]. The samples were irradiated from the bottom direction through the axis with the field size of almost 50 cm diameter in air at room temperature. As shown in table 2, the three different modes (Mode-1: thermal neutron rich, Mode-2: epi-thermal and fast neutron rich and Mode-3: the mixed modes) of neutron beams made by heavy water spectrum shifter and cadmium thermal-neutron filters were applied to each sample [14]. It is noted that the irradiation time is different in each mode.

MRI measurements were performed using a 1.5 T scanner (Siemens) with a quadrature torso SPEEDER coil. A multiple spin-echo sequence was applied using the following parameters, (TR / TE's / FOV / Slice Thickness / Matrix Size) = (5 s / 20 - 320 ms, 32 echos / 192 x 192 mm^2 / 10 mm / 192 x 192) with a resulting voxel size of 1.0 x 1.0 x 10.0 mm^3. The transverse relaxation rate ($R_2 = 1/T_2$) is determined by fitting equation, $S(TE) = S(0)\exp(-TE/T_2)$, where S(0) and S(TE) are the signal intensities at echo times, 0 and TE, respectively.

3. Results and discussion

Figure 1 shows the depth-R_2 profiles obtained from our polymer gel dosimeters exposed to neutron beams of the different energy spectrum modes. In the thermal neutron mode (Mode-1), both profiles of the gels with and without boron show monotonically decreasing with depth after the peak near the surface. It seems that the decrease for the gels with boron corresponds to decreasing of thermal neutron in tissue and the peak shifts to near the surface significantly due to the reaction with boron. In the epi-thermal and fast neutron mode (Mode-2), broad peaks are observed at deeper position, around from 20 to 25 mm of depth. It is suggested that the R_2 profile for the gel with boron corresponds to the distribution of the thermal neutron due to the moderation of epi-thermal neutron. But for the gels without boron, it seems that the decreases correspond to decreasing of gamma ray. As same as the profiles in Mode-1, the peak shift due to boron was also observed. The depth profiles of the mixed mode (Mode-3, not shown) were similar to that of Mode-1. It is suggested that the contributions of epi-thermal and fast neutrons are small compared to that of thermal neutron. These results indicate both MAGAT gel dosimeters with and without boron have the effective sensitivity on the thermal neutron rather than on fast neutron.

Table 1. The composition of MAGAT type gel prepared in this study. (Boric acid of 25mM approximately corresponds to 50 ppm of ^{10}B.)

Chemical	Concentration
Water	87 wt%
Gelatin	8 wt%
Methacrylic acid	5 wt%
Boric acid	0 and 25 mM
Tetrakis(hydroxymethyl)phosphonium chloride (THPC)	5 mM

Table 2. The dose rates composition (Gy/h) in the different irradiation modes (Mode1-3). Thermal, epi-thermal and fast neutron energies are less than 0.5 eV, from 0.5 eV to 10 keV, and more than 10keV, respectively.

Mode	Fast	Epi	Th	Gamma	Total	Irr. Time (min.)
1	0.05	0.00	0.20	0.20	0.44	15
2	0.33	0.04	0.00	0.12	0.50	30
3	0.33	0.05	0.48	0.66	1.52	5

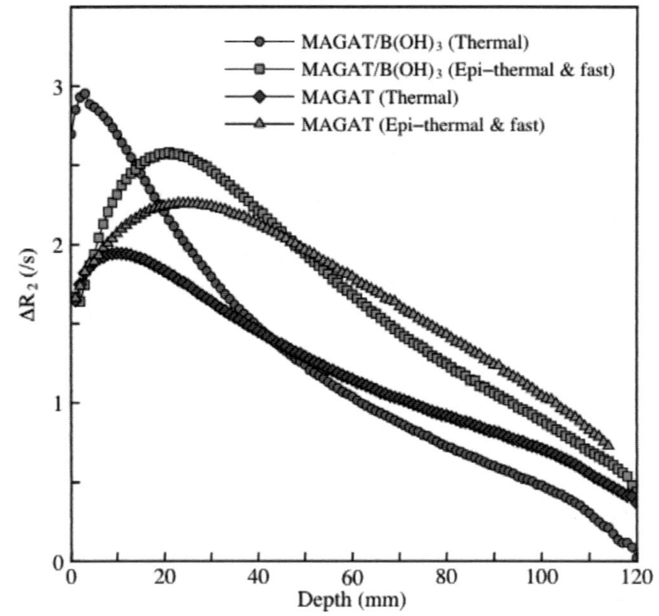

Figure 1. The depth from phantom surface vsΔR_2 [$=R_2 - R_{2,bg}$] responses in different energy spectra of MAGAT gel dosimeters with and without boron.

The thermal neutron reacts with nitrogen and hydrogen nuclei in polymer gel and produces proton [$^{14}N(n, p)^{14}C$] and prompt gamma ray [$^{1}H(n, \gamma)^{2}H$] from their nuclei as the secondary radiations. In this experiments, R_2 response of gel dosimeter without boron will mainly be due to them while the influence of boron-neutron capture [$^{10}B(n, \alpha)^{7}Li$] was obviously observed in the gel with boron. From these results the estimation of the dose distribution in the gel dosimeter with heterogeneous distribution of boron might be possible by using a gel-in-gel such as a gel container [15]. Moreover by adding some compounds containing neutron-capture-nuclei and by varying the concentrations, each dose component might be distinguished. The comparison with simulations is also needed in the next step.

Journal of Physics: Conference Series **573** (2015) 012074 doi:10.1088/1742-6596/573/1/012074

4. Conclusion

In this work, we have examined the depth-R_2 responses of MAGAT polymer gel dosimeter with and without boron on neutron beams with different energy spectra. In the results, it was shown that the gel dosimeter with boron had the effective sensitivity on thermal neutron. The studies on the fundamental characteristics, the effect of additive compounds with neutron-capture-nuclei and the comparison with Monte Carlo simulations and other dosimeters are under way.

5. Acknowledgement

We would like to thank Mr. Shinji Maeda of Hiroshima Prefectural Rehabilitation Center for his technical assistance with MRI. This work is supported by JSPS KAKENHI (Grant Number 26460736) and the Collaborative Research Program of Kyoto University Research Reactor Institute (Grant Number 26P14-29).

6. References

[1] Neutron Capture Therapy: Principles and Applications 2012. Sauerwein W A G, Wittig A, Moss R and Nakagawa Y (Eds). Springer Verlag (ISBN 978-3-642-31334-9)
[2] Boron and Gadolinium Neutron Capture Therapy for Cancer Treatment 2012. Hosmane N S, Maguire J A, Zhu Y and Takagaki M (Eds). World Scientific Publishing (ISBN 978-981-4338-67-7)
[3] Baldock C *et al* 2010 *Phys. Med. Biol.* **55** R1-63
[4] Hurley C *et al* 2005 *Appl. Radiat. Isotopes* **63** 443-56
[5] Baldock C 2006 *J. Phys.: Conf. Ser.* **56** 14-22
[6] Venning A J *et al* 2004 *J. Phys.: Conf. Ser.* **3** 155-8
[7] Brown S *et al* 2008 *Appl. Radiat. Isotopes* **66** 1206-12
[8] Farajollahi A R *et al* 2000 *Phys. Med. Biol.* **45** N9-14
[9] Gambarini G *et al* 2006 *J. Phys.:Conf. Ser.* **41** 275-81
[10] Uusi–Simola J *et al* 2007 *J. Appl. Clin. Med. Phys.* **8** 114-23
[11] Uusi-Simola J *et al* 2003 *Phys. Med. Biol.* **48** 2895-906
[12] Abtahi S M et al 2013 *IFMBE Proc.* **39** 1157-9
[13] Sakurai Y *et al* 2002 *Med. Phys.* **29** 2328-37
[14] Sakurai Y *et al* 2000 *Nucl. Instr. Meth. Phys. Res. A.* **453** 569-96
[15] Tominaga T *et al* 2014. Proceedings of 8th International Conference on 3D Radiation Dosimetry (IC3DDose) 4-7 September 2014, Ystad, Sweden

Assessment of the effects of CT dose in averaged x-ray CT images of a dose-sensitive polymer gel

T Kairn[1, 2], M B Kakakhel[3], H Johnston[4], A Jirasek[4] and J V Trapp[2]
[1]Genesis Cancer Care Queensland, Brisbane, Australia
[2]Science and Engineering Faculty, Queensland University of Technology, Brisbane, Australia
[3]DPAM, Pakistan Institute of Engineering and Applied Sciences, PO Nilore, Islamabad 45650, Pakistan
[4]Department of Physics and Astronomy, University of Victoria, Victoria BC, Canada

E-mail: t.kairn@gmail.com

Abstract. The signal-to-noise ratio achievable in x-ray computed tomography (CT) images of polymer gels can be increased by averaging over multiple scans of each sample. However, repeated scanning delivers a small additional dose to the gel which may compromise the accuracy of the dose measurement. In this study, a NIPAM-based polymer gel was irradiated and then CT scanned 25 times, with the resulting data used to derive an averaged image and a "zero-scan" image of the gel. Comparison between these two results and the first scan of the gel showed that the averaged and zero-scan images provided better contrast, higher contrast-to-noise and higher signal-to-noise than the initial scan. The pixel values (Hounsfield units, HU) in the averaged image were not noticeably elevated, compared to the zero-scan result and the gradients used in the linear extrapolation of the zero-scan images were small and symmetrically distributed around zero. These results indicate that the averaged image was not artificially lightened by the small, additional dose delivered during CT scanning. This work demonstrates the broader usefulness of the zero-scan method as a means to verify the dosimetric accuracy of gel images derived from averaged x-ray CT data.

1. Introduction

Given the obvious value of 3D gel dosimetry as a technique for verifying the accuracy and deliverability of radiotherapy treatment plans [1-3], the reading out of gel measurements using an imaging modality already available in many radiotherapy departments is highly desirable [4]. For this reason there is increasing interest in the use of x-ray computed tomography (CT) [5] to provide images of dosimetry gels [6-8].

While the small radiation-induced density changes in polymer gel dosimeters can be detected using clinical x-ray CT scanners, the low signal-to-noise ratio (SNR) in the resulting images means that the results are not immediately useful without further processing [7]. The main method for enhancing the SNR in an x-ray CT image of a dosimetry gel is to scan the gel multiple times and calculate the mean of each pixel value across the various scans, to give an averaged image [6, 7]. Obviously, each time the gel is CT scanned, it receives a small additional radiation dose which may affect the result of the measurement. For example, when a PAGAT gel [9] is exposed to 6 MV x-ray doses of up to 6.9 Gy and then CT scanned 50 times using a 120 kVp x-ray source, the resulting image pixel values can be shown to increase from one scan to the next, resulting in an averaged image that provides dose measurements 5% higher than the doses measured in the first CT image alone [10].

The inaccuracy caused by the gel's response to the CT dose can be mitigated by: optimising the gel for CT readout [11]; optimising CT scanning parameters to produce maximised signal-to-noise images with minimised dose [7, 12]; and by directly removing the effects of the CT dose by subtracting "background" CT scans of non-irradiated (but CT imaged) gel samples from the scans of the irradiated gel samples. When examining the results of applying such techniques, a dosimetrically reliable reference image is needed, against which the averaged image can be evaluated. We propose that Kakakhel *et al*'s zero-scan method [10, 13, 14] (described below) is well suited to this purpose.

2. Method

2.1 Gel fabrication and irradiation

Two identical containers of polymer gel were produced according to a recipe optimised for x-ray CT by Jirasek *et al* [11, 15]. The gel consisted of 15% NIPAM (TCI America, Portland, OR), 4.5% BIS (Sigma-Aldrich Canada, Oakville, ON), 5% gelatin (Sigma-Aldrich), 75.5% deionized water and 5 mM tetrakis-hydroxymethyl-phosphonium-chloride (THPC) (Sigma-Aldrich) as antioxidant [11].

One container of gel was left un-irradiated, to provide a background measurement, and the other was irradiated using three intersecting 3×3 cm^2 fields from a Varian Clinac 21EX linear accelerator (Varian Medical System, Palo Alto, USA), which delivered up to 27.6 Gy to the centre of the gel. (This treatment plan is illustrated in [15].)

2.2 X-ray CT image acquisition

X-ray CT images were acquired using a GE HiSpeed FX/i single-slice CT scanner (GE Medical Systems, Milwaukee, USA) with a 120 kVp source operating at 200 mAs. One 3 mm slice was scanned 25 times, in each container of gel, and reconstructed using a standard convolution kernel over a 25 cm field of view. Each image acquisition took approximately 2 seconds; scanning each sample took 50 seconds. No x-ray tube cooling breaks were required. The background gel was scanned three minutes after the irradiated gel.

2.3 X-ray CT image analysis

Prior to analysis, a background image from the un-irradiated gel was subtracted from each of the 25 images of the irradiated gel, to remove scanning artefacts [16, 17]. The values of each individual pixel value (Hounsfield unit, HU) in the resulting images were then averaged over the 25 scans to produce an averaged image [6, 7], and linearly extrapolated back to a theoretical "zero-scan" image of the gel without the effects of the small additional dose delivered during CT scanning [10, 13, 14].

The contrast (C), contrast-to-noise ratio (CNR) and SNR of the resulting images were then evaluated as [18]:

$$C = \frac{P_{high} - P_{low}}{P_{low}}, \quad CNR = \frac{P_{high} - P_{low}}{\sigma_{low}} \text{ and } SNR = \frac{P_{high}}{\sigma_{high}} \text{ or } \frac{P_{high}}{\sigma_{high}}$$

where P_{high} and P_{low} were the mean values in 1 cm diameter regions of interest in, respectively, high and low dose areas of each image and σ_{high} and σ_{low} were the standard deviations from those means.

3. Results

Qualitative examination of the images shown in figures 1(a)-(c) suggests that the averaged and zero-scan images are affected by less noise than the image from the first scan, with the averaged image being the least noisy of the three images. This observation is supported by the data shown in table 1.

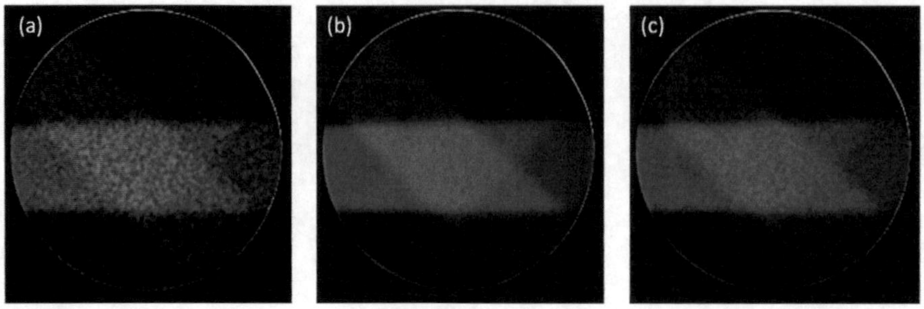

Figure 1. Transverse images of irradiated gel, derived from (a) the first of 25 sequential CT scans of the gel, (b) the average pixel values (HU) from all 25 scans of the gel and (c) the result of projecting the pixel values from the 25 scans back to the "zero scan" of the gel.

Table 1. Contrast, CNR and SNR, determined from images shown in figure 1.

Image	Contrast	CNR	High-dose SNR	Low-dose SNR
First scan	20.1	15.1	8.3	0.8
Average	21.2	67.9	33.4	3.2
Zero scan	21.8	33.9	21.4	1.6

The profiles in figures 2(a) and (b) show the same reduction in noise in the results achieved with the averaging and zero-scan analysis methods, while also demonstrating that the pixel values in the averaged image are not noticeably elevated, compared to the zero-scan result. This result suggests the averaged image has not been artificially lightened by the small, additional dose delivered during CT scanning.

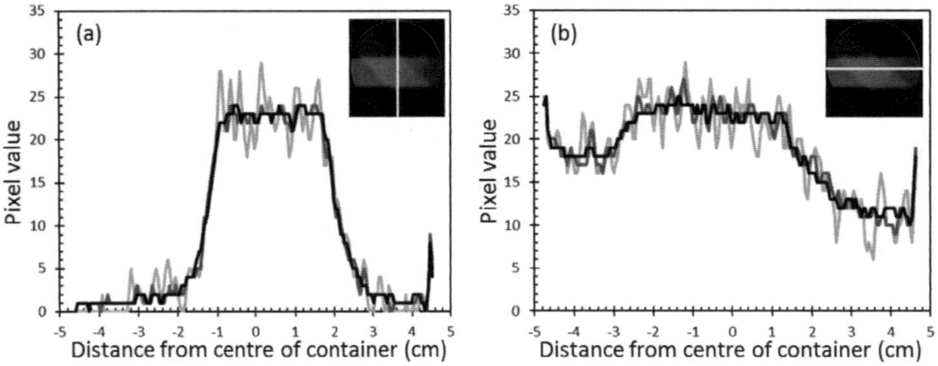

Figure 2. Profiles through the dose images of the irradiated gel, shown in figure 1, where the green (light grey) line represents data from the first scan, the black line represents data from the average of the scans and the red (dark grey) line represents data from the zero scan. Insets show positions of profiles.

Figure 3 shows a histogram of the gradients of the linear fits used to extrapolate the zero-scan pixel values, for all pixels in the transverse image of the gel, and indicates that all gradients are small (no value falls outside the range -0.30 < gradient < +0.30) and that values are symmetrically distributed around zero; pixels are neither becoming systematically lighter (as they would if the result was affected by radiation dose from the CT) nor systematically darker (as they would if the CT dose was being over-corrected), between one CT scan and the next.

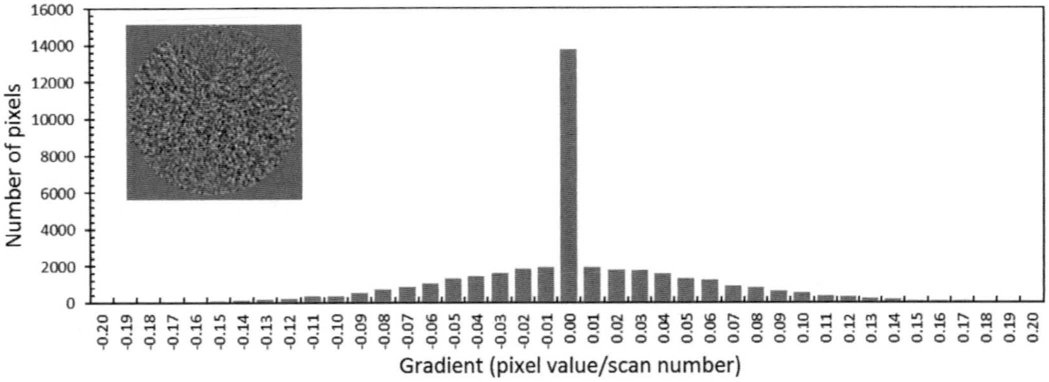

Figure 3. Histogram of gradients obtained from zero-scan extrapolation. Inset shows map of gradient values, with 0.0 appearing as mid-grey, values greater than 0.0 appearing as lighter grey (to white) and values less than 0.0 appearing as darker grey (to black).

4. Conclusion

This work used the zero-scan method to show that an averaged series of CT images of an irradiated gel were not affected by the dose delivered during the x-ray CT imaging process, and thereby demonstrated the broader usefulness of the zero-scan method as a means to verify the dosimetric accuracy of gel images derived from averaged x-ray CT data. A zero-scan image of a CT-scanned gel sample should be acquired and analysed each time the gel recipe or scanning protocol is altered, as well as whenever the averaging of CT scans of a dosimetric gel is being attempted for the first time.

5. Acknowledgements

TK's contribution to this study was supported by Premion (Genesis Cancer Care Queensland), in association with the Wesley Research Institute and the Queensland University of Technology, as part of Australian Research Council linkage project LP110100401. The authors wish to thank the Vancouver Island Cancer Centre for kindly providing access to their x-ray CT scanner.

6. References

[1] Baldock C *et al* 2010 *Phys. Med. Biol.* **55** R1-63
[2] Korreman S S 2013 *J. Phys.: Conf. Ser.* **444** 012007
[3] Kairn T *et al* 2012 *Phys. Med. Biol.* **57** 3359-69
[4] Baldock C 2006 *J. Phys.: Conf. Ser.* **56** 14-22
[5] Brown S *et al* 2008 *Appl. Radiat. Isotopes* **66** 1206-12
[6] Hilts M *et al* 2000 *Phys. Med. Biol.* **45** 2559-71
[7] Jirasek A 2013 *J. Phys.: Conf. Ser.* **444** 012005
[8] Hill B *et al* 2005 *Brit. J. Radol.* **78** 623-30
[9] Venning A J *et al* 2004 *J. Phys.: Conf. Ser.* **3** 155-8
[10] Kakakhel M B *et al* 2011 *Med. Phys.* **38** 5130-35
[11] Jirasek A *et al* 2012 *Phys. Med. Biol.* **57** 3137-53
[12] Hilts M *et al* 2005 *Phys. Med. Biol.* **50** 1727-45
[13] Kakakhel M B *et al* 2013 *J. Phys.: Conf. Ser.* **444** 012091
[14] Kakakhel M B *et al* 2014 *J. Appl. Clin. Med. Phys.* (In press)
[15] Johnston H *et al* 2012 *Phys. Med. Biol.* **57** 3155-75
[16] Trapp J V *et al* 2001 *Phys. Med. Biol.* **46** 2939-51
[17] Trapp J V *et al* 2004 *Phys. Med. Biol.* **49** N139-46
[18] Kairn T *et al* 2010 *Phys. Med. Biol.* **55** N533-45

Computational simulations of the influence of noise in optical CT reconstruction

Yves De Deene
Department of Engineering, Faculty of Science, Macquarie University, Sydney, Australia

E-mail: yves.dedeene@mq.edu.au

Abstract. In 3D radiation dosimetry with optical CT scanning readout, projections of transmitted light are recorded in either lines or planar. The projections are then transferred to optical density maps by use of filtered back-projection. Absolute dose maps can be derived from the optical density maps by calibration of the optical absorption coefficient to dose values. The transmission profiles will be subject to a certain level of detector noise and noise arriving from fluctuations in the light source. Different reconstruction filters in the frequency domain can be applied in the image reconstruction procedure. The noise level in the final reconstructed images is determined by the noise in the projections, the spatial resolution and the reconstruction algorithm. The reconstruction filters may also have an effect on the geometrical precision as a spatial frequency filter may also affect the spatial resolution. We here propose an easy method to assess both the noise sensitivity and the performance in terms of mapping dose distributions with various theoretical dose gradients.

1. Introduction

Three-dimensional radiation dosimetry has received growing interest with the implementation of highly conformal radiotherapy treatments [1]. Over the last two decades, much research has been conducted towards both the development of different radiation sensitive three-dimensional dosimeters and adequate scanning techniques. Although an acceptable precision and accuracy has been achieved with polymer gel dosimeters and MRI scanning [2-4], the procedure is relatively labor intensive and some expertise is needed to acquire reliable quantitative MR images. As an alternative to MR scanning, optical CT scanning has been proposed as an alternative [5]. Different staining dosimeters have been proposed as a better alternative to polymer gel dosimeters including polyurethane based dosimeters, PRESAGE™ [6] and micelle gel dosimeters [7, 8] and another system will be proposed at this conference. For a comprehensive discussion on optical CT scanning and different methods we refer to another communication and references herein [9]. An analysis of the propagation of noise in reconstructed images for different frequency filters has already been performed by Krstajic and Doran [10] for a uniform circular phantom.

In this study, we extend this approach to a circular phantom containing a theoretical square field. This allows not only to observe the effect of the filter on the noise propagation in the reconstructed images but also the effect of the frequency filter on the spatial integrity of resolving dose gradients that occur in the dose distribution. In addition to the study by Krstajic and Doran, we also note that the noise in the reconstructed image is not uniformly distributed but displays a radial dependence.

Journal of Physics: Conference Series **573** (2015) 012076 doi:10.1088/1742-6596/573/1/012076

2. Methods and Materials

2.1. SNR of reconstructed images

All image reconstructions were performed using Matlab R2008 using the built-in function "iradon" with the following spectral filters: Hamming, Hann, Ram-Lak and Cosine. A circular phantom with diameter 5 cm was considered. The number of projections from which the image was calculated was 180 over a demi-arc, corresponding to a radial increment of one degree. The attenuation through the phantom was considered to occur according to an inverse exponential power law. Additive stochastic white noise was added to the projection profiles prior to reconstruction that corresponds with the combined effect of read noise, thermal noise and shot noise of the detector.

The signal-to-noise in the projection is a function of the light intensity falling on the detector as described by equation 1.

$$SNR = SN = k\cdot \Phi \hbar v\cdot t\cdot A\cdot \eta \Phi \hbar v\cdot t\cdot A\cdot \eta + nCCD2 + nreadout2 = k\cdot SS + nCCD2 + nreadout2 \qquad (1)$$

where Φ is the light power falling on the detector (*W/m2)*, $\hbar v$ is the photon energy (*J*), *t* is the exposure time, *A* is the surface of the detector element, η is the quantum efficiency of the detector, *nCCD* is the CCD noise (electrons) and *nreadout* is the noise (electrons) due to read out of the CCD-image sensor (i.e. the electronics). The scaling constant k accounts for the binning and averaging during readout.

Several of these parameters are detector specific. In order to come up with some realistic values, we performed some noise measurements on the CCD camera that we used in house. To derive the relation of SNR versus signal intensity (S), 512 projection images where recorded with a resolution of 0.2 mm x 0.2 mm. The standard deviation and mean signal values over the 512 projection images were derived on a pixel-by-pixel basis and plotted in a scatter plot (SNR(S)). The relation was fitted to the function:

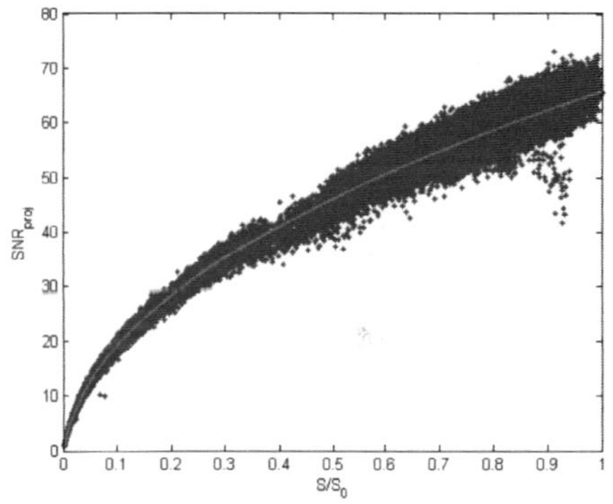

Figure 1. SNR_{proj} as a function of the detected normalized signal intensity S/S_0 in a realistic projection image obtained with a commercial CCD camera (DragonFly2 DR2-HIBW)

$$SNRproj = k.\ SS + ncam2 \qquad (2)$$

where *ncam* combines both *nCCD* and *nreadout* (i.e. *ncam2= nCCD2+nreadout2*).

Figure 1 shows the scatter plot of SNR_{proj} as a function of the detected signal (dots) with a fit against equation 2 (red line). The detected signal is normalized to the largest signal that was obtained before saturation occurs. The fitted values were: *k=66.5* and *ncam=0.1405*. The goodness of fit was: *r2=0.9992*. This functional relation is further used in the simulations.

2.2. Spatial integrity study

To study the effect of frequency filters in the reconstruction in the presence of noise on the measured penumbra, a synthetic circular phantom with square field was constructed (figure 2). The steepness of the gradient in the square field can be varied by only a single parameter. The theoretical equation for the optical density offset created by the square field is given by equation:

$$ODFx,y,a=ODR\cdot\sinh(\xi a)\cosh\xi x-x0+\cosh(\xi a)\cdot\sinh(\xi a)\cosh\xi y-y0+\cosh(\xi a) \qquad (3)$$

where ODR is the optical density in the flat (irradiated) region of the field, a is the field size, ξ is a parameter inversely correlated with the penumbra width and $x0$, $y0$ are the spatial coordinates of the offset of the field with respect to the geometrical centre. The theoretical equation allows to perform the performance study without the need of spatial discretization of the synthetic phantom which avoids any bias in the results caused by regridding along the projection lines.

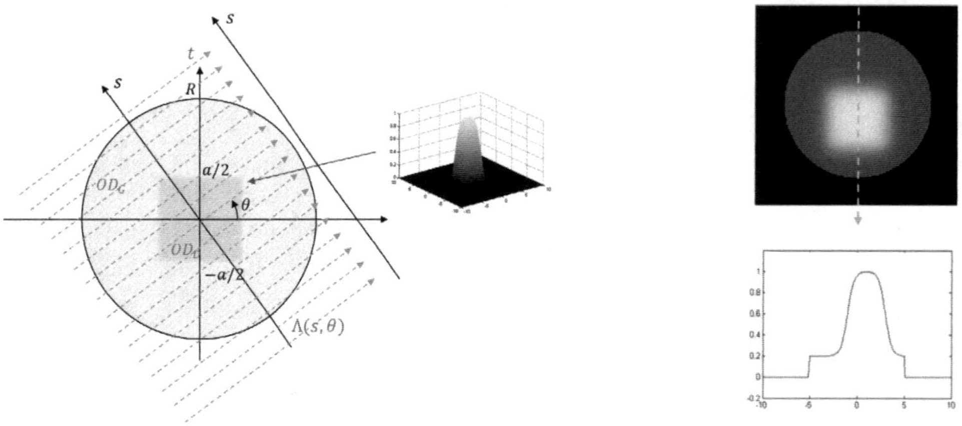

(a) (b)

Figure 2. A synthetic circular phantom with square field in the centre (a). Note that in the simulations the field was chosen acentric to generate a more general case (b).

3. Results and Discussion

3.1. SNR characteristics for different filters

The SNR as a function of optical absorbance coefficient is plotted for two different frequency filters in figure 3. All other parameters were kept the same, thus illustrating the strong influence of the frequency filter on the noise performance. A more than two fold increase in SNR is observed by applying a Hann filter instead of a Ram-Lak filter.

3.2. Spatial integrity of dose gradients

The spatial performance in terms of the ability to acquire the penumbra width is demonstrated in figure 4.

4. Conclusion

An easy way of simulating the performance of an optical scanner and the image reconstruction algorithm for 3D radiation dosimetry is described. These simulations can be used to predict and optimize the precision in terms of dose resolution, for any detector, number of projections, spatial binning and reconstruction algorithm. It is also helpful in describing the effect of the reconstruction software on the spatial integrity. It is shown that for a phantom of 10 cm diameter, the optimal average optical density of the dosimeter should be around 0.2 cm^{-1}. Note that this value is dependent on the size of the phantom. It is also found that the noise is not uniformly distributed in the reconstructed dose maps. From the four studied frequency filters, the Hann filter gives the best results, both in terms of precision and spatial integrity at steep dose gradients but it is likely that other filters can still improve the noise performance.

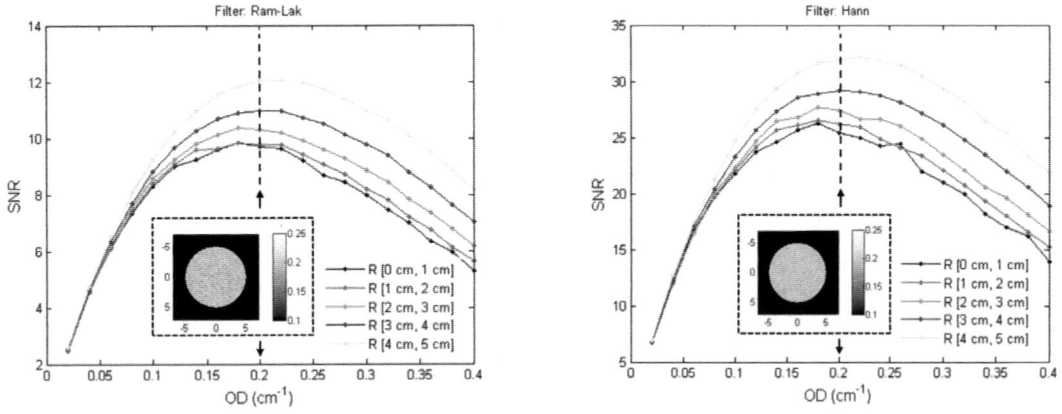

(a) (b)

Figure 3. SNR in OD maps reconstructed using a Ram-Lak filter (a) and a Hann filter (b). The SNR is plotted for different circular strips showing that the noise is non-uniform in the entire phantom. The different graphs are for different circular shells with inner and outer radius as indicated in the figure inset. A reconstructed image for OD = 0.2 cm^{-1} is also shown.

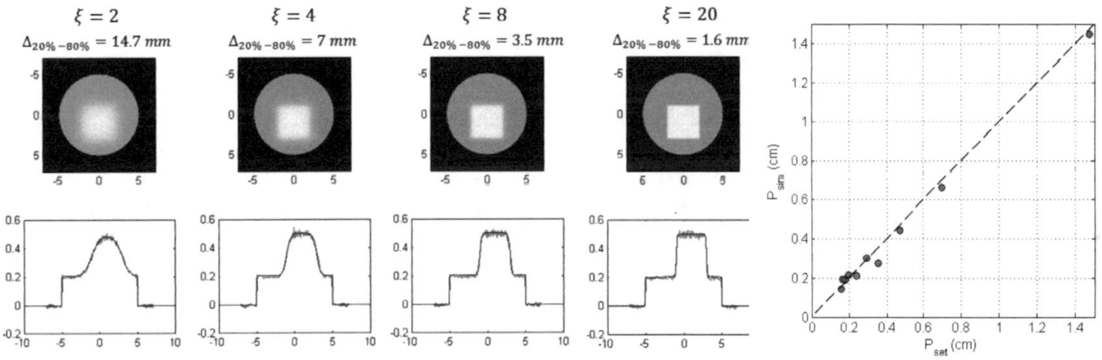

(a) (b)

Figure 4. Performance of filtered back projection (Hann filter) on the measured penumbra in the presence of noise. OD maps of simulated square fields with different dose gradients (a) and derived penumbra width as compared to the theoretical penumbra (b).

5. References

[1] Baldock C *et al* 2010 *Phys. Med. Biol.* **55** R1-63
[2] Vandecasteele J and De Deene Y 2013 *Phys. Med. Biol.* **58** 19-42
[3] Vandecasteele J and De Deene Y 2013 *Phys. Med. Biol.* **58** 43-61
[4] Vandecasteele J and De Deene Y *Phys. Med. Biol.* **58** 63-85
[5] Gore J C et al 1996 *Phys. Med. Biol.* **41** 2695-704
[6] Adamovics J A and Maryanski M 2004 *Med. Phys.* **30** 1349
[7] Jordan K and Avvakumov N 2009 *Phys. Med. Biol.* **54** 6773-89
[8] Vandecasteele J and De Deene Y 2013 *Phys. Med. Biol.* **58** 6241-62
[9] Doran S and Krstajic N 2006 *J. Phys. Conf. Series* **56** 45-57
[10] Krstajic N and Doran S 2007 *Phys. Med. Biol.* **52** 3693-713

Journal of Physics: Conference Series **573** (2015) 012077 doi:10.1088/1742-6596/573/1/012077

Is a quasi-3D dosimeter better than a 2D dosimeter for Tomotherapy delivery quality assurance?

Aitang Xing[1], Shrikant Deshpande[1,3], Sankar Arumugam[1], Armia George[1], Lois Holloway[1,2,3,4], Philip Vial[1,2] and Gary Goozee[1,4]

[1]Liverpool and Macarthur Cancer Therapy Centres and Ingham Institute, Liverpool Hospital, New South Wales, Australia

[2]Institute of Medical Physics, School of Physics, University of Sydney, New South Wales, Australia

[3]Centre for Medical Radiation Physics, University of Wollongong, New South Wales, Australia

[4]South West Clinical School, University of New South Wales, New South Wales, Australia

E-mail: aitang.xing@sswahs.nsw.gov.au

Abstract. Delivery quality assurance (DQA) has been performed for each Tomotherapy patient either using ArcCHECK or MatriXX Evolution in our clinic since 2012. ArcCHECK is a quasi-3D dosimeter whereas MatriXX is a 2D detector. A review of DQA results was performed for all patients in the last three years, a total of 221 DQA plans. These DQA plans came from 215 patients with a variety of treatment sites including head-neck, pelvis, and chest wall. The acceptable Gamma pass rate in our clinic is over 95% using 3mm and 3% of maximum planned dose with 10% dose threshold. The mean value and standard deviation of Gamma pass rates were 98.2% ± 1.98(1SD) for MatriXX and 98.5%±1.88 (1SD) for ArcCHECK. A paired t-test was also performed for the groups of patients whose DQA was performed with both the ArcCHECK and MatriXX. No statistical dependence was found in terms of the Gamma pass rate for ArcCHECK and MatriXX. The considered 3D and 2D dosimeters have achieved similar results in performing routine patient-specific DQA for patients treated on a TomoTherapy unit.

1. Introduction

It is recommended standard practice that the Tomotherapy plan for each patient be checked in the form of delivery quality assurance (DQA) prior to treatment [1]. DQA is a procedure in which the patient plan is copied and delivered to a phantom. The deliverability and accuracy of the patient plan is verified by comparing the dose calculated by the treatment planning system (TPS) and measured by one detector or detector array inserted or embedded in the phantom.

A single ionization chamber with small air cavity can be used to verify the dose at single points in a phantom [2]. To measure a dose distribution across different locations in the phantom, an array of detectors is required. Detector arrays along with the phantom can be classified as a 2D or 3D dosimeter according to how the detectors are distributed geometrically in space [3, 4]. If an array of detectors are arranged at different points in 3D space but not limited to a plane, the dosimeter is usually called a 3D dosimeter or quasi-3D depending on the density of detectors in space. Gel dosimeter is a true 3D dosimeter [5, 6], but it has not been commonly used for clinical routine QA. Most dosimeters used routinely in most radiotherapy centres are either 2D or quasi-3D detectors. The advantage of a 3D or quasi-3D dosimeter over a 2D dosimeter is that it is able to measure a 3D dose distribution in a phantom. In nature, the dose delivered to the phantom or patient is delivered throughout a volume not constrained to a plane. Both 3D and 2D dosimeters, can only measure the doses at a certain number of locations within a phantom. In practice these locations are chosen to be in high dose regions with low gradient. The agreement between the calculated and measured dose is usually quantified using the Gamma index [7].

There are several commercial dosimeters suitable for Tomotherapy DQA. Among them are ArcCHECK™ [8] (Sun Nuclear, Melbourne, FL) and MatriXX Evolution™ [9] (IBA, Schwarzenbruck, Germany). These two dosimeters have been used in our clinic for performing patient-specific quality assurance for Tomotherapy treatment since 2012. ArcCHECK is a quasi-3D dosimeter, whereas Matrix Evolution is a 2D dosimeter. The purpose of this study was to review the DQA results over the previous three years and to investigate if there was a statistical difference between DQA results performed with a quasi-3D dosimeter and a 2D dosimeter.

2. Materials and Methods

2.1. ArcCHECK versus MatriXX Evolution

The ArcCHECK 3D dosimeter is a cylindrical acrylic phantom with an array of 1386 diodes that are positioned on the surface of cylinder measuring 21 cm in diameter [8]. The diode array forms a spiral path of 1 cm and 1 cm inter-detector distance and there are 66 diodes for each turn. This cylindrical-spiral detector pattern was designed to reduce the directional dependence. The phantom also has an outer diameter of 26.6 cm and an inner holes diameter of 15.1 cm, the active area length is 21 cm and active detector size of each diode is 0.8 x 0.8 mm^2. Another advantage of this dosimeter is that it has a 15 cm diameter cavity in the phantom that can hold an insert with an ionization chamber for point dose measurement. It can simultaneously measure the exit and entrance dose in 3D space.

As a 2D dosimeter, MatriXX Evolution is composed of the MatriXX and Multicube. The Matrix device is an array of 1020 vented pixel ion chambers [9]. The chambers are divided into 32 rows and 32 columns lying in a plane. The active area is 24 x 24 cm^2. The chambers have low angular dependence. MultiCube is a two-part plastic water phantom. The matrix device can be easily inserted into the MultiCube for measuring 2D dose distributions within the MultiCube.

2.2. DQA with ArcCheck and MatriXX Evolution

ArcCheck and MatriXX Evolution were commissioned and calibrated during the period of commissioning TomoTherapy Hi-ART unit in June, 2012. The dosimetric and mechanical performance of the Tomo therapy machine was ensured to be its optimal status by yearly, monthly, weekly and daily QA following the recommendation by AAPM protocol [1]. Since then, the DQA of patient plans have been performed by different clinical physicists but following the established Tomo patient DQA protocol.

A DQA plan for each patient was created using the DQA software tool installed on the planning work station. This procedure is similar to ones described in other literature [8, 9]. Briefly, a patient plan is opened using Tomo DQA software, ArcCHECK or MatriXX CT image data sets can be selected, then an IVDT (intensity-to-density value) curve was applied to the CT images for late dose calculation and the Tomo couch was inserted into the phantom CT images. The critical point for establishing a DQA plan is to move the phantom to make sure the detector plan of ArcCHECK or MatriXX in the high dose region. The high dose region is usually chosen to be within the planning target volume (PTV) of the patient. Once the detector plane is in the correct position, the red lasers were moved to coincide with the external marks on the ArcCHECK or MatriXX Evolution. The phantom was then setup on the Tomo couch, scanned using MVCT and registered to the planning kV CT images. The position of the phantom was adjusted and the DQA plan was delivered to the phantom. Measured dose was then compared with TPS calculated ones. The Gamma map was calculated using 3mm distance tolerance and 3% of global maximum plan dose with threshold of 10%. This analysis was performed using the SNC software tool for ArcCHECK and OmniPro-I'mRT for MatriXX.

2.3. *Retrospective analysis of DQA results*

A total of 221 DQA plans were included in this study. These DQA plans corresponded to 215 patients treated in our clinic during the period from June 2012 to April, 2014. Tomotherapy was used to treat patients with a variety of diseases: Head-neck, breast wall and Pelvis. Over 90% of them were head-neck patients as Tomotherapy provided superior conformal dose to complicated targets while sparing the critical organs such as spinal cord, brain stem in head neck area, compared to other available options. For analysis, the DQA plans were divided into two subgroups: ArcCHECK subgroup and Matrix subgroup. A paired t-test for the gamma pass rate between the two subgroups was calculated.

3. Results and Discussion

Figure 1 and 2 showed the variation of Gamma pass rate for the patients measured with MatriXX and ArcCHECK, respectively. There were more patients measured with ArcCHECK than with Matrix due to efficiency considerations. If MatriXX was used for 2D dose measurements, another plan was created using the Tomotherapy cheese phantom for point dose measurement. The 2D dose map and point dose can be measured using one DQA plan for one patient by inserting the chamber and its holder into its cavity.

The Gamma pass rates were scattered in the same range between 92% and 100 % for both patients measured with MatriXX and ArcCHECK. The mean value and standard deviation were 98.2% ± 1.98 (1SD) for MatriXX and 98.5%±1.88 (1SD). There were several patients whose Gamma pass rates were below 95% for both MatriXX and ArcCHECK. Our local protocol for acceptable Gamma pass rate is no less than 95%. Further analysis indicated the detector planes for these patients were not in a high uniform dose region. Another DQA plan was created for these patients and all met local Gamma pass rate criteria.

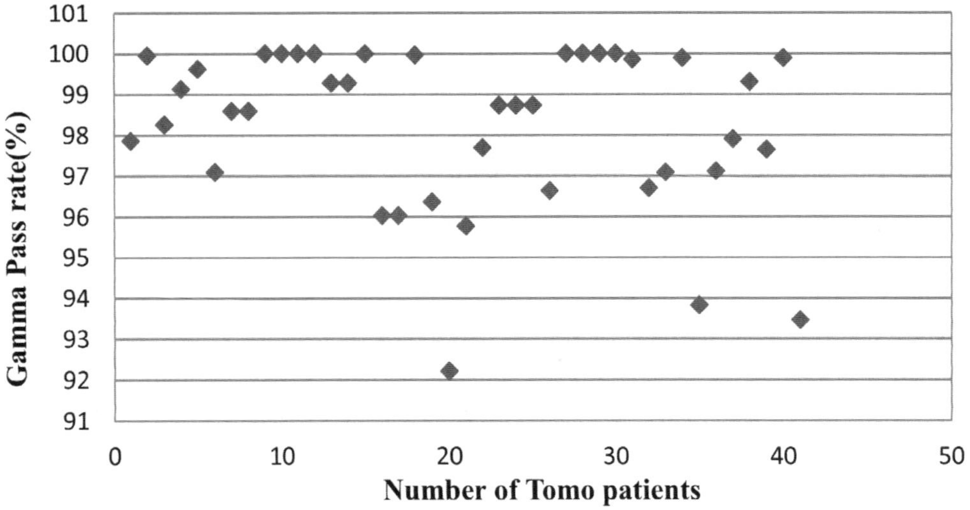

Figure 1. Gamma pass rates measured with MatriXX Evolution for each patient using 3mm/3% of global maximum planned dose with 10% dose threshold.

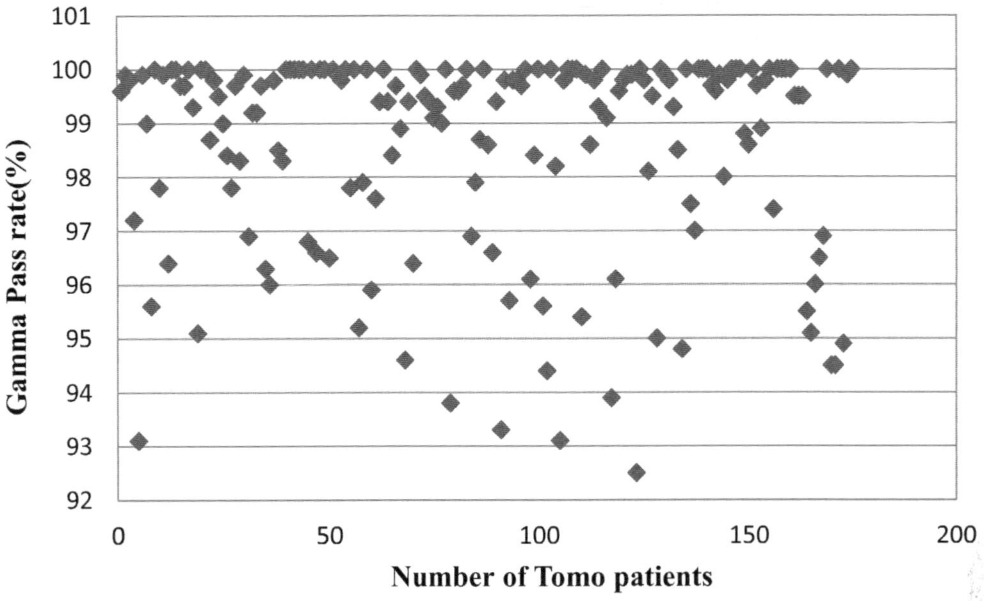

Figure 2. Gamma pass rates measured with ArcCHECK for each patient using 3mm/3% of global maximum planned dose with 10% dose threshold.

To evaluate the variability of the Gamma pass rate measured with ArcCHECK, we also analyzed the statistical dependence between these two patient DQA groups. A paired t-test was performed. In order to perform this test, the patient number in each DQA group has to be same. 41 patient DQA results were randomly selected from the whole ArcCHECK DQA results group. The t-test was run several times in this way. It was found the p-value for two-tail was greater than 0.09 and the p-value for one-tail is larger than 0.04. It was determined that there was no statistical difference for the Gamma pass rates measured with ArcCHECK and MatriXX.

4. Conclusion

Tomotherapy DQAs performed with a quasi-3D dosimeter or a 2D dosimeter for 215 patients in our centre were reviewed and analyzed. No statistical difference in Gamma pass rate values using these two types of dosimeters was found.

5. References

[1] Langen K M *et al* 2010 *Med. Phys.* **37** 4817-53
[2] Hill R *et al* 2009 *Med. Phys.* **36** 3971-81
[3] Greer P B *et al* 2007 *Med. Phys.* **34** 4389-98
[4] Vial P *et al* 2008 *Med. Phys.* **35** 1267-77
[5] Oldham M *et al* 2001 *Med. Phys.* **37** 1436-45
[6] Baldock C *et al* 2010 *Phys. Med. Biol.* **55** R1-63
[7] Low D A *et al* 2003 *Med. Phys.* **25** 656-61
[8] Neilson C *et al* 2013 *Med. Dosim.* **38** 77-80
[9] Xu S *et al* 2010 *Biomed. Imaging Interv. J.* **6** e24

Characterization of a commercially-produced chemically stable Fricke gel dosimeter

K I Penev[1] and K Mequanint[1,2]
[1]Department of Chemical and Biochemical Engineering, University of Western Ontario, London, Ontario, N6A 5B9, Canada
[2]Biomedical Engineering Program, University of Western Ontario, London, Ontario, N6A 5B9, Canada

E-mail: kmequani@uwo.ca

Abstract. The successful manufacturing of a modified Fricke gel with increased shelf life in a pilot scale (20-liter) commercial production facility is reported. The gels remained chemically stable and usable for over 30 days. Linear dose dependence was observed for cumulative doses up to 30 Gy. Dose-depth studies showed a good correlation between the experimental data and theoretical expectations, but the dose response appeared to be energy-dependent.

1. Introduction

The ferrous sulfate–gelatin dosimeter (Fricke gel) has been extensively studied over the past two and a half decades, but its applications remain mostly limited to academic research. Whereas this dosimeter is relatively easy and inexpensive to prepare, it has never been produced on a large scale or as a commercial product because of the perceived fast image deterioration due to fast iron diffusion post-irradiation [1, 2] and the short shelf-life of the Fricke gel [3]. Regarding the diffusion, a non-diffusing nanocomposite Fricke gel for ion dosimetry at high doses has recently been reported [4]. Furthermore, even when a diffusing Fricke gel is used, fast gel scanning modes such as cone beam optical computed tomography (CT) have been shown to yield reasonable dose distribution profiles [5]. Regarding the shelf-life, we have previously demonstrated that the ferrous iron auto-oxidation can be drastically reduced with appropriate additives [1]. This modified Fricke gel composition was licensed to Modus Medical Devices, Inc. (London, ON, Canada) in October 2013. Modus Medical has since assembled a pilot-scale, commercial gel manufacturing facility with a capacity of 20 liters per batch. Here, we report on the characterization of the dose response, stability and reusability of two of the first batches produced at the new facility.

2. Experimental

2.1. Gel manufacturing

A 20-litre jacketed process reactor was supplied by Chemglass Life Sciences (Vineland, NJ, USA), with built-in controlled stirring, aeration, temperature, and pH measurement. External peripherals, supplied by Cole-Parmer (Canada) and McMaster-Carr (USA), include: a high capacity heating/cooling circulating bath, liquid transfer peristaltic and gear pumps, tubing and miscellanea. In order to avoid contamination of the gels, all internal surfaces of the reactor and all other wetted parts were selected from acid- and oxidation-resistant medical- or food-grade materials: glass and polytetrafluoroethylene (Teflon®) for gel processing; polypropylene and polyethylene terephthalate (PET) for liquid/gel storage; flexible polyvinyl chloride, Noreprene® and polyvinylidene fluoride (Kynar®) for tubing and miscellanea; and stainless steel (grade 316L) in the pump head of the gear pump. The stainless steel had minimal contact with the gel solution with a liquid residence time less than two seconds. The gear pump was pre-conditioned in order to remove easily leachable surface materials by circulating warm dilute sulfuric acid (10 wt.%) for a period of two hours.

The gel manufacturing procedure is a multistep process which was optimized for high efficiency and batch-to-batch consistency. The final manufacturing step, before the dispensing of the product into PET jars, was filtration of the gel solution through a 0.45 μm membrane capsule filter (AcroPak™ 500, Pall Corp., USA) for improved optical clarity. The jars were placed into a custom built 100-liter water bath, cooled gradually to gelation, and then refrigerated at 4-6 °C until further use.

2.2. Gel compositions and testing

Two batches of the modified Fricke gel [1] were prepared using 300 Bloom bovine bone gelatin (Rousselot International, Peabody, MA, USA), identical with the Eastman gelatin used in [5]. Deionized water, sulfuric acid (SA), glyoxal (Gx) and ferrous ammonium sulfate hexahydrate (FAS) were supplied by Caledon Labs (Georgetown, ON, Canada), xylenol orange (XO) was purchased from TCI America (Portland, OR, USA), and 4-nitro-1,10-phenanthroline (Nn) from Alfa Aesar (Ward Hill, MA, USA). All chemicals were of analytical or higher grade and used without further purification. The first batch (lot no.: H09) contained in mmol/L: SA:FAS:XO:Nn:Gx = 30:0.10:0.1:0.07:2. The second batch (lot no: H10) contained in mmol/L: SA:FAS:XO:Nn:Gx = 30:0.12:0.1:0.07:2. The higher iron concentration in H10 was used to increase the sensitivity of the gel. Both batches had 4 wt.% gelatin and were filled into 1-liter PET jars.

All irradiations were performed at the London Regional Cancer Program (LRCP) in London, ON, Canada. The H09 gels were used in a longitudinal study of the dose sensitivity and gel stability over 32 days using a cobalt-60 source (Eldorado 6, Atomic Energy of Canada Ltd.), that had an initial dose rate of 136 $cGy.min^{-1}$. For each irradiation, a jar of gel was positioned vertically over the source, partially shielded with a 40 mm thick lead block and exposed to a 5 Gy dose. The jar was then rotated by 90° and exposed to a new 5 Gy dose (figure 1). On day 35 after preparation, five previously unused gels were exposed to uniform doses in the range from 1.25 to 20 Gy to confirm the linearity of the dose response.

The H10 gels were irradiated from above with square beams of photons (3×3 cm at 90 cm source-to-surface distance (SSD)) or electrons (4×4 cm at 100 cm SSD) on a Varian 21iX linear accelerator at the following energies: 6 and 10 MV (photons), 9, 12, 16 and 20 MeV (electrons). The centers of the beams were off-set from the center of the gel by 1 cm. The 6 MV dose depth curves were compared against the preliminary results from Monte-Carlo simulations of small field photon beams, courtesy of Kevin Jordan and Matt Mulligan at the LRCP. All scans were performed with a Vista15 cone beam optical scanner (Modus Medical Devices), with an amber LED source at 590±5 nm. Image processing was done using ImageJ 1.44 (National Institutes of Health, USA) and Matlab 7.9.0 (The Mathworks Inc.).

3. Results and discussions

Figure 1 shows the dose response behavior of the H09 gels over the course of 32 days, plotted as the change of attenuation (Δμ) versus the height of the gel. The left panel presents a typical example, where zone A received 10.0 Gy, zone D received the dose transmitted through the lead shield (approximately 0.8 Gy), while zones B and C received 5.0 Gy plus half of dose D. Note that at the bottom and top of the gels there are zones where the jar geometry and the scanner capabilities prevent exact attenuation measurements. Those zones are excluded from further analysis. Assuming linear dose dependence, the signals at A plus D should be equal to that of B plus C. The absolute error of linearity was generally less than $2\times10^{-3}cm^{-1}$, which is close to the noise level, measured as standard deviation at $(1\text{-}2)\times10^{-3}cm^{-1}$. Over the course of two weeks, the same gel was re-irradiated the same way three more times, maintaining linearity close to the noise level up to the third irradiation at day 11 (central panel). Six additional gels were irradiated at various time intervals up to 32 days after the batch was prepared (right panel). The error of linearity was close to the noise in all cases. The dose responses in the different gels were remarkably similar over the full course of the experiment, with the exception of Day 7. In that particular case, the temperature of the gel during irradiation may have played a role, as it had not been properly conditioned and cold Fricke gels are known to be less sensitive [5]. In addition, the re-irradiation of the first gel on the same day did not yield an increased sensitivity (central panel) as expected based on previous experience. Four of the gels were irradiated multiple times (only one is shown here) and generally, in the follow-up irradiations, the gels exhibited higher sensitivity but the linearity of the dose response was maintained up to a cumulative dose of 30 Gy (central panel).

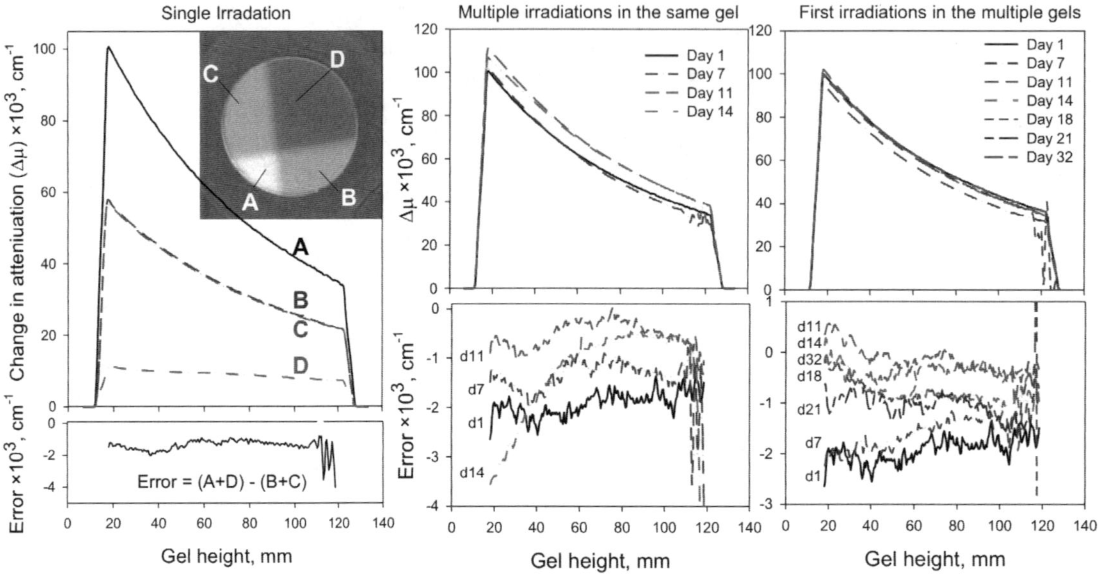

Figure 1. Dose responses of gels H09 over the course of a month. Left - single irradiation, center - multiple irradiations (10 Gy) of the same gel, right - first irradiations (10 Gy) of multiple gels.

On day 35 after the batch preparation, five gels were irradiated uniformly at increasing doses from 1.25 to 20 Gy. The light attenuation was in a linear dependence with the dose for all gel heights ($R^2 > 0.999$, data not shown). The dose sensitivity at 18 mm gel height was calculated to be 11.6×10^{-3} $Gy^{-1}cm^{-1}$ ($R^2 = 0.9999$). However, the absolute dose sensitivity was impossible to determine, as the surface at the bottom of the jar is concave and prevents accurate dose calculation when the gel is irradiated from below.

Dose response calibration and dose-depth curves were obtained from the H10 gels. Photon beams with known doses at 30 mm below the gel surface were used for calibration of the dose response (figure 2, left), obtaining a sensitivity of 19.5×10^{-3} $Gy^{-1}cm^{-1}$ with an excellent linearity ($R^2 > 0.9995$). Compared to H09 the sensitivity was higher, as expected due to the higher iron concentration. The experimental dose-depth distributions for the 6 MV photon beams correlated with the Monte Carlo simulations well with $R^2 > 0.99$ for all but the lowest dose (figure 2, center). It should be noted that the noise in the scans played a significant role, especially at the extreme top and bottom sections of the jars. Figure 2 (right) shows the dose-depth curves for the 10 MV photon beams and electrons beams at four energies. Unfortunately, by the time of the submission of this manuscript, we do not have complete modeling of the 10MV dose depth curves, and the 10MV data has not been correlated to modeling predictions. The calculated doses (calibrated with the 6 MV and 10MV photon beams) fit well with the 9 and 12 MeV electron beams, the 16 and 20 MeV beams showed higher than expected responses, which may indicate an energy-dependent dose response. Further examination of that effect is required before we can arrive at any definitive conclusions.

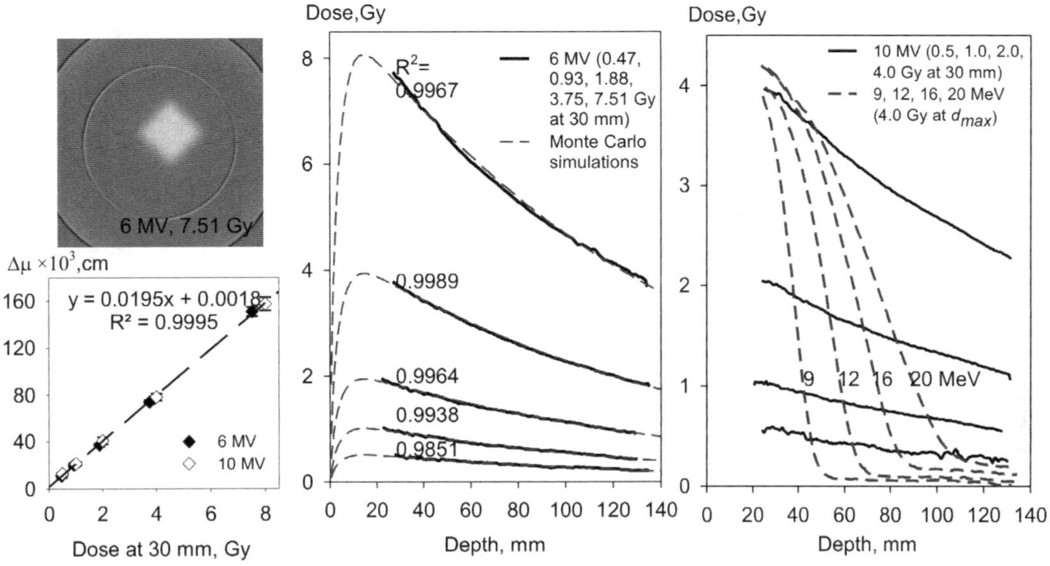

Figure 2. Dose response (left) and dose-depth curves in the H10 gels (center and right); d_{max} refers the depth of maximum dose.

4. Conclusions

Considering the prolonged stability and sensitivity of the reported product, we believe that the commercial production facility at Modus Medical has had a successful start of operation. The gels showed excellent dose linearity (relative dose response) and consistent sensitivity for over 30 days. However, the absolute dose response was less stable and is possibly energy-dependent. Future product development will focus on examining these effects, as well as minimizing the noise in the gel scans.

5. Acknowledgments

We would like to acknowledge the financial support of the Mitacs Elevate program, the Ontario Centres for Excellence and the Natural Sciences and Engineering Research Council of Canada, as well as Modus Medical Devices, Inc. Individual thanks are due to Jennifer Dietrich and John Miller from Modus Medical Devices Inc., and Kevin Jordan and Matt Mulligan from the London Regional Cancer Program.

6. References

[1] Harris P J *et al* 1996 *Phys. Med. Biol.* **41** 1745
[2] Baldock C *et al* 2001 *Australas. Phys. Eng. Sci.* **24** 19
[3] Penev K I *et al* 2013 *Phys. Med. Biol.* **58** 1823
[4] Maeyamaa T *et al* 2014 *Rad. Phys. Chem.* **96** 92
[5] Olding T *et al* 2010 *J. Phys.: Conf. Ser.* **250** 012028

Feasibility of polymer gel dosimetry measurements in a dynamic porcine lung phantom

P Mann[1], M Witte[1], S Armbruster[1], A Runz[1], C Lang[1], M Breithaupt[2], M Berger[2] J Biederer[3], C P Karger[1] and T Moser[1]

[1]Department of Medical Physics in Radiation Oncology, German Cancer Research Center, INF 280, 69120 Heidelberg, Germany

[2]Department Medical Physics in Radiology, Section of Pulmonary Imaging, University of Heidelberg, INF 280, 69120 Heidelberg, Germany

[3]Department of Diagnostic and Interventional Radiology, Section of Pulmonary Imaging, University of Heidelberg, INF 110, 69120 Heidelberg, Germany

E-mail: p.mann@dkfz.de

Abstract. A dynamic ex-vivo porcine lung phantom combined with polymer gel dosimetry is tested as a new tool to validate modern adaptive radiotherapy techniques (e.g. gating or tracking). The gel was inserted into the lung via a latex balloon to simulate a tumor. After irradiation, the location of the dose maximum was verified, however, the dose was higher than planned and a high background signal was seen. Potential reasons for this finding are the non-standard conditions of gel handling. These conditions were systematically studied. Besides temperature, the material of the balloon seems to be of special importance. The results identify open issues that have to be addressed in future studies.

1. Introduction

Modern adaptive radiotherapy techniques (e.g. gating or tracking) have the potential of significant normal tissue sparing [1]. This is especially relevant for lung tumors that may exhibit large displacements due to respiratory movement. Implementing these techniques for clinical applications, however, requires validation of the intended workflow that ideally includes 3D measurements of the delivered dose distribution. Although several phantom settings have been developed [2, 3], this issue is not satisfactorily solved yet as they usually do not reflect the complexity of real patient treatments and/or do not provide sufficient information on the actually delivered dose. An ideal validation phantom should provide (i) realistic motion patterns comparable to those of patients, (ii) anthropomorphic properties allowing for realistic condition for time-resolved radiological imaging, and (iii) the possibility of measuring the delivered dose distribution in 3D [4]. This study presents a new approach to meet these requirements by a combination of a dynamic porcine lung phantom with 3D polymer gel measurements.

2. Material and Methods

2.1. Dynamic lung phantom

The dynamic lung phantom consists of an artificial PMMA-thorax and contains a post-mortem explanted porcine lung [5]. Inflation of the lung is achieved by under-pressure within the thorax and a water-filled silicon balloon simulates the diaphragm. Modulation of the water volume over time with pre-defined temporal pattern allows to ventilate the lung under conditions comparable to those in patients and to generate arbitrary breathing patterns. The phantom is MR-compatible and has been previously used to study 4D motion patterns and radiological imaging protocols of artificial tumors produced by injecting Agarose-gel [6]. Thus, the phantom meets (i) and (ii) of the above requirements.

2.2. Polymer dosimetry gel

To measure the dose distribution in the phantom, the Agarose gel has to be replaced by a polymer dosimetry gel [7], e.g. the commercially available BANG-gel (Research, Inc. Madison, USA). Polymer gels include monomers that polymerize locally depending on the absorbed radiation dose. Measuring the relaxation rate R2 of the transverse magnetization with MRI techniques, the 3D-dose distribution can be determined after establishing a linear calibration curve [7]. For MRI-evaluation, a 32-echo multi spin echo sequence [8] with an inter-echo spacing of 18.5 ms was used. Gel production and storage after production (96 h) were performed as recommended by the manufacturer. To minimize the influence of oxygen and light on the calibration samples, brown borsilicate flasks (20 ml) were used and additionally covered by aluminum foil during storage.

2.3. Insertion of the gel into the lung

There are two major challenges in using the polymer gel in our experimental setting: (i) The gel has to be inserted into the lung and protected from contact with oxygen, and (ii) due to the limited durability of the lung, the recommended time between gel injection and irradiation (96 h) and irradiation and MRI evaluation (24 h) [7] had to be reduced to 3 h and 1 h, respectively.

To protect the dosimetry gel from oxygen after injection into the lung, a latex balloon attached on top of a urinary catheter by means of two o-rings was used. Prior to inserting the lung into the thorax, the catheter was guided into the pulmonary artery and the catheter-balloon located between the o-rings was inflated to fix it. Using a syringe, 50 ml of the gel were injected into the latex-balloon. As the gel was inserted after the recommended storage period and thus had to be reheated, the calibration samples were reheated as well to assure identical temperature histories.

2.4. Feasibility test with the lung phantom

As a first feasibility test of the intended experimental setting, the complete chain of insertion of the gel-tumor into the lung, irradiation and dosimetric evaluation was performed. Calibration samples were taken from the same batch and were irradiated in a water phantom under reference conditions with doses between 0 and 7 Gy correcting for the influence of the flask wall. In the treatment planning system (*VIRTUOS*) a small target volume (≈15 ml) within the measurement volume was defined and a 5-field treatment plan with a maximum target dose of 6 Gy was generated and subsequently irradiated using 6 MV beam (*Artiste*, Siemens). Due to the durability of the lung, MRI evaluation of the gel was performed 1 h after irradiation of the phantom and in a liquid-filled holding device for the calibration samples, respectively. The entire experiment was performed under static conditions.

2.5. Systematic investigation of gel response

As the feasibility test revealed several problems which are most likely related to the non-standard conditions of the gel handling, the gel response were systematically studied under these conditions.

2.5.2 Temperature effect

As temperature is a critical factor [8], the cooling behavior of our gel was measured placing the fiber of an MR-compatible optical thermometer (Fotemp Polytec 4.16) within the balloon. In addition, the response of unirradiated gel samples (latex balloon, 20ml flasks) was analyzed regarding its dependence on temperature during MRI evaluation. For this, the samples were measured 0, 43, 53, 66 and 104 minutes at room temperature (21°C) after removing it from storage at 12°C.

2.5.2 Oxygen permeability of the balloon

As continued oxygen contact may change the gel response to irradiation [8], the permeability of the latex balloon for oxygen was investigated. Four balloons and a set of calibration samples were filled with gel and stored light-shielded for either 3 or 24 hours at room temperature. For each storage time, one balloon was irradiated homogeneously with 4 Gy while the other remained unirradiated. During MRI measurement all samples were placed in a thermobox to assure constancy of temperature.

2.5.3 Water permeability of the balloon

As the influence of penetrating water may impact the dose response, it was investigated, if the thin-walled latex balloon is permeable for water in an aqueous environment. Two gel-filled balloons and a set of calibration samples were stored in water for 24 hours. One of the balloons was subsequently irradiated with 4 Gy. For constancy of temperature, all samples were placed inside a water filled thermobox 3 hours prior to the MRI measurement taking place 92 hours after irradiation.

3. Results

3.1 Lung experiment

Figure 1 shows the result of the lung experiment. Although the position of the dose maximum could be verified with the gel, there were two unexpected findings: (i) the maximum dose was measured to be 7.3 rather than the planned 6 Gy, and (ii) there is a significant background level of about 4.2 Gy. As these problems are likely to be related to the non-standard conditions of the gel handling potentially disturbing factors were further analyzed.

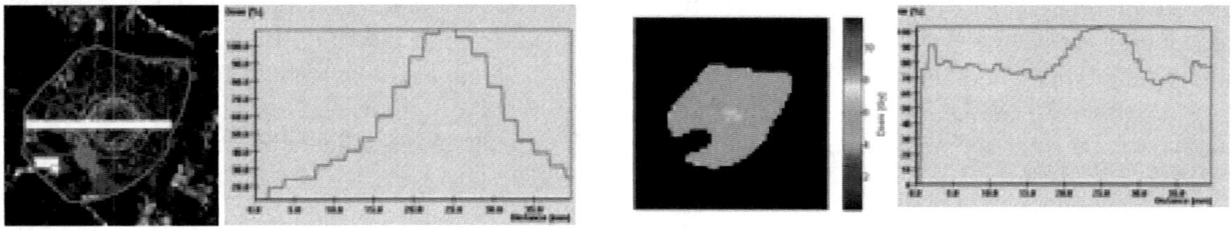

(a) (b) (c) (d)

Figure 1. CT-based dose-distribution (a) and normalized dose profile (b) along the white line. (c) and (d) show the respective results of the gel measurement.

3.2 Temperature effect

After reheating to 55°C, the gel cooled down and reached thermal equilibrium after 3h. Analyzing the temperature dependence of the MRI-signal over the first 104 min, an increase in T_2-values of 50% for the glass flasks and 33% for the balloon was found.

3.3 Oxygen permeability of the balloon

Figure 2 shows the dose maps determined with the calibration curve. For all measurements, the dose decreases towards the centre. Comparing the irradiated (figure 2c, d) with unirradiated balloons (figure 2a, b), an increased dose was found in the center, however, this hotspot occurred only after 3h (figure 2d). Doses after 3 and 24h differed by up to 1.7 Gy in corresponding regions. For irradiated calibration vessels (e.g. at 2 Gy, figure 2e and 3c) a homogeneous dose distribution is visible.

3.4 Water permeability of the balloon

Figure 3 shows the influence of the aqueous environment of the balloon on the dose response of the gel. Although homogeneously irradiated, the irradiated balloon shows a hot-spot of 4.4 Gy in the centre (figure 3a). The unirradiated balloon shows doses between 2.2 and 3.8 Gy. Excluding the hot spot, a similar dose distribution with an average dose of 3.2 Gy and 3.4 Gy is visible, respectively.

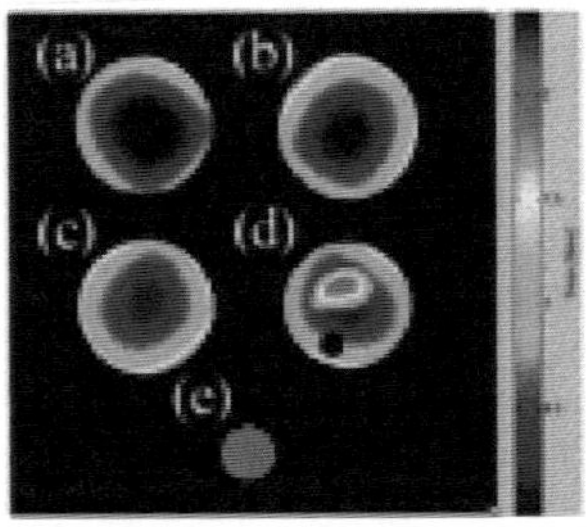

Figure. 2. Dose distribution measured in the gel-filled balloon stored in air for 24 h (a, c) and 3 h (b, d) prior to irradiation with 4 Gy (c, d) and no irradiation (a, b). The calibration sample (e) was irradiated with 2 Gy.

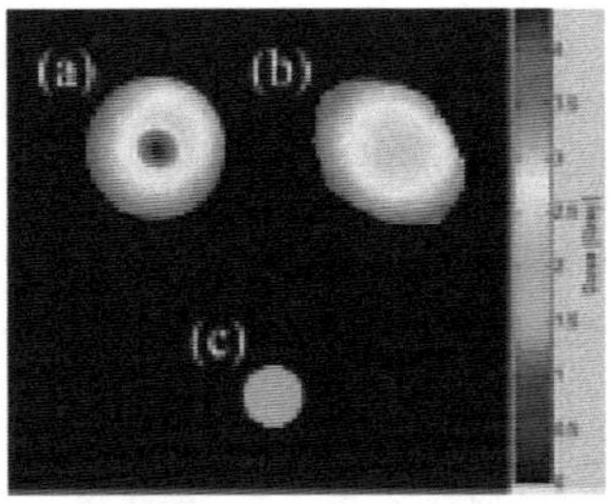

Figure 3. Dose distribution measured in the gel-filled balloon: (a) irradiated with 4 Gy (b) unirradiated. Prior to irradiation, the samples were stored in water for 24 h. The calibration sample (c) was irradiated with 2 Gy.

4. Discussion

The lung phantom was demonstrated to be a versatile tool for 4D-imaging studies [6]. Combining the phantom with polymer-based dosimetry-gel would allow to measure delivered dose distributions in radiotherapy and thus to validate complex treatment techniques like gating or tracking.

In the initial test of the gel in the lung phantom, the location of the dose maximum could be successfully verified, however, the measured maximum dose was markedly higher than planned, and there was a significant background signal. These results indicate the existence of additional uncontrolled factors that influence the gel response. Since the lung explant was stored in a freezer prior to preparation, temperature differences between the gel in the lung phantom and in the calibration flasks may occur. Thus, it is important to standardize the experimental procedures such that both pass the same temperature history. Other reasons may be related to deviations from the manufacturer's recommendations concerning the gel handling, which was necessary due to our experimental setting. With this respect, three aspects are important: (i) reheating of the gel prior to injecting it into the lung, (ii) reduced time between gel injection and irradiation as well as between irradiation and evaluation, and (iii) the use of balloons instead of glass to protect the gel from the environment.

While it can be expected that (i) and (ii) could be solved by carrying along calibration samples under the same conditions, it appears that oxygen and water may penetrate the latex-material. While the first might inhibit polymerization and thus reduce dose, the latter seems to increase the measured dose.

5. Conclusion

The combination of the dynamic anthropomorphic lung phantom with 3D gel dosimetry is a new promising approach to validate complex treatment techniques like gating or tracking geometrically as well as dosimetrically. As a first application of the phantom, the location of the dose maximum of a treatment plan could be verified. Regarding dosimetric measurements, the behavior of the gel under non-standard conditions is highly complex and although some influence factors have been identified in this study, further work is necessary to optimize protocols for use in the lung phantom.

6. Acknowledgements

This work was supported by the Federal Ministry of Education and Research (BMBF) Germany, grant number 01IB13001B. The authors would like to thank Prof. Dr. Andreas Berg for his advice on dosimetry gel handling.

7. References

[1] Glide-Hurst C K *et al* 2014 *J. Thorac. Dis.* **6** 303-18
[2] Vedam S S *et al* 2003 *Phys. Med. Biol.* **48** 45-62
[3] Ceberg S *et al* 2008 *Phys. Med. Biol.* **53** N387-396
[4] Baldock C 2006 *J. Phys.: Conf. Ser.* **56** 14-22

[5] Biederer J *et al* 2003 *Radiology* **226** 250-5
[6] Biederer J *et al* 2009 *Int. J. Radiat. Oncol. Biol. Phys.* **73** 919–26
[7] Ibbott G S *et al* 1997 *Int. J. Radiat. Oncol. Biol. Phys.* **38** 1097-103
[8] Baldock C *et al* 2010 *Phys. Med. Biol.* **55** R1-63

Investigating hydrogel dosimeter decomposition by chemical methods

Kevin Jordan
London Regional Cancer Program, London Health Sciences Centre, Department Medical Biophysics, Western University, London, Ontario, Canada

E-mail: kevin.jordan@lhsc.on.ca

Abstract. The chemical oxidative decomposition of leucocrystal violet micelle hydrogel dosimeters was investigated using the reaction of ferrous ions with hydrogen peroxide or sodium bicarbonate with hydrogen peroxide. The second reaction is more effective at dye decomposition in gelatin hydrogels. Additional chemical analysis is required to determine the decomposition products.

1. Introduction

The issue of hydrogel dosimeters disposal becomes important as larger volumes and more frequent usage becomes widespread. Materials with toxic reactants and products will require extra procedures for safe disposal. Preferably, non-toxic chemistries will be selected for gel preparation and safe and convenient disposal procedures will be available for clinical users. One aspect of *Green chemistry* is to replace toxic materials with non-toxic alternatives. An example, is the trend to replace halogenated chemistries in industrial and domestic environments. Genipin hydrogels were specifically developed because of this chemistry was used in medicine and the food industry. Most of the current, 3-D, radiochromic, hydrogel dosimeters are relatively safe to produce and handle. One disposal method involves breaking ferrous-xylenol orange or leuco-dye-micelle gels into small pieces and air drying. The solid film can then be disposed in a land-fill. This process is possible for a few litre volumes but as larger quantities are consumed this process becomes impractical. Also, molds may grow on the drying gels, contaminating the laboratory.

Globally dye contamination of water and soils is a common problem associated with textile industries. Many researchers are examining inexpensive methods to clean up dye contaminated water and soils. They tend to examine low cost approaches such as: solar photochemistry, locally available catalytic materials such as clays, sands and mining tailings. This literature is a resource for hydrogel dosimeter remediation ideas since low cost and low toxicity are starting criteria for the research. The radiochromic dosimeters often contain a dye or dye precursor such as a leucodye. Generally these specific compounds are chosen for their low toxicity. However, introduction of dyes into the environment should be minimized as a matter of principle. If these molecules can be degraded prior to leaving the laboratory it is preferred. Typically, these compounds are less than one thousandth of the total mass of a hydrogel dosimeter and gelatin represents 5% or more of the mass. For this reason, promising reactions in contaminated water based on generation of OH radicals may not work due to the scavenging effect of the gelatin [1]. Also, the radiochromic hydrogel dosimeters are acidic, with pH ~ 2 being typical. This allows acidic reactions to be examined and still follow the *atom economy principle of Green chemistry*. In this investigation, several methods of hydrogel degradation are examined and may stimulate ideas for practical gel decomposition methods.

This study is limited to the leuco-crystal violet micelle hydrogel dosimeters (LVC) and it is expected that results are applicable to similar triphenylmethane dyes, such as malachite green. The Fenton reaction involves generation of OH radicals by the breakdown of hydrogen peroxide (H_2O_2) by ferrous ions. This common reaction is used globally for treatment of waste water and sewage. The Fenton reaction for oxidative degradation of water contaminated with malachite green from textile industries was investigated [1]. Further studies examined the use of the Fenton reaction to decompose crystal violet in water [2] and the enhanced reaction with granular activated carbon [3]. Reaction rates can be increased using near UV (sunlight) to drive the corresponding photo-Fenton reaction, degrading chlorinated phenol [4].

Direct photochemical degradation of dyes in hydrogels may be possible for certain chemistries. However, the dyes crystal violet and malachite green have adequate photochemical stability to be useful when designing stable radiation dosimeter materials. In water, crystal violet is stable to exposure to simulated sunlight but decomposes with the addition of H_2O_2 [5]. However, the addition of mannitol quenched the decomposition reaction, indicating OH radical is involved. It is anticipated similar quenching results would be observed with hydrogels. Also, only a small fraction of the leucodye is converted to dye for doses below 100 Gy. The colourless leucodyes are not sensitive to visible wavelengths. UV exposure is an option, but it is anticipated that gelatin will preferentially absorb the UV light, limiting the effectiveness of this approach.

Another chemical degradation approach of organic pollutants involves the reaction of sodium bicarbonate and hydrogen peroxide to produce peroxymonocarbonate (HCO_4^-), superoxide ion ($O_2^{-}\cdot$) and singlet oxygen($^1O_2$) [6]. The formation of HCO_4^- , within metabolic conditions, is currently under investigation and it is interesting that cells would also use this mechanism for the disposal of metabolic waste [7].

In this study, several approaches to the decomposition of lueco-crystal violet and crystal violet in gelatin hydrogel micelle dosimeters were examined for extent of reaction. Direct, thermal decomposition at temperatures near the boiling point were not investigated because it is anticipated that few dosimeter users would have access to fume hoods for control of air quality.

2. Methods

Previously irradiated samples of leuco-crystal violet (LCV) micelle gelatin hydrogel were melted in a 50 C water bath. The gel composition was: 5 % by mass, gelatin, 25 mM trichloroacetic acid, 4 mM Triton X100 surfactant and 1 mM LCV. In the first sample, 1 mM ferrous ammonium sulphate was added with stirring. After 10 minutes, 100 mM H_2O_2 was added to the solution. The solution was divided for subsequent experiments. In one case the upper surface was illuminated with 365 nm UV light to a fluence of ~100 J while examining for photobleaching. The experiment was repeated with ~ 250 J of 630 nm red light. To another sample sodium bicarbonate ($NaHCO_4$) and additional H_2O_2 was added.

The second experiment involved the addition of 300 mM H_2O_2 to the melted gel followed by the slow addition of 50 mM $NaHCO_4$ to minimize foaming. The sample was maintained at 50 C and observed for two hours.

3. Results

The introduction of ferrous ammonium sulphate to the warm gelatin solutions resulted in a darkening of the material due to the formation of crystal violet and the addition of H_2O_2 accelerated the conversion to dye. However, even with excess H_2O_2 the samples did not fade to colourless solutions. UV light exposure also failed to visibly bleach the dark solutions. Red light exposure had minimal effects on colour change of the gel dosimeter at room temperature. However, the combination of $NaHCO_4$ and H_2O_2 in warm melted gels resulted in colour fading in less than one hour. Figure 1 shows a four month old sample of LCV micelle gel before and after the reactions with $NaHCO_4$ and H_2O_2.

Subsequent lowering of the pH to less than 7 resulted in the sample changing from an opalescent to a clear colourless solution.

Figure 1. Oxidative decomposition of leuco-crystal violet hydrogel dosimeter: left panel, before treatment, right panel, sample after two hour reaction with bicarbonate and hydrogen peroxide.

4. Discussion

Iron chemistry deserves further investigation because of its biological importance and because many hydrogel dosimeters employ similar reactions. It is a catalyst in the production of several oxidative species with the OH radical being the most reactive. In the case of gelatin hydrogels, the OH radical is scavenged primarily by the gelatin polymer, making it an inefficient species for oxidizing the low concentrations of dye and leuco-dye. In principle by adding larger amounts both the gelatin and dyes would be decomposed. But more efficient reactions may exist. Singlet oxygen chemistry also is expected to be helpful since it has a lower redox potential, leading to fewer possible reactions and fewer products to consider. In this study the addition of photosensitizer dyes was not examined in an attempt to minimize the number of additional chemicals added. Solar photochemistry should also be a very cost effective method of disposal under optimal conditions. Electrochemical methods may also be practical in a laboratory setting

It would be very helpful if the gelatin could be recovered from the hydrogel for reuse or separate disposal but this may not be feasible. In the case of LCV-micelle-gelatin hydrogels, a combination of sodium bicarbonate and hydrogen peroxide appears to be an effective method of decomposition requiring further validation by examining the oxidation products.

5. References

[1] Dutta K *et al* 2003 *J Environ. Sci. Health Part A* **A38** 1311-26
[2] Huan-Jung F *et al* 2009 *J. Hazard. Mater.* **171** 1032-44
[3] Chiing-Chang C *et al* 2011 *J. Hazard. Mater.* **196** 420-5
[4] Ruppert G *et al* 1993 *J. Photochem. Photobiol. A: Chem* **73** 75-8
[5] Denman S *et al* 1996 *Dyes and Pigments* **30** 67-72
[6] Xu A *et al* 2011 *Chemosphere* **82** 1190-5
[7] Bakhmutova-Albert E V *et al* 2010 *Inorg. Chem.* **49** 11287-96

Three dimensional dose verification of VMAT plans using the Octavius 4D dosimetric system

Sankar Arumugam[1], Aitang Xing[1], Tony Young[1], David Thwaites[2] and Lois Holloway[1,2,3,4]

[1]Liverpool and Macarthur Cancer Therapy Centres and Ingham Institute, New South Wales, Australia
[2]Institute of Medical Physics, School of Physics, University of Sydney, New South Wales, Australia
[3]Centre for Medical Radiation Physics, University of Wollongong, New South Wales, Australia
[4]University of New South Wales, New South Wales, Australia

E-mail: sankar.srumugam@sswahs.nsw.gov.au

Abstract. The Octavius 4D dosimetric system generates a 3D dose matrix based on a measured planar dose and user supplied Percentage Depth Dose (PDD) data. The accuracy of 3D dose matrices reconstructed by the Octavius 4D dosimetric system was systematically studied for an open static field, an open arc field and clinical VMAT plans. The Octavius reconstructed 3D dose matrices were compared with the Treatment Planning System (TPS) calculated 3D dose matrices using 3D gamma (γ) analysis with 2%/2mm and 3%/3mm tolerance criteria. The larger detector size in the 2D detector array of the Octavius system resulted in failed voxels in the high dose gradient regions. For the open arc fields mean (1σ) γ pass rates of 84.5(8.9) % and 94.2(4.5) % were observed with 2%/2mm and 3%/3mm tolerance criteria respectively and for clinical VMAT plans mean (1σ) γ pass rates of 86.8(3.5) % and 96.7(1.4) % were observed.

1. Introduction

Modern radiotherapy techniques are highly complex and Volumetric Modulated Arc Therapy (VMAT) follows the same trend [1, 2]. Three dimensional dose verification of VMAT plans is recommended to ensure the dosimetric accuracy of treatment delivery [3]. Gel dosimeters offer high resolution 3D information of delivered dose [4, 5]. However the requirement of extensive time for the preparation and readout of gel dosimeters limit their use for routine clinical use [6]. Recently electronic dosimeters have been introduced by many vendors that are able to provide 3D or quasi 3D dose information [7]. Octavius 4D (PTW, Freiburg, Germany), is one such system that provides measured 3D dose matrices of the VMAT plans. McGarry *et al* [8] studied the performance of the Octavius 4D system in the verification of rotational delivery using a Truebeam (Varian Medical Systems, Palo Alto, CA, USA) linear accelerator (linac). In their work the synchrony of the Octavius 4D rotational unit with Linac gantry was tested for discrete gantry angles. In this work we present the validation of the Octavius 4D system for the dosimetric validation of VMAT plans with an Elekta-Synergy linac. We also tested the synchrony of the Octavius 4D rotational unit with linac gantry throughout arc delivery at six different gantry speeds including minimum and maximum speed of the gantry.

2. Materials and Methods

2.1. Octavius 4D dosimetric system

The Octavius 4D dosimeter consists of a cylindrical unit of 32cm diameter and 34cm length that houses a 2D array detector. The measurement plane of the detector array aligns with the centre plane of the cylinder. The cylinder has the capability to rotate in synchrony with the linac gantry using input from the inclinometer. Two types of detector arrays, Octavius Detector 729 and Octavius Detector 1000SRS can be used for the measurement. All the measurements in this study were performed using the Octavius Detector 729 array that has 729 cubic ion chambers uniformly arranged in a 27 x 27 cm^2 area. Each ion chamber has dimensions 0.5cm x 0.5cm x 0.5cm. The Verisoft, v5.1, software was used for the measurement and dose analysis. During measurement the software integrates the frames for every 200ms and these dose planes in conjunction with gantry angle information from the inclinometer and user provided Percentage Depth Dose (PDD) measurements for field sizes ranging from $4x4cm^2$ to $26x26cm^2$ measured at 85cm Source to Surface Distance (SSD) are used for the reconstruction of a 3D dose matrix. The Octavius system was calibrated to measure the absolute dose as per the manufacturer recommended calibration procedure.

2.2. Planning and delivery system

A 6MV photon beam model for an Elekta –Synergy linac in the Pinnacle (Philips Ltd, USA) treatment planning system (TPS), v9.6, was used to generate all the treatment plans in this study. The Elekta – Synergy linac used in this study has the MLCi head and the VMAT plans were delivered with a continuously variable dose rate (CVDR) using Integrity, v1.1, console software.

2.3. Validation of synchronisation of cylinder and linac gantry rotation

The accuracy of the 3D dose matrix calculated by the Octavius 4D depends on the accurate synchronisation of the rotation of the cylindrical unit with the linac gantry. In order to test the synchronisation between the cylindrical unit and linac gantry, dose resulting from treatment arcs with an open field of 10x10 cm^2 and different gantry speeds ranging from 1°/min to 6°/min was measured. The gantry position for these arcs as a function of time was plotted using the service plot function in the Elekta_Synergy linac. Similar data was generated from the Octavius measured raw data using the sampling time interval and registered gantry angle. The synchronisation of the Octavius and linac gantry was assessed by comparing the angular position of the cylindrical unit and the linac gantry by comparing these data with respect to the radiation delivery start and detection time.

2.4. Validation of standard and clinical plans

The Octavius system reconstructs the 3D dose matrix based on measured planar dose and user supplied PDD information. The accuracy of the 3D dose matrix reconstructed by the Octavius system for open static and arc fields was compared with the Pinnacle calculated dose matrix. Thirteen clinical VMAT plans, consisting of prostate, prostate bed and head & neck, were also verified using the Octavius system. Some of the key characteristics of static open field, open arc field and clinical VMAT plans are shown in table1. The Octavius reconstructed and Pinnacle calculated dose matrices were compared using 3D gamma (γ) analysis with 2%/2mm and 3%/3mm tolerance criteria and 10% dose threshold [9].

Table 1. Key characteristics of the studied static field, arc and VMAT plans.

Parameter value	Static field	Arc	VMAT plans
Field size	$3x3m^2$ to $25x25cm^2$	$5x5cm^2$ to $25x25cm^2$	
Arc Length		180°	210° to 310°
Collimator angle	0°	0°	10°
Control-point spacing			2° to 4°

3. Results and Discussion

Figures 1a and 1b show the angular positions of the Synergy linac gantry and Octavius during the slowest and fastest possible arc delivery. The insert of figure 1b shows that linac gantry movement is slightly non linear at the start of the fastest arc. The cylindrical unit of Octavius closely follows the gantry at both fastest and slowest speeds.

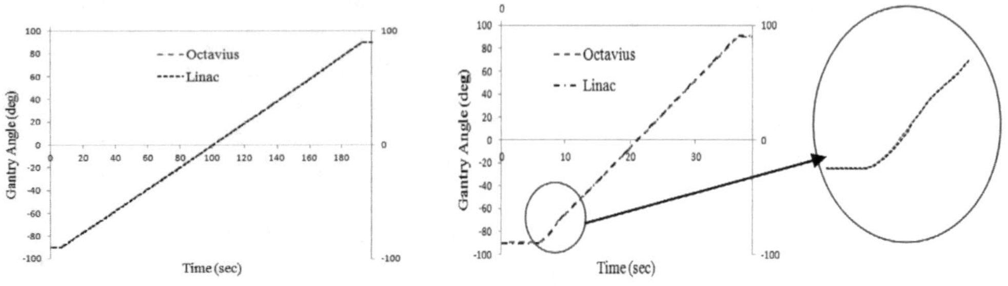

a b

Figure1. Synchronisation of the Octavius cylindrical unit with linac gantry for a slowest speed and b.fastest of gantry rotation speeds.

Figure 2 shows the gamma pass rates with 2%/2mm and 3%/3mm tolerance criteria for static open fields, open arc and clinical VMAT plans. In general, the absolute dose agreement between the 3D dose matrix reconstructed by Octavius and Pinnacle agreed very well. This ensures the correct implementation of PDDs used for the reconstruction of 3D dose and accurate field size response of Octavius phantom and detector array. However the Octavius reconstructed dose matrices showed broader shoulders at the field edges due to the relatively larger detector size. This resulted in reduced pass rates for some of the static open fields (figures 2 and 3) when the edge of the field falls on the active volume of the detectors. The dose reconstructed for open arc fields also showed failed gamma voxels at the edges of the field for some of the arcs (figure 2). Mean (1σ) γ pass rates of 84.5(8.9) % and 94.2(4.5) % was observed for 2%/2mm and 3%/3mm tolerance criteria for the studied open arc plans. Figure 4 shows Pinnacle and Octavius calculated dose matrices in axial, sagittal and coronal planes and the γ analysis results in corresponding planes. The clinical VMAT plans showed mean γ pass rates of 86.8(3.5) % and 96.7(1.4) % for 2%/2mm and 3%/3mm tolerance criteria.

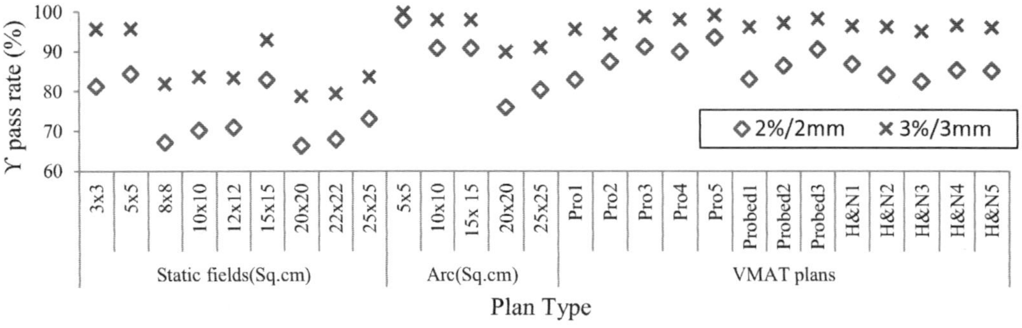

Figure 2. Gamma pass rate of static open fields, open arc and clinical VMAT plans.

The 3D dose information provided by the Octavius system enables the verification of delivered dose in the entire treatment volume (figure 4). The coarse resolution and relatively large active volume of the detectors in the 729 detector array resulted in failed voxels in high dose gradient regions of the measured static open field, arc field and clinical VMAT plans. The observed 3D γ pass rates for the clinical VMAT plans with 3%/3mm and 2%/2mm tolerance criteria are comparable with γ pass rates reported for VMAT plans using other detectors [7, 10, 11].

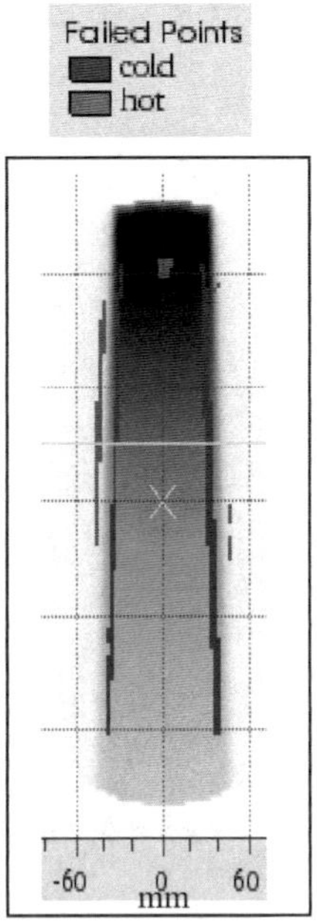

Figure 3. Gamma analysis results for 5x5 cm^2 static field.

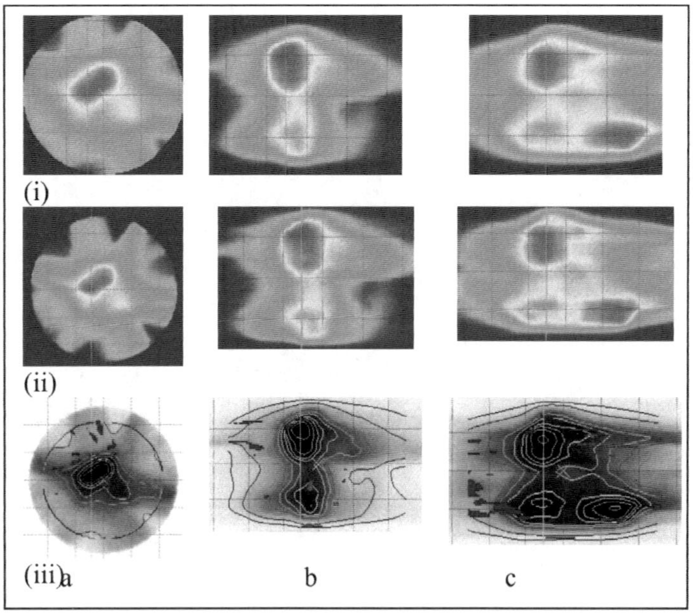

Figure 4. (i) Pinnacle and (ii) Octavius calculated dose matrices in a.axial, b.sagittal and c.coronal planes. (iii) Gamma analysis results with 3%/3mm criteria for a.axial, b.sagittal and c.coronal planes.

4. Conclusion

The Octavius 4D system has been shown to accurately predict 3D dose based on measurement and PDD information. The cylindrical unit of the Octavius 4D system has been shown to rotate in synchrony with the Elekta-Synergy gantry at both slowest and fastest speeds. However, due to the coarse resolution of the detectors and a relatively large detector size failed pixels were observed on the penumbra region of the conventional treatment fields and high dose gradient region of VMAT plans.

5. Acknowledgement

The authors are thankful to Mr Peter Douglas, Nucletron Pty Ltd, Australia for lending the Octavius 4D system to carry out this project.

6. References

[1] Otto K 2008 *Med. Phys.* **35** 310
[2] Webb S *et al* 2009 *Phys. Med. Biol.* **54** 4345
[3] Ezzell G A *et al* 2003 *Med. Phys.* **30** 2089
[4] Baldock C *et al* 2010 *Phys. Med. Biol.* **55** R1
[5] Venning A J *et al 2004 J. Phys.: Conf. Ser.* **3** 155-8
[6] Vial *et al* 2008 *Med. Phys.* **35** 1267
[7] Feygelman V *et al* 2011 *J. Appl. Clin. Med. Phys.* **12** 146
[8] McGarry C K *et al* 2013 *Med. Phys.* **40** 091707
[9] Low D A *et al* 2003 *Med. Phys.* **30** 2455
[10] Van Herk M *et al* 2007 *Int. J. Radiation Oncology Biol. Phys.* **67** 1568-77
[11] Arumugam S *et al* 2013 *Med. Phys.* **40** 071724

AUTHOR INDEX

AUTHOR INDEX

AUTHOR INDEX